THE FOURTEENTH
CENTURY

THE
FOURTEENTH
CENTURY
1307—1399

BY

MAY McKISACK

*Formerly Professor of History
at Westfield College
in the University of London
and Honorary Fellow of
Somerville College, Oxford*

Oxford   New York
OXFORD UNIVERSITY PRESS

Oxford University Press, Walton Street, Oxford OX2 6DP

Oxford New York
Athens Auckland Bangkok Bombay
Calcutta Cape Town Dar es Salaam Delhi
Florence Hong Kong Istanbul Karachi
Kuala Lumpur Madras Madrid Melbourne
Mexico City Nairobi Paris Singapore
Taipei Tokyo Toronto

and associated companies in
Berlin Ibadan

Oxford is a trade mark of Oxford University Press

First published 1959 as volume five of The Oxford History of England
First issued as an Oxford University Press paperback 1991

British Library Cataloguing in Publication Data

Data available

Library of Congress Cataloging in Publication Data
McKisack, May.
The fourteenth century, 1307-1399 / by May McKisack.
p. cm.
Reprint. Originally published: Oxford: Clarendon Press, 1959.
(The Oxford history of England; 5)
Includes bibliographical references and index.
1. Great Britain—History—14th century. I. Title.
942.01'7—dc20 DA225.M34 1991 91-11657
ISBN 0-19-285250-7

3 5 7 9 10 8 6 4

Printed in Great Britain by
Biddles Ltd
Guildford and King's Lynn

PREFACE

SOMETHING of my indebtedness to the many scholars whose labours have enriched our understanding of the fourteenth century is acknowledged below, in the footnotes and Bibliography. Among such scholars (though none of them is responsible for anything I have written amiss) I wish to thank especially Professor V. H. Galbraith, for his sustained interest in and enlivening criticism of my work; Dr. Rose Graham, for her kindness in reading and offering valuable comments on Chapter X; Professor Eleanora Carus-Wilson, for her unfailing readiness to offer guidance on problems of economic history; and three younger scholars—Dr. E. B. Fryde, Dr. J. R. L. Highfield, and Dr. G. A. Holmes—for their great generosity in allowing me to make use of some of their unpublished work. I also owe much to the learning and patience of the General Editor, Sir George Clark, and to the skill and courtesy of the staff of the Clarendon Press. The Constance Ann Lee Fellowship, awarded me by Somerville College for the academic year 1954-5, relieved me of most of my teaching responsibilities and has thereby enabled me to fulfil my contract at a much earlier date than would otherwise have been possible. I welcome this opportunity to express my profound gratitude to my college for its support and for countless benefits bestowed on me through many years, not least for the benefit of an incomparable tutor, the late Maude Clarke, whose book this should have been.

M. McK.

Westfield College, London
11 March 1959

CONTENTS

VI. EDWARD III AND ARCHBISHOP STRATFORD
(1330–43)

VII. PARLIAMENT, LAW, AND JUSTICE

XV. THE RULE AND FALL OF RICHARD II (1388–99)

XVI. LEARNING, LOLLARDY, AND LITERATURE

LIST OF MAPS

GENEALOGICAL TABLES

THE HOUSE OF PLANTAGENET

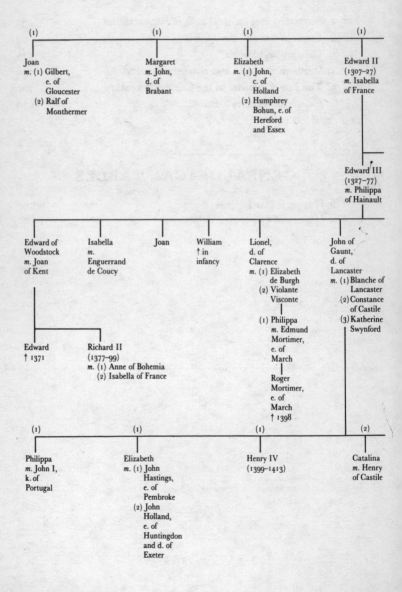

(1) Joan
m. (1) Gilbert, e. of Gloucester
(2) Ralf of Monthermer

(1) Margaret
m. John, d. of Brabant

(1) Elizabeth
m. (1) John, c. of Holland
(2) Humphrey Bohun, e. of Hereford and Essex

(1) Edward II
(1307–27)
m. Isabella of France

Edward III
(1327–77)
m. Philippa of Hainault

Edward of Woodstock
m. Joan of Kent

Isabella
m. Enguerrand de Coucy

Joan

William
† in infancy

Lionel, d. of Clarence
m. (1) Elizabeth de Burgh
(2) Violante Visconte

(1) Philippa
m. Edmund Mortimer, e. of March

Roger Mortimer, e. of March
† 1398

John of Gaunt, d. of Lancaster
m. (1) Blanche of Lancaster
(2) Constance of Castile
(3) Katherine Swynford

Edward
† 1371

Richard II
(1377–99)
m. (1) Anne of Bohemia
(2) Isabella of France

(1) Philippa
m. John I, k. of Portugal

(1) Elizabeth
m. (1) John Hastings, e. of Pembroke
(2) John Holland, e. of Huntingdon and d. of Exeter

(1) Henry IV
(1399–1413)

(2) Catalina
m. Henry of Castile

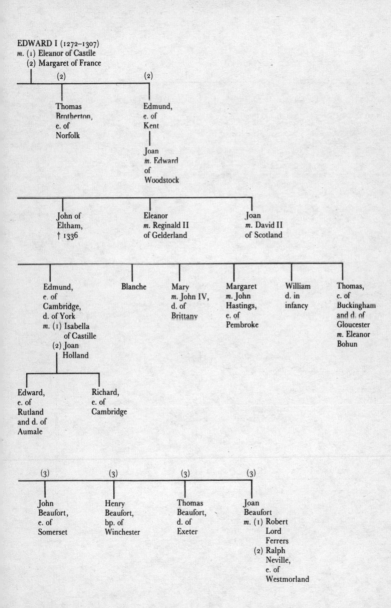

EDWARD I (1272–1307)
m. (1) Eleanor of Castile
   (2) Margaret of France

(2)
Thomas
Brotherton,
e. of
Norfolk

(2)
Edmund,
e. of
Kent

Joan
m. Edward
of
Woodstock

John of
Eltham,
† 1336

Eleanor
m. Reginald II
of Gelderland

Joan
m. David II
of Scotland

Edmund,
e. of
Cambridge,
d. of York
m. (1) Isabella
       of Castile
   (2) Joan
       Holland

Blanche

Mary
m. John IV,
d. of
Brittany

Margaret
m. John
Hastings,
e. of
Pembroke

William
d. in
infancy

Thomas,
e. of
Buckingham
and d. of
Gloucester
m. Eleanor
Bohun

Edward,
e. of
Rutland
and d. of
Aumale

Richard,
e. of
Cambridge

(3)
John
Beaufort,
e. of
Somerset

(3)
Henry
Beaufort,
bp. of
Winchester

(3)
Thomas
Beaufort,
d. of
Exeter

(3)
Joan
Beaufort
m. (1) Robert
       Lord
       Ferrers
   (2) Ralph
       Neville,
       e. of
       Westmorland

INTRODUCTION

K ING EDWARD II entered into a rich inheritance. England at the opening of the fourteenth century was a prosperous land, a land of expanding population, flourishing agriculture, fair cities, fine churches, rising universities and schools. The great King Edward I who had ruled this country for over thirty years had played his role magnificently, offending none of the conventions of his age. Immensely vigorous, both physically and mentally, he enjoyed the hawking, hunting, and mock fighting which were the approved relaxations of monarchy and was himself a soldier of distinction. He accepted the medieval ideal of a united Christendom and made his influence felt in Europe, while energetically maintaining what he conceived to be the rights and prerogatives of the English Crown. By the end of his reign he had conquered the principality of Wales and added the earldom of Cornwall to the royal demesnes. The great earldom of Gloucester was in the hands of his grandson, Gilbert de Clare, the earl of Hereford and Essex was his son-in-law, the earl of Surrey was his granddaughter's husband. The king's nephew, John of Brittany, was hereditary earl of Richmond; another nephew, Thomas of Lancaster, son of his brother Edmund, was earl of Lancaster and Leicester and had inherited a claim to the Ferrers earldom of Derby and, through his wife Alice, daughter and heiress of the king's friend, Henry de Lacy, to the reversion of the earldoms of Lincoln and Salisbury. Although Edward had run into serious difficulties in 1297, when he had been forced to make formal confirmation of the Great Charter and the Charter of the Forest, he had felt strong enough by 1305 to repudiate these concessions and to seek and obtain papal absolution from his oath. His most determined ecclesiastical opponents, Robert Winchelsey, archbishop of Canterbury, and Antony Bek, bishop of Durham, soon found themselves in exile; the king seemed to have insured himself against any renewal of the crisis of 1297, let alone of another Runnymede or Lewes. Meanwhile, a series of great statutes had amended and clarified the law of the land; the council in parliament was omnicompetent and the speed and equity of its judgements were attracting an increasing number of suitors; by summoning representatives of shires, cities, and boroughs, of cathedral and parochial clergy to his parliaments, Edward had enlarged the scope of the feudal assembly and laid the foundations of a system of parliamentary taxation. Much of his success must be ascribed to the efficiency of his ministers and of the administrative system which they controlled. Despite their inevitable unpopularity, both Robert Burnell, chancellor from 1274 to 1292, and Walter Langton, treasurer from 1295 until the end of the reign, served the monarchy well. In chancery and exchequer and in the household departments of wardrobe and chamber, Edward I had been able to rely on the loyalty and experience of a small army of well-trained clerks and officials who drafted his letters and directed the details of his policy. But, at least until the last years of his reign, he had been wise enough not to allow such men to usurp, or (what was more important) to appear to usurp the advisory functions proper to the hereditary aristocracy. For Edward I possessed in generous measure the political good sense in which both his predecessor, Henry III, and his successor, Edward II, were conspicuously deficient. High-handed, overbearing, and often unscrupulous, he might provoke men's resentment, even their hatred, but seldom their contempt. If there were some who suspected him of too great dependence on the officers of state and household, none charged him with subordinating the public interest to his private affections. The pattern of skilful ruling which he bequeathed to his heir might have been turned to good account by a wiser man than Edward of Carnarvon.

The cloud on the horizon was Scotland. For in 1307 England and Scotland wereat war and the new king's most damaging liability was the in flexible hostilityof the Scots whose struggle for independence had already persisted through a decade. By 1306 Robert Bruce of Annandale was conspiring secretly with certain of the Scottish magnates to have himself accepted as king. When his principal Scottish rival, John Comyn the Red, was murdered at Dumfries in February, Bruce, who was suspected of complicity in the crime, was forced to take to the hills; but he declared himself the champion of national independence, renewed his claim to the Crown, and a few weeks later secured his own coronation at Scone. Though subsequently driven into exile, he reappeared in Scotland early in the following year. Edward I spent the last winter of his life at the priory of

Lanercost, near Carlisle, and died as he was moving towards the Border to renew the attack on Bruce. English determination to maintain the vassal status of Scotland was thus still at grips with the Scottish will to resistance; a Scottish war with all its implications, military, financial, and political, was the unenviable legacy of Edward II. Militarily—though the lesson had not been digested—experience had shown that the Scots could be defeated in battle and their country temporarily overrun, but that to hold them in permanent subjection was a task beyond England's resources. The Scottish campaigns of Edward I had strained these resources severely; and, though his credit operations look trifling if measured against those of Edward III at the opening of the Hundred Years War, debts amounting to over £60,000 remained unpaid at his death and exchequer accounts were in chaos. The customs were mortgaged to the Italian banking-house of Frescobaldi and money was owing to magnates, troops, courtiers, tradesmen, and clerks. Politically, the Scottish *rapprochement* with France during the Anglo-French conflict of 1294–7 had been among the most sinister developments of the war of independence. At the end of the reign, England and France were at peace. But the vital question outstanding between them, the question of the precise status of the king of England in his capacity of duke of Aquitaine, remained unresolved to threaten the peace of western Europe.

I

EDWARD II AND THE ORDAINERS
(1307–13)

WHEN, in July 1307, the great Edward I lay dying at the Cumberland village of Burgh-upon-the-Sands, it may well have seemed to some of his principal subjects that he had lived too long. Around the formidable old king, whose last campaign was undertaken in his sixty-eighth year, there had been growing up a circle of much younger barons, many of them linked to the royal house by ties of blood or marriage. Even Henry de Lacy of Lincoln, veteran among the earls, was Edward's junior by twelve years; and, of the rest, only John of Brittany, earl of Richmond, and the earls of Oxford and Ulster were over forty. Aymer de Valence, shortly to assume the title of Pembroke, may have been thirty-seven;[1] but Humphrey Bohun of Hereford and Essex, Thomas of Lancaster, Leicester, and Derby, Guy Beauchamp of Warwick, Edmund Fitzalan of Arundel, and John de Warenne of Surrey were all young men in their twenties or early thirties; while the new king's nephew, Gilbert de Clare of Gloucester, was a boy of sixteen. There were grounds for rejoicing in the accession of a prince in his twenty-fourth year, 'fair of body and great of strength', whose education had been such as befitted his rank.[2] If his household was unruly and his habits extravagant there was little in such youthful excesses to call for comment.[3] For many years he had been suitably betrothed to a French princess, Isabella, daughter of Philip IV; he was duke of Aquitaine and lord of Ponthieu-Montreuil; he had played his part unremarkably in four Scottish campaigns, had acted as regent for his father during his absences

---

[1] He was born probably c. 1270 and assumed the title on the death of his mother in April 1308. *Complete Peerage*, x. 382–4.

[2] According to Robert of Reading (*Flores*, iii. 137), Edward II was acclaimed *cum ingenti laetitia*. The author of the *Vita Edwardi Secundi* (ed. N. Denholm-Young, 1957, p. 40) also refers to the popular favour he enjoyed at the beginning of the reign.

[3] Edward's household as prince of Wales is the first of its kind of which we have detailed knowledge. See T. F. Tout, *Chapters in Medieval Administrative History*, ii. 165–87. His early life is the subject of a monograph by Hilda Johnstone, *Edward of Carnarvon* (1946).

abroad, and had attended him in parliament and on other state occasions. Long sojourns at his Buckinghamshire manor of Langley may have developed his taste for country pursuits and there is evidence that he was already addicted to swimming and boating; but he was by no means a boor. He enjoyed play-acting and minstrels were much in his company; his interest in architecture and shipbuilding was more than perfunctory;[1] he borrowed (and failed to return) lives of St. Anselm and St. Thomas from Christ Church, Canterbury;[2] and if he was indeed the author of the Anglo-Norman lament ascribed to him, he knew something of versification.[3] The members of his household can hardly have been blind to his emotional instability; but there is little to suggest any general awareness of its dangers at the time when the new king hastened to Carlisle to receive the homage of his barons.

Even the immediate recall of Piers Gaveston does not appear to have awakened serious apprehension. It is not easy for the modern reader to penetrate the vituperations of the chroniclers so as to gain a clear picture of this unfortunate man. The younger son of a Béarnais knight, he had been a member of the household of the Prince of Wales and had enjoyed the privileges of a royal ward. In his early years he may have been guilty of nothing worse than indiscretion: but the prince's infatuation for him had been the occasion of a stormy scene in the last winter of the old king's reign. Edward was determined to treat Gaveston publicly as a brother and demanded for him some especial mark of royal favour, the county of Ponthieu, according to one author, the earldom of Cornwall, according to another. A request which threatened the hard-won solidarity of the royal estates roused the choleric temper of the elder Edward, who drove his son from his presence with bitter reproaches and banished the favourite for a time—though with such careful provision for his maintenance as to suggest that he was regarded as the victim rather than the instigator of the prince's folly.[4] The new king lost no time in

[1] Certain works carried out at Westminster early in the reign and repairs to the ship which brought Queen Isabella from France are said to have been *per proprium . . . divisamentum Regis* (P.R.O. E 101/468/15.) I am indebted for these references to the courtesy of Mr. H. M. Colvin.

[2] M. R. James, *Ancient Libraries of Canterbury and Dover*, pp. xlv–xlvi, 148.

[3] P. Studer, who discovered and edited the poem, accepts the ascription (*Mod. Lang. Rev.* xvi (1921), 34–46). For a sceptical comment see V. H. Galbraith, 'The Literacy of the Medieval English Kings', *Proc. Brit. Acad.* xxi (1935), 231, n. 6.

[4] Johnstone, *Edward of Carnarvon*, pp. 122–3.

proffering amends. On 6 August, by a charter sealed in the presence of the earls, Gaveston was put in possession of the royal earldom of Cornwall, with all the lands, manors, castles, vills, hundreds, and honours appurtenant to it; and arrangements were made for his marriage to the king's niece, Margaret de Clare, sister of the earl of Gloucester. His elevation was a gesture, not only of affection for the living, but of defiance to the dead: it heralded a general reversal of royal policy. Within a month of his accession, Edward II, 'for the honour of God and St. Cuthbert and on account of the special affection in which we have long held him', had restored Antony Bek to the see of Durham of which he had been deprived by Edward I. By the end of the year he was even writing to Clement V to demand the reinstatement of his father's most determined opponent, Archbishop Winchelsey. Walter Langton, bishop of Lichfield and treasurer of England, against whom the new king cherished many old grudges, had been jettisoned already; orders for the seizure of his temporalities and the payment of his revenues into the royal chamber were issued to the sheriffs in September.[1] It may have been Gaveston who urged strong measures:[2] but Langton had many enemies at court and in the country and the appointment of a commission to investigate charges of peculation aroused no protest. Edward put the former keeper of his wardrobe, Walter Reynolds, in Langton's place at the treasury and began to press for his elevation to the see of Worcester. Custody of the privy seal was entrusted to William Melton, already controller of the wardrobe; for he was not unwilling to avail himself of the expertize of those of his father's old servants who did not look dangerous politically; both the new chancellor, John Langton, and the new keeper of the wardrobe, John Benstead, had been in the service of Edward I. The abandonment of the Scottish campaign was a yielding to necessity and, with winter approaching, could hardly fail to be popular; but Edward was careful not to incur the charge of indifference to the safety of the northern borders. Two of the most responsible of the magnates—the earl of Richmond and Aymer de Valence—were appointed his

[1] Langton retained office as treasurer until 22 Aug., but entries in the Receipt Roll (E 401/167) suggest that he had ceased to act by 19 July. (I am indebted to Dr. E. B. Fryde for these references.)

[2] As the St. Paul's annalist suggests (*Chrons. Edward I and II*, i. 257). See also *The Chronicle of Walter of Guisborough* (Hemingburgh), ed. H. Rothwell (Camden Series lxxxix, 1957), pp. 382–3.

lieutenants in Scotland; and special keepers of the peace were
assigned to the northern counties, the sheriffs being ordered to
assist them when necessary, with horses, arms, and the entire
posse comitatus.[1] Money and supplies were hurried to the Border
where the king himself remained until the beginning of September. He could not stay longer, for urgent and costly matters
demanded his attention—a state funeral, a royal marriage, and
a coronation—and it was necessary to hold a parliament. Lords,
clergy, and commons were summoned to meet the king at Northampton on his way south; and this, the first parliament of the
new reign, granted a fifteenth from the clergy and a fifteenth and
a twentieth from the laity. With the prospect of a subsidy to
offset the 40,000 marks already swallowed up by his father's
exequies and the expenses of his own household, Edward was
able to repay a loan of £500 from the earl of Cornwall to whom,
a few weeks later, he granted custody of the Audley lands.[2] After
keeping the feast of Christmas with his friend, the king appointed
him keeper of the realm, while he himself went to France to
fetch his twelve-year-old bride whom he married at Boulogne on
25 January 1308, having first done homage to her father for the
duchy of Aquitaine and the county of Ponthieu. During his
absence Gaveston was, in effect, ruler of England, with power
to issue licences to elect, to grant royal assents, restore temporalities and presentations, and to deal with wardships and
marriages.[3] It may have been malicious rumour which suggested
that the best of the royal wedding-presents went to the earl of
Cornwall; but there could be no mistaking the implications of
the place accorded to him at the coronation on 25 February
when, 'so decked out that he more resembled the god Mars than
an ordinary mortal', he bore the crown and the sword of St.
Edward in procession before the king.[4] His position, in short,
was creating a major political dilemma and destroying the
atmosphere of general amity in which the reign had opened; and
tension was not eased when Gaveston and his friends worsted
the earls at a tournament at Wallingford.[5]

The coronation of Edward II has been provocative of
much learned debate.[6] Some historians, following the St. Paul's

[1] *Cal. Pat. Rolls 1307-13*, p. 14.
[2] *Col. Chancery Warrants*, i. 266; *Issues of the Exchequer*, p. 119; *Foedera*, ii. 16.
[3] *Foedera*, ii. 28.   [4] *Ann. Paul*, pp. 258-62.
[5] *Vita Edwardi*, p. 2.   [6] See Bibliography below, p. 555.

annalist, believe that the postponement of the ceremony for a week was a consequence of baronial exasperation and that Edward had to promise to dismiss Gaveston and to concede a number of other demands before the magnates would allow him to be crowned. But this version of events is peculiar to one chronicle and it seems almost certain that the true reason for the postponement was Edward's desire to have himself crowned by Winchelsey.[1] When this hope failed, hurried arrangements had to be made for the bishop of Winchester to act instead. Jealousy of Gaveston must have been simmering violently below the surface, but it does not appear to have erupted until after the king and queen had been crowned. A new *ordo* with elaborated rubrics and expanded ceremonial was in use for the occasion;[2] and the king took the oath, not in Latin but in French. Stubbs used this fact as evidence that Edward II was the type of un- learned king whom his remote ancestor, Fulk the Good, had declared to be no better than a crowned ass;[3] but it has since been shown that the vernacular oath was no novelty and that Edward made use of it, not because he was illiterate, but because French was the common idiom of most of those present at the ceremony. What made the occasion memorable was not the use of the vernacular but the fact that Edward II took a fourfold instead of the traditional threefold oath.[4] His second and third promises —to maintain peace and to do justice—corresponded to the first and second clauses of the oath taken by Richard I, his first—to maintain the laws and customs allowed by former kings— amounted to a general undertaking to uphold the whole body of English law and custom; but interpretation of the fourth promise is more difficult, for this looked to the future as well as to the past, the king swearing to maintain and keep 'les leyes et les custumes droitureles les quiels la communaute de vostre roiaume aura eslu'. The most plausible explanation of these much dis- cussed words is that they were intended to cover any changes in law and custom which might arise during the reign as a result

[1] He wrote to the archbishop from Dover on 9 Feb. urging his speedy return (*Foedera*, ii. 32). See H. G. Richardson, 'Clement V and the See of Canterbury', *Eng. Hist. Rev.* lvi (1941), 97–103.

[2] P. E. Schramm, *A History of the English Coronation*, tr. L. G. Wickham Legg (1937), pp. 74–78.

[3] *Constitutional History of England* (4th ed. 1906), ii. 332.

[4] The oath has come down to us in two forms, the Latin in the Coronation Office, the French (which is preferable) in the chancery enrolment at the end of the *Liber Regalis* of Edward II, in what appears to be an official record of the coronation.

of enactments promulgated by the king with the consent of the community; and it is likely that they were prompted by memories of Edward I's violations of his own enactments, on the plea that he had been coerced. The concession is substantial but by no means revolutionary; for the subordination of the king to the law was the pure gospel of Bracton; and though by common consent the king was the supreme lawgiver, very ancient custom required that when the law was shown to be defective it should be amended with the consent of the great men of the realm.[1] It seems impossible to believe that a coronation *ordo* which laid special stress on the priestly character of the king as manifest in his power of healing, should have included an oath designed to place him 'below the level of his subjects', or that he should have renounced an ancient prerogative right, by binding himself in advance to accept laws chosen by his subjects, or by any section of them.[2] The fact that the oath was used later as a weapon in the wordy warfare between Edward and the earls should not mislead us into crediting its framers with revolutionary intentions.[3]

Though there is little reason to suppose that the fourth clause was framed with reference to Gaveston, it was none the less evident that, once the coronation was over, he would have to be dealt with. The queen's kinsmen had returned indignantly to France declaring that the king loved Gaveston more than his wife;[4] the English magnates were closing their ranks against the favourite, and the return of Winchelsey—his intransigence in no way mitigated by the new king's efforts on his behalf—greatly strengthened their hands. Only Richmond and Gloucester still remained neutral; and Henry of Lincoln assumed leadership of a movement which had the support of Lancaster, Warwick, Hereford, Pembroke, and Warenne. At the April parliament the barons appeared in arms and forced the king to promise to banish Gaveston who was said to have disinherited the Crown,

---

[1] The point is put succinctly in the alleged answer of the earls to the manifesto against the Ordinances drawn up by French jurists in 1312: 'England is not governed by written law but by ancient laws and customs used and approved in the time of the king's forebears: and where these are in any point deficient the king and his prelates, earls and barons, on the complaint of the people (*vulgi*) are bound to amend them.' *Ann. Lond.*, pp. 210-14.

[2] As suggested by Professor Wilkinson (*Const. History*, ii. 12).

[3] Mr. Richardson has argued convincingly that the oath was probably drafted some months earlier and before the *ordo*. *Speculum*, xxiv. 58-59.

[4] *Ann. Paul*, p. 262.

alienated the king from his magnates, and bound confederates to himself by oath. Two points of interest emerge from this so-called 'Declaration of 1308'.[1] According to the *Gesta Edwardi*, written by a canon of Bridlington, the earls invoked the fourth clause of the coronation oath in support of their demand for Gaveston's exile, interpreting it to mean that the king was thereby bound to accept their demands. If the canon's version is correct, it seems that the oath was already being used—and distorted—in the interests of current controversy. Moreover, both Bridlington and the London annalist assert that the magnates, at the same time, drew a significant distinction, declaring that homage and allegiance were due to the Crown, rather than to the person of the king. Their statements are puzzling; for if the 'doctrine of capacities' had been so clearly formulated at this early date, in relation to the king and his office, it is strange, to say the least, that the Ordainers should have failed to make use of it; and it seems possible that the chroniclers, writing in their knowledge of after events, may have been reading back into the crisis of 1308 a justification of the political programme of 1321. However this may be, it was not by verbiage but by show of arms that the barons attained their ends. Edward was compelled to issue letters patent promising to suffer no impediment to Gaveston's departure;[2] while Winchelsey pronounced him excommunicate should he fail to leave England on the appointed day (25 June) or should he attempt to return.

Then, as always, Edward's promises meant little. His true purpose is revealed by the series of negotiations on which he embarked at once, wire-pulling for his friend and offering concessions to his enemies and to others whose co-operation might be useful. On 9 May he bestowed the hereditary stewardship of England on his cousin Thomas of Lancaster, a young man who had not as yet taken a conspicuous part in the baronial opposition and whom he may have believed it possible to conciliate.[3] A peremptory order to Amerigo dei Frescobaldi to cease interference with the archbishop's dies and moneyers at Canterbury may have been intended as a sop to the formidable Winchelsey.[4] Henry Percy was allowed to take up his residence in Scarborough castle; Queen Isabella was given the counties of Ponthieu and

---

[1] *Ann. Lond.*, pp. 153–4; Bridlington, p. 33.
[2] *Foedera*, ii. 44.
[3] Ibid.
[4] Ibid., p. 45.

Montreuil for her personal expenses.[1] To her father, Edward sent an exposition of his difficulties with a request that he would use his influence to allay the discords which had arisen. Clement V and the cardinals were appealed to in similar terms and asked to annul the conditional sentence of excommunication.[2] Meanwhile, every precaution was taken to ensure that Gaveston should be maintained in comfort. Castles and manors in England were granted to him on 7 June, together with lands in Aquitaine to the value of 3,000 marks a year; on 16 June he was appointed king's lieutenant in Ireland[3] and, according to the Lanercost chronicler, was provided with a supply of blank charters sealed with the king's great seal, to use as he wished.[4] The appearance among the witnesses to some of these grants and letters, of Richmond, Gloucester, and Henry Percy, alongside such officials as Hugh Despenser and William Melton, suggests that Edward's policy of conciliation was not proving wholly ineffective. When the day of departure came, he accompanied Gaveston to Bristol and watched him embark, with a great household, for Ireland. Here Gaveston showed himself open-handed and seems to have made no bad impression;[5] but the king endured his absence with difficulty, while he sent letters and embassies to the French court and to the Curia, made fresh concessions to Gaveston in Gascony, and looked after the interests of his property at home.[6] The king's attitude was an open secret and provocative of bitter resentment. There were demands for the removal of officials and for redress of other grievances, embodied in a long schedule which magnates and commons presented in the parliament of April 1309. Ready as always to strike a bargain in the interests of Gaveston, Edward boldly proposed his recall as the price of concession; but the suggestion was bluntly rejected and the petition shelved. Before the next parliament met at Stamford in July, the king had improved his position. He had secured papal annulment of the sentence of excommunication; and Lincoln and Pembroke seem to have been won over, doubtless because they hoped that both Edward and the favourite had now learned their lesson.

---

[1] *Cal. Pat. Rolls 1307-13*, pp. 68, 74.   [2] *Foedera*, ii. 49-50.
[3] *Cal. Pat. Rolls 1307-13*, pp. 74, 78, 83.   [4] *Chronicon de Lanercost*, p. 212.
[5] Murimuth (p. 12) declares that the Irish loved him because of his lavish gifts, the London annalist (p. 156) that he introduced new manners and led a life of wonderful variety.
[6] e.g. the order dated 26 Feb. 1309 for the punishment of trespassers on Gaveston's property in the Isle of Wight. *Foedera*, ii. 67.

Their support emboldened the king to reverse the sentence of exile and appear in parliament with Gaveston at his side, prepared to concede all the magnates' demands in return for his restoration. The success of this barefaced manœuvre makes it apparent how deeply divided were the counsels of the magnates, on whose fickle character the author of the *Vita Edwardi* passes bitter judgement.[1] Only Winchelsey remained irreconcilable, refusing to take any part in the business of parliament[2] and continuing to press for the release of Langton while he drew up a long schedule of clerical *gravamina* for presentation to the papal curia.

The paper victory of the Statute of Stamford (which, in itself, amounted to little more than a reissue of the *Articuli super Cartas* of 1300), was thus won at the price of Gaveston's return and full restoration to his English honours and titles. It soon became evident that his character was not changed nor the king's infatuation cured. The chroniclers complain that though he had promised to behave well and to live in peace, he soon began to treat the earls and barons as his servants and to breed discord between them and the king. It was at this time that he began to bestow *turpia cognomina* upon the earls;[3] and it was his insolence and the ejection of a Lancastrian dependant from court through his agency that finally drove Thomas of Lancaster to the side of the opposition.[4] When the king summoned a council to York in October, five of the earls refused to attend 'because of Peter'.[5] By December it had become necessary to order the arrest of scandalmongers and to prohibit unauthorized gatherings of armed men; and when parliament was summoned for February 1310 the earls again refused to attend while Gaveston was present. Edward had been sparing no pains to retain the loyalty of his supposed adherents, particularly of the wealthy young earl of Gloucester on whom he bestowed both lands and money;[6]

---

[1] *Vita Edwardi*, pp. 6–7.

[2] He tried to ensure the absence of some of the other prelates by arranging for the consecration of Droxford as bishop of Bath and Wells to coincide with the opening of parliament; see Kathleen Edwards, 'The Political Importance of the English Bishops during the Reign of Edward II', *Eng. Hist. Rev.* lix (1944), 318.

[3] *Flores*, iii. 152. The only one for which there is contemporary authority is 'black dog of Arden' applied to Warwick.

[4] *Vita Edwardi*, p. 8.   [5] Guisborough, p. 384.

[6] Gloucester received the royal manors of Fakenham, Causton, Aylsham, and Dalham in Oct. (*Cal. Charter Rolls*, iii. 130); in Nov. the king made him a gift of 3,000 marks (*Cal. Chancery Warrants*, i. 305).

and he now made Gloucester responsible, with Richmond, Lincoln, and Warenne, for ensuring that none should appear armed in parliament. It was thus an act of open defiance when the earls attended in full military array and laid a statement of their grievances before the king. The substance of their complaint was that, led by evil counsellors, he was so far impoverished as to have to live by extortion, that the grants made to him had been wasted, and that by the loss of Scotland he had dismembered his Crown. A commission of reform was demanded forthwith, Edward's attempts at prevarication were swept aside, and he was forced to concede what was asked. By letters patent dated 16 March 1310 he agreed to the appointment of Ordainers with full power to reform his realm and household; on the day following, the lords promised that the commission should not be held to establish a precedent. The Ordainers were chosen in the Painted Chamber of Westminster palace by a process of indirect election reminiscent of 1258.[1] There were twenty-one of them and they constituted a fairly representative body, intractable prelates like Winchelsey being balanced by a curialist like Langton of Chichester, the most intransigent of the earls by moderates like Lincoln and royalists like Richmond and Gloucester; among the barons were Robert Clifford and William Marshall, both of whom had royalist leanings.[2] They began their work without delay, issuing six preliminary ordinances forthwith; but the drafting of the main document proved a lengthy business and it was not complete until August 1311, a fact which suggests that the Ordainers may have been at pains to collect evidence.[3] In the interval thus afforded him, Edward (who had already dispatched Gaveston to the north) did what he could to consolidate his own position.

On 30 April he sought to assert his control over the executive. The chancellor (himself one of the Ordainers) was ordered to

[1] In 1258 the Council of Fifteen was elected by 4 persons, themselves elected by a body of 24, 12 of whom had been nominated by the king, 12 by the barons. In 1310 the bishops elected 2 earls, the earls 2 bishops; these 4 elected 2 barons; and the 6 co-opted 15 others.

[2] The Ordainers were: the archbishop of Canterbury, the bishops of London, Salisbury, Chichester, Norwich, St. Davids, and Llandaff; the earls of Lincoln, Pembroke, Gloucester, Lancaster, Hereford, Richmond, Warwick, and Arundel; Hugh de Vere, Robert Clifford, Hugh Courtenay, William Marshall, and William Martin.

[3] The interval between the Provisions of 1258 and 1259 afforded an obvious precedent.

come to the king at Windsor and henceforward to follow him
wherever he went 'with the great seal and all the Chancery, as
the chancellors of the king's ancestors did', and to command
the justices of the king's bench to follow him likewise.[1] When
Langton proved unamenable he was replaced by Walter Reynolds 'against the will of the community of England'.[2] The keeper
of the wardrobe (Ingelard Warley) was sent into the city of
London with a writ of privy seal prohibiting injuries to Gascons.[3]
In June Edward made a bid for the support of Warenne by
granting him the farm of the castle and honour of High Peak for
life, while fairs and markets were allowed to the loyal Richmond.[4] Grants to Gaveston on various dates indicated the king's
determination to stand by the favourite;[5] and in September he
went north after making a vain attempt, obeyed only by
Gloucester and Warwick, to force the earls to follow him.
Lincoln was made keeper of the realm and by October the king
was in Scotland. From Linlithgow he sent instructions to the
chancellor to join him at Newcastle on 1 December and to make
arrangements for the exchequer to be at York after Easter; and
he remained in the north, chiefly at Berwick, until July 1311.
Little was achieved against the Scots, except, indeed, by Gaveston who succeeded in penetrating to Perth and harrying the
country between the Forth and the Grampians; but the king
was determined to keep out of harm's way for as long as possible.
When Henry of Lincoln died in February, Gloucester was
appointed regent in his place. The old earl's death was a serious
misfortune for the king, not only because it deprived him of a
loyal servant, but also because it put Lincoln's son-in-law,
Thomas of Lancaster, in possession of his earldoms of Lincoln
and Salisbury and made him forthwith the richest of the earls.[6]
Meanwhile, Edward's financial difficulties had been increasing
and soon after Lincoln's death he was offering his executors
some of the crown jewels as security for a loan of 4,000 marks;[7]
a little later he was demanding a tax on the spiritualities from

---

[1] *Cal. Chancery Warrants*, i. 314.   [2] *Ann. Paul.*, p. 269.
[3] *Cal. Letter-Book D*, pp. 228–9.
[4] *Cal. Fine Rolls*, ii. 63; *Cal. Charter Rolls*, iii. 137; *Foedera*, ii. 109.
[5] e.g. a grant of free warren in some of his demesnes, July 1310 (*Cal. Charter Rolls*, iii. 138); the office of justice of the Forest south of Trent, Oct. 1310 (*Cal. Fine Rolls*, ii. 73).
[6] He did not assume the titles, however: see R. Somerville, *History of the Duchy of Lancaster* (1953), i. 22.   [7] *Parl. Writs*, II. i. 58.

the prelates.[1] In June 1311 the regent and all the great officers of state were ordered to take into the king's hands all the customs of England, Ireland, and Scotland received since Whitsun, so that he be 'hastily served with the moneys for the great businesses touching him'.[2] But the results were disappointing; and at last it was becoming evident that the day of reckoning could not be deferred much longer. On 16 June writs were issued summoning magnates, knights, burgesses, and clerical proctors to a parliament at Westminster on 9 August, so that the Ordinances made for the common weal might be completed and confirmed. A few weeks later, having lodged Gaveston in the rocky stronghold of Bamborough, Edward came reluctantly south. He achieved the maximum delay by going on pilgrimage to Canterbury and failing to appear at Westminster until several days after the parliament was due to begin; and even then, he made a desperate and unseemly effort to save Gaveston by an offer to accept any ordinance touching himself provided that the favourite were left unharmed. But the barons stood firm and the king's intimate counsellors advised him to yield rather than face civil war.[3] When the royal assent had been won, the Ordinances were published in parliament; five days later the king ordered their proclamation throughout the land. Meanwhile, they were published by the lords at Paul's Cross, and Winchelsey threatened any who should violate them with excommunication.

The demands of the Ordainers were arranged under forty-one heads.[4] A number of individuals were singled out for condemnation, among whom Gaveston inevitably took first place. The indictment against him is of formidable length. He is said to have given the king bad counsel and led him into evil ways; to have laid hands on the royal treasure and taken it out of the realm to Ireland and Gascony; to have 'accroached' (i.e. usurped) royal power to himself and estranged the king's heart from his people; to have replaced royal ministers by those of his own 'covin', some of them aliens; to have forced the king to alienate Crown lands to the great damage of his estate; to have led him to war without the assent of the baronage and to have put him in personal danger; and to have caused him to seal

---

[1] *Foedera*, ii. 132.   [2] *Cal. Chancery Warrants*, i. 369.
[3] *Vita Edwardi*, pp. 17–18.
[4] The Ordinances are printed in *Rot. Parl.* i. 281–6 and in *Statutes of the Realm*, i. 157–67.

blank charters under the great seal.[1] It is recalled that Edward I
had him banished and alleged that his return was never generally approved. His sentence is renewed banishment, as an open
enemy of king and people, from all the king's dominions; he is
to sail from Dover before All Saints' Day (1 November) and, if
he fails to go or if he returns, he is to be treated as a public
enemy. It was easy enough to construct a case against the
favourite whose greed and arrogance were not in question.
Edward's infatuation, however interpreted, was derogatory to
the dignity of the Crown. Extravagant grants from the royal
demesne were indefensible in the existing state of the national
finances. Yet the explanation of Gaveston's unpopularity offered
by the author of the *Vita Edwardi* reveals the large element of
jealousy underlying the attack on him. He was a Gascon who
rose too fast; his pride was so intolerable that he thought none
but the king his equal; when earl of Cornwall (a dignity intended
by Edward I for one of the king's sons) he forgot that he had
once been a humble squire; and the king favoured him to excess
and would not speak with the barons save in his presence.[2] The
charge that he had led Edward into war was absurd and hardly
consistent with baronial complaints of neglect of the Scottish
danger; and the suggestion that his recall in 1307 was not
approved reads like a lame attempt at justification after the
event. No doubt Gaveston was an intolerable nuisance; but the
Ordainers were laying on his shoulders full responsibility for
Edward II's extravagance and want of judgement. He was the
first of a series of scapegoats who for twenty years stood between
the king and the consequences of his own folly.

All the three other individuals singled out for condemnation
were aliens. Henry de Beaumont, a cousin of the king's on his
mother's side, had served in his household as Prince of Wales
and had been rewarded by grants of manors in Lincolnshire and,
in May 1311, of the Isle of Man.[3] His sister, Isabella de Vescy,
had held the custody of Bamborough castle since the beginning
of the reign.[4] The hostility of the Ordainers towards two persons

---

[1] This charge probably refers to Edward's use of the great seal before Gaveston's
departure to Ireland (above, p. 8). See Jocelyne G. Dickinson, 'Blanks and Blank
Charters in the Fourteenth and Fifteenth Centuries', *Eng. Hist. Rev.* lxvi (1951),
375–87.
[2] *Vita Edwardi*, pp. 14–16.   [3] *Cal. Pat. Rolls 1307–13*, p. 300.
[4] She received custody of Bamborough in Nov. 1307 (ibid., p. 26). Lady de
Vescy was a member of the queen's household, a fact which throws some doubt on

who do not appear to have played any very conspicuous part in the political intrigues of the preceding years is probably to be explained by their control of points of strategic importance for the Scottish war.

More complex motives prompted the attack on the Italian banker, Amerigo dei Frescobaldi. He and his company were ordered to render their accounts by mid-October and, in the meantime, they were to be arrested and their goods seized.[1] Edward I had left behind him a debt to all creditors amounting to at least £60,000 and his son's financial embarrassments had forced him into continued dependence on the foreign bankers, particularly the Frescobaldi who had been receivers of the customs since 1304. Amerigo dei Frescobaldi was also constable of Bordeaux, a post which brought him into contact with the friends and relatives of Gaveston, conspicuous among whom was the banking family of Calhau. Bertrand de Calhau, a nephew of Gaveston's, held the post of valet of the household and acted as financial agent to his uncle. The St. Albans chronicler believed that Gaveston was in the habit of transmitting to these aliens the huge treasure which he amassed at the expense of Englishmen; a royal licence permitting Calhau to go to Ireland during Gaveston's exile there lends some colour to these rumours.[2] Financial power brought political influence in its train: one Calhau was mayor of Bordeaux, another was a knight of the royal household and seneschal of Saintonge. Thus, both Italian and Gascon bankers were laying hands on affairs of state and were believed to be establishing a stranglehold on the king, whose financial dealings with the Frescobaldi were matters of public knowledge. In the summer of 1309 all the wool customs of Scotland and Ireland had been granted to them, with the proviso that they were not to answer for them except at the king's order under writ of privy seal; and they had profited considerably by the fall of Walter Langton whose debtors, many of them Italians, had been instructed to pay what they owed to the Frescobaldi. Yet the king's own debts to them, estimated in 1310 at £21,635, were not of unprecedented magnitude; as Dr.

the chroniclers' hints of a rift already threatening between Isabella and her husband. One of the charges against her was that she had used her influence to secure the issue of writs of privy seal, against the law.

[1] An earlier order (6 July 1311) for the arrest of the Frescobaldi had been countermanded by the king three weeks later. *Cal. Letter-Book D*, pp. 268–9.

[2] J. de Trokelowe, *Annales*, pp. 64, 68; *Cal. Pat. Rolls 1307–13*, p. 94.

Fryde has pointed out, they could have been met from the proceeds of a single subsidy; and the system of regular Crown borrowing from major financiers, which was already customary, seems to have been working reasonably well.[1] The unpopularity of the Frescobaldi with the baronage was due less to the size of their loans to the king than to the financial independence which such loans allowed him and to the belief that the Italians were politically mischievous.

Other clauses in the Ordinances reveal the barons' desire to force a policy of retrenchment upon the king and to establish control of the royal finances. The Crown is said to be 'abased and dismembered' by alienations of land and other gifts. All such are to be annulled and none made in future, until the king's debts are paid and his financial estate restored. The issues of the customs and all other royal revenues are to be paid entire into the exchequer; from them the exchequer officers are to make allowance for the maintenance of the king's household so that he may 'live of his own without recourse to prises other than those due and accustomed'. In future, there are to be no prises of corn, merchandise, or other goods, whether under colour of purveyance or otherwise, against the will of the owner and without due payment. New customs and *maltoltes* levied since the coronation of Edward I are to be abolished.[2] But in thus forbidding increases in the customs while at the same time cutting off the supply of foreign credit, the Ordainers were making solution of the financial problem impossible; for it was only by allowing the king access to the wealth of the native merchants that his dependence on foreign creditors could be reduced. Nor could he help himself by tinkering with the coinage, for monetary changes henceforth needed the consent of the baronage in parliament.

The Ordinances restrict the king's freedom of action in many other ways. He is not to quit the realm, nor engage in foreign war, nor appoint a keeper of the realm, without the consent of the baronage in parliament. If he does engage in war without

[1] I am greatly indebted to Dr. Fryde for allowing me to see a draft of his forthcoming chapters in the *Cambridge Economic History*, vol. iii.

[2] Edward I had made unsuccessful attempts to persuade English merchants to pay the *nova custuma* imposed on aliens in 1303. In 1309 at the Stamford parliament the barons had persuaded Edward II to suspend the new custom, except on wool, woolfells, and leather; but it was restored within the year. See N. S. B. Gras, *The Early English Customs System* (1918), pp. 66–70.

the barons' consent their military service will be denied him. His power to choose his own servants is severely restricted, for the appointment of all the chief officers of state and of the household needs the consent of the baronage in parliament.[1] Any appointments which the king may have made already are to be reviewed in parliament. The sheriffs are to be appointed, either by the chancellor, treasurer, and council, or by the treasurer, the barons of the exchequer, and the chief justices; the collectors of the customs are to be natives. All officials are to take oath on appointment to maintain the Ordinances. There are restraints on the privy seal, used by the king for his personal correspondence, particularly with the chancellor. One ordinance states that the law of the land is often interfered with by letters under the privy seal and others insist on the use of chancery writs for forest indictments and commissions to sheriffs and officials. Restraints on the activity of the controller of the wardrobe, who normally acted as its keeper, meant further restraints on the use of the seal. Though the Ordinances nowhere state specifically that the controller is not to keep the seal, the Ordainers' demand for a 'suitable clerk' to keep it amounted to the institution of a new office, that of keeper of the privy seal.[2] The king is also denied the right to issue charters of protection or pardon to open malefactors, though he is still free to exercise his ancient prerogative of showing mercy. Several clauses take up the question of abuses in the administration of the law. In future, litigants may not plead letters from the king to excuse their non-appearance in court or to hold up the hearing of pleas. The jurisdiction of the exchequer is confined to matters touching the king and his ministers and such common-law actions as debt and trespass are excluded from it.[3] The household courts of the steward and the marshal are likewise restrained from entertaining pleas which belong to the courts of common law. Accused persons, if they are of good fame and can find sureties, ought not

---

[1] These officers included the chancellor, treasurer, and chief justices; the chancellor and chief baron of the exchequer; the steward of the household, the keeper and controller of the wardrobe, and the keeper of the privy seal; the wardens of the forests, the two royal escheators, the clerk of the common pleas, the warden of the Cinque Ports, and 'good and sufficient ministers' for Gascony, Scotland, and Ireland. The only exceptions are the chamberlain (who was dealt with in a later ordinance) and officers appointed as delegates by magnates holding offices in fee.

[2] Tout, *Chapters*, ii. 285.

[3] Virtually a recapitulation of cap. 4 of *Articuli super Cartas* (1300): 'Nul commun plai ne seit desoremes tenu a l'eschequer countre la forme de la grant chartre.'

to be imprisoned pending trial; and in several other ways the Ordinances seek to protect such persons from the consequences of malicious appeals. Abuses in the administration of the forest law are also remedied.

Under the Ordinances, parliaments must be held once or, if need be, twice a year in some convenient place. Parties to suits in the courts are said to suffer delay because their opponents declare that answer ought not to be made without the king, i.e. until the next parliament. All statutes are to be maintained if they are contrary neither to the Charters nor to the Ordinances; and the Ordainers claim for themselves the right to interpret doubtful points in both the Great Charter and the Charter of the Forest. In each parliament there is to be appointed a committee of lords to hear complaints against the king's ministers. The final clause provides that the Ordinances shall be maintained in every particular, that the king shall publish them under the great seal, and shall send copies to every county in England and to the Cinque Ports so that they be proclaimed and kept, both within the liberties and without.

Learned opinion on the significance of this famous document has suffered many vicissitudes since the days of Stubbs, who saw it as 'a summary of old grievances and, in all respects but one, of new principles of government by restraint of the royal power'.[1] Where it disappointed him was in its failure to make any provision for the action of the commons; in this respect, the Ordainers seemed to him to be behind, rather than in advance of their time. Scholars of later generations have seen the Ordainers as aiming at efficiency on conservative lines, as anticipating 'the Whig ideal of a constitutional king whose authority was in practice wielded by a united aristocracy'.[2] The difference between the Provisions of Oxford and the Ordinances, it has been said, lay in the 'immense particularity with which the men of 1311 pressed for the radical reform of the royal household'.[3] Ministers of the court as well as ministers of state are subject to purge. To some historians, this insistence on the reform of the household has looked like an attack on a system, and its promoters like radicals trying to introduce new principles of government.[4] Others see little in the Ordinances which might not have

[1] *Constitutional History*, ii. 346.   [2] Tout, *Chapters*, ii. 194.
[3] Tout, *The Place of the Reign of Edward II in English History*, 2nd ed. (1936), p. 83.
[4] J. Conway Davies, *The Baronial Opposition to Edward II* (1918), pp. 60, 356.

been taken from earlier programmes of opposition under Henry III:[1] they regard the attack on the household officers as nothing more than an echo of ancient grudges against royal *familiares* concentrated in this instance on abuses of the system of purveyance and misuse of the privy seal.[2]

Stubbs was unquestionably right in viewing the movement of 1310-11 as aristocratic rather than popular. The Ordinances are the work of the baronage and of the baronage alone. Though they contain a demand for annual parliaments, there is no hint that such parliaments must include representatives of the commons and the object of the demand is stated explicitly to be the furthering of petitions. The commons were asked for their approval of the Ordinances in the parliament of August 1311, but it is doubtful whether this was in any sense a formal ratification and it seems likely that it was on account of their usefulness as agents of propaganda that they were bidden to attend. It is obvious to us, as it was not obvious to Stubbs, that they were not yet an integral part of parliament. None the less, this is what they were fast becoming; and there is substance in the argument that the Ordinances may have been a turning-point in their development.[3] Edward II's reluctance to accept the Ordinances was overcome with difficulty and he yielded only because of the general fear of civil war; his character made it likely that he would retract his promises at the first opportunity; and for this reason the barons seem to have been at pains to publicize their Ordinances and to secure widespread popular assent. Winchelsey explained to the bishops that the Ordinances had been drafted by the magnates and after publication commonly accepted and approved, *quasi ab omnibus*. They were read at Paul's Cross and we hear of oaths to maintain them being taken there and elsewhere by knights and burgesses as well as magnates. Simon de Montfort had recognized the value of middle-class support at a time of political crisis; neither Edward II nor his enemies can have been blind to its uses. It may be more than coincidence that the fairly regular summons of knights and burgesses to parliament dates from about this time.

The Act of 1311 is, then, a baronial document seeking to base

[1] J. E. A. Joliffe, *The Constitutional History of Medieval England* (1937), p. 363.
[2] B. Wilkinson, *Studies in the Constitutional History of the 13th and 14th Centuries* (1937), pp. 227-46.
[3] Maude Clarke, *Medieval Representation and Consent* (1936), p. 161.

itself on some form of popular approval, but at no point suggesting new constitutional powers for the commons nor any desire
to take the people into partnership in controlling the government. It is, indeed, highly doubtful whether the main concern
of the Ordainers was with matters which can properly be
described as constitutional, or even administrative. What they
most desired was the removal of Gaveston and his associates, to
prevent which the king had been prepared to concede all their
other demands. The personal element in medieval state conflicts
is the most transitory of all, the least readily recoverable by
posterity. Yet it is impossible to read the chronicles without being
made aware of the supreme importance of personalities in these
conflicts; and the chronicles of Edward II's reign make it very
plain that it was loathing and jealousy of Gaveston which
brought the whole business to a head. The removal of Gaveston
was not, however, all that the earls desired. They wished also to
be rid of those who were under his influence and might be regarded as his agents, and they aimed at making impossible the
recurrence of a situation in which a favourite, through his
influence with the king, was able to dominate the whole
administration. This kind of outcry against favourites, *familiares*,
has, of course, many precedents. Such jealousies were inevitable
in a feudal society where the great landowners, by virtue of the
king's claim on them for suit of court, were regarded as his
natural counsellors, where the intrusion of upstarts and foreigners
into the king's intimate circle must awaken resentment. The
course of English history since the end of the twelfth century had
intensified and consolidated these resentments, partly as a
result of the increasing self-consciousness of the baronial order
which developed during the long absences of Richard I, the
struggle for the Charters, the periodic conflicts of Henry III's
reign, and in the crisis under Edward I, which taught them to
describe themselves as *universitas* or *communitas regni* and was soon
to lead to the use of the term 'peers of the realm'; and partly as
a result of the increasing complexity of the business of administration, demanding the services of highly trained lawyers and
bureaucrats, of a permanent civil service, masters of special tech-
niques and difficult to control from without. Increasingly and
inevitably, these bureaucrats became the day-to-day advisers of
the king, the real framers of his policy; the advisory functions of
the baronage tend to disappear into the background, to become

a formality. The reason for the specific demands for reform and control of the royal household may be found, primarily, indeed, in the provocative folly of Edward II, but also in the departmental developments of the past half-century and in the baronial conviction that the best way to secure permanent control of policy was to make the heads of these departments answerable to a parliament dominated by the lords. In 1258, when another foolish king had provoked a political crisis, the great officers of state had been made answerable to a Council of Fifteen; in 1311, not only the great officers, but the lesser also, are made answerable to parliament. There is little to suggest that the Ordainers drew a sharp distinction between the two types of officers or that they thought of them as controlling two rival systems of administration, though no doubt they were aware that disproportionate influence was being exercised by such a person as the controller of the wardrobe through his misuse of the privy seal. Thus, the Ordinances are conservative in so far as their general objective—control of the king by his 'natural' advisers, the barons—is the same as in 1264 or in 1258. The appearance of novelty derives from the greater precision with which the details are specified and this in itself suggests a new awareness on the part of the barons of the real difficulties confronting them. But the Ordinances could provide no permanent solution; for the greater complexity of the administration did not eliminate its personal character. To name a single example, it can be seen how chimerical was the notion of parliamentary control of ministers when the king, going north to fight the Scots, thought nothing of bodily removal, not only of the wardrobe and privy seal, but also of chancery, exchequer, and common-law courts, so that he himself might remain in contact with them.

In so far as they were believed to spring from misconduct of officials or their subordinates, many of the points raised in the Ordinances can be linked with their general objective; and questions of foreign war, defence and finance, which probably appeared to the Ordainers to be second only in importance to the removal of evil counsellors, were closely associated with it. The king's political unwisdom and their own inveterate hostility to aliens provoked the attack on the Frescobaldi; but nowhere do the barons reveal themselves as more conservative than in their handling of the financial problem. Before their eyes was the outmoded and unrealistic concept of a king 'living of his own'; and, apart from economy and the resumption of alienated Crown property, they have no constructive suggestions to make. While they are eager to be rid of the foreign capitalists, to abolish the new customs, to prevent unwarranted purveyances, they nowhere suggest balancing the losses which the king must incur with a regular series of parliamentary grants, loans from native merchants, or any other kind of subvention. Here was, perhaps, their greatest mistake. For it was the financial predicament of late medieval kings which drove them to dependence, not only on the alien bankers, but also on the skilled administrators of the household as the only people able and willing to try to keep them solvent. The result was an impasse; for the magnates bitterly resented the king's dependence on his ministers and, as will be seen later, they were to single out as their victims men like Stapledon and Baldock who were seriously attempting administrative and financial reform. At the same time, by their own reluctance to sanction increases in the customs or to offer adequate parliamentary subsidies, they were making such dependence and the search for new sources of credit inevitable. Many of the grievances voiced in the Ordinances were, however, of long standing. They formed part of a solid deposit of abuses and they need not have come to a head so early in a new reign if intense jealousy of Gaveston had not served to stimulate a general sense of discontent and to put the barons in a mood for cataloguing complaints. Gaveston himself, so far as can be discovered, was innocent of any notions about government: the household arrangements were much as they had been in the latter years of Edward I and so was the state of the royal finances. The overmastering impulse behind the Ordinances was the desire to end the thraldom in which a weak king was held by a greedy and foolish favourite. It was this, not any desire to broaden the basis of the constitution or to alter the structure of government, which underlay the whole movement and gives it its essentially conservative character.

The sequel to the publication of the Ordinances affords a striking example of the gulf commonly dividing the enactment of a medieval statute from its enforcement as the law of the land. The Frescobaldi fled the country and the royal debt to them was never repaid; this, in conjunction with their misfortunes abroad, led to their speedy bankruptcy. A London citizen, John of

Lincoln, succeeded Amerigo dei Frescobaldi as keeper of the exchanges of London and Canterbury; Gaveston's adherent, John Hotham, was replaced by Henry Percy as warden of the forests north of Trent. But the earls evidently felt that more remained to be done, for they continued to meet and deliberate. A parliament of magnates was held in November 1311 and a second set of Ordinances was there promulgated, specifying certain individuals who were to be removed from court; they included the chamberlain, John Charlton, the keeper of the wardrobe, Ingelard Warley, and the cofferer, John Ockham. These changes were highly unwelcome to the king and when parliament was dissolved on 18 December he betook himself to Windsor declaring that he was being treated like an idiot.[1] As for Gaveston, one of Edward's first moves after the publication of the August Ordinances had been to provide him with a safe-conduct, 'coming to the king at his command', until All Saints' Day when his exile was due to begin.[2] They remained together until Gaveston sailed from the Thames on 3 November. His movements thereafter are exceedingly mysterious. The author of the *Vita Edwardi* says that he sailed to Flanders and that after that no one knew where he went.[3] His original destination seems to have been Brabant, for Edward had already written to his brother-in-law, Duke John, asking him to receive the earl of Cornwall 'quem in quadam peculiari praerogativâ dilectionis hucusque concepimus'.[4] The St. Paul's annalist says that he went to Bruges, Trokelowe that he went first to France but fled to Flanders on being driven out by Philip IV, the Lanercost chronicle that he was driven from Flanders through the influence of the French king. Wherever he went, it is clear that he soon returned. On 30 November the barons forced the king to commission Hugh Courtenay and William Martin 'to search for Peter de Gaveston supposed to be hiding and wandering from place to place in the counties of Cornwall, Devon, Somerset and Dorset'.[5] According to the *Vita Edwardi*, he was thought to be lurking, 'now in the king's apartments, now at Wallingford, now in the castle of Tintagel'.[6] With characteristic effrontery he

---

[1] *Vita Edwardi*, p. 21.   [2] *Foedera*, ii. 143. He is described as *dilectum et fidelem nostrum*.

[3] *Vita Edwardi*, p. 21.

[4] *Foedera*, ii. 144. Edward wrote also to the duchess and to the duke's lieutenant.

[5] Ibid., p. 151.

[6] *Vita Edwardi*, p. 21. The earldom of Cornwall was in the hands of Gaveston's kinsman, Bertrand de Calhau.

joined the king openly before Christmas and they kept the feast together at Windsor. From this time onwards, Edward's main objective was the building up of a party to overthrow the Ordinances.

As in the spring of 1310, his first move was to secure control of the great seal. Until nearly the end of the year, writs had been sealed under the eye of the Ordainers, but on 30 December some chancery clerks brought the seal to the king at Windsor. A few days later, Edward and Gaveston set out for the north accompanied by a number of the household officers.[1] By 7 January 1910 the officers of the chancery had received instructions to remove to York; they obeyed without delay and, appearing before the king there on 20 January, took up their quarters in St. Mary's abbey. The chancery thus secured, Edward turned his attention to the exchequer and conceived the bold plan of availing himself of the skill and experience of his father's old counsellor, Walter Langton. The sheriffs were ordered to restore his property and Edward wrote to the pope on his behalf; before the end of the month, his appointment as treasurer was announced. But the barons were determined not to admit him. Pembroke, Hereford, and John Botetourt confronted him within the walls of the exchequer itself and threatened him with the penalties of perjury, since he, like everyone else, was sworn to maintain the Ordinances.[2] Meanwhile, Edward had retained both Warley and Ockham and was taking steps to restore the position of Gaveston. On 18 January he himself drafted orders to the sheriffs (when they had been sealed in his presence, he laid them on his bed), to proclaim that the earl of Cornwall, who had been exiled against the laws and customs of the kingdom, which the king by his oath was bound to maintain, had now returned to the king by his command.[3] It was not only the barons who could play at the game of manipulating the coronation oath to suit their own purposes. On 20 January a second series of writs, also sealed in the king's presence, announced the restoration to Gaveston of all his lands and castles.[4] In March he was assigned the customs in the port of Berwick, in payment of a debt owing to him from the king; in April he

[1] A daughter was born to Gaveston in Jan., Bridlington, p. 42.
[2] Conway Davies, pp. 551-52. The barons sent an account of this incident to the king.
[3] *Foedera*, ii. 153.
[4] Ibid., p. 154.

was given the office of justice of the forest, south of Trent.[1] The
Londoners were ordered to see to the defences of their city and
to hold it for the king; the sheriffs were told to make weekly
proclamation of the king's peace and of the rights of the Crown.
A commission, consisting of the bishop of Norwich, John Salmon
(an Ordainer of royalist sympathies), six knights, and six clerks,
was appointed to treat with the Ordainers for the revision of such
parts of the Ordinances as were prejudicial to the king or to any
other person. The custody of Bamborough, which had been
given to Henry Percy, was restored to Lady de Vescy; the Isle
of Man was regranted to Henry de Beaumont; loans were
raised and purveyances taken;[2] fortresses were placed in the
hands of the royalists. Armed conflict had become inevitable;
and the terms on which the king granted custody of Scarborough
castle to Gaveston show that he was prepared for the worst.
Gaveston's orders were to deliver the castle to no one but the
king himself and, if it should happen that the king were brought
a prisoner there, not to deliver the castle to him or to anyone
else. In the event of the king's death, the castle was to remain to
Gaveston and his heirs.[3] The author of the *Vita Edwardi* repeats
a current rumour that Edward, knowing Gaveston to be in
danger both in his own dominions and in France, planned to
send him to Scotland and that he entered into negotiations with
Bruce to whom he offered the Crown of Scotland in perpetuity;
but Bruce replied, 'How shall the king of England keep faith
with me when he does not observe the sworn promises made to
his own liegemen?'[4] Tales of this kind were bound to circulate
when the king and Gaveston were known to be perambulating
the north parts, probably in the hope of raising a sufficient army.

Meanwhile, the Ordainers had taken up the king's challenge.
Gaveston's return and restoration to favour were tantamount to
a declaration of war. Winchelsey stood firmly by the barons and
declared Gaveston excommunicate, while Lancaster, Pembroke,
Hereford, Arundel, and Warwick formed a confederacy for the
defence of the Ordinances and even Gloucester offered his help.
A plan of defence was framed. Gloucester was to hold London
and the south; Hereford, Essex and the eastern counties;

---

[1] *Cal. Pat. Rolls 1307-13*, pp. 449, 450.
[2] It was about this time that a new Italian creditor, the Genoese Antonio
Pessagno, began to supply Edward with loans and with goods on credit.
[3] *Cal. Pat. Rolls 1307-13*, p. 454.   [4] *Vita Edwardi*, p. 22.

Lancaster, the west of England and north Wales, 'lest there be a tumult among the people'.[1] Robert Clifford and Henry Percy were to bar the possible ways of escape across the northern border, while Pembroke and Warenne were to go to the king and seize Gaveston. Warlike preparations were begun under the convenient cover of tournaments; and, since the archbishop could take no part in such activities, leadership of the opposition group began to pass to Thomas of Lancaster, as the richest and most powerful of the earls. It was about this time that he sent a letter to the queen telling her that he would not rest until he had rid her of the presence of Gaveston.[2] On 4 May 1312, when Edward and Gaveston were at Newcastle, they were surprised by the news that Lancaster, with a large force, was descending on the town. King and favourite escaped down the river to Tyne-mouth just in time to avoid capture; next day they made for Scarborough where Gaveston took refuge, while Edward retired to York. The people of Newcastle had shown no disposition to put up a fight and town and castle fell without a siege. Lancaster seized the royal servants, arms, treasure, and horses and kept his army in being so as to intercept Gaveston should he try to escape. Meanwhile, Pembroke and Warenne were carrying out their appointed task of advancing on Scarborough to seize him. Within a fortnight, lack of supplies had forced Gaveston to come to terms. Robert of Reading, who clearly thought the terms too lenient, suggests that Pembroke had been bribed by the king;[3] for Gaveston's personal immunity was assured. His men were to be allowed to hold Scarborough until the following August and, if no agreement as to his fate had then been reached in parliament, he was to be allowed to return to the castle. This settlement, dated 19 May 1312, was attested by Pembroke, Warenne, and Henry Percy, all of whom swore on the Host to maintain it.[4]

Pembroke's intention was to take Gaveston to his own castle of Wallingford; but when they reached Deddington in Oxford-shire the prisoner begged for a rest, and the company put up in the house of the local rector. As his countess was at the neigh-bouring manor of Bampton, Pembroke took the opportunity to

[1] *Ann. Lond.*, p. 204.
[2] Trokelowe, *Annales*, pp. 75-76.
[3] *Flores*, iii. 150-1. Pembroke is said to have been offered £1,000. See *Vita Edwardi* (p. 24) where it is stated that Pembroke acted on his own authority and pledged his lands and tenements for Gaveston's safety.
[4] *Ann. Lond.*, p. 205; Bridlington, p. 43.

visit her, leaving Gaveston in charge of his servants. This proved to be a fatal error of judgement; for news of Gaveston's proximity reached Guy Beauchamp at Warwick and very early in the morning of Saturday, 10 June, he appeared with a small force and surrounded the rectory. According to the chroniclers, Gaveston saw them approaching when they were yet a little way off and looked about him wildly in the hope of finding a way of escape; but he was trapped. Warwick approached the house shouting, 'Get up, traitor, you are taken!' and Gaveston was forced to throw on some clothes and descend the stairs. Barefoot and bareheaded, brought forth, not as an earl but as a common thief, he had to set out on foot; only when the company was half-way to Warwick, was he mounted on a mare to hasten his progress. Followed by a jeering crowd he reached the castle and was thrown into one of its dungeons to await events.[1]

The news of Warwick's violent action filled Pembroke with dismay. According to the *Vita Edwardi*, he at once sought out the earl of Gloucester and begged his help in saving his name from dishonour and his property from forfeiture; receiving no encouragement, he then made for Oxford and laid his case before the clerks and burgesses; but as they were unwilling to undertake the siege of Warwick castle, the second appeal met with no better fortune than the first.[2] Lancaster, Hereford, and Arundel were all at Kenilworth, a few miles distant from Warwick, and after a brief consultation they made up their minds. On Monday, 19 June, they took Gaveston out of Warwick castle and brought him to Blacklow Hill, in Lancaster's territory, on the road between Warwick and Kenilworth. Here, in the presence of the three earls (for Warwick remained prudently in his castle), Gaveston was executed by two Welshmen of Lancaster's retinue who reserved his head for their master.[3] Four cobblers bore the headless body back to Warwick on a ladder; but the earl refused to receive it and it was left to some Oxford Dominicans to afford it a temporary refuge in their

---

[1] The London annalist (pp. 206-7) and Robert of Reading (*Flores*, iii. 151-2) supply the liveliest accounts of Gaveston's arrest.

[2] It is tempting to see in Pembroke's action an early instance of an appeal to academic arbitration; but as the appeal was not to the masters but to the clerks and burgesses, it is likely that military aid was his objective.

[3] According to the *Vita Edwardi* (p. 26) it was Gaveston's kinship with Gloucester that saved him from hanging.

convent. They dared not bury it in consecrated ground as Gaveston had died excommunicate.

Stubbs felt no hesitation in declaring that Gaveston was killed 'illegally, if not unrighteously';[1] but the legal position was by no means clear. If the Ordinances were still in force, a plausible case could be made out for the summary treatment accorded to an outlaw who had been designated a public enemy, should he return from exile, and so might be killed without thought of trial. In the subsequent peace negotiations, Lancaster and Warwick countered the king's offer of a general pardon for the death of Gaveston by the argument that such a pardon would imply that Gaveston was under the law at the time of his execution. If this were so, they could be proceeded against for homicide, a suggestion which they repudiated; for they claimed that they had acted lawfully against an outlaw and an enemy of the king and the kingdom, one so designated under the Ordinances.[2] The canon of Bridlington, in a version of events peculiar to himself, tells us that Gaveston was brought before two justices of jail delivery at Warwick and that, by authority of the Ordinances, *which they did not know the king had revoked,* they held that he should be beheaded as a traitor. The canon adds that this execution of an earl, the peers of the land being neither summoned nor present, aroused undying hatred between the king and the magnates. His line of argument seems to be that, if the Ordinances were still in force, Gaveston could be executed summarily; had the justices known that the king had revoked them, they would have advised differently for, if that were so, Gaveston was an earl, entitled to formal judgement by his peers.[3] The author of the *Vita Edwardi,* who also argues the point, stresses the liability of Lancaster on whose land Gaveston was killed and to whom his head was sent. In killing Gaveston the earls, he thinks, took a great responsibility upon themselves, for they slew a great lord whom the king had adopted as a brother

[1] *Constitutional History,* ii. 348.

[2] See E. A. Roberts, 'Edward II, the Lords Ordainers and Piers Gaveston's Jewels and Horses, 1312–13', *Camden Miscellany,* xv (1929), 16.

[3] Bridlington, pp. 43–44. The two justices were not, of course, *trying* Gaveston who, on this view, was entitled to no trial: they were giving what can best be described as a legal opinion. Some plausibility is lent to this tale by the fact that Bridlington supplies the names of the justices—William Inge and Henry Spigurnel —for although, as Sir James Ramsay pointed out (*Genesis of Lancaster,* i. 46, n. 1), these two held their commissions, not in Warwick, but in Leicestershire and Nottinghamshire, it is conceivable that their advice was sought by the earls.

and cherished as a son, who was his companion and friend. Such a deed needed to be defended by a great man and Lancaster, as the most nobly-born and powerful of them all, took the charge upon himself and ordered Gaveston who had been thrice exiled, given, as it were, three lawful monitions, to be executed.[1] In other words, an unprecedented emergency demanded unprecedented action and it was lawful for the greatest of the king's subjects to take the law into his own hands. The earls, we must suppose, held the Ordinances to be still valid; but Edward II had specifically repealed the clause relating to Gaveston and had restored his lands and titles. His right to do so might be questioned; but it may be recalled that, thirty years later, Edward III did not hesitate to annul a statute by letters close to the sheriffs, on the plea that he had conceded it under coercion.[2] Edward II's action could be construed as a breach of his coronation oath: but all the evidence suggests that most contemporaries felt no certainty that, in June 1312, the ordinance touching Gaveston was legally enforceable. Whatever view is taken of its legality, however, there can be no doubt that the summary execution of one who might at least claim to be an earl established a most dangerous precedent; for it is the first of the long series of quasi-judicial murders which darkens the history of the fourteenth century.[3] Pembroke does not appear to have charged Lancaster formally with perjury, but clearly he felt that he had played him false. The leading Ordainers were members of a confederacy and, on any view, Lancaster, Arundel, Hereford, and Warwick should have held themselves bound by the solemn pledge which Pembroke, Warenne, and Percy had given at Scarborough.

The immediate effect of the execution of Gaveston was the dissolution of this confederacy. Pembroke and Warenne went over to the king's side and began to work with the men they had formerly despised.[4] The courtiers urged Edward to collect an

[1] *Vita Edwardi*, p. 28.
[2] Below, p. 177.
[3] Cf. the comment on Gaveston's murder made by an anonymous northern chronicler writing *c.* 1327: 'O viri execrabiles, mors execranda! O mors nepharii quam nephanda! O mors impii impiissima! O mors scelerati sceleratissima! Nec immerito est mors eius tam vilis et reproba censenda, cuius pretextu tantus cruor et tam preciosus falso et maliciose effunditur subsequenter.' G. L. Haskins, 'A Chronicle of the Civil Wars of Edward II', *Speculum*, xiv (1939), 76.
[4] *Ann. Lond.*, p. 208: Lanercost, p. 219. Warenne received a grant of markets so early as 24 July; Pembroke was given the New Temple and other property of the

army, and within a few weeks of the murder of his friend he
felt strong enough to return to London, where the mayor had
already been ordered to take the city into the king's hands and
to seize war-horses and armour for his use.[1] Some of Edward's
counsellors, however, were still of opinion that civil war must
at all costs be avoided; for if the king were taken the triumph of
the Scots and the destruction of the realm must follow.[2] Though
the barons were still in arms, these wiser counsels prevailed.
The king was not really in a position to fight, for the wardrobe
receipts had fallen heavily and these losses were only partly
offset by the forfeited revenues of the Templars; and his grief
for Gaveston was in some degree mitigated by the birth of his
elder son on 13 November.[3] Thus, the way was open for compromise; and Cardinal Arnold and the bishop of Poitiers, whom
the pope had sent to England in response to an appeal from
Edward II, supported by the queen's uncle Prince Louis of
Évreux, the earl of Gloucester, and some of the bishops, set to
work to arrange a settlement. Negotiations between the two
parties were pursued under a series of letters of safe-conduct and
there were long debates in the parliament which sat throughout
the autumn of 1312.[4] Hereford took a leading part in these, but
neither Lancaster nor Warwick associated themselves with him
and it remained very doubtful whether any agreement that
might be reached could be enforced. What was optimistically
termed a 'final peace' was proclaimed before Christmas, the
barons to offer humble apology to the king, in return for a
general amnesty. Lancaster and Warwick reacted to this announcement by sending in a list of objections arranged under
twenty heads; and, though the jewels taken from Gaveston
were restored in February, negotiations dragged on for the best
part of another year. Their early stages can be followed in detail
in the careful report which the cardinal and the bishop of
Poitiers submitted to Clement V at the end of their mission,

Templars in December, *Cal. Charter Rolls*, iii. 194, 203. Royal writs prohibiting
conventicles and confederations 'to live and die together' were issued on 20 July,
Foedera, ii. 172.

[1] *Cal. Letter-Book D*, p. 290.   [2] *Vita Edwardi*, pp. 30–32.

[3] Gaveston's castle of Knaresborough was granted to the infant prince in Dec.
1312. According to Geoffrey le Baker (p. 6) Edward was merely concealing his
grief for Gaveston at this time.

[4] One of the most difficult questions proved to be the return to the king of the
jewels, horses, and arms seized from Gaveston at Newcastle. For these negotiations
see *Camden Miscellany*, xv.

revealing their own persistence in the cause of peace and the generous support lent them by Gloucester. Meanwhile the nation was kept in a state of nervous tension by constant proclamations and cancellations of tournaments, by orders to the earls not to move about the country in arms and by their refusal to attend councils because they were unwilling to trust the king's safe-conducts. Though there were neither battles nor sieges, there was something very like a cold war. Ineffective parliaments met in the spring and summer of 1313;[1] Edward followed the example of his grandfather Henry III and escaped from the scene of his troubles to pay two visits to France. The death of Winchelsey in May removed one of his most dangerous enemies. His recall in 1308 had been a crass blunder; and the king was determined to ensure that the next archbishop should be a man of different calibre. Clement V had already reserved the see, but within a fortnight of Winchelsey's death the Canterbury monks elected Thomas Cobham, a man of good family, learning, and rectitude. Edward, however, persuaded the pope to overset the election and, by means of an undoubtedly simoniacal transaction, Walter Reynolds was appointed to the primacy.[2] With a friend at Canterbury the king's position was greatly strengthened; and in October 1313 the recalcitrant lords at last consented to make public apology in Westminster Hall. Two days later pardons were issued to some 500 lesser offenders, a high proportion of whom were Lancastrian dependants.[3] Neither side was sincere in its expressions of amity; but there can be no doubt that the settlement represented a substantial victory for the king. The Ordainers had to be content with terms which contained no reference at all to the Ordinances; and by accepting the king's pardon they could be held to have conceded his point that the Ordinances were invalid when Gaveston was slain and thereby to have admitted their guilt in the matter of his death. Like Henry III before him, Edward II emerged from the struggle with his barons bound by no new restraints on his liberty of

[1] The one positive achievement of these months was the Ordinance of the Staple. Below, pp. 350-1.
[2] According to the canon of Bridlington, Reynolds gave 1,000 marks for the see (p. 45); another source puts the figure at 32,000 (Chron. Mon. de Melsa, ii. 329). Reynolds himself spoke of 'non modica et intolerabilis pecunia' (Eng. Hist. Rev. lvi. 99).
[3] Their names are printed in Foedera, ii. 230. About the same time Pembroke released to Lancaster some of the lands formerly the property of the Templars, perhaps as a gesture of reconciliation. See Somerville, Duchy of Lancaster, p. 24.

action. None of these points was stated explicitly; but the record of the negotiations makes it clear that both parties were well aware of them. Lancaster had suffered what amounted to a major humiliation, and it was exceedingly unlikely that a man of his temper would be prepared to accept such a settlement as final.

Summing up the record of Edward II's first six years, his observant biographer can find little solid achievement to record.

> He has achieved nothing laudable or memorable, save that he has married royally and has begotten an heir to his crown. . . . If our King Edward . . . had not accepted the counsels of evil men, none of his ancestors would have been nobler than he. For God had endowed him with every gift. . . . At the beginning of his reign he was rich, with a populous land and the goodwill of his people. He became son-in-law to the King of France, first cousin to the King of Spain. If he had followed the advice of the barons he would have humiliated the Scots with ease. . . . If he had devoted to arms the labour that he expended on rustic pursuits, he would have raised England aloft: his name would have resounded throughout the land.[1]

The author's comment is probably a fair enough reflection of public opinion at the time. What was agitating the public mind was the question of Scotland virtually forgotten, it would seem, in the course of the dreary wranglings between the king and his subjects. The nation had no desire to see Edward II evince bellicose qualities in a civil war. What was wanted was that he should display his prowess in the conflict with the Scots which could not be postponed much longer.

[1] *Vita Edwardi*, pp. 39–40.

II

FROM BANNOCKBURN TO
BOROUGHBRIDGE (1314–22)

EDWARD II had not been many months on the throne before
it became apparent that English policy towards Scotland
would no longer be guided by any consistent or statesman-
like purpose. Edward I had placed the subjugation of Scotland
in the forefront of his ambitions: for his son, Scottish affairs and
the defence of the northern border would always be subordinate
to his private loves and hates. England's weakness was Bruce's
opportunity; and in the years between Edward's accession and
the battle of Bannockburn he concentrated his strategy on three
main objectives. He had to secure the open country, large areas
of which were still under the control of his Scottish opponents;
he had to fill his war-treasury, and the method which commended
itself to him was to subject the north of England to the devastating raids which forced the inhabitants of the border counties to
pay large sums for temporary truces; and he had to recover the
castles of Scotland from their English garrisons. Before he met
the English at Bannockburn, Bruce had gone far towards achieving all his aims. He defeated John Comyn on the Aberdeenshire
coast in December 1307 and a few months later routed him at
Inverury in his hereditary earldom of Buchan and recovered
most of the east Highlands. Argyll was overrun that autumn and
Edward Bruce recovered Galloway. In 1311, while Edward II
was meeting the Ordainers, Robert Bruce crossed the Solway
and swung eastward into Tynedale; a month later, he had
burned Corbridge and the men of Northumberland were offering
him £2,000 for a truce. Next year Bruce embarked upon an even
more daring raid. The towns of Hexham and Durham were
gutted and James Douglas penetrated to Hartlepool, returning
with booty and prisoners. The men of the Border, knowing that
they could look for little help from their own king, paid immense
sums for a further series of short truces.[1] Meanwhile, one by one,

---

[1] The men of Northumberland were unable to contribute anything to the subsidy
of 1309; Northumberland, Cumberland, and Westmorland all had to be exempted
from contribution in 1313 and for the rest of the reign; see J. F. Willard, *The Scotch*

the castles were being recovered. A rash attempt on Berwick met with failure, but Linlithgow fell to the Scots in 1311, Perth, Roxburgh, and Edinburgh in the early months of 1313. The spring of that year, in which Edward Bruce began the fateful siege of Stirling, saw the restoration of Scottish rule in the Isle of Man. Robert Bruce now had behind him something like a united Scotland. The kindred and friends of the murdered Comyn had submitted; many of the Anglo-Norman barons had been won over; and at Dundee, in February 1310, the Scottish clergy had accorded him formal recognition as king.

English reactions to these events were vacillating and hesitant. Edward II had gone south for his coronation, leaving Bruce to fight his Scottish enemies without their English allies. An elaborate series of writs which went out in the spring of 1308 to the sheriffs and others to furnish supplies for a Scottish campaign was cancelled within a couple of months. Since Edward was unwilling to leave England, he was ready to listen to suggestions from Clement V and Philip IV of France, advising the conclusion of a truce. In 1309 there was talk of an expedition against truce-breakers; but when Sir John Segrave went to the Border, he agreed privately with Bruce for a suspension of hostilities and advised Edward to postpone operations until the summer. It was a winter of severe frost and the problem of finding fodder for the horses was only one of many obstacles to a campaign in Scotland at this season.[1] Next year, when the Ordainers were in control of affairs in England, Edward was glad enough of a pretext to go north. Ingelard Warley was dispatched 'throughout the kingdom' to arrange for the provision of supplies for a war in Scotland. He raised a loan of 3,000 marks from the earl of Lincoln and substantial contributions from some of the religious houses, though the king professed himself disappointed at the poor response from others.[2] Moreover, most of the earls ignored Edward's summons to follow him and his long sojourn on the Border achieved little, while the castles were largely neglected. An order for supplies to go to Perth in October 1312 had not been fulfilled by December and the town and castle fell a few weeks later. A small sum was sent to the governor of

Raids and the Fourteenth Century Taxation of Northern England (Univ. of Colorado Studies v, 1908), pp. 238–39.
[1] Lanercost, pp. 213–14.
[2] *Cal. Pat. Rolls 1307–13*, pp. 280, 286; *Cal. Close Rolls 1307–13*, pp. 260, 278.

Stirling, Sir Philip Mowbray;[1] but it was shortage of supplies which led Mowbray to come to an agreement with Edward Bruce in the summer of 1313.

Built in 1288, the stone castle of Stirling stood to the north-west of the medieval town on a precipitous rock overlooking the Forth. Its unique strategic importance derived from its command of the road running north to the Highlands from Falkirk and of the fords over the river. It had ready access to the sea and to reduce it by starvation should have been an almost impossible feat. None the less, Bruce was determined to attempt the capture of this key fortress and, after establishing the blockade, he left his brother with a force sufficient to guard the landward approaches and to maintain the siege. Supplies soon ran low within the castle and Sir Philip Mowbray persuaded Edward Bruce to agree to a covenant to raise the siege on condition that, if an English army had not appeared within three leagues of the town by Midsummer Day 1314, the castle would be surrendered. Even Edward II could hardly ignore such a challenge and Robert Bruce fully recognized that by the compact the Scots had been 'set in jeopardy';[2] for his brother had committed him to the pitched battle with the English which he had wished at all costs to avoid. He was well aware of the inferiority of his cavalry and archers to those of his opponents and he feared that a single battle might cost him the fruits of all his labours, his Crown, and the freedom of his country.

Mowbray, meanwhile, hurried south to impress the strategic importance of Stirling upon Edward II who professed himself eager for a trial of strength, and in February 1314 announced that he was determined to invade Scotland soon after Easter.[3] Brushing aside the protests of the earls who reminded him that, under the Ordinances, he should not engage in war without consent of parliament, he sent out orders for the mustering of a great army and enormous quantities of supplies. The number of troops raised cannot be estimated, however, with anything approaching precision. Barbour's figure of 100,000 infantry, 40,000 cavalry, and 50,000 archers is clearly incredible, but some modern attempts to reduce the size of the English army to between 6,000 and 7,000 infantry with 450 to 500 cavalry, are

[1] £100 worth of victuals, *Cal. Close Rolls 1307-13*, p. 399.
[2] Barbour's *Bruce* (Early Eng. Text Soc.), ii. 259.
[3] *Cal. Chancery Warrants*, i. 395.

hardly less so.[1] All the chroniclers make it plain that it was a very impressive force which left for the Border in May 1314; the figure of 20,000 may not be far wide of the mark.[2] Many of these were seasoned troops. The best of the infantry were highly skilled archers from Wales and the marches; combined with them were levies from the midlands and north-west, who had seen service under Edward I, and, possibly, some Irish. It was a force well adapted to the country in which it would have to fight, for Edward had already proclaimed that he had heard that 'a great part of the exploit will come to footmen'.[3] Among the magnates who accompanied the king were the earls of Gloucester, Pembroke, and Hereford. The army with its immense train of baggage-wagons was at Berwick by 12 June: and here Edward took occasion to confer the Scottish lands of John Graham and Thomas Randolph upon Hugh Despenser the younger.[4] The Tweed was crossed on 17 June and four days later the English were in Edinburgh where they paused to collect stores from their ships. Time was running short, however, and on 22 June the advance was resumed. It was a body of tired troops which reached Falkirk in the evening after a march of twenty miles.

Reconstruction of the events of the next twenty-four hours cannot be other than conjectural. Of those who described the battle of Bannockburn only one was an eyewitness[5] and all but one were ecclesiastics. John Barbour, archdeacon of Aberdeen, whose *Bruce* preserves by far the most graphic and circumstantial account, did not write until 1375. Sir Thomas Gray's *Scalacronica*, the work of a soldier whose father had been present at the engagement, was composed in Edinburgh jail in 1355. The English monastic chroniclers wrote far from the event and with little understanding of tactics. Thus, there is room for almost unlimited speculation, and that, not only on points of detail, but on such a central matter as the site of the battle itself. The result has been a learned debate of some acerbity.[6] There is

[1] *Complete Peerage*, xi, App. B, pp. 11–13.

[2] W. Mackay Mackenzie, *The Battle of Bannockburn* (1913), p. 30.

[3] *Cal. Chancery Warrants*, i. 398.   [4] *Cal. Documents relating to Scotland*, iii. 69.

[5] Robert Baston, an English Carmelite, whom Edward II took with him to celebrate the victory in verse. Taken prisoner by the Scots, Baston was forced to devote his unremarkable talent to writing up their triumph in verses which add nothing to our knowledge of the battle. They were published by W. D. Macray in *Eng. Hist. Rev.* xix (1904), 507–8.

[6] See W. Mackay Mackenzie, *Barbour's Bruce* (1909); *The Battle of Bannockburn*

general agreement, however, that Bruce had taken up his position on the plateau north of the Bannock burn, expecting that the English would try to force their way along the so-called Roman road from Falkirk to Stirling. In anticipation of this, he ordered the digging of a number of small holes, or 'pottes', the depth of a man's knee, on either side of the road: covered with turf and branches, they were intended to serve as a deterrent to the English horsemen.[1] If, however, the English should prefer a route farther to the east, the Scots could swing round to face them, with no great difficulty. Bruce divided his army, which may have numbered between 6,000 and 10,000, into four infantry 'battles', with a small cavalry reserve under Sir Robert Keith, and retired into camp in the wood known as the New Park, out of sight of the approaching enemy. It was his intention to fight a defensive action on foot, on ground of his own choosing,

> And wreik on thame the mekill Ill
> That thai and tharis has done us till.[2]

The English army approached the Bannock on the afternoon of Sunday, 23 June, and halted to the south of it. Much of the upper course of the burn then ran through a deep gorge, the lower course ran through pools and marshes. Mowbray came down to meet the king and reminded him that, as he was now within three leagues of Stirling, he need come no farther for, under the terms of the covenant, the castle was relieved and the enemy, he said, had blocked all the narrow roads through the New Park. But the younger English captains were not thus to be baulked of the victory which seemed to lie within their grasp and two bodies of cavalry were sent forward to reconnoitre. The first of these, led by Sir Humphrey Bohun, advanced up the Roman road; at the top of the slope they encountered Bruce, mounted on a grey palfrey and wearing his crown over a leather helmet. The clash was brief but decisive. Bohun charged the Scottish king, only to be felled by a stroke of his battle-axe and the rest of his party then withdrew. Lord Robert Clifford and

(1913); *The Bannockburn Myth* (1932); J. E. Morris, *Bannockburn* (1914); Sir H. Maxwell, 'The Battle of Bannockburn', *Sc. Hist. Rev.* xi (1914), 233–51; Sir C. Oman, *Art of War in the Middle Ages* (2nd ed. 1924), ii. 84–100; Rev. T. Miller, *Site of the Battle of Bannockburn* (1931); Major A. F. Becke, 'The Battle of Bannockburn', *Complete Peerage*, xi (1949), App. B; and review of the last by J. D. Mackie, *Sc. Hist. Rev.* xxix (1950), 207–10.

[1] In the event, the famous potholes played very little part in the battle.
[2] *Bruce*, ii. 291.

Henry de Beaumont were meanwhile feeling their way round to
the east of the Scottish position, probably with the object of
reinforcing the garrison of the castle, for their numbers were too
small for them to have attempted to surround the enemy camp.

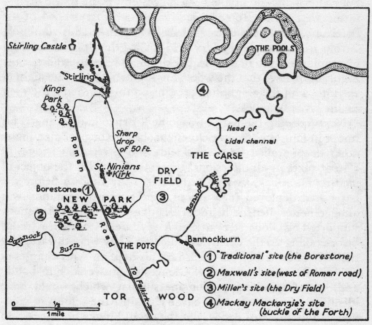

FIG. 1. Suggested sites for the Battle of Bannockburn.

Somewhere in the vicinity of St. Ninian's kirk they encountered
the earl of Moray who repulsed them after a violent exchange;
among those taken prisoner was Sir Thomas Gray, the father of
the chronicler. These two reverses had a bad effect on English
morale. The passage by the Roman road was now judged too
dangerous and the generals were faced with the problem of
finding an alternative crossing. When and where they found it,
where the English army spent the night of 23–24 June, and
where the battle itself was fought, cannot be determined with
certainty. It seems likely that the English chose to cross about a
mile below Bannockburn village, near the point where the river
turns north; and that somewhere in this 'evil, deep, wet marsh'
they pitched their camp on Midsummer Eve. The baggage-trains

can hardly have been brought across but most of the troops
may have forded the river before nightfall. It is possible, however, that the whole army spent the night south of the burn, the
troops crossing next morning, only in order to fight the battle.[1]
Four alternative sites have been suggested for the battle itself—
to the west of the Roman road; in the neighbourhood of the
Borestone, where Bruce is said to have planted his standard;
farther to the east in the 'dry field' below the slope down from
St. Ninian's kirk; or in the buckle of the Forth where some
historians believe that the whole English army spent the night of
23 June, with their right flank resting on the Forth and their left
on the river Bannock.[2]

Next morning the battle was joined. Bruce had arranged his
troops in four 'schiltroms', closely packed oblongs of men many
ranks deep, walled by their shields and presenting hedges of
18-foot pikes in all directions. It seems that, with the object of
inciting an attack, they moved towards the English before the
latter had deployed their troops. The Scottish formation was
altogether unsuited for an initial assault and the English veterans,
supported by Gloucester, advised a day's delay to rest the troops
before taking up the enemy's challenge. They were thwarted by
the folly of the king who turned on his nephew with charges of
prevarication and treachery. Gloucester answered him boldly
and, without waiting to don the surcoat which would have
identified him, plunged his cavalry against the schiltrom commanded by Edward Bruce. But they could make no impression
on the hedge of pikes and the young earl himself fell early in the
fight. Meanwhile, Moray and Douglas engaged the remainder
of the front-line cavalry, fighting desperately on a narrow front.[3]
Behind the cavalry the English infantry were almost immobilized:
when the archers tried to assail the Scots they found themselves
shooting their own men in the back; and the Scots succeeded in
using their light horse under Sir Robert Keith to launch a flank
attack on the archers. Soon, both armies were locked in an

[1] The statement of the *Scalacronica* (p. 142), that the English lay 'en un plain
deuers leau de Forth outre Bannokburn', seems to me, however, to indicate beyond
reasonable doubt that they camped *north* of the Bannock. The suggestion that the
author was repeating what he had heard from his father who, as a prisoner of the
Scots, would be looking at the battle from the north and that 'outre Bannokburn'
therefore means *south* of the burn seems over-ingenious and irreconcilable with the
allusion to the Forth.   [2] See p. 70 below.
[3] Barbour stresses the 'gret stratnes of the plase' (*Bruce*, ii. 299).

appalling mêlée. The English showed no lack of courage; but, when some Scottish camp-followers and guerillas began to stream down a hill to the west, behind the Scots, they took them for a reserve force and their resistance crumbled. With shouts of 'On them! On them! On them! they fail!', the Scots pressed home their advantage, thrusting the enemy into the marsh and river where great numbers were drowned or suffocated. Edward himself made for Stirling, but the governor refused him admission, reminding him that the castle must now be surrendered to the Scots. Making a long détour and hotly pursued by Douglas, the king and his friends headed east for Dunbar, whence they fled by sea to Berwick.

Seldom, if ever, had English arms suffered a disaster comparable to Bannockburn, and the dismayed and bewildered chroniclers are hard put to it to find an explanation. For most of them the defeat appeared as a moral judgement. Edward II had robbed the monasteries on his way north; the English had spent the night before the battle in carousing, while the Scots were fasting and praying; it was the *peccata enormia* of the vanquished that brought about their downfall. The Lanercost chronicler sees the absence of Lancaster and some of his fellow-earls as contributing to the defeat; Trokelowe suggests more reasonably that the English lost the battle because they attacked with tired men and beasts in difficult country. Only the author of the *Vita Edwardi* comes near to perceiving the military significance of the battle when he compares Bannockburn with Courtrai where, twelve years earlier, the townsmen of Flanders had shattered the feudal chivalry of France.[1] In 1315, at Morgarten, the pikemen of the Swiss cantons were to inflict a comparable defeat on the feudal armies of the Habsburgs. These three great battles opened a new period in the history of war. On the battlefields of France the English were soon to show that the lesson of Bannockburn had not gone unrecognized. Henceforward, the archers, the core of the army, were deployed in such a way as to protect, not to impede, the cavalry and infantry; and the long bow, effectively used, was to render obsolete the methods of fighting which had been common in western Europe since the eleventh century.

[1] Lanercost, pp. 224–5; Trokelowe, p. 87; *Ann. Lond.*, p. 231; Baker, p. 7; *Vita Edwardi*, p. 56.

More immediately, the results of Bannockburn were many and disastrous. The most obvious was the casualty-list. Though figures are not available, it seems clear that the English army as a fighting-force was temporarily destroyed. The death of the gallant young earl of Gloucester entailed the break-up of the great Honour of Clare, *altera columna Angliae*, among his three sisters, and the scramble for its spoils led directly to the civil wars of Edward II's latter years. The earl of Hereford was taken prisoner, as was Roger Northburgh, keeper of the privy seal, with all his clerks and records and the seal itself; the steward of the household, Sir Edward Mauley, perished in the Forth. Altogether, some forty knights were killed or captured;[1] while the Scottish casualties seem to have been relatively light. Robert Bruce became forthwith a national hero. Sentence of forfeiture was passed on all who refused to recognize his authority; abroad, he was regarded as an ally much better worth cultivating than his English opponent. Although he had to wait some years for official recognition in Europe and for the lifting of the ban of the Church, he was, says the Lanercost chronicler, 'after his victory . . . commonly called King of Scotland by all men because he had acquired Scotland by force of arms'.[2] Bannockburn proved the most important single factor in ensuring the ultimate independence of Scotland and saving her from the fate of Ireland.

For the inhabitants of the northern counties of England the defeat spelt ruin. The Scots now took the initiative, harrying the border shires in a series of devastating raids and penetrating farther and farther into England. Within six weeks of their victory, Edward Bruce and James Douglas broke into Northumberland and Durham while some of their followers crossed the Tees into Yorkshire and even ventured south of Richmond. The advent of winter afforded no respite, for the Scots were again on the Tyne by Christmas. The raids continued throughout 1315; in 1316 the Scots were once more in the heart of Yorkshire, and Lancashire was wasted as far west as Furness. Next year the north was the scene of an even more sensational outrage, when the Cardinals Gaucelin of Eauze and Luca Fieschi, sent by John XXII to make peace between the rival factions in England and to consecrate the bishop of Durham, were attacked by a gang of desperadoes led by Sir Gilbert

[1] *Ann. Lond.*, p. 231.   [2] Lanercost, p. 228.

Middleton whose castle at Milford on the Wansbeck had become a centre of highway robbery. According to the *Scalacronica* his activities were expressive of the disgust felt in the north at the king's failure to relieve the miseries of the Border. Middleton's act of sacrilege had, indeed, the effect of galvanizing the English government into one of its rare bursts of activity. Proclamations were issued for the arrest of the offenders, and Middleton suffered execution as a traitor. Bruce, who had refused to admit the legates over the Border, was excommunicated afresh and Scotland was laid under interdict. None the less, by April 1318 the Scots had attained one of their principal objectives in the capture of Berwick. These disasters in the north added to the general confusion and weakening of morale consequent on Bannockburn and helped to undermine the authority and prestige of Edward II still further. Nor was it only in England that the English could now be attacked; his triumph at Bannockburn opened the way for Bruce to Ireland.

Although Edward I never set foot in Ireland, he was well served there by his justiciar, Sir John Wogan, who had been twelve years in office when the old king died. About three-quarters of Ireland was nominally subject to English control and under Edward I this control was more of a reality than ever before.[1] Apart from the short interruption of Gaveston's term as lieutenant, Wogan continued as justiciar until 1312, his main preoccupation being the enlistment of Irish mercenaries for the Scottish war. These troops were under the command of Richard de Burgh, the 'Red Earl' of Ulster, a vigorous and able ruler who was restoring some degree of order in his turbulent earldoms of Ulster and Connaught. On the eve of Bannockburn, the prospects for the Anglo-Irish colony probably looked less gloomy than at any time since the landing of Strongbow. But underlying the apparent strength of the English government were deep and dangerous divisions. There were rivalries among the native Irish themselves and among the Anglo-Irish lords, exemplified, the one by the succession wars of the O'Conors in Connaught, the other by conflicts between the Lacys and the Verdons in Meath and Louth. The old-established 'conquest families' were suspicious of such absentee landlords as Gilbert of Gloucester,

[1] For the condition of Ireland under Edward I see Sir Maurice Powicke, *The Thirteenth Century* (vol. iv of the present *History*), pp. 560 ff.

hereditary lord of Thomond, and Roger Mortimer of Wigmore who arrived in 1308 to claim his liberties of Trim and Leix. More significant still was the great resurgence of the Celtic peoples of Ireland, Scotland, and Wales against their English conquerors. Given fresh impetus by the Edwardian conquest of Wales and by the Scottish war of independence, and openly encouraged in Ireland by Franciscans, Cistercians, and secular clergy, this was rapidly assuming the guise of a great confederacy. The *lingua Scotica* still meant the common Gaelic tongue of both Ireland and Scotland, and the Irish heard with joy of the triumphs of the Scots. Thus, when Donal O'Neill, king of Tyrone, offered his shadowy claim to the high kingship of Ireland to Edward Bruce, the prospect of an invasion from Scotland aroused excitement in many quarters. A Scottish victory might mean the recovery of Ireland by the Irish; but it might also mean the recovery of Meath by the Lacys, of Louth by the Verdons, of the lordship of the Antrim glens by the Byset rivals of de Burgh.

Though not themselves bred in the Gaelic tradition, the Bruces were ready enough to make use of it.[1] Edward Bruce was a product of Anglo-Norman feudalism, more alien in outlook to the 'mere Irish' than were the lords of the English Pale. But he was notoriously ambitious, and Robert Bruce was quick to perceive the advantage to them both of an Irish kingdom for the man who

> Thoucht that scotland to litill were
> Till his brothir and him alsua.[2]

Moreover, the Bruces were no strangers to Ireland. Robert Bruce's wife was the daughter of the Red Earl, and it must have been with his father-in-law's connivance, if not at his suggestion, that Bruce had sheltered in Rathlin during the winter of 1306–7. He had inherited the earldom of Carrick from his mother and now passed this title to his brother, who was thereby provided with ancestral claims along the Antrim coast and with a convenient landing-place. When it became obvious that an invasion of Ireland was imminent, the bishop of Ely was sent as a special envoy to consult with the Anglo-Irish magnates and Edmund

---

[1] The author of the *Vita Edwardi* (p. 61) repeats a rumour that, had he been successful in Ireland, Edward Bruce intended to proceed to the conquest of Wales. John XXII, in Mar. 1317, warned 'Robert de Brus and his abettors' to desist from invading and occupying the lands of England, Wales, and Ireland. *Cal. Pap. Reg.* (*Letters*), ii. 138.   [2] *Bruce*, ii. 335.

Butler was appointed justiciar; but it was two years before a relieving army arrived from England.

Edward Bruce landed in Larne harbour on Lady Day 1315.[1] With him was another of the victors of Bannockburn, Thomas Randolph, earl of Moray, and several knights of high renown. The army comprised some 6,000 mail-clad warriors, veterans trained in the hard school of the war of independence and heartened by their recent triumphs. Supported by mobile native auxiliaries familiar with the countryside, the Scots were manifestly superior to their opponents, for the Anglo-Irish colonists of the eastern region had little experience of serious fighting. Their troops consisted mainly of local levies, 'a gadering of the cuntre', who could not stand against trained soldiers. Within a year of his landing, Edward Bruce had won a series of spectacular victories. Dundalk fell to him in June and, in September, the earl of Ulster (who, despite the family connexion had not been persuaded to join the invaders) was defeated in Antrim. Bruce then invaded Meath, and after he had worsted Roger Mortimer at Kells he was joined by Hugh and Walter Lacy who guided him through the midlands and brought over many of the gentry to his standard. A few months later their combined forces routed the justiciar's army at Skerries (co. Kildare); and in May 1316, on a hill near Dundalk, in the presence of a great host of his allies, Edward Bruce was crowned king of Ireland. From Wicklow to Sligo the Irish people rose in support of the invader, who proclaimed that his objective was the overthrow of the English conquerors and the restoration of the ancient Irish kingdoms to their hereditary rulers, under himself as high king. The fall of Carrickfergus in September after a protracted siege seemed to set the seal on a triumph which was signalized by the arrival of Robert Bruce with a force of 'gallowglasses' at the end of 1316. Together the brothers ranged through the midlands, spreading fire, slaughter, and devastation over a countryside already wasted by the famine, murrain, and plague which assailed much of Europe in these years.

[1] The fullest accounts of the Bruce invasion are to be found in Barbour's *Bruce* and Fordun's *Scotichronicon* (ed. Goodall). The Irish sources are much less informative but brief descriptions of the invasion are included in most of the Irish annals, e.g. of Ulster, Clonmacnoise, and Loch Cé. The best modern authorities are G. H. Orpen, *Ireland under the Normans* (1920), iv. 160–206, and E. Curtis, *Medieval Ireland* (1938), pp. 178–201. See also Olive Armstrong, *Edward Bruce's Invasion of Ireland* (1923), and R. Dunlop, 'Some Notes on Barbour's Bruce, Books XIV–XVI and XVIII', *Essays presented to T. F. Tout* (1925), pp. 277–90.

But the tide was already turning against the invaders and their friends. At Athenry in Connaught, a great native host had been defeated after a fierce contest by William de Burgh and Richard de Bermingham; and the citizens of Dublin beat off a determined attack, forcing the Scots, who lacked artillery, to abandon the siege of the capital. Robert Bruce withdrew to Scotland; Mortimer of Wigmore, lately appointed justiciar, landed with an army in the spring of 1317 and drove the Lacys out of Meath, while Edward Bruce retreated to the north. When Mortimer was recalled to England a year later, the back of the invasion had been broken; and, at the hill of Faughart, near Dundalk, on 14 October 1318, Bruce was defeated by an Anglo-Irish force under John de Bermingham and lost his life in action. The failure of his grandiose scheme was doubtless inevitable, for he had to deal with a people susceptible, indeed, to nationalist propaganda, but also passionately jealous for local and individual rights. Moreover, it seems evident that in the course of their sojourn in Ireland the Scots had alienated many sympathizers. They brought to the Irish midlands the violence and terror with which the men of northern England had long been familiar; and a note of relief at their departure is clearly audible in the so-called O'Madden Tract,[1] which designates the Scots invaders as worse than their English predecessors, and in the comment of the native annalist on the slaying of Edward Bruce:

There was not done from the beginning of the world a deed that was better for the Men of Ireland than that deed. For there came dearth and loss of people during his time in all Ireland in general for the space of three years and a half and people undoubtedly used to eat each other throughout Ireland.[2]

While the struggle was at its height, the native princes had addressed to Pope John XXII their famous remonstrance embodying a damning indictment of English rule in Ireland.[3] The Irish, so the princes allege, 'have been depraved, not

---

[1] This was a local tract describing the purpose of a charter issued by Mortimer, at the instance of the earl of Ulster, to Eoghan O'Madden of Hy Many, his brothers, and their heirs (Rot. Pat. Canc. Hib., p. 28).

[2] Annals of Ulster (Rolls Series), ii. 433.

[3] The text is found only in Scotichronicon (ii. 259-67). The remonstrance was dispatched to Avignon by the two cardinal nuncios who were in England 1317-18. Olive Armstrong (pp. 112-15) believes it to have been the work of Donal O'Neill and suggests that it should be used with caution as a picture of the state of Ireland under English rule.

improved, by intercourse with the English who have deprived them of their ancient written laws ... aliens from us in language, circumstances and actions, all hope of maintaining peace with them is out of the question'. John XXII gave the rebels no encouragement and excommunicated their clerical supporters; but he was sufficiently impressed by their appeal to transmit its content to Edward II together with an adjuration to apply the necessary remedies.[1] A great statesman might well have seized the opportunity offered by the defeat of the Scots to issue a general charter of liberties and to initiate a policy of reform; but great statesmen were not bred at the court of Edward II. A few grudging concessions to Irishmen desirous of obtaining the benefits of English law and an abortive attempt to set up a university in Dublin[2] represented the sum of the reforms. The shaken authority of the English government was not restored; the disintegrating forces already at work were stimulated, not stifled; and the Bruce invasion has justly been regarded as marking the first stage in the decline of the medieval colony in Ireland.

For Edward II himself, the most fatal consequence of Bannockburn was that it put him at the mercy of his domestic enemies. From the time of Gaveston's murder, the king had relied mainly on the support of the earl of Pembroke whose advice he sought on every detail of government. Pembroke acted as president of the council, and the numerous writs addressed to him show that much of the king's executive power was in his hands.[3] Now the Scots had played into the hands of Edward's old enemy, Thomas of Lancaster, the chief of the Ordainers. Though we may discount rumours that, at this early date, Lancaster had been intriguing with Bruce, there is no doubt that the Scottish king had proved useful to him. Had Edward returned from Scotland in triumph, he might well have turned his victorious army against his rivals at home. Instead, Lancaster was able to use the force which he himself had assembled at

---

[1] *Cal. Pap. Reg.* (*Letters*), ii. 440 (May 1318).

[2] Four masters were created in 1320 at the instance of Alexander Bicknor, archbishop of Dublin, who next year issued an 'ordinance for the University of Dublin'; but when he fell into disgrace as a partisan of Mortimer, the infant university languished.

[3] Conway Davies (pp. 323 ff.) points out that the informal character of some of these writs emphasizes Pembroke's unofficial position as a person acting on the king's behalf.

Pontefract to overcome the king and to secure his own objectives. Edward's biographer has left us a clear account of his movements after his panic-stricken flight had brought him safely to Berwick:

> On the advice of his friends the king . . . made for York and there took counsel with the earl of Lancaster and the other magnates and sought a remedy for his misfortune. The earls said that the Ordinances had not been observed and therefore events had turned out badly for the king: both because the king had sworn to stand by the Ordinances and because the archbishop had excommunicated all those contravening them: so that no good could come unless the Ordinances were fully observed. The king replied that he was willing to do anything ordained for the common good and he promised to observe the Ordinances in good faith. The earls said that . . . if the Ordinances ought to be observed it was necessary to ask for their execution. The king granted their execution: he denied the earls nothing. Therefore, in accordance with the Ordinances, the chancellor, treasurer, sheriffs, and other officers were removed and new ones substituted. The earls also willed that Hugh Despenser, Henry Beaumont, and certain others should leave the king's court until they should answer certain objections put to them . . . but at the king's instance this was deferred. Hugh Despenser was compelled, however, to retire.[1]

This version of events is substantially accurate. The York parliament of 1314 saw Walter Reynolds displaced from the chancery in favour of the treasurer, John Sandale, whose office passed to Sir Walter Norwich; and Lancaster extorted pardons for breaches of the peace for himself and some hundred of his followers. During the autumn, nearly all the sheriffs were changed, and the king's helplessness made it possible for Lancaster and his friends to begin the purging of the household at which they had been aiming for the past three years. Bannockburn had removed the keeper of the privy seal and the steward of the household. The keeper of the wardrobe, Ingelard Warley, was now replaced by William Melton, soon to become archbishop of York, one of the very few officials who enjoyed the confidence of all parties; and with Warley disappeared his colleague, the wardrobe cofferer, John Ockham, who had been stigmatized as an undesirable in 1311. In February 1315 the king had to submit to the removal from the council of Walter Langton and of the elder Despenser, with whom Pembroke had been accustomed

[1] *Vita Edwardi*, pp. 57-58.

to act. Edward was reduced to morbid brooding on the past. Soon after Christmas he ordered that the embalmed body of Gaveston (whom he had hoped to see avenged before he was buried) should be taken from Oxford to his favourite manor of King's Langley where he had already made generous gifts to the Dominicans in whose church the body was to be interred. The last rites were performed by the king's friend, Archbishop Reynolds, and four other bishops in the presence of a large congregation of abbots and religious; but few lay magnates attended the ceremony.

Gloucester had fallen at Bannockburn; Pembroke's influence was much diminished; Warwick was still a force to be reckoned with,[1] but when he died in August 1315 Lancaster was left in a position of unchallenged supremacy. The majority of the barons were his supporters, and clergy and people looked to him as the guardian of their liberties. Throughout most of the year 1315 the whole administration of the country was under his direct control. His instructions to the chancellor and the keeper of the wardrobe were given in written schedules; orders and pardons were issued at his request and appointments made at his instance.[2] Yet he himself preferred to keep aloof from ordinary meetings of the council, thereby underlining his unique position and forcing Edward to treat him as virtually an independent prince. In the Lincoln parliament of 1316 the seal was set upon his triumph.[3] Edward had spent Christmas in congenial fashion, rowing in the Cambridge fens 'with a great company of simple people',[4] before he hastened to Lincoln for the opening of the

[1] According to one chronicler, Guy of Warwick, early in 1315, was made 'the king's principal counsellor', *Ann. Lond.*, p. 232.

[2] For examples see Conway Davies, pp. 396–9. In June 1315 the chancellor went as envoy from the king at Westminster to Lancaster at Kenilworth. In October the king sent Melton and Hugh Audley to Lancaster at Donnington to seek his advice on certain administrative matters; Lancaster embodied this in a written schedule which he sent to the king who forwarded it to the principal councillors. Safe-conducts were given at his instance (e.g. to the outlawed men of Bristol) and petitions were addressed to him (e.g. from the merchants of Chester).

[3] Memoranda of the proceedings of this parliament were compiled by William Airmyn, clerk of the chancery, who was specially nominated by the king for the purpose (*Rot. Parl.* i. 350). Tout calls this 'the first full and intelligible record of the proceedings of a parliament' (*Place of Edward II*, p. 166). For a full description of the roll see Richardson and Sayles, 'Early Records of the English Parliament', *Bulletin Inst. Hist. Res.* vi (1929), 141–2, 151. These authors do not regard the roll as a precedent for others, or as a true journal, but as an *ad hoc* production written up later from notes.

[4] *Flores*, iii. 173.

session on 28 January. With characteristic insolence, Lancaster delayed his arrival and, since little of importance could be done without him, the king grew weary of waiting and made preparations to leave. But on 12 February the earl arrived and magnates and commons were then informed that the principal business before them was the problem of the Scottish war on which they agreed to deliberate apart. A few days later the bishop of Norwich, by the king's command, addressed the magnates. The king, he said, wished to observe the Ordinances and also to remove a certain *dubietas* which the earl of Lancaster was believed to harbour in regard to himself. The earl could rest assured that the king bore good will towards him and all the magnates who were the objects of his special benevolence; he wished that Lancaster should be chief of his council and arrange all his affairs. Lancaster thanked the king and deferred his reply. Afterwards, says the roll, he was sworn of the king's council; and the form of his oath shows his intention to make his own terms. He consented for love of the king and for the sake of the Ordinances; but if the king would not follow his advice, he must be at liberty to discharge himself. In other words, the king was to be deprived of all freedom of action and Pembroke and his friends of all effective influence. The council, with Lancaster at its head, was to assent in advance to every administrative act, and his control of it was assured by the condition that any member who gave 'bad counsel' might be removed in parliament where Lancaster knew that he could command support. Grants were then made for the Scottish war and a muster at Newcastle was ordered for July.

After the dissolution of the Lincoln parliament there was a serious attempt to enforce the Ordinances. Writs to the sheriffs ordered their proclamation and observance in every shire and instructions were issued relevant to particular clauses, notably those touching prises and customs. The king continued to defer to Lancaster who seems to have been present at several meetings of the council. Yet his control of his colleagues cannot have been easy to maintain for they included, besides the permanent officials whose sympathies always tended to be royalist, the royalist archbishop of Canterbury and the royalist earl of Richmond, as well as the discomfited earl of Pembroke. Probably, it was only by a Scottish campaign vigorously pursued that Lancaster could have secured the co-operation of the whole council. He had replaced Pembroke as king's lieutenant and chief captain of all the forces against the Scots on the Border, but so little was he inclined to take his duties seriously that rumours of an understanding between himself and Bruce continued to circulate. In consequence, his work as chief councillor proved almost completely abortive. He drew up schemes of reform but could not secure their acceptance and it was not long before he reverted to his former policy of abstention from meetings of council and parliament. The plots of the courtiers, he said, made it dangerous for him to approach the court; and it is true that Edward was now plucking up courage to restore certain victims of the opposition purge. Thomas Charlton was controller of the wardrobe by July 1316; within the next twelve months both Ingelard Warley and John Ockham became barons of the exchequer; and Hotham of Ely, formerly a confidant of Gaveston's, replaced Walter Norwich as treasurer. The officials hated Lancaster and his men, and the author of the *Vita* represents the situation as one of undeclared war between two rival households:

Whatever pleases the lord king the earl's servants try to upset; and whatever pleases the earl the king's servants call treachery: and so, at the suggestion of the Devil, the *familiares* of each put themselves in the way and their lords, by whom the land ought to be defended, are not allowed to be of one accord.[1]

The result was paralysis of all effective government at a time when a series of natural calamities and political upheavals made such an impasse peculiarly dangerous.

For the English people, the year following their defeat at Bannockburn opened a period of acute physical distress. Torrential rains in 1315 ruined the harvests of Europe, and famine spread from the Pyrenees to Russia and from Scotland to Italy.[2] The canon of Bridlington speaks of misery 'such as our age has never seen'.[3] Men were reduced to eating horses and dogs and charges of cannibalism were not confined to the Irish. Wheat had to be dried in ovens before it could be ground and there was no nutriment in the bread when it was baked. In the open country men were murdered by others desperate for food, and in the towns many died of hunger. When the king visited St.

[1] *Vita Edwardi*, p. 75.
[2] H. S. Lucas, 'The Great European Famine of 1315, 1316, and 1317', *Speculum*, v (1930), 343–77.   [3] Bridlington, p. 48.

Albans in the autumn of 1316 it was scarcely possible to buy bread for his household. Wheat rose to six or even eight times its normal price, fetching 30s. or 40s. a quarter in the summer of 1315; that year the Leicester chronicler saw it sold in Leicester market for 44s.[1] The price of peas, beans, oats, malt, and barley rose proportionately and salt—an indispensable commodity—cost as much as wheat. Government attempts at price control proved futile. Their only effect was to cause dealers to withdraw their goods from the market and, as the chronicler remarks, it was better to buy dear than not to be able to buy at all.[2] The ordinance fixing a schedule of maximum prices was repealed a year later. It had achieved nothing and was regarded by some as a violation of natural law. 'How contrary to reason', exclaims the canon of Bridlington, 'is an ordinance on prices, when the fruitfulness or sterility of all living things are in the power of God alone, from which it follows that the fertility of the soil and not the will of man must determine the price.'[3]

Against this background of social wretchedness, unrest developed in many parts of the country. The Scots were harassing the Border counties and raiding Yorkshire. On the upper Severn the courtier John Charlton contested the lordship of upper Powys with one of Lancaster's adherents, Griffith of Welshpool. Llewellyn Bren rebelled in Glamorgan where his attack on Caerphilly castle forced the Anglo-Norman lords of south Wales to combine to repress him. In Bristol the mass of the burgesses rose against the governing oligarchy and drove them out of the town; the government had to bring up siege-engines to compel their surrender.[4] The keeper of the king's horses complained that he and his men had been assaulted and robbed while staying at Oseney abbey and that the offenders were being maintained by the burgesses of Abingdon who 'have made a common collection to which each gives a certain sum of money each week and they put it in a box'.[5] In his own county of Lancaster, Earl Thomas found himself faced with serious trouble. Adam Banaster, one of his military tenants, headed a revolt against his chief household official and special confidant, Sir Robert Holland. Holland suppressed the rising with vindictive savagery

---

[1] Knighton, i. 411.   [2] *Vita Edwardi*, p. 69.   [3] Bridlington, pp. 47-48.
[4] For the Bristol disturbances see S. Seyer, *Memoirs of Bristol*, ii (1823), 88-106, and E. A. Fuller, 'The Tallage of 1312 and the Bristol Riots', *Trans. Bristol and Glos. Arch. Soc.* xix (1894-5), 172-278.
[5] *Cal. Chancery Warrants*, i. 412-13.

and for some years his faction pillaged and murdered with impunity, pleading Lancaster's protection for their acts of brigandage.[1] Nor was the earl any more successful in maintaining harmony within the circle of his own family. His wife Alice Lacy, from whom he derived much of his property, left him in 1317 and sought asylum with Earl Warenne, thereby precipitating a private war in Yorkshire.[2] The king, who was said to have connived at her action, went twice to the north on the pretext of enforcing the peace. When the royal army was ordered to muster at York, Lancaster caused the bridges to be broken and did all in his power to prevent the passage of men and arms. It is in this context that we meet the first explicit reference to Lancaster's desire to make political capital out of his office of hereditary steward of England. 'He asserted that he acted thus because he was Steward of England and it was his duty to see to the welfare of the realm and if the king wished to take up arms against anyone he ought first to warn the Steward.'[3] Edward's reply was to march past Pontefract with an imposing display of arms as he took his road to the south. An acrimonious exchange of letters, in which Edward remonstrated at Lancaster's private gatherings of magnates and people and retention of troops at excessive wages and the earl accused the king of ignoring the Ordinances and failing to discuss his grievances in parliament, indicated the growing tension between them.[4]

Such conditions opened the way for the emergence of the new political group to which historians have given the name of the Middle Party. Its leader was Aymer de Valence, earl of Pembroke, who, throughout his public career, remained a firm supporter of the Ordinances which he himself had helped to frame. But to support the Ordinances was not necessarily to support Lancaster, whom Pembroke probably had never forgiven for his murder of Gaveston, and by whom he had been supplanted after Bannockburn. The ineffectiveness of the government between 1312 and 1314, when Pembroke was the king's

---

[1] Banaster's revolt is described fully by G. H. Tupling, *South Lancashire in the Reign of Edward II* (Chetham Soc. 3rd ser. I, 1949).

[2] Alice Lacy went to live with a lame squire named Ebulo L'Estrange whom she married in 1322. Warenne himself had lately made an unsuccessful attempt to obtain a divorce from his wife, Joan of Bar, the king's niece. See I. S. Leadam and J. F. Baldwin, *Select Cases before the King's Council* (Seldon Soc. xxxv, 1918), pp. lxvi–lxix, 27–32.   [3] *Vita Edwardi*, p. 81.

[4] Bridlington, pp. 50–52.

right-hand man, suggests that his abilities were in no way outstanding. But his ambitions were less grandiose than Lancaster's,
he was a moderate man and an experienced diplomatist, and no
great qualities of statesmanship were needed to enable anyone
to perceive the disastrous effects of the cold war between the
king and his cousin, both at home and abroad. Pembroke's outlook was shared by many of the prelates, whose general aim was
compromise; and it seems likely that the Middle Party was born
during a mission to Avignon at the end of 1316. The envoys
were Pembroke himself, John Salmon, bishop of Norwich, John
Hotham, bishop of Ely, and Bartholomew Lord Badlesmere, a
Kentish tenant of the archbishop of Canterbury; and it was soon
afterwards that the two cardinals, Gaucelin of Eauze and Luca
Fieschi, came on their peace mission to England. Other prelates
who joined the party were the king's friend, Walter Reynolds,
archbishop of Canterbury, John Langton of Chichester, a
veteran survivor from Edward I's chancery, Walter Stapledon
of Exeter, and John Sandale, who had obtained the see of Winchester through Pembroke's influence. Among Pembroke's lay
associates the most conspicuous was Badlesmere, a baron of the
second rank whose principal affiliations had been with Gloucester and Hereford. Humphrey Bohun of Hereford, hereditary
constable of England, was a supporter of the party, as was also
Thomas Brotherton of Norfolk, the king's half-brother, to whom
the office of marshal had been granted in tail at the Lincoln
parliament. Warenne's private feud with Lancaster kept him
in Pembroke's group, in company with his brother-in-law,
Edmund Fitzalan, earl of Arundel, another of the Ordainers;
and they were soon joined by a powerful body of marcher lords,
including Roger Mortimer of Chirk, his nephew, Roger
Mortimer of Wigmore, and the husbands of the three coheiresses
of Gloucester, Hugh Despenser, Roger Damory, and Hugh
Audley. The household officers rallied to the new party and
Pembroke soon found himself at the head of a very powerful
combination. That the association was more than casual is
proved by the survival of a secret indenture, dated 24 November
1317, between Pembroke and Badlesmere on the one hand and
Roger Damory on the other, whereby Damory undertook to use
his utmost diligence to induce the king to be governed by their
counsel and to place faith in them before all other men, and to be
so governed himself; he took oath on the Host and pledged the

very large sum of £10,000 as security. Pembroke and Badlesmere, for their part, swore to warrant and maintain Damory against all other persons, saving their allegiance to the king.[1] Whether this indenture was unique or remains as the sole survivor of a series, we do not know; but it affords interesting evidence of the aims of Pembroke and Badlesmere and of the lengths to which they were prepared to go to attain them.

What were the objects of the Middle Party? Maintenance of the Ordinances was always in the forefront of their programme and, though this may have been lip-service on the part of the courtiers, there is no reason to doubt that Pembroke and the bishops were sincere in their desire for administrative reform. On the other hand, the large official and courtier element in the party precludes the possibility that it was, in any sense, a combination against the king. Magnates and officials alike had had full opportunity, however, to take the measure of Edward II. They must have been aware of his instability, his readiness to be led by his personal friends. If Pembroke could recover the key position which he had lost after Bannockburn, if the king could be induced to place faith in him and his colleagues 'before all other men', then there was reason to hope that restraint of Edward's rash impulses could be combined with the maintenance of some degree of royal dignity. The officials would be free to pursue their departmental activities unhampered, better order might be kept in the land, and those responsible for its defences might be able either to come to terms with the Scots or to rally the nation in a great effort against them. So much is obvious enough. What is more difficult to determine is the attitude of the party towards Thomas of Lancaster, who was likely to prove the most serious obstacle to the fulfilment of their plans. The inveterate hostility of some of the courtiers to Lancaster may be taken for granted; but Pembroke and his episcopal colleagues must have been alive to the impossibility of ousting Lancaster altogether without resorting to the methods of violence which they wished at all costs to avoid. What they probably wanted from Lancaster was at least a show of co-operation and abandonment of his claim to a uniquely privileged position *vis-à-vis* both the king and his fellow-barons. Since the marshal and constable were members of the new group, an attempt might be made to dispute the superiority of the steward, and the progress made by

[1] Conway Davies, pp. 433-4.

the Middle Party in 1318 was encouraging. A series of parleys opened with a conference at Leicester in April between the leaders of the party and Lancaster, or his representative.[1] Here agreement was reached that the Ordinances should be maintained, evil counsellors removed, Lancaster and his friends pardoned all trespasses, and that he should meet the king at some convenient place and under proper guarantees. Two months later the Middle Party leaders met a group of the king's councillors in the exchequer. Here they seem to have agreed to invite Lancaster to attend the next parliament as a peer of the realm 'but without accroaching sovereignty towards the others'. He was to be promised redress of his grievances and told that if he would not co-operate the magnates would act without him. Immediately afterwards Edward made a solemn declaration at St. Paul's that he would confirm the Ordinances, make peace with Lancaster, and rely on the help and counsel of his barons. But it was now clear to Lancaster that the Middle Party, or some of its members, was moving to the right, and he refused to consent to a general pacification until he had been satisfied on two points—that lands alienated and gifts made in contravention of the Ordinances should be resumed, and that evil counsellors should be removed, so that he might approach the king with safety. These demands entailed further parleys in the course of which Lancaster suggested the expedient of a standing baronial council, and it was not until 9 August 1318 that a treaty, or indenture, was finally sealed at Leake in Nottinghamshire, between Lancaster on the one hand and the prelates and barons of the Middle Party on the other. This provided that the Ordinances should be maintained; that Lancaster and his associates should be pardoned; that a parliament should be summoned; and that a standing council—one member to be a banneret nominated by Lancaster—should be established, without consent of which the king was not to perform any of the normal acts of sovereignty.[2] Five days later the king and Lancaster met to exchange the kiss of peace and letters of pardon were issued to more than 600 of the earl's associates.

The terms of the treaty of Leake make plain the straits to which the monarchy, in the person of Edward II, had now been

---

[1] J. G. Edwards, 'The Negotiating of the Treaty of Leake', *Essays . . . presented to R. L. Poole* (1927), pp. 360–78; B. Wilkinson, 'The Negotiations preceding the Treaty of Leake', *Studies . . . presented to F. M. Powicke* (1948), pp. 333–53.

[2] *Rot. Parl.* i. 453–5.

reduced; and it is hardly surprising that rumours that he was a supposititious child should have begun to circulate about this time.[1] The strength of Lancaster's position seemed assured; a compact between a single earl and the leading barons of England speaking for the king, was without precedent in English history. Yet, in fact, the treaty had given Lancaster little more than a pardon for himself and his friends and the satisfaction of seeing the Ordinances formally approved. Whether by his own wish or that of his opponents, he was not to be a member of the standing council; and no positive steps had yet been taken to remove 'evil counsellors' or to resume gifts made in contravention of the Ordinances. Lancaster doubtless hoped to see these matters settled in the next parliament; but when it met at York in October he was no longer in a position to make his own terms. Some satisfaction may have been afforded him by the prominence given to the Ordinances, which were coupled with the Great Charter and thereby accorded the status of fundamental law, and by the appointment of a committee to amend the king's household. Officials high and low were subjected to systematic review, but, apart from another clean sweep of the sheriffs, the only important change effected was the substitution of Badlesmere for Sir William Montagu (promoted to be seneschal of Gascony) as steward of the household. This appointment provoked Lancaster's sole recorded intervention in the business of parliament. It seems certain that he had planned to use his position as steward of England to control the *domus regis* and so rid the court of those whom he held to be enemies of his cause; for he now declared that, as hereditary steward of England, he was entitled to nominate the steward of the household as his deputy, in the same way as the marshal of England nominated the marshal of the household. His claim was met by an offer to search the records and report in six months' time; and Badlesmere meanwhile took possession of the office, thereby securing for Pembroke just the kind of influence which Lancaster had coveted for himself. Such a rebuff was not offset by the Household Ordinance of York which, though it attempted to remove some long-standing financial abuses and provided for certain resumptions of demesne, embodied no really new policy.[2] Pembroke and his party were much more deeply concerned to use their newly acquired influence to deal with the problem of

[1] *Vita Edwardi*, p. 86. [2] For the text see Tout, *Place of Edward II*, pp. 244–81.

Scotland. When, at the York parliament of May 1319, Lancaster renewed his demand in more extreme form, praying that the king 'should grant him the stewardship of his household which appertains to him by reason of his Honour of Leicester',[1] he was once again put off by a promise of further searches in the records. Whether or not this had been their original intention, Pembroke and Badlesmere had now deprived him of all effective influence, except such as his representative might be able to exercise on the standing council which was formally established in this parliament.

Meanwhile, Berwick had fallen to the Scots and Edward II's new advisers held it to be a matter of national honour to attempt its recovery. The muster at Newcastle arranged for June 1319 represents the nearest approach to a national effort that the reign can show; for even Lancaster was there, with Henry his brother and heir-presumptive, their old enemy Warenne, Pembroke, Hereford, Arundel, Despenser, and

> all the Erllis als that war
> In yngland worthy for to ficht,
> And baronis als of mekill mycht.[2]

Towards the end of July the army moved forward to the attack. Though the walls of Berwick were low, both sides were furnished with elaborate and ingenious siege-appliances. Neither had made much headway when Bruce wisely resolved to attempt a diversion. He sent Sir James Douglas with the main Scots force right into Yorkshire where they are said to have been guided by an English knight named Sir Edmund Darel, who offered to lead them to York, where the queen, the exchequer, and the courts were lodged. Isabella escaped by water; and archbishop Melton, at the head of a motley force of priests, monks, farmers, and townsmen, made a valiant attempt to repel the invaders, whom he encountered, probably on 20 September, at Myton-in-Swaledale. Kindling great fires of hay to conceal their advance, the Scots bore down upon the Yorkshiremen, scattering them far and wide. Many were drowned in the Swale and the archbishop himself had a narrow escape,

> Tharfor that bargane callit ware
> 'The chaptour of mytoune'; for thare
> Slayn sa mony prestis ware.[3]

---

[1] Cole, *Documents illustrative of English History*, p. 48.
[2] *Bruce*, ii. 416–17. (Barbour describes the siege with a wealth of detail.)
[3] *Bruce*, ii. 428. William Airmyn was among those taken by the Scots. See A. D. H.

When the battle was over the Scots pressed on towards Lancaster's castle at Pontefract; and, as Bruce had hoped, the news of the disaster split the English army before Berwick. Lancaster and the northern earls insisted on returning to protect their own lands and, while the English were wrangling, Douglas and his army made good their escape. There was nothing for it but to agree to a two-years' truce and the high hopes of the Middle Party were shattered. As for the king, Robert of Reading declares that from this time onwards his infamy began to be notorious, his torpor, his cowardice, his indifference to his great inheritance.[1]

Lancaster's brief essay in co-operation was ended, and when parliament met again at York in January 1320 he refused to attend. 'It was not fitting, he said, that a parliament should be held *in cameris*: for the king and his associates (*collaterales*) were suspect by him and he had openly proclaimed them his enemies.'[2] What he meant by this speech is not altogether easy to determine. The words *in cameris* suggest some kind of secret assembly, but there was nothing secret about the York parliament to which the magnates had been summoned in due form. No commons had been summoned, probably because of the physical difficulties attendant on a mid-winter session in the north of England; but it is most unlikely that the absence of the commons was the ground of Lancaster's complaint.[3] More probably, he meant to imply that this parliament would be at the mercy of the courtiers —all the more so because of being held in the north—that it had been framed in advance, so to say, and that it was useless, therefore, for him to attend it. No doubt he was still smarting from the rebuff he had received over the household stewardship; and from his own point of view he was probably wise not to attend, for the ministerial changes effected at York amounted merely to a reshuffle of offices among the members of the dominant party. Salmon of Norwich succeeded Hotham of Ely as chancellor and Stapledon of Exeter stepped into the office of treasurer left vacant by the death of Sandale of Winchester. Robert Baldock, archdeacon of Middlesex, who became keeper of the privy seal, was known only as a well-trained wardrobe clerk. When Edward went to France in the summer, he left Pembroke as keeper of

Leadman, 'The Battle of Myton', *Yorks., Arch. and Top. Journal*, viii (1883–4), 117–22.
    [1] *Flores*, iii. 192–3.   [2] *Vita Edwardi*, p. 104.
    [3] See Helen Cam, 'From Witness of the Shire to Full Parliament', *Trans. Royal Hist. Soc.*, 4th ser. xxvi (1944), 31.

the realm, while Badlesmere became constable of Dover castle and warden of the Cinque Ports. Despite the disasters at Berwick and Myton, the Middle Party was still in control.

Meanwhile, the two Despensers by whose ambitions the party was ultimately to be shipwrecked, were moving to the centre of the political stage. Hugh the elder, a son of the justiciar of 1260, was an experienced official of strongly royalist sympathies whose abilities had often proved useful to Edward II. He was the only baron to stand firmly by Gaveston during the movement against him in 1308 and after the murder he represented the king in the negotiations with the Ordainers. He is found acting as the king's creditor, assuming responsibility for his debts, helping to direct the minor officials of the household, receiving in return such profitable offices as keeper of the royal castles of Devizes and Marlborough. He accompanied Edward on his flight from Bannockburn and, though dismissed in 1315, he returned to court in association with. Pembroke and thereafter worked in support of his son who was rapidly becoming the dominant figure in the partnership. The younger Hugh was a contemporary of Edward II and had been a member of his household as prince of Wales; the king had given him the hand of his niece Eleanor, eldest of the three sisters of Gilbert de Clare. After a brief period of association with the Ordainers, Despenser moved towards Pembroke's party and his appointment as chamberlain in 1318 was generally approved.[1] In the meantime, the death of his brother-in-law at Bannockburn had vastly improved his prospects. Less than a month later, the escheators had been ordered to take all the Gloucester estates (valued at over £6,000)[2] into the king's hands; but the widowed countess held up their partition by feigning pregnancy and it was not until November 1317 that orders were issued for delivery of the portions of the inheritance.[3] To Despenser, as husband of the eldest of the coheiresses, fell the important lordship of Glamorgan with adjacent lands in Wales, valued at £1,276. 6s. 9¾d.[4] Hugh

---

[1] 'de cuius diligencia et fidelitate magnam gerebant fiduciam tunc temporis proceres et magnati', 'Chronicle of the Civil Wars of Edward II', *Speculum*, xiv. 77.

[2] J. Conway Davies's figure of £15,000 has been corrected by G. A. Holmes, *The Estates of the Higher Nobility in XIV century England* (1957), p. 36, n. 1.

[3] J. Conway Davies, 'The Despenser War in Glamorgan', *Trans. Royal Hist. Soc.*, 3rd ser. ix (1915), 21-64.

[4] E. B. Fryde, 'The Deposits of Hugh Despenser the Younger with Italian Bankers', *Econ. Hist. Rev.*, 2nd ser. iii (1951), 344.

Audley who had married Gaveston's widow, the second of the Clare sisters, received Newport, Netherwent, and the county of Gwennllwyg, while the lordship of Usk went to Roger Damory, husband of the youngest sister. Despenser was far from satisfied with these arrangements; for he wished to acquire the whole of the Clare property in Wales and to secure for himself the title of earl of Gloucester. The earldom of Norfolk had been revived for Thomas of Brotherton, that of Kent for Edmund of Woodstock, and there were strong arguments for a revival of the great earldom of Gloucester. Both Norfolk and Kent, however, were of the blood royal. It would be a very different matter to bestow one of the richest of the English earldoms on the son of a secondary baron, especially when this must entail overriding the claims of both Audley and Damory. Yet Despenser was determinedly ambitious and by 1320 he had achieved some initial success. He had encroached on the county of Gwennllwyg and forced Audley to surrender both this and his lordship of Newport in exchange for other lands in the south-east. In Carmarthen he had secured Cantrefmawr, the old stronghold of the princes of south Wales, and the castle and town of Drusslan. In Glamorgan he had obtained a grant of full sovereignty as complete as any earl of Gloucester had ever possessed; and his acquisition of Lundy island was clearly designed to give him control of the Bristol Channel. But to round off his territories in the west he needed the lordship of Gower which would establish the river Loughor as the boundary between his lands and the Lancaster lordship of Kidwelly. Gower was the property of William de Braose, an impecunious baron who was prepared to sell it to the highest bidder; but when Braose died in 1320 his son-in-law, John Mowbray, seized both the barony of Gower and the lordship of Swansea. Despenser now decided to show his hand. He persuaded Edward II to declare that Gower had escheated to the Crown, since it had been alienated without royal licence, thereby raising a vital question of marcher privilege. The suggestion that the lords of the march were subject to the English law of alienation sufficed to bring their mounting resentment to a head, and a powerful confederation of marchers, determined to resist the acquisitiveness of Despenser, was soon in being.[1] Besides Audley, Damory, and Mowbray, it included Humphrey Bohun of

---

[1] W. H. Stevenson, 'A letter of the younger Despenser on the eve of the Barons' rebellion, 21 March 1321', *Eng. Hist. Rev.* xii (1897), 755–61.

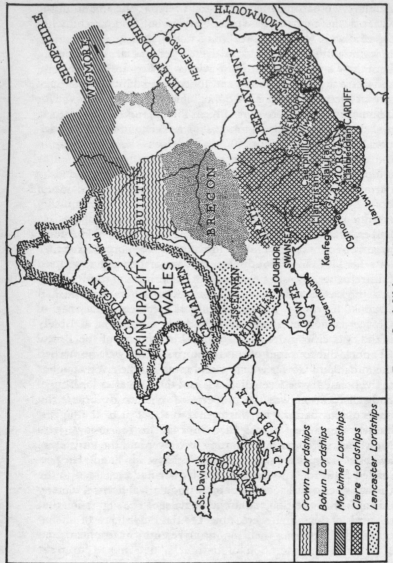

SHROPSHIRE

WIGMORE

HEREFORDSHIRE

• HEREFORD

MONMOUTH

USK
Sef oog

NEWPORT
Grumhlyn
Caerphilly
Llanrumney
Llantrisant
St. Fagan's
Llanbleddian
Ogmore
CARDIFF

GLAMORGA

NEATH

ABERGAVENNY

BRECON

BUILTH

Painscastle •

PRINCIPALITY
OF
WALES

CARDIGAN

CARMARTHEN

ISCENNEN

KIDWELLY

GOWER

LOUGHOR •

SWANSEA •

Kenfeg

Oystermouth •

Llanwm

CARDIGAN

PEMBROKE

HAV...

• St. David's

Fig. 2. South Wales and the march.

	Crown Lordships
	Bohun Lordships
	Mortimer Lordships
	Clare Lordships
	Lancaster Lordships

Hereford whose lordship of Brecon lay next to Glamorgan, John Giffard[1] whose tenants in Cantref Bychan had been at odds with Despenser's in Cantrefmawr, Mortimer of Chirk whose position as justice of Wales was threatened by Despenser's encroachments in Carmarthen, and Mortimer of Wigmore who supported his uncle; behind it were many of the native Welsh who had suffered at the hands of Despenser and held him responsible for the execution of Llewellyn Bren. Among the great landowners of the march, only Pembroke and Arundel continued to stand by the king.[2] In January 1321 Edward was ordering Hereford and other barons to refrain from discussing the affairs of the realm; in March the king came to Gloucester and forbade assemblies of men in the Welsh marches, but his instructions were ignored. The confederate lords seized Newport and overran Glamorgan, where no one would defend the castles. Despenser was completely defeated and Edward was forced to agree to produce him for judgement in the next parliament.

As lord of Kidwelly, Thomas of Lancaster could not be indifferent to Despenser's activities; but it was characteristic of him that he took no active part in the war, preferring to pursue a line of his own. His method was to summon a number of peculiar assemblies in the north of England, the first of which met at Pontefract on 24 May 1321, when in the chapter-house of the priory there assembled a group of northern magnates, including Multon of Gilsland, Furnival of Sheffield, Deyncourt, Percy, and Fauconberg, all of whom took oath to defend their lands and one another should any of them be attacked. No mention was made of the Despensers and it is possible that the confederates were thinking primarily of the danger from the Scots.[3] Before committing themselves to further action, they asked for the advice of the prelates, a request which may suggest a desire to evade any more extreme demands from Lancaster. The earl then wrote to the archbishop of York and to the bishops of Durham and Carlisle, asking them to meet him at Sherburn-

---

[1] For Giffard see Ruth Butler, 'The Last of the Brimpsfield Giffards and the Rising of 1321–2', *Trans. Bristol and Glos. Arch. Soc.*, lxxvi (1957), 75–97.

[2] Pembroke went to France in Nov. 1320 to prosecute his marriage with Marie de St. Pol which took place in Paris on 5 July 1321.

[3] There was a precedent of a kind. In Apr. 1315, at the urgent request of Lancaster, Archbishop Greenfield had summoned his clergy to attend a council convened at Doncaster to discuss measures of defence against the Scots. *Letters from the Northern Registers*, p. 245.

in-Elmet on 28 June.[1] He invited also many barons, bannerets, and knights, southerners as well as northerners; both Bohun of Hereford and Mortimer of Wigmore were there. When all had assembled Sir John de Bek proclaimed the cause of summons. Matters to be discussed included the bad character of the king's ministers, forfeitures, and banishments decreed without judgement of peers, the inequities of the staple system, and imprudent treaties made with foreign nations. Lancaster then asked the prelates to retire to the house of the rector to consider their answer, while the lay lords deliberated apart. The bishops replied that they were willing to aid the earl to the utmost in defending the realm against the Scots, but asked that consideration of the matters raised in Bek's speech should be deferred until the next parliament. Nothing is known of the laymen's deliberations. But documents preserved at Christ Church, Canterbury, and at Lambeth contain the text of an indenture drawn up in the name of some sixty magnates whereby they pledge themselves to secure the destruction of the Despensers.

Stubbs, who was followed by Tout, thought that at Sherburn Lancaster was trying to assemble 'a parliament of his own', summoning prelates, barons, and knights and even parodying parliamentary procedure by using Sir John de Bek in the role of chancellor and arranging for the withdrawal of one of the estates for separate deliberation; and Miss M. V. Clarke suggested that it may have been with some confused idea of his rights as steward that Lancaster relied upon what she calls 'the irregular counter-Parliaments that he organised in the North'.[2] Most historians have seen the assembly as symptomatic of the way in which the greed of the Despensers had played into the hands of Lancaster, the meeting at Sherburn paving the way for the union of the forces of north and west, in a true parliament, against the Despensers and putting Lancaster once again at the head of an almost completely united baronage against an almost completely isolated king.[3] Recent investigation suggests, however, that the triumph of Lancaster and the unity of the baronage were much less complete than has been supposed.

---

[1] Both Melton of York and Beaumont of Durham were well disposed to the king; Halton of Carlisle was an aged prelate who had been preoccupied with the Scottish menace for thirty years.

[2] Stubbs, *Chronicles of the Reigns of Edward I and II* (Rolls Series), ii. lxxxvii-lxxxviii; Tout, *Place of Edward II*, p. 129; Maude Clarke, *Medieval Representation and Consent*, p. 163.   [3] Conway Davies, pp. 478-9.

Professor Wilkinson has pointed out that the Lancastrian sphere of influence was essentially north midland as distinct from northern. Lancaster never controlled the great families of Northumbria, the Percies, Mowbrays, and Cliffords, by whom, indeed—and, as events were to show, with good reason—he was suspected of too great friendliness towards the Scots. As to the marcher lords, some of the greatest of them, notably Arundel and Pembroke, had not yet declared themselves. The northern lords, Professor Wilkinson believes, never set their seals to the Sherburn indenture at all; only the marchers, and not all of them, were concerned with it.[1] For this reason, although opposition to the Despensers was coherent enough to secure sentence against them in parliament, the king found it possible to secure certain concessions.

On their way to the parliament which had been summoned to meet at Westminster in July, certain of the lords, so the St. Paul's annalist tells us, had put in writing 'a certain tract ordained and approved according to ancient custom'.[2] Conclusive evidence as to the nature of this tract is wanting, but Miss Clarke believed it to have been the famous treatise on the office of the steward which embodies and expounds the Lancastrian claim that it was the steward's function, after the king, to supervise and regulate the whole realm of England, all the officers of the law, in war and peace, to punish negligent officials and, if necessary, to dismiss and punish evil counsellors.[3] Such a tract would have afforded a fitting prelude to the great attack on the Despensers which was about to be launched, though it is unlikely that such pure Lancastrian doctrine can have been accepted by more than a section of the baronage. Unfortunately we have no official record of the proceedings of this parliament, but the chroniclers make it clear that agreement was reached only after considerable dispute. The author of the *Vita Edwardi* tells us that at first the king tried to refuse the lords a hearing and that he yielded only when a message was

[1] B. Wilkinson, 'The Sherburn Indenture and the Attack on the Despensers, 1321', *Eng. Hist. Rev.* lxiii (1948), 1–28. Several versions of the indenture are extant, suggesting successive drafts, the product of difficult negotiations: one version omits the most conspicuous northern names. It is pointed out that the indenture was not mentioned in the subsequent proceedings against Lancaster and that Edward II never accused the northern lords of supporting him.

[2] *Ann. Paul.*, p. 293.

[3] Maude Clarke, op. cit., p. 242. For the text of the tract see L. W. Vernon Harcourt, *His Grace the Steward and Trial of Peers* (1907), pp. 164–7.

sent through Pembroke and others warning him that if he
would not hear them they would withdraw their homage and
appoint a better ruler.[1] Edward may well have been shaken by
threats coming from such a source, and when Pembroke urged
him to save his throne by conceding the banishment of the
Despensers he reluctantly agreed to a compromise. The Despensers were banished in perpetuity as 'evil counsellors'; but
the magnates withdrew their demand for a statute against them
and for official recognition of a doctrine distinguishing between
the Crown and the person of the king.[2] Formal pardons were
issued to Lancaster and some 500 of his adherents for their
action in resisting the king and the Despensers.

The break-up of the Middle Party was now fairly complete.
By the summer of 1321 one of its leading members, Lord Badlesmere, whose action in accepting the stewardship of the household had been a principal cause of Lancaster's withdrawal, was
so far identified with the Lancastrian group as to need one of the
king's pardons. But his reconciliation with Lancaster was no
more than superficial, and as the king was unwilling to forgive
his treachery he drifted into an isolated position. Edward soon
found occasion to attack him openly. On her way to Canterbury
in October, Queen Isabella sought hospitality at the royal
castle of Leeds in Kent of which Badlesmere was constable. He
himself was not in residence but his wife refused the queen
admittance on the grounds that she could receive no one without her husband's permission. The king made this insult the
pretext for raising a large army and the support lent him by Pembroke, Arundel, Warenne, and many other lords is sufficiently
indicative of the lack of unanimity among the peers. The few
marcher lords who made the gesture of raising a force to help
Badlesmere reached Kingston-on-Thames only to learn that
Leeds castle had fallen to the king.[3] Lancaster had withdrawn
to the north where he summoned another of his 'counter-
parliaments' to meet at Doncaster in November. Edward II
forbade the meeting and we have no direct evidence that it took
place; but there is preserved in the British Museum the draft of

---

[1] *Vita Edwardi*, p. 112.

[2] Professor Wilkinson's argument (*Const. Hist.* ii. 136-9) on this difficult point
seems to me conclusive. It offers a reasonable interpretation of the otherwise incomprehensible statement in the official record of 1322 that the Despensers were
condemned *inter alia* for *holding* this doctrine.

[3] Lady Badlesmere was taken and kept prisoner in the Tower until Nov. 1322.

a petition which is likely to have emanated from some such assembly.[1] This so-called Doncaster petition opens with a recapitulation of the unlawful acts of the Despensers and goes on to state that, despite his banishment, the younger Despenser is still being maintained by the king in a career of piracy. Men are still being disinherited without judgement, and if redress is not granted the magnates will be forced to defend their rights by show of arms. The facts were correct enough. All the chroniclers agree that, whereas the elder Despenser had accepted his sentence and gone abroad, the younger had embarked on a career of piracy in the Channel, robbing merchant-ships and basing his activities on the Cinque Ports, with the king's full connivance. But threats of the kind contained in the Doncaster petition afforded Edward ample justification for his resolve to crush Lancaster once and for all. He had a strong army mustered and the prospects were good. Before the end of the year both the Despensers had been recalled and through the medium of Reynolds the king had secured from the convocation of Canterbury a condemnation of the process against them.

From his base at Cirencester Edward marched north and crossed the Severn at Shrewsbury in January 1322. The Mortimers, probably because they were uncertain of Lancaster's intentions, submitted without a fight and most of the other leaders followed their example. Hereford, Mowbray, and Clifford escaped to join Lancaster at Pontefract; but Badlesmere, who accompanied them, 'could find no favour with him'.[2] Edward swung round through the middle and southern marches, occupying the main strongholds of his enemies without meeting any resistance. On the feast of the Epiphany he had extended his peace formally to the Despensers and by the end of the month he was assembling levies at Coventry and proclaiming that he was about to march to the north where 'certain magnates of the realm' had leagued themselves with the earl of Lancaster and were awaiting the arrival of the Scots with whom they had conspired.[3] Lancaster missed the opportunity of attacking the king's army while it was in process of formation,

[1] G. L. Haskins, 'The Doncaster Petition, 1321', *Eng. Hist. Rev.* liii (1938), 478–85.

[2] 'A Chronicle of the Civil Wars of Edward II', *Speculum*, xiv. 78.

[3] The charge of conspiracy was well founded. In Dec. 1321 James Douglas sealed a safe-conduct for two of Lancaster's agents and in Feb. 1322 Mowbray and Clifford were invited to confer with Randolph in Scotland (*Foedera*, ii. 463, 472).

preferring to concentrate his forces before the royal castle of
Tickhill.[1] After wasting three weeks there he moved south and
occupied the bridge at Burton-on-Trent in the hope of checking
the royalist advance. But Edward forded the Trent at a point
above Burton and, faced with the danger of encirclement,
Lancaster set fire to the town and retreated hastily to Pontefract,
some of his men plundering and others deserting as they went.
It seems to have been at this juncture that Sir Robert Holland,
whom Lancaster had summoned to his aid, decided to abandon
his master's cause.[2] Lancaster wished to make a stand at Pontefract but Hereford and others insisted on withdrawal towards
the Border, probably in the hope of receiving help from the
Scots with whom a formal alliance had now been concluded.
The king seized Tutbury where, on 12 March, Lancaster was
proclaimed a rebel; and the earls of Kent and Surrey were sent
to pursue him and to lay siege to Pontefract castle. By 16 March
the Lancastrians had reached Boroughbridge, on the south
bank of the river Ure. At the northern end of its narrow wooden
bridge there waited the levies of Cumberland and Westmorland under the command of Sir Andrew Harclay, a capable
veteran of the Scottish wars. Harclay had dismounted all his
knights and men-at-arms and arranged them with some supporting pikemen on the north bank of the river.[3] Faced with
this opposition, the Lancastrian army divided, one band under
Hereford and Clifford, concentrating on the bridge, another
under Lancaster, making for a nearby ford. Neither achieved
anything. Hereford was killed on the bridge by a Welsh pikeman
who had concealed himself beneath it, Clifford was severely
wounded, and the discouraged troops made no attempt to cross.
Lancaster's force had meanwhile lost its nerve under the fire of
Harclay's archers and he was driven to seek an armistice until
the morning. During the night there were many desertions and
the arrival next day of the sheriff of York from the south cut off

[1] A passage from the Lichfield Chapter Acts (Bodleian MS. Ashmole 794, f. 2d)
is significant of the state of general alarm at this time. Master William of Eiton,
mason, with his seven masons, swears to help the Chapter in the defence of the
close and to forewarn them if they should hear of any impending harm or danger
(5 Feb. 1322); see J. Harvey, *Gothic England* (1947), p. 172, app. ii.
[2] G. H. Tupling, *South Lancashire in the Reign of Edward II*, p. xxxiii.
[3] T. F. Tout, 'The Tactics of the Battles of Boroughbridge and Morlaix', *Eng.
Hist. Rev.* xix (1904), 711-15. Tout suggests that Boroughbridge may have been the
real starting-point of the new tactics of combining archers with dismounted men-at-
arms.

his last hope of retreat. Together with Mowbray, Clifford, and the other captains, he surrendered to Harclay who took all his prisoners to York. Edward II saw at last the opportunity of vengeance for which he had waited so long. He ordered the prisoners to be brought before him at Pontefract and there, in the hall of his own castle, Lancaster listened while the king 'recorded' the operations leading to the battle, the earl's war-like array with banners flying, the conflict, and his capture. These facts were presented to the seven earls and other unnamed barons who were present, as manifest and notorious and there-fore admitting of no answer.[1] Lancaster was sentenced to death as a traitor and a rebel by these magnates, only his rank and his royal blood saving him from the most degrading penalties of treason. On the morning of 22 March he was led out from the castle in penitential dress, 'on a lene white Jade with owt Bridil', and beheaded in the presence of a jeering crowd.[2] With the king's leave, the monks of Pontefract priory then removed the body and buried it, though with little ceremony, before the high altar of their church.

Until the final débâcle, the earl's stormy career seems to have had no adverse effect on the smooth running of the great Lancastrian estates.[3] Though little surplus capital was available for investment in land, Thomas maintained his father's policy of adding small parcels of lands and rents as opportunity offered, and on more than one occasion he was able to turn a political crisis to advantage in this way. Thus, in the York parliament of 1314, Pembroke yielded him the manor of New Temple and the former Templars' lands outside Temple Bar; at the conclusion of their private war, Lancaster obtained Warenne's Yorkshire lands and effected an exchange of manors and hundreds which enlarged his resources in Wiltshire, Somerset, and Dorset. Building operations, not exclusively military in character, added to the value of his many castles. Pickering was furnished by him with a new hall and chamber, Kenilworth with a chapel, Ponte-fract with a tower intended, it was said, to serve as a prison for the king. The scanty records which survive show that for

[1] T. F. T. Plucknett, 'The Origin of Impeachment', *Trans. Royal Hist. Soc.*, 4th ser. xxiv (1942), 57. The earls present were Kent, Richmond, Pembroke, Warenne, Arundel, Athol, and Angus (*Rot. Parl.*, ii. 3).

[2] Leland, *Collectanea*, ii. 465.

[3] Somerville, *Hist. of the Duchy of Lancaster*, i, chs. ii, iv.

administrative purposes the wide and scattered estates had largely been integrated. By a method which made for administrative unity, the reeves paid in their dues to receivers at centres like Kenilworth and Pontefract, irrespective of the earldom to which the particular manor was attached. Though there is as yet no trace of a single receiver-general, two auditors helped to unify the organization, at the head of which stood the earl's council. For the defence of his estates and for other purposes Earl Thomas could rely on a considerable private army. More than 500 retainers wore his livery of argent and azure and for the muster at Newcastle in 1319 he was asked to find 2,000 footmen. Yet, unlike some of his successors, he made little or no use of his great territorial powers to win special liberties for his lands. Unfortunately for himself, since he lacked political capacity, his ambitions were essentially political rather than territorial.

The clue to Lancaster's policy seems to lie in his conception of himself as standing in a unique relationship both to the king and to his fellow-barons. It is not without significance that in his correspondence with Bruce he should have adopted the pseudonym of 'King Arthur'.[1] For Thomas was not only the richest landowner in England, after the king; he was nephew to Edward I, first cousin to Edward II; his mother had been queen of Navarre; he was brother-in-law to Philip IV, uncle to the queen of England and to three kings of France. Such a man might well expect to occupy a place of special confidence at court and, as the irresponsibility of Edward II became increasingly apparent, to dominate the whole administration. When it seemed that neither his illustrious connexions nor his territorial influence would suffice to win him the position he desired, Thomas fell back upon the powers inherent, or believed to be inherent, in the office of steward, appurtenant to his earldom of Leicester. In so far as he had a 'constitutional programme' his aim was to see a puppet-king, guided and controlled by a baronial council under the presidency of the steward and served by a household, the head of which would be his deputy and the officers his nominees, occasional recourse being had to parliaments likewise amenable to his control. Thwarted in the execution of this programme, first by Gaveston, then by Pembroke and, finally, by the Despensers, Lancaster almost inevitably adopted a policy of non co-operation. He was isolated

[1] *Cal. Close Rolls 1318–23*, p. 526.

by his own ambitions which drove him to maintain a private army, to summon private 'parliaments', and to the final treachery of negotiations with the king of Scots. Had he been a man of different character, capable of winning and holding loyalty, he might have anticipated the action of his descendant and ended by overthrowing and supplanting the reigning king. As it was, his nobly-born wife left him for a squire of low degree; his principal official and many of his retainers deserted him before Boroughbridge; his constables yielded up his castles without so much as a show of a fight; and his fellow-barons acquiesced in his condemnation to a traitor's death. He may be seen as the supreme example of the over-mighty subject whose end must be either to destroy or be destroyed.

Little of this was apparent, however, to the mass of Earl Thomas's contemporaries. His early association with Winchelsey (commonly regarded as a second Becket) had not been forgotten and in some ecclesiastical circles his credit was high. The prior and convent of Christ Church sought his assistance in their efforts to secure the canonization of Winchelsey;[1] the monastic chroniclers record the death-bed speech in which Henry of Lincoln was believed to have exhorted his son-in-law to protect the Church from the oppression of the Romans and the exactions of kings, to defend and deliver the people, to persuade the king to remove evil counsellors, to choose English servants, and to observe the Great Charter;[2] they write of Earl Thomas as a martyr in the cause of Church and State.[3] Not all these eulogies can be dismissed as the fruit of sedulous propaganda. Edward II had to place an armed guard round Pontefract priory to keep off the weeping crowds, who then transferred their devotion to the place of Lancaster's execution where a chapel was built later with the aid of funds collected from all over England. There were similar scenes before the tablet which Lancaster had caused to be set in St. Paul's to commemorate the Ordinances.[4] Unaware of the real nature of his aspirations, men were ready to see Lancaster as a second de Montfort, the

---

[1] Conway Davies, p. 108.   [2] Trokelowe, pp. 72–73.
[3] 'pro justitia ecclesiae et regni' (Knighton, i. 426); 'pro ecclesiae jure et statu regni (*Flores*, iii. 204).
[4] It has been suggested that a devotional plaque recently acquired by the British Museum, which depicts scenes from the life and downfall of Lancaster, may have been an offering at this shrine. See H. Tait, 'Pilgrim Signs and Thomas Earl of Lancaster', *British Museum Quarterly*, xx. 2 (1955), 39–46.

Ordinances which he had championed as a second Great Charter. They remembered that he had died in conflict with a king whom none could respect and with courtiers whose avarice was soon to unite the whole nation against them. For the unhappy subjects of Edward II, Earl Thomas thus became a symbol of resistance to the powers of darkness, the upholder of ancient liberties against new-fangled tyranny. His reputation far exceeded his deserts; but the five years which followed his death were to afford ample proof of the reality of the evils with which he had contended.

Note on the Site of the Battle of Bannockburn

Objections can be raised to all the suggested sites. Sir H. Maxwell's view that the battle was fought to the west of the Roman road appears to be irreconcilable with Barbour's statement that the English were in close touch with Stirling throughout, with his emphasis on the marshy character of the ground, and with the subsequent flight of the English to the Forth. If the battle was fought round the Borestone, as was long assumed, or in the Dry Field favoured by Mr. Miller, it is equally hard to understand how contact with the castle was maintained and why or how Edward II fled there after the battle. Dr. Mackay Mackenzie's reconstruction, which moves the whole action into the buckle of the Forth, disposes of these difficulties but raises others. It necessitates the assumption that a great army, and probably heavy wagons as well, were taken over the Bannock and its marshes after an exhausting march and across the flank of an active enemy; thus, it forces us to credit the English leaders with astonishingly bad generalship. Edward and many of his captains must have been familiar with the country, yet we are invited to believe that they made deliberate choice of the marshes of the Carse for their camp. Gray, however, clearly thought that the English were badly led; and Major Becke may be right in supposing that most of the baggage was left behind at Berwick, though there is no contemporary warrant for this suggestion. It is conceivable that some of it was left behind at Edinburgh where the pause made by the army was so short as hardly to allow time for taking in a full quantity of supplies. The case for this site seems, on the whole, to be rather stronger than that put forward for any of the other three.

III

REACTION AND REVOLUTION
(1322-30)

ON 2 May 1322, six weeks after the execution of Lancaster, there assembled at York a parliament which included—besides magnates, knights, and burgesses—clerical proctors, representatives of the Cinque Ports, and twenty-four discreet men empowered to act for the *communitas* of the principality of Wales, lately the scene of civil war. The main business of this famous parliament was to restore the king to his full dignity by means of a repeal of the Ordinances. If there was opposition to this measure, we know nothing of it; for the chroniclers are uninformative and no official record of the proceedings has survived. All that is left to us is the text of a statute, repealing the Ordinances and ending with the ambiguous sentences which have raised such a dust of controversy among modern scholars:[1]

All the things ordained by the said Ordainers, [the statute concludes] and contained in the said Ordinances shall henceforth and forever cease . . . statutes and establishments duly made by our lord the king and his ancestors shall remain in force. And that henceforth . . . all manner of ordinances or provisions, made under any authority or commission whatsoever, by the subjects of our lord the king or his heirs, concerning the royal power of our lord the king or of his heirs, or contrary to the estate of our lord the king or his heirs, or contrary to the estate of the crown, shall be null and of no validity or force. But things which are to be established for the estate of the king and his heirs and for the estate of the realm and people, shall be treated, granted and established in parliament by our lord the king and with the consent of the prelates, earls and barons and of the commonalty of the realm, as has been hitherto accustomed.[2]

Thus, the statue is designed to administer a rebuke to the baronial cabal which had engineered the Ordinances and forced them upon the king; those who drafted it may have had in mind also Lancaster's pseudo-parliaments and the marchers' confederation of 1321. The intention seems to be to point a

[1] See Bibliography, pp. 556-7 below.
[2] 'Revocacio novarum Ordinationum'. *Statutes of the Realm*, 15 Edw. II.

contrast between ordinances improperly initiated by subjects and traditional methods of legislation by the king in parliament. Whether the statute will bear the weight of further meaning which has been imposed upon it remains open to question. The final phrase, 'as has been hitherto accustomed', would appear to preclude the possibility that it was any part of the purpose of the statute to introduce new methods of legislation, or to confer new rights on 'prelates, earls, barons and the commonalty of the realm'. Many scholars read 'commonalty of the realm' as a synonym for 'commons in parliament' and, in Miss Clarke's view, the ambiguity of the wording was deliberate. At this date the position of the clerical proctors was still in doubt and to define the estates of the commons more precisely would have been to draw attention to a current controversy, irrelevant to the main issue. Miss Clarke linked the statute with the doctrine of parliamentary authority set out in the *Modus Tenendi Parliamentum* and thought that its purpose was 'to integrate the estates of the realm in parliament'.[1] Yet in the text, the inclusion of the commons, if they are included, appears to be almost incidental; the words 'commonalty of the realm' read like the kind of phrase which runs readily off the pen of a well-trained clerk, and it is possible that the intention is merely to reinforce the notion of universal consent dear to medieval political theorists.[2] Nor is it easy to discover in the entourage of Edward II the far-sighted statesmen who, in Mr. Lapsley's view, sought by means of this statute to deflect the current of political dispute from the battlefield to parliament. The framers of the statute, we may well believe, were thinking mainly of the past. They desired to demonstrate the irregularity of the conduct of the Ordainers and of Lancaster by putting a statute in the way of any who might try to repeat their offence. In so far as the future was in their minds at all, it was probably the immediate future; and their aim was less to provide constitutional machinery for the settlement of future disputes on the prerogative, than to acquire a weapon with which to defeat any attempt at resuscitation on the part of surviving Ordainers, or any attempt at a movement similar to theirs. It is indeed significant that the royalist party in their hour of triumph should have chosen to emphasize the

---

[1] *Medieval Representation and Consent*, p. 139.
[2] As suggested by J. R. Strayer [*Amer. Hist. Rev.* xlvii (1941), 22], with whose view of the statute I am in general agreement.

dignity of the representative parliament by selecting it as the most effective setting for their revenge upon their enemies, and that an act of parliament should have appealed to them as affording the best guarantee of future security; but the importance of the Statute of York in other respects seems to have been greatly exaggerated.

For many contemporaries, the most significant act of the York parliament was not the statute but the measures taken to punish the king's enemies and to reward his friends.[1] The process against Lancaster was solemnly recorded: the lands of Damory, who had died at Tutbury, and of Hereford, who had fallen at Boroughbridge, were declared forfeit. Clifford, Dayville, and Mowbray were hanged at York, Badlesmere at Canterbury, Sir Henry Tyes in London; and some fifteen other knights were drawn and hanged. Both the Mortimers were imprisoned in the Tower and Hugh Audley in Berkhamsted castle, while those who escaped the avengers fled to France.[2] The process against the Despensers was revoked, on the grounds that they had not been called to answer in 1321, that the award, having been made without the consent of the prelates 'who are peers of the realm in parliament', was in violation of cap. 39 of Magna Carta, and that the magnates had framed the sentence without the king, before they came (in arms) to parliament and had thereby accroached royal power. Sir Andrew Harclay was rewarded for his services at Boroughbridge by elevation to the rank of earl of Carlisle and the elder Despenser became earl of Winchester; but even with his enemies at his mercy, it seems that Edward dared not revive the earldom of Gloucester for the younger Hugh. To both Despensers the royalist victory brought access of solid territorial power. Between them they acquired some 175 knights' fees from Lancaster's widow and many castles and manors; the elder secured Lancaster's north Welsh stronghold of Denbigh, the younger his manors of Donnington and Bisham. From Mowbray the younger Hugh obtained the castle and town of Swansea and the castles of Oystermouth, Pennard, and Loughor in Gower, from Damory's widow, the lordship of

---

[1] See 'A Chronicle of the Civil Wars of Edward II' (*Speculum*, xiv. 79). *Primus tractatus parliamenti* was the annulment of the sentence on the Despensers, *secundus punctus* was the recording of the sentence on Lancaster. Like most of the other chroniclers, this anonymous northern writer does not mention the statute.

[2] Mortimer of Chirk died in the Tower in 1326.

Usk,[1] from Mortimer of Chirk the castles of Blaen Llyfni and Bwlch y Dinas and the land of Talgarth. The lordship of Iskennin, gained as a result of the forfeiture of John Giffard, enabled him to bridge a gap between some of his south Welsh territories; and his influence was extended eastward when Norfolk made him a grant for life of the honour and castle of Chepstow and of all the old Bigod lands beyond the Severn. Thus, the two Despensers were virtually in control of the whole of south Wales; and in the north and centre their only rival was Arundel, who combined the office of justiciar with custody of most of the Mortimer estates. As time went on, it seemed that they had less and less to fear. Pembroke, who might have rivalled them in influence, had associated himself with the judgement on Lancaster and had acquiesced in the restoration of the Despensers without recorded protest. If he showed no tendency to dispute their subsequent pre-eminence at court, this was probably because they had already seen to it that he should be rendered harmless. About a month after the dissolution of the York parliament, because the king 'was aggrieved against him for certain reasons' (doubtless for the part he had played in the parliament of 1321), Pembroke had been made to pledge his body, lands, and goods to obey and aid the king in war and peace and not to ally with anyone against him.[2] Thereafter, he seems to be identified with the court party, associating himself with the younger Despenser in negotiations with Scotland and, possibly, in some of his administrative reforms. The times were not easy for a man of the middle way and it may even be that he felt some affection for the wayward king.[3] But, though he was not a strong man, it is improbable that the Despensers grieved overmuch when he died suddenly in France in the summer of 1324, leaving no issue. No one else seemed likely to challenge their power. Neither Arundel nor Warenne was trusted by his fellows; John of Brittany, earl of Richmond, was a foreigner and closely aligned with

---

[1] Elizabeth Damory was allowed Gower in exchange but within two years, by a piece of legal chicanery which throws a sinister light on his methods, Despenser had succeeded in depriving her of it again; see G. A. Holmes, 'A Protest against the Despensers, 1326', *Speculum*, xxx (1955), 207–12.

[2] *Cal. Close Rolls 1318–23*, pp. 563–4.

[3] In a petition relating to the funeral arrangements, the Countess of Pembroke referred to her husband's devotion to the king, 'a qui il fust si procheyn et vous ad servi cum vous savetz'. Marie de St. Pol remained in control of Pembroke's large estates until her death in 1377; see H. Jenkinson, 'Mary de Sancto Paulo, Foundress of Pembroke College, Cambridge', *Archaeologia*, lxvi (1914–15), 435.

the court,[1] as were the king's two half-brothers, Thomas of
Norfolk and Edmund of Kent; the heirs of Hereford and Warwick were minors. Henry of Lancaster, brother and heir to
Earl Thomas, was a moderate man and so little suspect by the
king that in 1324 he was allowed possession of the earldom of
Leicester; but there was not much prospect that he could
further repair the broken fortunes of his house.

Like all their predecessors in power, the Despensers found
themselves faced with the problem of Scotland. When the truce
expired in the summer of 1322 the Scots swarmed over the
Border and ravaged the north-west of England down to Preston.
Edward attempted retaliation by wasting the Lothians, but the
Scots refused to give battle and within a few weeks he withdrew,
his troops decimated by famine and pestilence. Bruce determined to pursue him. A Scottish force made its way into the
heart of Yorkshire and Edward II himself narrowly escaped
capture in a skirmish between Byland and Rievaulx. All Yorkshire was stricken with panic and the canon of Bridlington tells
us that many northerners, his own brethren among them,
despairing of protection from Edward were beginning to look to
the king of Scots.[2] Such impulses were not confined to the rank
and file. The archbishop of York issued formal letters empowering the heads of religious houses to treat severally with the Scots
for terms of peace; Edward II's favourite, the bishop of Durham,
Louis de Beaumont, entered into direct negotiations with them.
Most sinister of all was the defection of the hero of Boroughbridge, Andrew Harclay, earl of Carlisle. After the skirmish
near Byland, the earl dismissed his troops, sought out Bruce in
person, and agreed with him for the establishment of a committee of six Scottish and six English lords with power to conclude
a peace between the two countries, on the basis of recognition
of Bruce as king of Scots. Harclay was taken and brought to
London to suffer the penalties of treason; but his action had
made it plain to all that the war with Scotland was lost.[3]
Permanent peace was out of the question so long as Edward II
refused to recognize Bruce; but a thirteen years' truce, negotiated

---

[1] Richmond was a prisoner in the hands of the Scots, 1322–4. See I. Lubimenko,
Jean de Bretagne (1908), pp. 66–70.   [2] Bridlington, p. 81.
[3] According to Murimuth (p. 39) Harclay wished to marry the sister of Robert
Bruce. See J. E. Morris, 'Cumberland and Westmoreland Levies in the time of
Edward I and Edward II', *Trans. Cumb. and West. Arch. Soc.*, N.S. iii (1903), 324 ff.

in March 1323, by Pembroke, Baldock, and the younger Despenser, brought the war to an end for the rest of the reign. The English undertook to do nothing to hinder the Scots from coming to terms with the papacy, and papal recognition of Bruce's title was soon obtained. Other clauses in the truce were designed to give some relief to the unhappy northerners. The march laws were to be maintained and harbour allowed to ships of either nation driven into one another's ports in times of stress. Thus, to the Despensers fell the opportunity denied to all Edward II's former advisers, of ruling a country free from the menace of war in the north.

They were by no means incompetent for the task, being much more than greedy and irresponsible favourites. Tout has demonstrated their pursuit of a clear-cut policy of developing the inner fastnesses of the royal household in such a way as to render the king virtually independent of the great offices of state which tended to be subject to baronial or parliamentary control.[1] Hugh the younger concentrated his attention on the chamber, revival of which had begun under John Charlton as chamberlain, when the custodians of the forfeited lands of the Templars and of Walter Langton had been ordered to account for them to him, instead of to the officers of the exchequer. The Ordainers had tried to arrest this process by demanding Charlton's removal and the payment of all the issues of the land into the exchequer; but it did not prove difficult to evade these restrictions and when Hugh Despenser became chamberlain in 1318, he resumed Charlton's policy. The chamber was used as a source of private revenue to the king and as a means for the acquisition of manors; and under Despenser the system of reserving lands to the chamber and exempting them from the jurisdiction of the exchequer was greatly expanded. Edward II's favourite residence of King's Langley, where Gaveston lay buried, was one such manor, Burstwick-in-Holderness was its northern counterpart. Furnished with its own seal, the *sigillum secretum*, the chamber became the source of writs and letters of a kind that in an earlier age would have emanated from chancery. Reforms in other departments were promoted by Despenser's principal associates. Walter Stapledon, bishop of Exeter, during his two periods of office as treasurer (1320-1 and 1322-5), made a heroic effort to introduce some order and system into the unwieldy

[1] *Place of Edward II*, pp. 142 ff.; *Chapters*, ii, 314-60.

mass of exchequer records.[1] A series of ordinances issued by
the king in council, the Cowick Ordinances of 1323, and the
Westminster Ordinances of 1324 and 1326, provided for their
classification and calendaring. Strict rules were laid down for
the yearly compilation of the Pipe Roll, large additions were
made to the exchequer staff, the work of the two remembrancers
was clearly distinguished, and four auditors were appointed. In
1324 the two escheators, one for the north and one for the south
of the Trent, were replaced by nine, each in charge of a much
smaller group of counties, a system which lasted until the end of
the reign and was revived again under Edward III. But Stapledon's outlook was not purely departmental. He aimed at a comprehensive survey of all the administrative records of the Crown
outside the chancery, including those of the wardrobe and such
as were scattered in castles and elsewhere throughout the country.
Robert Baldock, chancellor from 1323 to 1326, was also an
active reformer. Before his promotion Baldock had combined,
in what had become the traditional fashion, the offices of keeper
of the privy seal and controller of the wardrobe, and as chancellor he sought to retain control of the privy seal. The result was
the separation of the latter from the wardrobe and the limitation
of the wardrobe's activities which becomes evident in the latter
years of Edward II. A series of ordinances issued between 1318
and 1324 lopped off from the wardrobe a number of its former
responsibilities. The king's butler, the receivers of stores, the
keeper of the king's horses, and other royal messengers were
ordered to account directly to the exchequer. Thus, the reign
of Edward II sees both the culmination of the wardrobe's
activities and the beginnings of a decline which was to be pro-
gressive throughout the century. Credit must also be given to
the Despensers and their associates for the decision to retain
some part of the reform programme of the Ordainers; it may
not be altogether fanciful to see the hand of Pembroke here. In
the very parliament which formally repealed the Ordinances,
Edward II issued 'establishments', later embodied in a statute,
whereby he confirmed the rights of the Church as set forth in
the Great Charter and other statutes, the amendments to the
Forest Law and the limitations on the jurisdiction of household

---

[1] In Tout's view, 'it is hardly too much to say that we owe to Stapledon, more
than to any other one person, the fact that our vast collection of exchequer records
before 1323 is still preserved to us' (*Place of Edward II*, p. 173).

officers enacted by the Ordinances, and the act of 1316 touching the power of the sheriffs, combining with these provisions for the better keeping of the peace, sumptuary legislation, and standardizing of weights and measures. Moreover, the younger Despenser was 'the principal mover with the king and council' in securing the Ordinance of Kenilworth (May 1326), which substituted home for foreign staples.[1]

Yet for all their administrative ability, the Despensers lacked the political acumen which should have enabled them to appreciate the significance of their defeat in 1321. Their avarice was notorious and gave rise to mounting discontent. Recent investigations have revealed something of the vast resources of the younger Hugh.[2] At the beginning of September 1324 he had deposits amounting to nearly £6,000 with the Florentine firms of Bardi and Peruzzi; between November 1324 and Michaelmas 1326 he deposited a further £5,735 with the Peruzzi alone. He was almost certainly the Italian bankers' most important English client, his deposits with them exceeding even those of the papal collectors; yet these deposits constituted only a part of the total reserve funds at his disposal. The bulk of Despenser's wealth derived from his landed estates; but he did not scruple to increase it by exploiting his position at court and by methods even more dubious. The bishop-elect of Rochester was not allowed livery of his temporalities until 'contrary to all justice and the custom of England', he had made the chamberlain a present of £10.[3] One John de Sutton was held in prison until he had surrendered the castle of Dudley;[4] a wealthy heiress, Elizabeth Comyn, was kidnapped and detained for more than a year until she had made over a number of her estates to Despenser and his father.[5] Such things could not be hid. In 1324 the younger Hugh complained to the pope that a necromancer at Coventry was hatching a plot to kill the king, his father, and himself.[6] More ominous still was the growing tendency to hark back to the events of 1321-2 and to make a hero of Thomas of Lancaster. Men and women continued to resort to Pontefract and St. Paul's;

---

[1] *Cal. Pat. Rolls 1324-27*, p. 274. See p. 351 below.
[2] E. B. Fryde, 'The Deposits of Hugh Despenser the Younger with the Italian Bankers', *Econ. Hist. Rev.*, 2nd ser. iii (1951), 344-62.
[3] *Hist. Roffensis (Anglia Sacra* i), p. 361.   [4] *Complete Peerage*, iv. 266.
[5] *Econ. Hist. Rev.*, 2nd ser. iii. 348, n. 7. Despenser was also believed to be inducing the king to exact forced oaths from his subjects: the oath taken by Pembroke (above p. 74) may be a case in point.   [6] *Complete Peerage*, iv. 268, n. *f.*

at Barton Regis, near Bristol, where two of the lesser rebels had
been hanged, there were rumours of miracles; a certain Sir
Reginald de Montfort was said to have bribed a poor child in
the district to pronounce to the people that his sight had been
restored.[1] In the parliament of 1324 magnates and prelates
petitioned that the rebels' bodies still hanging on the gallows
might be taken down and given Christian burial. It was doubtless in order to check such tendencies that the king ordered the
record of the proceedings against Lancaster and his associates
to be read before the magnates in this parliament and formally
enrolled.[2] Some precautions were very necessary; for an opposition party was forming and at the heart of it stood the queen.

Little evidence is available to show by what road Isabella of
France had arrived at this position. Her rare interventions in
politics hitherto had been mainly pacific in character. After the
murder of Gaveston she had supported the bishops in their
efforts to promote peace between Edward and Lancaster; she
had been active in procuring the settlement at Leake;[3] and in
1321 she joined Pembroke and Richmond in begging the king to
have mercy on his people. While Gaveston was in the ascendant
she was still very young; and the birth of her first child within a
few months of his death may well have led her to hope that the
worst of her troubles were over and that better times lay ahead.[4]
In the years that followed, her relations with her husband cannot
have been wholly unhappy. Three more children were born to
them—John of Eltham, Eleanor of Woodstock, and Joan of the
Tower.[5] The queen was furnished with a household appropriate
to her station, its upper ranks numbering at least seventy persons,
the attendant throng of laundresses, watchmen, messengers,
servants, carters, grooms, and pages, bringing the total of her
staff to about 180.[6] Grants to her included the important honours

[1] Cal. Chancery Warrants, i. 543; Cal. Pat. Rolls 1321–24, pp. 378, 442.
[2] G. O. Sayles, 'The Formal Judgements on the Traitors of 1322', Speculum, xvi
(1941), 57–63.
[3] 'procurante indies et viriliter instante Regina' (Trivet. Ann. Cont., ed. A. Hall,
1722, p. 27).
[4] According to Trokelowe (p. 68) Isabella had complained bitterly to her father
of Gaveston's avarice.
[5] John of Eltham, b. 15 Aug. 1316, d. 13 Sept. 1336, created earl of Cornwall
Oct. 1328; Eleanor of Woodstock, b. 18 June 1318, d. 22 Apr. 1355, m. May
1332, Reginald, count of Gelderland; Joan of the Tower, b. 5 July 1321, d. 7 Sept.
1362, m. July 1328, David, afterwards David II, king of Scotland.
[6] Hilda Johnstone, 'The Queen's Household' (Tout, Chapters, v. 241–50).

of Wallingford and St. Valéry. Edward entrusted her with a
mission to France in 1314;[1] he dropped his own nominee in
deference to her wish to secure the see of Durham for Louis de
Beaumont; and although the king and queen began by supporting different candidates for Rochester in 1316, it was the king
who was persuaded to change his mind, 'whereat the cardinals
marvelled and put it down to his inconstancy'.[2] It is true that on
more than one occasion, notably in 1319 and 1322, Edward had
shown himself oddly indifferent to his wife's safety; but the
sources afford no suggestion that she had given him cause for
jealousy or that there was any serious breach between them up
to the time of Boroughbridge. Almost certainly, it was the
position accorded to the Despensers after the royalist victory
which finally alienated Isabella from Edward II; and it may
have been not long afterwards that she began to attach to herself
the little group of bishops and lay magnates who were to be her
principal supporters for the rest of the reign.

In a letter written to the pope after Boroughbridge, Edward
declared that there were three bishops whom he could no longer
tolerate in his kingdom—Burghersh of Lincoln, Orleton of
Hereford, and Droxford of Bath and Wells.[3] The first two were
to become prominent members of the queen's party. Burghersh,
who may have been in some degree indebted to Isabella for his
see, was a nephew of Badlesmere and deeply implicated in his
fall. His temporalities were seized by the king, but John XXII
refused to accede to his translation and exhorted the other
bishops and magnates to intercede for him. Nor would the pope
agree to move Orleton of Hereford who, in 1321, had thrown in
his lot with his friend and patron, Roger Mortimer of Wigmore.
These two bishops—*alumpni Iezabele*, as Baker terms them—were
the leaders of the new group, but others soon joined them. The
new bishop of Winchester, John Stratford, who was in disgrace
for accepting the see in defiance of the king's wishes and had
had to buy his peace at the price of a grant of £1,000 to the
younger Hugh, had no love for the Despensers. Among the lay
magnates, the queen could count on the support of her uncle,
Henry of Lancaster, in any movement directed against the
Despensers; while both Thomas of Norfolk and Edmund of Kent

---

[1] *Cal. Pat. Rolls 1313-17*, p. 85.   [2] *Hist. Roffensis*, p. 359.
[3] See Kathleen Edwards, 'The Political Importance of the English Bishops during the Reign of Edward II', *Eng. Hist. Rev.* lix (1944), 335 ff.

were being alienated by their arrogance. Henry de Beaumont, a member of both the great and the secret council, refused his advice to the king, and when told to leave replied that he would rather go than stay.[1] Most important of all, Roger Mortimer of Wigmore assisted by Orleton and by two influential Londoners, John de Gisors and Richard de Bettoyne, had escaped from the Tower in 1323[2] and made his way to France where he offered his services to Charles IV for the war in Gascony.[3]

Though there were rumours afoot that the younger Despenser was trying to engineer an annulment of Edward II's marriage at the curia, the first open move against Isabella was the sequestration of her estates in September 1324, on the pretext of danger of a French invasion. It is possible to exaggerate the significance of this incident, for the queen-mother's castles had been sequestrated on similar grounds in 1317; and the allowance of 2,920 marks a year made to Isabella was adequate.[4] Yet the loss of her property was certainly unwelcome to one whose whole history shows her to have been more than ordinarily acquisitive; and that he had deprived the queen of her dower was among the charges subsequently brought against the younger Despenser. Thus, when a way of escape offered, Isabella was quick to take advantage of it. An English embassy in Paris was failing to achieve a settlement of the Gascon question and it was proposed by the papal nuncios that Isabella should undertake a personal mission and try to bring about an understanding between her husband and her brother.[5] Burghersh, Orleton, and the ambassadors already in France warmly supported the proposal, Edward with some reluctance gave his consent, and on 9 March 1325 Isabella, accompanied by many members of her household, sailed for France.[6] Her intervention proved remarkably effective. Agreement was reached that Edward II's French possessions (except for the Agenais and La Réole whose fate was to be decided later) should be returned to him undiminished so soon as

[1] Speculum, xiv. 61. (The breach was not permanent, however, for Beaumont was a member of the embassy to France.)

[2] The date has been established by E. L. G. Stones, Eng. Hist. Rev. lxvi (1951), 97–98.   [3] Baker, p. 16.

[4] Tout, Place of Edward II, p. 140, n. 1; Chapters, v. 245.

[5] The ambassadors were the bishops of Norwich and Winchester, the earl of Richmond, and Henry de Beaumont; see pp. 109–10 below.

[6] Isabella's huntsmen and her pack of hounds were left in charge of the prior of Canterbury who later complained bitterly to Despenser that they were eating him out of house and home (Lit. Cant. i. 164–70).

he had performed his homage; in the meantime, the duchy of Gascony was to be put in the custody of a French seneschal. The great council very properly decided that the king should do homage in person; but the Despensers were determined that he should neither be parted from them nor allowed to rejoin the queen; and on 24 August Edward declared himself too ill to travel. The Despensers persuaded him to ignore expressions of uneasiness at home and to adopt a plan which the papal nuncios had already mooted—that Prince Edward, now thirteen years of age, should be invested with the duchy of Gascony and the county of Ponthieu and that he should perform the homage in his father's stead.[1] Accompanied by Walter Stapledon, one of the few bishops whom the king still felt able to trust, Prince Edward sailed accordingly to France and on 21 September did homage to Charles IV at Bois-de-Vincennes. With the heir-apparent at her side, Isabella held a trump card. Stapledon was so much alarmed by what he saw and heard in France that he returned secretly by night; and Edward began to press earnestly for his wife's return. After putting him off with excuses for a time, she came out at last into the open; a letter entrusted to Stratford informed Edward that neither the queen nor the prince would return to his court so long as the younger Despenser was there; and Isabella told her brother that her marriage had been broken and that she must live as a widow until the Despensers had been removed from power.[2] When the news of her decision became known in England, there was widespread alarm. Invasion scares were in the air throughout the spring of 1326, though what seems to have been anticipated was a joint invasion by the king of France and the queen of England,[3] supported by the influential group of exiles then in Paris—Kent, Richmond, Beaumont, Stratford of Winchester, Airmyn of Norwich, and, above all, Roger Mortimer of Wigmore, to whom Edward threatened death should he return.[4] When renewed appeals to the queen proved fruitless, Edward and his friends began to take some

---

[1] 'pater et filius Despenser . . . dederunt consilium et praevaluerunt, ut filius regis transiret' (Murimuth, p. 44).

[2] Vita Edwardi, p. 143.

[3] This is apparent from the correspondence between the prior and the archbishop of Canterbury (Lit. Cant. i. 172–4).

[4] Hist. Roffensis, p. 365. Edward blamed Airmyn for the terms of the treaty with France and had refused to restore his temporalities, thereby forcing him into alliance with the queen; see J. L. Grassi, 'William Airmyn and the Bishopric of Norwich', Eng. Hist. Rev. lxx (1955), 550–61.

belated precautions. The Tower and other castles were put in a
state of repair and orders were given for the guarding of the
coasts; commissions of array were issued and Frenchmen in
England put under arrest; the cautious Despenser withdrew
£2,000 in cash from his bankers. But all was not going smoothly
for Isabella in France. Her liaison with Mortimer was beginning
to cause scandal,[1] Despenser was using his influence against her,[2]
and Charles IV was being urged by the pope to countenance
the guilty couple no further. Since her brother was not disposed
to help her, Isabella found it prudent to withdraw to the Low
Countries and to strike a bargain with William II, count of
Hainault, Holland, and Zeeland, who, in return for the betrothal
of his daughter, Philippa, to Prince Edward, was prepared to
put a force of Hainaulters at her disposal. The ports of Holland
and Zeeland offered good facilities for embarkation; and, though
Edward II ordered the muster of a fleet, the sailors of England
are said to have refused to fight 'because of the great wrath they
had towards Sir Hugh Despenser'.[3] Thus, when the queen and
her company sailed from Dordrecht on 23 September 1326, they
met with no resistance at sea. Next day they anchored in the little
Suffolk port of Orwell, prepared to annihilate the Despensers and,
if necessary, to secure the forcible removal of Edward II himself.

The invaders spent their first night at Walton-on-the-Naze,
moving thence to Bury St. Edmunds and Cambridge, where the
queen stayed for some days at Barnwell priory; when she
reached Dunstable she was only thirty-three miles from London.
The invading force cannot have been large, the Hainaulters,
under the command of the Count's brother John and of Roger
Mortimer, numbering only some 700 men; but the queen, says
Knighton, found favour with all.[4] The gentry of East Anglia
joined her at once; Thomas of Norfolk, Henry of Leicester, and
a number of the bishops followed suit. Isabella was already in
communication with London and Edward's efforts to rouse the
city in his defence proved unavailing.[5] On 29 September he

---

[1] 'Tunc secretissimus atque principalis de privata familia regine', Baker, p. 21.
[2] *Scalacronica*, p. 151.
[3] *French Chronicle of London* (Camden Soc. xxviii), p. 51.   [4] Knighton, i. 432.
[5] She wrote to London for assistance immediately on landing, but no answer was
returned 'for fear of the king' (*French Chron.*, p. 51). Froissart, ed. Kervyn de
Lettenhove (xvi. 159), has a plausible story that a group of London citizens had
helped to procure the invasion by assuring Isabella that if she were to land she
would find widespread support.

offered a reward for Mortimer's head; next day, he caused a papal Bull denouncing invaders (which had been provided for use against the Scots) to be read aloud at Paul's Cross. But the reactions of the citizens offered him no encouragement and, recognizing that London was becoming too hot for him and that 'the whole community of the realm' adhered to the queen's party, he left the city for the last time and fled to the west accompanied by the Despensers, Arundel, Warenne, and Baldock. A week later a copy of a letter from Isabella, asking particularly for the destruction of the Despensers, was posted in Cheapside and others soon began to appear in the windows of private houses. Their effect was inflammatory. On 15 October the citizens flocked to the Gildhall and forced the mayor, Hamo de Chigwell, to declare publicly that enemies of the queen and the prince would remain in the city at their peril. Immediately afterwards the mob took up arms and seized and beheaded a certain John the Marshal, believed to be a spy of Despenser's. Worse was to follow. As the bishop of Exeter was riding home across the city from an episcopal conference (hurriedly assembled at Lambeth in the hope of offering mediation) a cry of treason was raised against him. Attempting to reach sanctuary in St. Paul's, he was intercepted, dragged from his horse and through the churchyard to Cheapside where, with two of his escort, he was beheaded with a butcher's knife. The more sober citizens were probably appalled by the outrage, for no one interfered with the minor canons and vicars choral of St. Paul's when they came out into the streets in procession and brought the headless body of the bishop into the cathedral.[1] Next day the corpse was taken to St. Clement Dane's, a church in the gift of the bishop of Exeter, but the rector refused admission to his patron's mortal remains. It was left to a woman in the crowd to throw a sheet over the body of this notable benefactor of religion and learning,[2] before it was hastily interred 'without office of priest or clerk', in a place nearby which came to be known as 'lawless church'. The whole city was now given over to looting and violence. Baldock's manors and the house of the Bardi were robbed and, an armed mob, assembled on Cornhill, forced the constable of the Tower to hand over young Prince John of Eltham, Roger Mortimer's children, and a number of other prisoners, the bishop

[1] The head was sent to the queen at Gloucester (Murimuth, p. 48).
[2] See below, pp. 299, 500.

of Lincoln among them. Oaths to live and die with the citizens
in their quarrel and to maintain their peace were exacted from
the released prisoners, the dean of St. Paul's, the abbot of Westminster and other religious, and from justices and clerks. On 17
October the tablet commemorative of the Ordinances was replaced in St. Paul's, whence it had been removed by the king's
orders. Much of the normal business of the city was suspended,
no pleas being heard in the church courts for fear of renewing
the riots. Order was not fully restored until 15 November, when
the bishop of Winchester appeared with letters from the queen
authorizing the election of the new mayor. London's adherence
to her cause was finally assured by the election of Mortimer's
partisan, Richard de Bettoyne, 'who had been greatly persecuted
by the king and Despenser', and by the appointment of John de
Gisors as constable of the Tower.[1]

While London was in tumult, the queen had been pressing
westward in pursuit of Edward and the Despensers. She was in
Wallingford on the day of Stapledon's murder and advanced
from there to Oxford, where she heard Orleton preach to the
university from the apposite text, 'I will put enmity between
thee and the woman and between thy seed and her seed'.[2] At
Gloucester, she was joined by Percy, Wake, and other magnates
from the north and from the marches. Bristol, where the earl of
Winchester was in command, was her next objective and here,
as in London, the sympathies of the burgesses were so strongly
with her that Despenser was forced to yield the town without
even a show of resistance. On 27 October he was brought before
a group of magnates, including Norfolk, Kent, Leicester, and
Mortimer, and sentenced, *sub lingua gallica*, in these words:

Sir Hugh, this court denies you any right of answer, because you
yourself made a law that a man could be condemned without right
of answer and this law shall now apply to you and your adherents.
You are an attainted traitor, for you were formerly banished as such,
by assent of the king and the whole baronage, and have never been
reconciled. By force, and against the law of the land and accroaching
to yourself royal power, you counselled the king to disinherit and
undo his lieges, and notably Thomas of Lancaster whom you had put
to death for no cause. You are a robber and by your cruelty you have
robbed this land, wherefore all the people cry vengeance upon you.

---

[1] The best accounts of the London riots are in the *French Chronicle* (pp. 51–55)
and *Annales Paulini* (pp. 315–17).

[2] Twysden, *Decem Scriptores*, col. 2765.

You have traitorously counselled the king to undo the prelates of Holy Church, not allowing the Church her due liberties. Wherefore, the court awards that you be drawn for treason, hanged for robbery, beheaded for misdeeds against the Church: and that your head be sent to Winchester, of which place, against law and reason, you were made Earl . . . and because your deeds have dishonoured the order of chivalry, the court awards that you be hanged in a surcoat quartered with your arms and that your arms be destroyed forever.[1]

The 'trial' of the elder Despenser was clearly intended as a parody of Thomas of Lancaster's; but conviction by the king's record being impossible in the absence of the king, it was necessary to substitute the procedure of 'conviction by notoriety' behind which lies the ancient notion of manifest ill-fame. And, though desire for a savage vengeance clearly inspired the sentence, the participation of a number of magnates who might be regarded as the peers of the accused, points to the anxiety of the queen's advisers to cast a veneer of legality over this latest example of judicial murder.[2]

The rebel army now moved to Hereford and divided. Isabella and Mortimer remained in the town, while Henry of Leicester guided by a Welsh clerk named Rhys ap Howel, went to hunt down the king and the younger Despenser. They had fled from Bristol to Chepstow with the idea of taking refuge in Despenser's island of Lundy, whence escape to Ireland might have been possible. But storms in the Bristol Channel prevented them from embarking and they made instead for Glamorgan in the hope of obtaining aid from Despenser's tenants. On 16 November Lancaster found them in Neath abbey, together with Robert Baldock and a household clerk, named Simon of Reading. Edward was left in the custody of his cousin who took him to Kenilworth, by way of Monmouth and Ledbury, while the queen and her friends proceeded to complete their task of vengeance. On 17 November Arundel was taken by John Charlton and executed, apparently without any kind of formal judgement; Simon of Reading was drawn and hanged; two other clerks were executed, 'at the instance of Mortimer who hated them and whose counsel the queen followed in all things'.[3] Some show of formality, however, attended the sentence on the arch-enemy, Hugh Despenser

---

[1] *Ann. Paul.*, pp. 317-18.
[2] Murimuth (p. 49) says that Despenser was hanged 'by clamour of the people', the exact phrase used fifty years later in the proceedings against William of Wykeham.   [3] Murimuth, p. 50.

the younger. The canon of Bridlington speaks of a 'tribunal' of judges before whom he was brought at Hereford on 24 November, and reports verbatim the speech addressed to the prisoner by William Trussell, a former adherent of Thomas of Lancaster, who presided.[1] All the chroniclers refer to Trussell as *justiciarius* and it is just possible that he held an (unrecorded) commission under the privy seal of the prince of Wales; but the judgement is said to have been awarded by 'totes les bones gentz du Roialme, greindres et meindres, riches et poures, par commun assent'— words again suggestive of conviction by notoriety and of the desire to spread responsibility which was later to be so obvious in the proceedings against the king himself.[2] Despenser was reminded that he had been banished from the realm *per assensum communem*, never to return, unless by judgement of the king and the magnates in parliament. He and his father, accroaching to themselves royal power, had spoiled and murdered many noblemen at Boroughbridge, had procured the execution of the noble earl of Lancaster and other barons, had led the king out of the realm to fight the Scots and lost more than 20,000 men by their incompetence and had returned in shame with nothing achieved. The younger Despenser, on his return, had advised the king to abandon the queen at Tynemouth, in peril of her life. He had despoiled the bishops of Lincoln, Ely, and Norwich of their plate and treasures, induced the king to confer the earldom of Winchester on his father and that of Carlisle on the notorious traitor, Andrew Harclay, while he himself had acquired many lands pertaining to the Crown. He had been the means of depriving the queen of her dower and when she and her son were abroad, he had tried by means of bribery to prevent their return. He had compelled the king to withdraw himself from them when they did return and had led him out of the realm, to the great disgrace of his person and estate, taking with him the great seal and much treasure. This ingenious tissue of fact and fiction sufficed to procure for the younger Despenser a sentence almost identical with his father's, which was carried out forthwith.[3]

---

[1] Bridlington, pp. 87–89.

[2] See G. A. Holmes, 'Judgement on the Younger Despenser, 1326', *Eng. Hist. Rev.* lxx (1955), 261–7. For the best text of the judgement see J. Taylor, 'The Judgement on Hugh Despenser the Younger', *Medievalia et Humanistica*, xii (1958), 70–77.

[3] Despenser's head was placed subsequently on London Bridge, 'with much tumult and the sound of horns in the presence of the mayor and commonalty' (*Ann. Paul.*, p. 322).

There remained only Baldock, whom the bishop of Hereford, claiming benefit of clergy, imprisoned in his London house. But he was withdrawn by force, the bailiffs asserting that 'the bishop could have no private prison within the walls of the city of London', and thrust into Newgate where shortly afterwards he died.[1]

So perished the chief of the queen's enemies.[2] Meanwhile, the need to make some provision for the government of the realm had not been altogether overlooked. On 26 October the magnates assembled at Bristol had proclaimed Prince Edward keeper of the realm which the king had deserted.[3] Two days later the prince issued writs in his father's name, under his privy seal as earl of Chester and duke of Aquitaine, for a parliament to meet at Westminster on 15 December. The writs stated that the king would then be absent from the kingdom but that business would be transacted before the queen and her son, as keeper of the realm.[4] On 20 November the bishop of Hereford was sent to demand the great seal from Edward II who was then at Monmouth; on 26 November he brought it to the queen and, four days later, it was put in the custody of the bishop of Norwich. As it was no longer possible to maintain the fiction that the prince was acting as regent in his father's absence, the rebels substituted the fiction that Edward personally had resumed the government. New writs, deferring the meeting of parliament to the morrow of the Epiphany (7 January), were then issued in the normal form.

Like the parliament of 1322, that of 1327 was very fully representative of the commons. The Cinque Ports and the principality of Wales were again represented and, by a process of double election which may reflect the disturbed condition of the city, London chose six persons, two of whom were to attend the parliament. When it met on the morrow of the Epiphany, the London mob crowded into Westminster Hall.[5] The *Historia*

---

[1] *Ann. Paul.*, pp. 320–1; *Decem Scriptores*, col. 2763.
[2] For reasons which are not altogether clear, but probably because his support was thought to be valuable, Warenne was allowed to make his peace with the queen.   [3] *Foedera*, ii. 646.   [4] *Parl. Writs*, ii. i. 350.
[5] The absence of an official record and the conflict between the two best unofficial sources (*Historia Roffensis* and Lanercost) make it impossible to establish the chronology of this momentous assembly with precision. For the conjectural reconstruction which follows I am much indebted to Professor B. Wilkinson ('The Deposition of Richard II and the Accession of Henry IV', *Eng. Hist. Rev.* liv (1939), 225,

Roffensis makes it plain that there was division of opinion among the bishops, and the first two days of the session may have been taken up with efforts to silence contrariants. When a measure of agreement had been reached, Mortimer's party opened a campaign for popular support. The bishops of Lincoln and Winchester had already been sent to the king now at Kenilworth, with a formal request that he should come to London; but 'he utterly refused to comply therewith; nay, he cursed them contemptuously, declaring that he would not come among his enemies, or, rather, his traitors'.[1] On Monday (12 January) this reply was recited publicly before the clergy and people; and on the same day, the mayor, aldermen, and commonalty of London sent a letter to the magnates (with whom they must have been already negotiating), asking whether they were willing to be in accord with the city, to swear to maintain the cause of Queen Isabella and her son, to crown the latter and to depose his father for his frequent offences against his oath and his Crown.[2] Next day (Tuesday, 13 January) the bishop of Hereford, preaching from the text, 'A foolish king shall ruin his people', stirred up a popular clamour, 'We will no more have this man to reign over us!' Meantime a large and formal cavalcade had set out for the Gildhall where representatives of all the estates of the realm took a confederate oath to support the queen and her son to the death, to keep the ordinances made, or to be made, in parliament, and to maintain the freedom of the city. The swearing-in of these deputies who included bishops, abbots, priests, clerks, earls, barons, judges, lawyers, knights, barons of the Cinque Ports, and burgesses, occupied three days. While it was in progress, the people heard two more sermons, one on Wednesday (14 January) from the bishop of Winchester on the text 'My head is sick'; the other on Thursday (15 January) from the archbishop of Canterbury who, taking *vox populi, vox dei* as his text, announced that by unanimous consent of all the magnates, clergy, and people, King Edward was deposed from his royal dignity, never more to reign and govern the people of England; and that the magnates, clergy, and people had unanimously agreed that his first-born son, the Lord Edward, should succeed

n. 4), whose chronology seems to me more convincing, on the whole, than that suggested by Miss Clarke (*Representation and Consent*, pp. 178 ff.).
[1] Lanercost, p. 257.
[2] *Cal. Plea and Memoranda Rolls of the City of London, 1323–64*, pp. 11–12.

to the Crown.[1] If the so-called Articles of the Deposition, which may have been compiled by Stratford, were read aloud, the reading may have taken place on this day.[2] The Articles charged Edward with being incompetent to govern, unwilling to listen to wise counsel, destroying the Church and many noble men of the land, losing Scotland, Ireland, and Gascony, breaking his coronation oath to do justice to all, stripping his realm and, by his cruelty and weakness, showing himself incorrigible and without hope of amendment. It was either on this day or the next (Friday, 16 January) that a formal deputation set out for Kenilworth to announce the decision of the nation to the king. Its composition, like that of the deputation to the Gildhall, shows it to have been fully representative of the different estates of freemen in the realm. There were two earls, Leicester and the renegade Warenne; three bishops, Winchester, Hereford, and, perhaps, Ely; four barons, Courtenay, Gray, Rhos, and Percy; the abbot of Glastonbury and the prior of Dover; four friars, two of them Dominicans and two Carmelites;[3] two barons of the Cinque Ports; four knights (possibly two from each side of Trent); three Londoners; and perhaps one or two representatives of lesser towns.[4] The spokesman was William Trussell, the Lancastrian knight who had pronounced sentence on the younger Despenser. He is variously described as 'procurator of all in the land of England and of the whole parliament', as speaking 'in the name of all the earls and barons of the realm of England' and as 'procurator of the prelates, earls and barons named in his procuration'. Geoffrey le Baker is our best authority for the scenes at Kenilworth. According to him, the main deputation was preceded by the two bishops who had already visited Edward the week before, to invite him to come to London. They now urged him to resign the Crown, promising that he should be maintained in a state of royal dignity and threatening that, if he refused to abdicate, the *populus* would repudiate both him and his sons and would choose as king some other person, not of the blood royal (doubtless a thinly veiled allusion to Mortimer).

---

[1] Lanercost, p. 258.

[2] *Foedera*, ii. 650. The Articles appear only in Orleton's *Apology* (1334) and may have been compiled merely for purposes of propaganda. The draftsman was Stratford's clerk, William Mees.

[3] According to Lanercost (p. 258), the Franciscans were excused at the queen's request.

[4] On the composition of the deputation see Clarke, pp. 186 ff.

Defeated by these threats, the king, with tears and sighs, consented. Clad in a black gown and half-fainting, he was then brought out to meet the main deputation. The bishop of Hereford (Orleton) addressed him harshly, repeating his threat to reject his children if he would not resign. Edward replied, with tears, that it grieved him that his people should be so far exasperated with him as to wish to repudiate his rule, but that he would bow to their will if his son were accepted in his stead. The next day, William Trussell, *ex parte tocius regni*, renounced all homage and allegiance to Edward of Carnarvon. Sir Thomas Blount, the steward of the household, broke his staff of office and announced that the royal household was dissolved. The deputation then returned to parliament, and the new reign was held to begin on 25 January 1327.[1]

The deposition of Edward II raises many problems. Was the assembly of 7 January 1327 a true parliament? Stubbs described it as such and he has been followed by some other scholars; but Professor Wilkinson asks the pertinent question, whether a true parliament was possible without the presence of the king.[2] Similar doubts may have troubled intelligent contemporaries, like Prior Eastry of Canterbury who urged the desirability of a renewed request to Edward II to come to London. Some of the chroniclers described the assembly throughout as *parliamentum*, but Lanercost speaks of *clerus et populus* and the Lichfield chronicle of *concilium generale tocius cleri et populi*. The writs had been issued in due form in the king's name and, as he was represented by his son and heir, we may be justified in describing the assembly, at least in its initial stages, as a parliament. But, whatever the legality or otherwise of the summons, it seems evident that before very long the assembly must resolve itself into a revolutionary convention. Isabella and Mortimer were, indeed, faced with a knotty problem; for they had to find an answer to a question which had not been asked seriously in England since the Norman Conquest. By what means might an undoubted king lawfully be removed? The constitution, as Stubbs observed, had no rule or real precedent for deposing a worthless king; and, though it is unlikely that any of the revolutionaries fully appreciated the momentous significance of the task on which they were engaged, the act was fraught with sufficient danger to make them go

[1] Baker, pp. 26-28.   [2] *Eng. Hist. Rev.* liv. 223 ff.

warily and to do all in their power to give the proceedings an
air of legality. It might have appeased some uneasy consciences
if the king had been allowed the opportunity to defend himself;
but it seems certain that the revolutionaries of 1327, like their
successors in 1399, were afraid to produce their victim before a
national assembly, and it is unlikely that Burghersh and Stratford made any serious attempt to persuade Edward to come to
London. The deputations to the Gildhall and to Kenilworth
suggest a nervous desire to spread responsibility as widely as
possible. Miss Clarke saw in the deputations an attempt, im-
perfectly executed, to put into practice the rules laid down in the
Modus Tenendi Parliamentum for dealing with difficult cases and
judgements.[1] But the connexion is purely inferential and it seems
more probable that the deputations were conceived of as representing, not the estates of parliament, but the estates of the
realm. The participation of the Londoners was vital and not
hard to obtain, for, as the author of the *Historia Roffensis* points
out, they were concerned to secure immunity from the consequences of Stapledon's murder which had committed them
irretrievably to the queen's cause.[2] By means of these representative deputations the repudiation of Edward II could be
made to look like a national act. Even so, it was felt to be insufficient to declare him deposed; hence the manifest anxiety of
the rebels to obtain from the king some show of voluntary resignation. Their chief concern, no doubt, was to secure themselves
against future trouble; but the pitiful scenes at Kenilworth
suggest that some of them may have felt the kind of scruple
which troubled the mind of the bishop of Rochester, who took
the oath at the Gildhall under protest, saving his order and
everything contained in the Great Charter. Efforts were certainly made to slur over the ugly fact of coercion. Two days
before he was crowned the new king issued a proclamation
stating that his father, 'of his own good will and by common
consent of the prelates, earls, barons and other nobles and the
community of the realm', had removed himself from the government and willed that it should devolve upon his heir who had,
therefore, undertaken the task of ruling the kingdom.[3]

[1] *Representation and Consent*, pp. 189 ff.
[2] A sense of uneasiness is seen in the treatment accorded to Archbishop Reynolds
when he came to take the oath. He had to buy his peace with 50 tuns of wine and
was badly mauled on leaving the hall (*Hist. Roff.*, p. 367).
[3] *Foedera*, ii. 683.

The rapid and complete success of the revolution calls for some explanation. Mortimer had never been popular and the conduct of the queen was not such as to inspire confidence. Yet those who stood by the king, or attempted to mediate, were few indeed. Among the latter, to their credit, we may number the two archbishops. Walter Reynolds, who owed his see to Edward II, had been genuinely scandalized by the murder of Stapledon and through the autumn of 1326 he used his influence to promote a policy of appeasement, despite the fact that he himself had a number of solid grievances against the king.[1] But he was not a man of strong character and in the end he bowed before the storm.[2] Archbishop Melton of York and the bishops of London and Rochester had the courage to protest publicly in London, in the presence of Stapledon's murderers. But, as has been seen, the queen was assured of the support of a powerful section of the episcopate and the other bishops remained neutral. Among the lay magnates there were some, like the king's half-brothers of Kent and Norfolk, whose hatred of the Despensers was stronger than their distrust of Mortimer; some, like Henry of Leicester, who felt bound to avenge the victims of Borough-bridge; some, like Henry de Beaumont, who had private grudges against the court; and, presumably, a large number whose defection is to be explained by disgust at the incompetence shown by the king and his friends in regard to the safety of the Border and the conduct of the Scottish war. It is easy to understand the reluctance of the northern magnates to risk anything for Edward II; they had even less reason than had the lords of the Welsh march to love the Despensers or the king who had made himself their tool. Of the officials, only the chancellor, Robert Baldock, and the controller of the wardrobe, Robert Holden, stood by Edward. The remarkable continuity of the administration through the autumn and winter of 1326–7 affords decisive proof that the rank and file of the civil servants were playing for safety. As to the mass of the people they, like the magnates, were the victims of the Despensers' greed and the king's incapacity; and their natural loyalty to their lawful ruler had been further undermined by propaganda measures, assuring

[1] As shown by Miss Edwards, *Eng. Hist. Rev.* lix. 340.
[2] Eastry wrote to Reynolds in December, 'juxta consilium Apostoli, cautius est ambulandum' and excused himself from attendance at the Epiphany parliament (*Lit. Cant.* i. 202–3).

them that the pope absolved them from their allegiance, that the queen was an injured wife and the king a degenerate, an idiot, or a changeling.

None the less, the king, while he lived, could still be dangerous.[1] In April 1327 Mortimer had him removed from Kenilworth to Berkeley castle, where he was put in the custody of two reliable jailers, Thomas Lord Berkeley and Sir John Maltravers. They were allowed an ample sum for his maintenance in a reasonable degree of state; but after his removal to Berkeley we find ourselves in a realm of mystery and surmise. It is clear, however, that there were two attempts to rescue Edward and that the first of these achieved some temporary success. In the disturbed country of the lower Severn, a Dominican named Thomas Dunhead led a band of conspirators who raided Berkeley and rescued the late king from his dungeon; they may have taken him to Corfe, in Dorset. He was soon recaptured, however, and Dunhead suffered a horrible death in some unnamed cell. The raid on Berkeley must have been highly alarming to Isabella and Mortimer, but it was the second of the two plots which sealed Edward's fate. This was organized in Wales by a certain Sir Rhys ap Griffith who was probably inspired by the hatred of Mortimer which he shared with many Welshmen, rather than by love of Edward II. Early in September 1327 Mortimer learned from one of his agents that this new plot was hatching and that there was serious danger that the king might once again be rescued. Through one William Ogle he forthwith transmitted the news, together with a covering letter hinting at the obvious remedy, to Maltravers and to a Somerset knight named Thomas Gurney. A fortnight later it was announced that Edward of Carnarvon was dead. He was only forty-three and of robust constitution; the weight of the evidence seems to place it beyond doubt that he was murdered. Three years later, when somewhat perfunctory judicial proceedings were undertaken against Ogle, Gurney, and Maltravers, the first two were found guilty of the murder and Maltravers of being an accessory; all three escaped their punishment by flight. Gurney died abroad and nothing more is heard of Ogle; but in 1351 Maltravers had

---

[1] For the history of Edward II after his deposition, see F. J. Tanquerey, 'The Conspiracy of Thomas Dunheved, 1327', *Eng. Hist. Rev.* xxi (1916), 119-24, and T. F. Tout, 'The Captivity and Death of Edward of Carnarvon', *Collected Papers*, iii (1920), 145-90.

his sentence of outlawry remitted. His estates were restored and he sat in parliament, dying at a ripe old age in 1364. The honour of Edward III, as Tout justly remarks, shines no brighter for his complaisance towards his father's murderers; and the elaborate ordinances issued for the late king's funeral make the conduct of Isabella appear peculiarly revolting. At Gloucester abbey, Edward's magnificent tomb with its two-storied canopy, soon became the centre of a popular cult; the pilgrims' offerings enabled the monks to recast their Norman choir in the new 'perpendicular' style, soon to become fashionable all over England.[1]

His dark and dreadful death raises Edward II, in his last hours, to the high plane of tragedy; and the instinct to canonize him was natural enough. 'Neverthelesse kepynge in prison, vilenes and obprobrious dethe cause not a martir, but if the holynesse of lyfe afore be correspondent',[2] and seldom, if ever, have contemporaries written of an English king with such unmitigated contempt, or modern historians shown such unanimity in adverse judgement. We may, indeed, pity the weak-willed prince, successor to a famous father, to whom fell not only the administrative problems endemic in the medieval state, but also the *damnosa hereditas* of a hostile Scotland, financial chaos, and an over-mighty cousin. We may take a less hard view than did the chroniclers of Edward's rustic tastes, his preference for the company of simple people, his delight in music and acting.[3] Yet underlying such contemporary criticisms was the sound instinct that the second Edward lacked altogether the dignity and high seriousness demanded of a king. Our sources afford us no evidence that at any time he tried to rise to his responsibilities or to learn from his misfortunes and mistakes. Alike in his relations with his friends and with his enemies, he showed himself a weakling and a fool. His mishandling of the Scottish war and his neglect of the safety of the north proved him wanting, not only in military capacity, but also in imagination, energy, and common sense. Reforms achieved in household and exchequer were

---

[1] *Hist. Mon. Sancti Petri Gloucestriae* (Rolls Series), i. 46.

[2] *Polychronicon*, viii. 325–7.

[3] Criticism of Edward's plebeian tastes was not confined to the chroniclers. A messenger in the royal household got into trouble for telling a friend that the king's way of spending his time was 'vacare et intendere circa fossata facienda et ad fodendum et eciam ad alia indecencia' (Hilda Johnstone, 'The Eccentricities of Edward II', *Eng. Hist. Rev.* xlviii (1933), 264–7).

the work of capable officials, not of the lazy and indifferent king.
Edward lived a life devoid of noble purpose or of laudable
ambition. He lowered the reputation of his country abroad and
at home he was the means of bringing the monarchy into the
most serious crisis that had faced it since 1066. It was his own
folly which delivered him into the hands of his cruel foes; and
the consequences of his deposition reached far beyond his own
generation. Memories of it were to haunt the dreams of his
great-grandson, Richard II; it smoothed the path of the revolutionaries of 1399 and it opened the way for dynastic conflict
and the decline of the medieval monarchy. More immediately,
his fall exposed the nation to the rule of the greedy and disreputable couple under whose aegis Edward III succeeded to
the throne.

The fourteen-year-old king was crowned by Walter Reynolds
on 1 February 1327 and Henry of Lancaster was made chief of
the council of regency. Mortimer did not claim a place on the
council, being content to see his interests represented by Hotham
of Ely as chancellor, Orleton of Hereford as treasurer, Sir Oliver
Ingham and Sir Simon Bereford. In the first parliament of the
new reign, Henry of Lancaster entered a petition for reversal of
the sentence on his brother. He argued that the condemnation
was erroneous because, although it was admittedly a time of
peace (the chancery and other courts being open), Thomas had
not been allowed to answer the charges against him; whereas a
peer of the realm should have been judged by his peers and not
on the king's record.[1] Henry's concern thus seems to have been
to counter the notion that summary judgement could be defended as a form of military justice, permissible in time of war.
Boroughbridge might have been regarded as an interruption of
the time of peace; the continuing function of the courts is cited
as proof to the contrary.[2] His plea was admitted and he was
reinstated in a great part of the Lancastrian inheritance and
allowed the title of earl of Lancaster.[3] Reversal of sentence of

---

[1] Rot. Parl. ii. 3-5.
[2] See T. F. T. Plucknett, 'The Rise of the English State Trial', Politica, ii (1937),
547 ff. Some of Lancaster's fellow-rebels, notably Damory, had been sentenced in
the court of the constable and marshal, i.e. by military law.
[3] The most important exception was the earldom of Lincoln and the Lacy inheritance which Thomas had enjoyed in his wife's right. The Lacy lands were recovered gradually by the fourth earl and by John of Gaunt (Somerville, Duchy of
Lancaster, i. 32-36).

forfeiture was also granted to those who were *de querela Comitis Lancastriae*; gifts of land to Arundel, Baldock, Stapledon, and others of their party were annulled. Queen Isabella's dower, originally worth £4,500, was restored to her, with additions which raised its annual value to £13,333;[1] and the Bardi proving indispensable, their services were retained.[2] Mortimer's policy seems to have been to rule indirectly through the queen, while putting as much responsibility as possible on Henry of Lancaster. But there was nothing indirect about his acquisitiveness. From the parliament of January 1327 he too procured reversal of the sentences passed on his family and obtained restoration of his father's property. A few months later he received an enormous grant of forfeited property, including the Despenser lordship of Denbigh and a number of lordships and manors which had been taken from Arundel; he seems also to have gained possession of the lands of his uncle of Chirk.[3] His tenure of the office of justice of Wales gave him a power greater even than his uncle's over the Principality and the march. The climax of his ambitions was reached when, in 1328, he assumed the title of earl of March; such an earldom, says the St. Paul's annalist, as was never before heard of in England.[4] Mortimer lived in a style corresponding to his wealth. In 1328 he followed the example of his grandfather by holding at Bedford a tournament, designated as a 'Round Table', possibly to serve as a reminder of the Mortimer claim to descent from Arthur and Brutus.[5] When Edward III and Isabella visited him at Hereford on the occasion of his daughter's marriage, they were offered another tournament of unparalleled magnificence; a few months later, the king was lavishly entertained at the family strongholds of Ludlow and Wigmore. Mortimer's avarice was as blatant as the Despensers'

---

[1] The movables, plate, and jewels of the younger Despenser were granted to Isabella in Jan. 1327 (*Complete Peerage*, iv. 270). A striking example of the queen's relentless rapacity was her confiscation of the books on canon and civil law, valued at £10, which Edward II had presented to the Master of King's Hall, Cambridge. See J. B. Mullinger, *The University of Cambridge* (1873), i. 253.

[2] E. B. Fryde, 'Loans to the English Crown, 1328–31', *Eng. Hist. Rev.* lxx (1955), 198–211.

[3] G. A. Holmes, *The Estates of the Higher Nobility in XIV century England*, p. 13.

[4] Mortimer's choice of this title in preference to Shrewsbury may have been dictated by his desire to commemorate his descent from one of the co-heirs of the ancient counts of La Marche (*Complete Peerage*, v. 634, n. *b*); but he can hardly have been blind to its implications in regard to his position on the marches of Wales.

[5] Knighton, i. 449. See Mary Giffin, 'Cadwalader, Arthur and Brutus in the Wigmore Manuscript', *Speculum*, xvi (1941), 109–20.

and it soon began to threaten the superficial unity of the revolutionaries, already menaced by ugly rumours concerning the fate
of Edward II.

Meanwhile, the trend of events in the north did nothing to
enhance the popularity of the new government. In 1327 the truce
with the Scots had still several years to run, but neither side was
in a mood to maintain it. There were complaints of minor
breaches, Bruce began to move troops towards the Border and,
soon after the coronation, the regents were constrained to prepare for war. A campaign in Scotland offered the prospect of
useful employment for the mercenaries from Hainault who, after
the manner of their kind, were already beginning to prove an
embarrassment to those who had hired them. But an unfortunate
brawl between them and the citizens of York while the army
was on its way north, made it necessary to send them home
again. This was an ominous beginning; and the Scots soon took
the initiative and advanced over the Border into Northumberland. The two armies approached each other in the neighbourhood of Stanhope Park; but the English, despite their superior
numbers, seem to have been afraid to offer battle and withdrew
to Newcastle with nothing achieved. The young king felt the
retreat as a personal humiliation; according to the chroniclers,
it was with difficulty that Isabella and Mortimer persuaded him
to agree to the final peace on which they were now determined.
They had many preoccupations elsewhere; they were disinclined
to spend on the Scottish wars the revenues which they had earmarked for their own advantage; and Bruce was in a mood for
conciliation. But peace was to be had only on the Scots' own
terms. One hundred Scottish knights were invited to attend a
parliament which met at York on 7 February 1328, for the
express purpose of making peace with Scotland; but the documents which embodied the settlement were drawn up at
Holyrood abbey, in the presence of Bruce himself, and the proceedings at Northampton (from which the treaty took its name)
were merely a formal ratification.[1] The treaty of Northampton

---

[1] E. L. G. Stones, 'The English Mission to Edinburgh in 1328', *Sc. Hist. Rev.*
xxviii (1949), 121-32; 'An Addition to the Rotuli Scotiae', ibid. xxix (1950), 23-51;
'The Anglo-Scottish Negotiations of 1327', ibid. xxx (1951), 49-54; 'The Treaty
of Northampton, 1328', *History*, N.S. xxxviii (1953), 54-61. The English ambassadors
to Scotland were the bishops of Lincoln and Norwich, Sir Geoffrey Scrope C.J. of
the King's Bench, Henry Percy of Alnwick, warden of the marches, and Sir William
Zouche of Ashby.

conceded to the Scots everything for which they had fought. It recognized Robert Bruce as king of an independent realm; it restored the frontier as it had been in the time of Alexander III; it cancelled all obligations or treaties implying any subjection to England; and it provided for a marriage between the four-year-old David Bruce and Edward III's young sister, Joan of the Tower, and for an offensive and defensive alliance against all parties, except the French. Bruce undertook to pay £20,000 in three yearly instalments, *pro bono pacis*, the English to restore all *obligationes, conventiones et pacta* touching the subjection of Scotland. The terms of peace contain no reference to relics or trophies, such as the Stone of Scone. It is the chroniclers who tell us, probably correctly, that the return of the Stone was contemplated but that the abbot of Westminster and the people of London refused to allow its removal.[1] It was not the retention of the Stone, but of the Ragman Roll, recording the homage due from individual Scottish barons to Edward I, which constituted a violation of the treaty. Nor did the terms include any reference to forfeited estates, apart from a clause safeguarding the rights of the Church, which was interpreted as a promise to restore ecclesiastical lands. At the time when the treaty was drafted Bruce was strongly opposed to any general policy of restoration, but, perhaps because of failing health, he seems to have withdrawn his opposition a few months later and to have sanctioned an agreement to restore the Scottish lands of Percy, Beaumont, and Wake.

The chroniclers' condemnation of the *turpis pax* is readily understandable. It followed hard upon a humiliating failure of English arms; it was regarded as the work of an unscrupulous adventurer whom even his supporters were beginning to mistrust, and of a notorious queen; it was concluded for reasons of private gain, rather than of national honour.[2] Yet, however base the motives which inspired the treaty, the modern historian cannot well deny that conclusion of the protracted warfare which had brought *mala innumerabilia* to both countries, was in itself desirable and right. The English were fighting a losing battle against Scottish independence. Refusal of recognition to a king

[1] The Black Rood of Scotland must have been restored, however, for the Scots lost it again at Neville's Cross.

[2] Isabella appropriated a large share of the £20,000 paid by the Scots, which may have been intended to be the repayment of sums exacted by way of blackmail from the north of England. See Jean Scammell, 'Robert I and the North of England', *Eng. Hist. Rev.* lxxiv (1958), 385–403.

de facto could benefit nobody and must expose the border counties to a renewal of invasion. But the bitter hostilities engendered during the long War of Independence were not to be allayed by diplomacy, and the memory of Bannockburn still rankled. With a warrior king on the throne of England and with Scotland soon to come under the rule of a minor, there was little hope that peace could be permanently maintained.

Henry of Lancaster and his friends offered no opposition to the settlement with the Scots. But six months after its conclusion, they were complaining that the council of regency was being disregarded, that the king had not enough to live of his own, and that the queen should enjoy her dower and not burden the people.[1] Lancaster refused to attend the parliament at Salisbury in October, where the title of March was conferred on Mortimer; his supporters, including some of the bishops, withdrew from it without licence. Lancaster even gathered troops at Winchester and it looked as though the country were threatened with a renewal of civil war; but, after the failure of some attempts at negotiation, Mortimer invaded the earldom of Leicester in January 1329, occupied the town, and forced Henry to patch up a peace.[2] The weakness of Edmund of Kent, who deserted Lancaster at this juncture, contributed to Mortimer's triumph. None the less, Isabella and her lover had already decided that Kent was dangerous. They set agents to work to persuade him that Edward II was still alive and the foolish earl fell into their trap and became party to a supposed plot to rescue the late king. This was enough to procure his doom. In the Winchester parliament of March 1330, he found himself suddenly arrested. His own speeches and letters were quoted against him and, after a hurried trial by the peers, allowed him, doubtless, because his guilt was undeniable, he was sentenced to death as a traitor and executed outside the walls of the city.

It was now clear to Lancaster that unless he took prompt and vigorous action, he and his friends would suffer a like fate; and he determined that, by some means, he must gain the ear of the king himself. Two members of the royal household were ready to help him—Richard Bury, keeper of the privy seal, and a

[1] It was later alleged that Mortimer had made peace with the Scots in order to destroy Lancaster, *Cal. Plea and Memoranda Rolls 1323-64*, p. 80.

[2] See G. A. Holmes, 'The Rebellion of the Earl of Lancaster, 1328-9', *Bull. Inst. Hist. Res.* xxviii (1955), 84-89.

yeoman of the household, named William Montagu. Between
them, they set to work to persuade Edward to break free of his
leading-strings. Montagu who, with Bartholomew Burghersh,
had already been on a mission to Avignon, sought a private
interview with John XXII and explained to him the humiliating
position in which the king was placed. The pope heard Montagu
sympathetically and asked him to arrange for some secret sign
whereby he might distinguish those letters which represented
the king's personal wishes from those dictated by Mortimer and
his associates. On Montagu's return to England, Bury sent a
letter to the Curia, explaining that the use of the words *pater
sancte* in any communication under the privy or secret seal,
would afford the necessary evidence; a specimen of the king's
handwriting was appended.[1] Edward informed the pope that
knowledge of the intrigue against Mortimer was confined to
Bury and Montagu, in whose discretion he placed full confidence.
Thus encouraged by both pope and king, secure in the goodwill
of Lancaster, and aided by some of his own contemporaries,
Montagu laid a plot. When the great council met at Nottingham
in October, Mortimer was clearly uneasy and suspects were
closely interrogated; Montagu advised the king to strike before
he was struck. Mortimer and Isabella shut themselves in the
castle and ordered the gates to be locked and the walls guarded;
but the constable showed Montagu a secret entrance through
an underground passage into the castle yard, where Edward
himself joined the conspirators. They made their way up to
Mortimer's chamber, struck down the two knights who were
guarding the door, and laid hold of their enemy. Hearing the
noise of a scuffle, the queen rushed into the room and entreated
her son to 'have pity on gentle Mortimer'; but her cries went
unheeded and the earl was taken out of the castle and hustled
with all speed to London. A few weeks later, on the ground of
crimes declared to be notorious, his peers in parliament sentenced him to be drawn and hanged as a traitor.[2] His ignominious end seemed to mark the final collapse of the hardy marcher
family which could trace an unbroken line of male descent from
that Ralph de Montemer on whom the Conqueror had bestowed

[1] C. G. Crump, 'The Arrest of Roger Mortimer and Queen Isabel', *Eng. Hist.
Rev.* xxvi (1911), 331–2. The letter, which was discovered in the Vatican archives,
contains what seems to be the earliest surviving autograph of an English sovereign.
See V. H. Galbraith, 'The Literacy of Medieval English Kings', *Proc. Brit. Acad.*,
xxi (1935), 223, 236, n. 47.   [2] 29 Nov. 1330.

the insignificant lordship of Wigmore. The earldom of March was now extinct, the great Mortimer estates were in the hand of the Crown. But, of Mortimer's associates, only Sir Simon Bereford shared his fate. Oliver Ingham and the bishops were pardoned; and Isabella was fortunate far beyond her deserts. Edward III decided to ignore her liaison with Mortimer and the only charge laid against him in connexion with her was of having caused discord between the queen and the late king. Though she had to surrender her ill-gotten gains, Isabella was endowed with the ample allowance of £3,000 a year and was allowed to move freely about the country. At her favourite residence of Castle Rising, she maintained a considerable degree of state, and presents to her constituted a steady drain on the purses of the neighbouring burgesses of Lynn.[1] She amused herself with hawking, reading or hearing romances, and collecting relics; and it may be that in her old age she suffered a change of heart, for she took the habit of the Poor Clares and was buried in the Franciscan church at Newgate.[2]

With the fall of Roger Mortimer, we reach the appropriate climax of a long tale of violent political vicissitudes. High claims have been made for the reign of Edward II as the breeding-ground of new political ideas, and it remains to consider something of its significance in this respect. The self-consciousness of the baronial order, already well developed in 1307, was inevitably enhanced by a period of protracted and sustained conflict with the Crown. More and more the magnates tend to conceive of themselves, socially and legally, as a class apart. Gaveston is denounced as an outsider; the elder Despenser is reminded that he has dishonoured the order of chivalry. With increasing frequency, the terms 'peers of the land', 'peers of the realm', 'peers of parliament', appear in the records.[3] The

[1] e.g. £6. 0s. 2d. for meat and swans in 1333; £4. 6s. 10d. for meat, lampreys, and wine in 1334; £9. 12s. 9d. for a pipe of wine and a barrel of sturgeon and their carriage to Rising in 1351 (*Hist. MSS. Comm.* 11th Rep., App. pt. iii, pp. 214–18).

[2] At her death in 1358, the inventory of Isabella's possessions included romances, some of the Charlemagne cycle, some of the Trojan war, and a few of the Arthurian group, as well as a gradual, an ordinal, and a book of homilies. (Tout, *Chapters*, v. 249; cf. H. Johnstone, 'Isabella, the She-Wolf of France', *History*, xxi (1936), 208–18.) We may discount John of Reading's jealous suggestion (*Chronicle*, pp. 128–9) that Isabella had been 'seduced' by the Grey Friars to change her will and seek burial among them.

[3] The earliest official use of the term 'peers of the land' appears to be in the treaty of Leake of 1318. Cf. A. F. Pollard, *The Evolution of Parliament* (2nd ed. 1926), p. 93.

thirty-ninth clause of Magna Carta is being made the basis of a claim to exclusive privilege, specifically asserted in the parliament of 1330 when the lords, after judging March, declared that, as peers, they were under no obligation to judge those who were not their peers. Here, undoubtedly, was a new and explicit concept of peerage, pregnant with possibilities for the future. Yet the fate of Mortimer and of almost all his aristocratic predecessors on the scaffold, makes it abundantly plain that, from the standpoint of the individual peer, the privilege was not worth a scrap of paper. Murimuth remarked with truth that, from the death of the earl of Lancaster to the death of the earl of March, it had not been customary for great men accused of treason to be given opportunity to defend themselves;[1] and at the close of this period an accused peer enjoyed no guarantee that he would be allowed to put in an answer to the charges against him, or even to be present at his so-called trial. Moreover, treason was undefined by statute and vague charges of 'accroaching the royal power' were sufficient to expose any man to its terrible penalties. Little wonder, then, that in a world fraught with so much insecurity and peril, individual magnates should have concentrated their energies on the stabilizing of their personal fortunes and that common political action proved hard to achieve and harder still to maintain. All the barons' devices for securing control of the king and his advisers ended in failure. The 'doctrine of capacities', adumbrated, possibly in 1308, certainly in 1321, had to be abandoned; the Ordinances received their death-blow in 1322 and were never resurrected in their original form; the standing baronial committee set up by the treaty of Leake foundered in the tempest of personal rivalries; and Lancaster's claims for the stewardship perished with him. Isabella and Mortimer had no constitutional programme at all. They triumphed, and Edward II lost his throne, not because they entertained principles of limited monarchy and he of absolutism, but because they were astute enough to take advantage of the fact that he was the prey of unpopular counsellors and was himself feeble, incompetent, and irresponsible. Among the magnates there were a few, like Pembroke and Archbishop Melton, whose pursuit of their private advantage did not blind them to the need for sound government, adequate frontier defences, and an intelligent foreign policy. But at heart

[1] Murimuth, p. 62.

they were all conservatives, looking back to Simon de Montfort, to the Great Charter, to the good old days. Such radical propensities as the period affords are to be found, not in the circle of an embryonic peerage, but in the departments of state and of the household, where officials like Stapledon and the younger Despenser were attempting administrative reorganization on lines which might have led ultimately to a unification of all the secretarial offices such as took place in France.[1]

As Stubbs perceived long ago, the most significant constitutional development of the reign lies in the growing importance of parliament as a political assembly and, within parliament, of the knights and burgesses. He and other scholars may have read too much into the Statute of York; but the fact remains that slowly, almost imperceptibly, and largely because of their usefulness as agents of propaganda and as vehicles of public opinion, the knights from the counties and the citizens and burgesses from the towns are establishing themselves, in this reign, as an integral part of parliament. No such claim could have been made for them in 1307. By 1330 parliament (restored to its rightful dignity by the Statute of York), never meets without them. Legislation is beginning to be founded on their petitions; sixteen of the articles which they present in the parliament of 1330 give rise to statutes. The role of the commons is still a minor one and the representative assembly of the nation is still in its infancy; but if those scholars are right who date the *Modus Tenendi Parliamentum* to this early period, there was someone alive in the England of Edward II who cherished a vision of parliament instinct with sound prophecy.

[1] Tout, *Place of Edward II*, p. 151.

IV

THE ORIGINS OF THE HUNDRED YEARS WAR

Anglo-French differences over Gascony afford the natural starting-point for an inquiry into the origins of the great war between the two nations which was to persist, albeit very intermittently, for over a hundred years. Yet by 1337 the 'Gascon question' was already of long standing; and, although it must be reckoned among the fundamental causes of conflict, reference to it does not altogether explain why the war should have begun in this particular decade. To understand why the war came when it did, it is necessary to try to distinguish the confusing and interacting causes of friction which arose, or became intensified, in the immediately pre-war years, to look, not only at Gascony, but also at the dynastic situation in France, at Scotland, the Low Countries, the Rhineland, the course of papal diplomacy, and the projects for a crusade. The motives and intentions of Edward III and the other war leaders have also to be reckoned with; and it is unfortunate that the material, abundant though this is, allows us so little insight into their psychology, letters and manifestoes written for propaganda purposes being often designed to conceal the motives which they profess to reveal. Such hints as the evidence may yield cannot be left out of account; but much remains obscure and any conclusions reached must necessarily be tentative. All that is certain is that the origins of a great international conflict are not to be found in one area alone, nor in the ambitions of a single individual, or of a single people.

The Gascon dilemma is familiar. The treaty of Paris of 1259 and the subsequent agreements under Edward I gave rise to a situation which was both delicate and dangerous. The rights of suzerainty which the treaties conferred upon the French king were manifestly difficult to exercise fully over a vassal who was also a sovereign; the duties of vassalage became increasingly painful for a king to fulfil; and the treaties had bequeathed to posterity a host of uncertainties as to the mutual obligations of the king of France and the duke of Aquitaine and the precise

THE HOUSES OF CAPET AND VALOIS

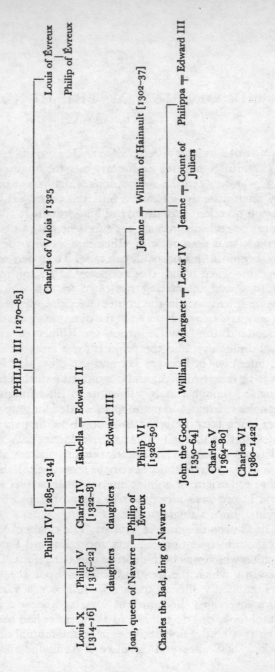

boundaries of their respective possessions. By 1303, when Edward I and Philip IV reached agreement at Paris, the status of Gascony, as Sir Maurice Powicke has demonstrated, 'had become involved in a network of juristic learning' and the old disputes had not been settled. The treaties had produced 'an interminable wrangle' and abundant causes of conflict remained when Edward II succeeded his father.[1] There is no reason to suppose that the Capetian kings hoped to deprive the Plantagenets of the duchy of Aquitaine to which they had an undisputed title; but they wished to make the most of their rights as overlords and particularly of their right to hear appeals from the Gascon vassals of the king of England. As for the Plantagenets, there could be no question of their relinquishing the rich land of Gascony which in 1306–7 had yielded a revenue greater than that of the English Crown;[2] their aim, from the time of Edward I onwards, was to escape from the obligations to France imposed on them by the treaty of 1259 and to establish their right to hold Gascony as an alod, in full sovereignty. Such a conflict of aims, it may be argued, was bound to lead ultimately to war. But wars were expensive and it is significant that under Edward I and Edward II only two minor ones broke the peace between England and France. Maintenance of peace under Edward I was due to his desire to uphold his cause by peaceful means and to the reluctance of both Philip III and Philip IV to press their claims against him to the point of complete intransigence; and the treaty of 1303 strengthened the family ties between Plantagenet and Capet. Edward II was son-in-law to Philip IV and brother-in-law to his three successors. His own tastes were not martial, his disposition was indolent and his impulses fitful; and the unhappy relations persisting between him and his barons throughout most of the reign would have made organization of large-scale continental enterprises hazardous, if not impossible. The war in Scotland strained English resources to their limit and the result of Bannockburn was not encouraging. On the French side, too, there were solid reasons for avoiding war. Philip IV's campaign of aggression in Flanders

[1] Sir Maurice Powicke's penetrating analysis of the Gascon question under Edward I (*The Thirteenth Century*, vol. iv of the present *History*, pp. 270–318; 644–54) may be supplemented by P. Chaplais, 'Le Duché-Pairie de Guyenne', *Annales du Midi*, lxix (1957), 5–38.

[2] See G. P. Cuttino, 'Historical Revision: the Causes of the Hundred Years War', *Speculum*, xxxi (1956), 468–9.

was a major preoccupation; and he seems to have been genuinely anxious not to provoke an open rupture with his daughter's husband. The short reigns of Louis X (1314–16) and Philip V (1316–22) were unfavourable to elaborate military undertakings. Thus, the encroachments of Philip IV's officials in Gascony culminated, not in war, but in the peace conference known as the Process of Périgueux (April 1311). Here (as at Montreuil in 1306), English and French plenipotentiaries set out their respective demands in a series of complex articles and debated them at length.[1] As neither side was willing to admit the claims of the other, the result was a deadlock; but still there was no war. Nor, though they gave dangerous prominence to the inflammable subject of the homage, did the deaths of three French kings during Edward II's reign produce an outbreak of hostilities. Edward was able to secure deferment of his homage to Louis X until it was too late to perform it at all; and he succeeded in delaying his homage to Philip V until within two years of the latter's death. When he did at last appear before the high altar of Amiens cathedral on 20 June 1320, there was exacerbation of feelings on both sides. Some of Philip's counsellors took occasion to declare the homage inadequate since it was unaccompanied by an oath of personal fealty and Edward countered their claim with some heat; but in the end it was withdrawn.[2] No serious threat to peace arose until after the accession of the last Capetian king, Charles IV; and the history of his relations with Edward II is instructive.[3]

Like all his predecessors, Charles IV was concerned to secure the homage for Gascony; but he showed no desire to profit by the embarrassments of his brother-in-law. It was not until September 1323, when the thirteen years' truce with the Scots had been achieved and Thomas of Lancaster finally disposed of, that the first summons to fulfil his obligations reached Edward II. The letter, which was conciliatory in tone, invited Edward to present himself at Amiens between Candlemas and Easter 1324. After consultation with his council, Edward sent ambassadors to represent the dangers that might follow from his leaving

[1] G. P. Cuttino, *English Diplomatic Administration, 1259–1339* (1940), pp. 12–14.
[2] E. Pole Stuart, 'The Interview between Philip V and Edward II at Amiens in 1320', *Eng. Hist. Rev.* xli (1926), 412–15.
[3] This was carefully worked out in the eighteenth century by de Bréquigny, 'Mémoire sur les différends entre France et Angleterre sous Charles le Bel', *Mém. Acad. Inscr. et Belles-Lettres*, xli (1780), 641–92.

the country and Charles, still willing to accommodate him, agreed to a postponement until 1 July. In the meantime, however, trouble was brewing within Gascony itself. At Saint-Sardos, a priory in the Agenais dependent on the Benedictine house of Sarlat,[1] Charles IV decided to authorize the building of a new fortified town, a *bastide*. His right to do so was questioned by the seneschal of Gascony, Sir Ralph Basset, and the place was attacked and burnt in November 1323, and a French serjeant hanged on gallows bearing the royal arms, among those implicated in the outrage being Raymond Bernard, lord of Montpezat. Edward II's protests of ignorance of the whole matter were accepted by the French king; but he summoned Basset, Bernard de Montpezat, and other English officials to appear before him at Toulouse in January; when they had twice defaulted they were declared banished and their possessions forfeit. In this serious dilemma, Edward II's ambassadors (the earl of Kent and the archbishop of Dublin) acted with great ineptitude. They first agreed to surrender the castle of Montpezat and then went back on their word, while at the same time requesting a new postponement of the homage, thus making it appear that Edward II intended both to condone the acts of his Gascon subjects and to default on his own obligations towards his overlord. In declaring the duchy of Gascony and the county of Ponthieu confiscate, Charles IV was taking the only step possible in the circumstances.[2]

Edward II had been making some half-hearted preparations for war but he was in no position to offer effective resistance to the French; and Kent, who had been put in command in Gascony, shut himself up in the fortress of La Réole when Charles of Valois entered the duchy in August 1324. On 22 September, urged on, it was said, by the archbishop of Dublin, a bitter enemy of the Despensers, Kent capitulated and agreed to a six months' truce. Edward dared not dispatch reinforcements to Gascony because the Channel was infested with French ships and, though indignant at Kent's surrender, he allowed himself to be persuaded to send the bishops of Winchester and Norwich at the head of an embassy to the French court. John XXII was now intervening actively and when Charles IV

---

[1] Sarlat itself had been incorporated by charter in the French royal demense.
[2] See P. Chaplais, *The War of Saint-Sardos, 1323–1325* (Camden 3rd ser. lxxxvii, 1954).

demanded the surrender of Gascony as a preliminary to negotiations, it was the papal nuncios who suggested that Queen Isabella should be sent to assist the peace-makers and that Prince Edward might perform the homage in his father's stead. By the end of March 1325 the queen was in France and by May Edward II had accepted a truce. When it became clear that the Despensers would not allow the king to leave the country, Prince Edward was invested with the duchy and the county for which he did homage in September, undertaking to pay a relief of £60,000.[1] Charles IV insisted, however, on retaining the Agenais as indemnity for the losses he had suffered in the war. Isabella and the prince refused to return and Edward, feeling himself deceived, pronounced himself 'governor and administrator' of all his son's lands; in July 1326 he even made the empty gesture of declaring war. But by this time Isabella and Mortimer were in the Low Countries and the revolution which was to overthrow Edward II was already taking shape.

The war of Saint-Sardos affords decisive proof, if proof be needed, that at this date confiscation of the Plantagenet territories in France formed no part of the considered policy of the Capetian kings.[2] Had this been his objective, Charles IV could have attained it with little difficulty. He could have seized the advantage afforded him by the crisis in England at the time of his own accession and refused to admit Edward's title to the duchy until he had done homage in person; above all, he could have profited by the presence of Isabella and the exiles on French soil by lending full support to the invasion of September 1326. On the contrary, Charles delayed his demand for homage until Edward II appeared to be clear of his embarrassments and he did not make even the outrage at Saint-Sardos an excuse for declaring war. It was only when Edward defaulted on the homage that the overlord took the perfectly legitimate step of confiscating the vassal's fiefs; and, though no doubt Charles IV was not averse from taking a nibble at the duchy and hoped that the Agenais might be transferred permanently to the domains of the French Crown and that his right to entertain appeals from Gascony might be extended, he was ready enough to invest Prince Edward with the bulk of his father's lands. He

---

[1] *War of Saint-Sardos*, pp. 241–5.

[2] As suggested, for example, by W. I. Lowe, 'The Considerations which induced Edward III to assume the title of King of France', *Annual Report of Amer. Hist. Assocn.* i (1900), 538.

was not prepared to take active steps to deliver his nephew or his sister into the hands of the Despensers, but neither was he ready to lift a finger to help Isabella and her lover to dethrone the king. While standing by his undoubted rights, he was ready to listen to papal counsels of moderation and to respect the rights of others. In short, if French policy had been determinedly aggressive, aggression seldom stood a better chance of success than in the closing years of Edward II. That war to the death for Gascony did not break out in these years is to be ascribed, less to the incapacity of Edward II, than to the reluctance of Charles IV and to papal intervention.

For the first four years of the new reign, English foreign policy was directed by Isabella and Mortimer. Charles IV had no choice but to accept the *fait accompli* of the revolution which he survived by little more than a year, and Isabella hastened to patch up a truce whereby her brother undertook to restore Gascony on payment of a war indemnity of 50,000 marks in addition to the £60,000 relief already promised, and to offer an amnesty to all the Gascon rebels, with the exception of eight who were to be banished and their castles destroyed. But the premature death of Charles IV in February 1328 raised a new issue in Anglo-French relations. For only the second time in more than three centuries,[1] there was no heir apparent to the throne of France. If women were barred from the succession, the nearest heir in the male line was Philip of Valois, a nephew of Philip IV and cousin to the late king. Contemporary feudal practice and the prevailing uncertainty about the law of succession made it almost inevitable that in these circumstances a claim should be proffered on behalf of Edward III, a direct descendant, through his mother, of Philip IV. Failure to make such a claim would have been tantamount to letting Edward's case go by default; and Isabella's overtures for alliances in Gelderland, Brabant, and Castile suggest that she may have had some notion of taking action at a later date.[2] In the meantime, neither she nor Mortimer was in any position to lead the nation into war. Philip VI was crowned at Rheims in May 1328 and a year later, though reservations were expressed on both sides, Edward III did

---

[1] The posthumous son of Louis X died within a few days of his birth (1316) and the throne was seized by the late king's brother, Philip V.

[2] Isabella is said to have told the French ambassadors that her son, who was the son of a king, should never do homage to the son of a count, *Les Grandes Chroniques de France*, v. 323–4.

simple homage at Amiens.[1] Rumours of war in 1330 proved
baseless; a convention was achieved at Bois-de-Vincennes in
May of that year, and, so soon as his hands were freed, Edward
III hastened to remove the causes of misunderstanding. In March
1331, 'so that there be no contention nor discord in regard to
the said homage', he acknowledged his obligation to perform
the liege homage which bound the vassal personally and carried
with it the specific duty of military service to the overlord. The
form of the oath was specified with precision. 'The king of
England, the duke of Guienne, shall place his hands between
the hands of the king of France whose spokesman shall thus
address the king of England:

> "Sir, will you become the liegeman of the king of France, as
> duke of Guienne and a peer of France and will you promise to
> bear faith and loyalty to him?"

The king of England shall answer,

> "I will."

And then the king of France shall receive the king of England to
the said liege homage and fealty with the kiss of peace, saving
his right.' The same form was observed for Ponthieu and Montreuil and Edward undertook, if the French king so desired, to
issue it in letters patent under the great seal.[2] A few days later,
perhaps because he feared to offend the nationalist suscepti-
bilities of his own subjects if he should appear over complaisant
towards Philip, Edward visited France in the disguise of a
merchant and the two kings met secretly at Pont-Sainte-Maxence.
Here, they affirmed their common desire to settle the difficulties
still outstanding about Gascony and discussed the possibilities of
a marriage alliance between their two houses. On the face of it,
the prospects for a durable peace had never looked brighter; but
some delicate problems still awaited solution.

The settlement of 1327 had reduced English domination in
Gascony to the coastal strip between the Charente and the
Pyrenees, pending payment of the large sums already promised
as relief and indemnity. The French held the Agenais, south of
the Dordogne and the Bazadais, south of the Garonne. Under
the convention of Bois-de-Vincennes (1330), commissions of

---

[1] The Bardi contributed over £3,600 towards the expenses of this visit (E. B.
Fryde, 'Loans to the English Crown, 1328–31', *Eng. Hist. Rev.* lxx (1955), 203).
[2] *Foedera*, ii. 813.

experts had been set up to examine the relevant evidence and to settle all questions in dispute touching the duchy, and from 1331 to 1334 the tradition of diplomatic negotiation established

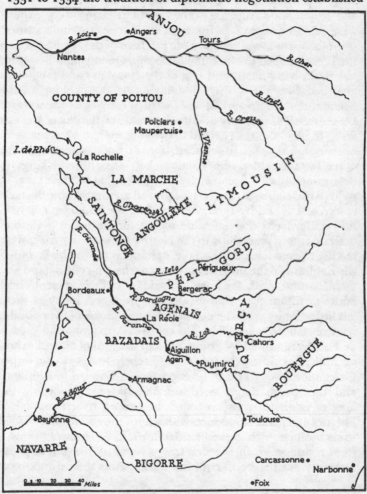

FIG. 3. Gascony and Poitou.

by the Processes of Montreuil and Périgueux was maintained by the Process of Agen. There was no lack of documentary evidence for, in the previous reign, the initiative of Bishop Stapledon, ably assisted by three household clerks, John Hildesle, Elias Joneston, and Henry of Canterbury, had led to the compilation of two great calendars of documents relating to Gascony, while both kings were well served by skilled jurists, like John Shoreditch and Raymond d'Aubenas.[1] The questions facing these experts were the old questions of frontiers, currencies, and indemnities; conspicuous among them was the problem of the Agenais which the French hoped to add to the Limousin, Querçy, and Périgord, territories lying north and east of the Garonne, and long since absorbed into the Capetian domain. At the opening of the negotiations, Philip and Edward were in conciliatory mood and the seneschal of Gascony and the constable of Bordeaux went so far as to proclaim that accord between them had been reached. Such optimism was ill-founded; for, as time wore on and as other factors caused the relations between the two kings to deteriorate, it soon became clear that no final settlement was in sight. Rival claims to the Agenais proved an insuperable obstacle to agreement and by 1336, when an impasse had been reached, Edward ordered the seneschal to put his castles in a state of defence and to send ships to the Norman coast. His orders were readily obeyed for, on the whole, opinion in the duchy favoured the English. To the majority of Gascons the king of England was their 'natural lord', the lineal descendant of their ancient ducal house to whom their first loyalty was due; and many of them felt little affinity with the northern French. Some of the nobility, it is true, tended to be fickle. The count of Armagnac and Gaston de Foix were bribed by Philip VI, and the lord of Navailles laid an appeal against Edward in the French *parlement*. But other Gascon lords were said to be eager to join Edward in Scotland, and the merchants of Bordeaux and Bayonne looked to the king of England whose ancestors had created their municipal liberties and to the Londoners with whom they were accustomed to do business;[2] while even a small town like Puymirol put up a stout resistance to the French troops when they began to penetrate beyond the borders of the duchy. And it was important

---

[1] See V. H. Galbraith, 'The Tower as an Exchequer Record Office in the Reign of Edward II', *Essays . . . presented to T. F. Tout* (1925), pp. 231–47; G. P. Cuttino, 'An unidentified Gascon Register', *Eng. Hist. Rev.* liv (1939), 293–9; 'Henry of Canterbury', ibid. lvii (1942), 298–311; 'The Process of Agen', *Speculum*, xix (1944), 161–78; *The Gascon Calendar of 1322* (Camden 3rd ser. lxx, 1949).

[2] See Margaret K. James, 'Les activités commerciales des négociants en vins gascons en Angleterre durant la fin du moyen âge', *Annales du Midi*, lxv (1953), 35–48, and below, pp. 360–3.

that throughout the crucial period of the negotiations and the opening of the war, Edward could count on the loyalty and experience of his seneschal Sir Oliver Ingham, an old servant of Edward II, who had been sent to assist Kent in 1324 and was appointed seneschal two years later. Though subsequently a partisan of Mortimer and a victim of the *coup d'état* of 1330, Ingham had been pardoned 'for his good service in Aquitaine' and restored to office forthwith. Such acts of judicious clemency were to prove characteristic of Edward III and none paid him better than this one. Ingham held the fort for him in Gascony, and when Philip took the final step the duchy was not prepared. On 24 May 1337 he declared Gascony confiscate to France 'on account of the many excesses, rebellions and acts of disobedience committed against us and our royal majesty by the king of England, duke of Aquitaine'. It is noteworthy that the specific reason given for this action was the support accorded to Robert of Artois by Edward III; the manifesto contains no reference to the points at issue within the duchy itself.[1] These issues were not new; and in confiscating Gascony Philip VI was merely following the example of Philip IV and Charles IV. The earlier confiscations had not been productive of a long-drawn-out conflict; and no major war need have followed Philip's action had it not been for other causes of Anglo-French hostility, foremost among which was the question of Scotland.

The treaty of Northampton of 1328 had been no more popular with Edward III than with his subjects and, once he had shaken off the tutelage of his mother and Mortimer, Scotland offered the most promising outlet for his martial energies. Robert Bruce had died in the summer of 1329 and the government of Scotland then fell to Randolph, earl of Moray, acting as regent for the five-year-old David II who was Edward's brother-in-law. To Moray, Edward addressed a number of formal requests that he should honour Bruce's undertaking to restore the earldoms of Buchan and Angus and the barony of Liddell to Beaumont, Umfraville, and Wake. When these requests were ignored, Edward Balliol, the pretender to the Scottish throne who had already been received at the English court, resolved to lead a group of malcontents to try their luck in Scotland. Edward III acted

---

[1] For the text of this document see *Chroniques de Froissart*, ed. Kervyn de Lettenhove, xviii. 34–37.

at first with the utmost propriety. He was not prepared to
involve himself in another disaster such as had helped to ruin his
father and he issued a proclamation affirming his entire regard
for the treaty of Northampton and forbidding armed men to
cross the Scottish border without his permission;[1] but at the
same time he turned a blind eye to the equipment of a small

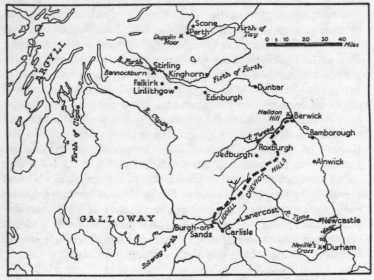

FIG. 4. Southern Scotland and Northern England.

force in the remote waters of the Humber. The adventurers
(known to history as the Disinherited) sailed to the Forth, disembarked at Kinghorn, and, after a series of hazardous enterprises, defeated the Scots, led by the earl of Mar, at Dupplin
Moor near Perth, on 11 August 1332, a few weeks after the death
of Moray, last survivor of Robert Bruce's great companions in
arms. Perth was occupied and on 24 September Balliol was
crowned at Scone as Edward I of Scotland; at Roxburgh, in
November, he acknowledged Edward III as his liege lord and
undertook to give him extensive territories in the lowlands and
on the Border. Edward must have been privy to these arrangements; but he maintained relations with the Scottish national
government and wrote to the pope disclaiming all knowledge of

[1] *Foedera*, ii. 810, 837.

and responsibility for, Balliol's actions.[1] Meanwhile, however, the levies of four shires were called out for the defence of the Border and Edward moved his government to York which was to be his administrative capital for the next five years. At the parliament which met there in December he showed his hand and asked his magnates for what was, in effect, a mandate for a Scottish war. Never much attracted by the prospect of campaigning among the northern hills, the lay lords asked for time to consult the prelates, many of whom were absent; and during the recess the whole picture changed. The nationalists rallied their forces and drove Balliol out of Scotland; and when parliament reassembled the lords declared themselves unable to agree. Edward thanked them for their attendance and announced that he would consult the pope and the king of France; but, in fact, he consulted neither and began to prepare for war without delay.[2] No doubt he feared lest the Scots, well aware of his partiality for Balliol, might themselves take the initiative. Berwick was invested and, on 19 July 1333, the Scottish force which attempted to relieve it was decisively defeated at Halidon Hill by a combination of archers with dismounted men-at-arms occupying a defensive position. Berwick capitulated, Balliol did homage for Scotland and handed over eight of the lowland counties, while David and Joan, accompanied by some of the Scottish bishops, fled to France, where they landed in May 1334.

Edward III and his advisers had not been blind to the possibility that English aggression in Scotland might lead to a resuscitation of the Franco-Scottish alliance. Edward's plan seems to have been to keep Anglo-French negotiations actively simmering, as earnest of his desire for a permanent settlement with Philip and in order to distract the French king's attention from events in the north. In December 1332 the bishops of Norwich and Worcester were in Paris where they were joined by Bartholomew Burghersh, now seneschal of Ponthieu and Montreuil and one of the most important English officials on the continent. Three days before Halidon Hill, English sailors were warned not to provoke the French without reason, so as to avoid the risk of war; Edward had already written to Philip VI, apologizing for the interruption of negotiations by unforeseen events and reiterating his desire for a final settlement with France at the

[1] *Foedera*, ii. 849 (15 Dec. 1332).   [2] *Rot. Parl.* ii. 67–69.

earliest opportunity.[1] The bishop of Norwich was sent again to
Paris, supported by the learned jurist, Thomas Sampson, and
the bishop of Worcester was ordered to go quickly from Avignon
to join them. Nothing significant emerged from these negotiations; but when a fresh embassy headed by the new archbishop
of Canterbury, John Stratford, arrived in Paris in April 1334,
rumours spread that peace had been achieved. It was at this
point—just after the arrival of David and Joan in France—
that Philip dropped his bombshell. He informed the English
emissaries of his intention that King David of Scotland and all
the Scots should be included in any treaty between England and
France. In short, the Franco-Scottish alliance was to stand and
Edward III was faced with the choice of abandoning his pre-
tensions in Scotland or risking active French intervention which
might transform a local conflict into a general war. From this
time onwards, the support lent by Philip VI to Scotland remained as a major obstacle to the conclusion of an Anglo-French
agreement and as a dangerous threat to the peace of western
Europe. Philip's attitude gave substantial encouragement to the
Scottish nationalists, who rallied under the leadership of Andrew
Moray and Robert the Steward, while the Disinherited quarrelled
among themselves and Edward, like his grandfather before him,
found himself committed to a series of costly and inconclusive
campaigns in the north. Before Christmas 1334 he had overrun
the lowlands in a short campaign which cost him over £12,000
in soldiers' wages; the following summer he and Balliol launched
a more ambitious expedition in the hope of achieving final victory
over the nationalists. A great host marched through southern
Scotland in two columns and converged on Perth; but still the
enemy was not brought to heel.[2] Meanwhile, ambassadors were
active. The papal nuncios were trying to achieve peace between
England and France, and representatives of Philip VI appeared
at York to offer unwelcome mediation between England and
Scotland. Benedict XII, who sensed the dangers inherent in
Philip's attempts to act as go-between, was bent on supplanting
him and in November 1335 a short truce was secured; but
Philip was unwilling to be ousted. By the following summer
Edward was informing the archbishop of York that the king of

---

[1] E. Déprez, *Les Préliminaires de la guerre de cent ans* (1902), p. 92.
[2] Professor E. A. Prince concluded that, with two possible exceptions, this was
the largest force ever assembled by Edward III. See *Eng. Hist. Rev.* xlvi (1931), 356.

France and David Bruce between them had rejected all his reasonable offers of peace, and that Philip had declared openly that his sympathies lay with the Scots, had promised to assist them to the utmost of his power, and was assembling a fleet for the invasion of England. It was true enough that acts of piracy were on the increase. In May 1335 the Scots had taken an English merchant ship at the mouth of the Seine and David Bruce's fleet even made an abortive attack on Jersey and Guernsey; but it was the transference of the fleet intended for the crusade from Marseilles to Normandy which finally convinced Edward that Philip intended large-scale intervention on behalf of the Scots. In the manifesto which he issued for propaganda purposes when war was seen to be inevitable (28 August 1337), Edward accused Philip of seeking every occasion to aid and maintain the Scots and of trying to keep him at war in Scotland so that he might not pursue his rights elsewhere; French sailors and men-at-arms had already committed acts of war in support of the enemies of England.[1] Here, in short, was a major cause of Anglo-French hostility, hardly a hint of which had been present when Edward III promised liege homage to Philip VI six years before.

Germany and the Low Countries played a less significant part in the political drama which preceded the war. The princes and burghers of the Netherlands and the emperor Lewis IV appear on the stage only in the final act and Edward III's diplomatic manœuvres in these regions were less a cause of the war than a consequence of his recognition that war had become inevitable. He had many affinities, both inherited and acquired, in the Low Countries. Edward I had enlisted the support of the Flemings in his quarrel with Philip IV and it might se em that Flanders was as natural an ally for the English as was Scotland for the French. The count of Gelderland was Edward's brother-in-law and the margrave of Juliers was married to his wife's sister. By his own marriage Edward had acquired an intimate connexion with the most powerful of the Netherland princes, William count of Hainault, Holland, and Zeeland; and after his victories at Halidon and Berwick, Edward's military renown began to attract adventurers from across the Channel to his court. William of Juliers and Guy count of Namur, with two of his brothers,

[1] *Foedera*, ii. 994–5.

joined Edward in the north in 1335; a yeoman from Cologne, John de Thrandeston, offered his services and became a valued member of the household.[1] Moreover, an uncle of the young count of Namur was becoming a familiar figure at the English court about this time. He was Philip VI's kinsman, Robert of Artois, who had been convicted of using forged documents in support of his claim to Artois and was strongly suspected of having poisoned the aunt in whose favour his claim had been rejected. Stripped of his lands and honours and condemned as a traitor, Robert fled to England where, or so many contemporaries believed, he concentrated his energies on persuading Edward III to renew his claim to the French Crown.

Yet despite these useful contacts, all was not plain sailing for Edward in the Netherlands. Alliance with the count of Flanders, though he is believed to have tried for it, proved impossible.[2] Philip of Valois had helped Louis de Nevers to defeat his rebellious subjects at Cassel in 1328 and the count was not now to be deflected from his allegiance. Edward's only hope lay in the strained relations persisting between the count and the cloth towns of Flanders. He held the whip hand over the burgesses of Ghent, Bruges, and Ypres who depended on English wool for their livelihood, whereas other markets, both at home and abroad, were open to the English growers and purveyors. Flanders was important to Edward at this stage, less for economic than for military and political reasons. He wished to prevent the Flemish ports being used against him in time of war and Flanders afforded a reservoir of mercenary troops and a convenient base from which to launch operations on French territory. For the moment, however, he could make little headway. An embargo laid on the export of wool to the Netherlands in August 1336 was intended, *inter alia*, to put pressure on the count; but its only immediate effect was to lead de Nevers to imprison English merchants and to order the seizure of English goods. In the spring of 1337 Flemish and other foreign clothworkers were invited to England; but the loyalty of Louis de Nevers to Philip VI served to frustrate most of these manœuvres. Philip could count also on the friendship of the bishop of Liège and of the two Luxemburgers, John of Bohemia and his son, Charles of Moravia.

[1] *Cal. Pat. Rolls 1334–38*, p. 167.
[2] H. S. Lucas, *The Low Countries and the Hundred Years War 1326–1347* (1929), p. 190.

Such was the situation when, in May 1337, an elaborately equipped embassy led by the bishop of Lincoln and the earls of Salisbury and Huntingdon, took up its quarters in Valenciennes, the capital city of Hainault, and began to angle for allies. The

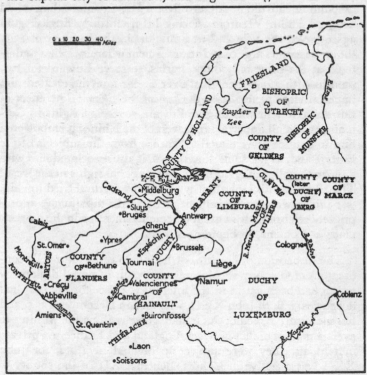

FIG. 5. The Low Countries in the Fourteenth Century.

magnificence of the ambassadors and their apparently unlimited resources soon won them what they sought; and it was not long before the counts of Hainault, Gelderland, Berg, Cleves, and Marck, the count palatine of the Rhine, the margrave of Juliers, and the elector of Brandenburg had promised their support—at a price—in the war with Philip which was now seen to be inevitable. Duke John of Brabant and Limburg proved rather less easy to woo, but he was won over by the colossal bribe of £60,000, coupled with a promise to establish the wool staple at Antwerp. Acting on their master's instructions, the ambassadors aimed higher still. Edward was doubtless aware of the difficulties which he was likely to encounter in controlling his new allies and he desired a position which would give him some authority over them. The emperor Lewis IV was the English queen's brother-in-law and the fact that he was also a heretic, schismatic, and excommunicate did not deter Edward III, any more than it had deterred Philip VI, from seeking his friendship. English gold again paved the way and at Frankfurt in the summer of 1337 the ambassadors negotiated an agreement whereby Lewis undertook, at the price of 300,000 florins, to serve Edward for two months with 2,000 men. Just over a year later he created him imperial vicar-general *per Alemanniam et Galliam*.[1] The effect of this appointment was to give Edward sovereign rights in Germany and in all imperial lands west of the Rhine; it empowered him to enforce service against France from the subjects of the Empire and, as Dr. Offler has shown, it also associated him with a German policy of revindication of imperial rights in the west.[2] The alliance with Lewis, for which Philip VI had tried in vain, was the crowning achievement of Edward's diplomacy, appropriately celebrated by a great ceremonial meeting in the market-place at Coblenz, in September 1338.

The suspension of hostilities which followed Philip VI's confiscation of Gascony was largely the result of papal diplomacy which had been active in the interests of peace for the past half dozen years. Both John XXII, who died in December 1334, and his successor, Benedict XII, had good reasons for wishing to avert a major conflict between England and France. As private individuals, they were subject to divided loyalties, for John XXII was a native of Cahors, Benedict XII of the county of Foix; and, as popes, they were bound to deplore the outbreak of war between the two great sovereigns of the west, whose orthodoxy was never in question, while the throne of Germany was occupied by the notorious Lewis IV and the Turks were advancing in the Mediterranean. Nor need they be supposed to have been indifferent to the inevitable consequences of such a war, to the destruction of churches and houses of religion, to the sufferings of the innocent, the shedding of Christian blood by Christians, the growth of rancour and vainglory and the harden-

---

[1] For the text of Lewis's letters to the *échevins* of Ypres, announcing Edward's elevation to the imperial vicariate, see Froissart, ii. 551–2.

[2] H. S. Offler, 'England and Germany at the Beginning of the Hundred Years War', *Eng. Hist. Rev.* liv (1939), 608–13.

ing of men's hearts. For John XXII, moreover, the crusade took precedence of all other causes. A great crusading army under the leadership of France was to recover the Holy Land and at the same time put an end to the depredations of the Seljuk Turks in the Aegean and to the attacks of the Syrian Mamluks on Cyprus. Unrealistic as such schemes may look in retrospect, they should not be dismissed as anachronistic nonsense. Contemporaries had no reason to believe that the age of the crusades was over; they still drew inspiration from tales of the heroic exploits of Godfrey de Bouillon and St. Louis in the east; and for Edward III and Philip VI, both of whom were essentially men of their own time, 'hoping to purchase Immortal Honour by their Deeds of Arms and Noble Chevalry',[1] the prospect of tasting adventure and winning renown under the banner of the Cross, while at the same time gaining access to the wealth of their own clergy, was no less alluring than it had been to Edward I and Philip IV. Brother Roger of Stavegni and Marino Sanudo could count on a cordial reception for the crusading treatises which they offered to the two kings;[2] and it seemed as though plans for a joint enterprise would serve to consolidate the friendly compact of March 1331. Edward made presents to the Armenian envoys who appeared in London in the spring of the following year; and the bishop of Winchester explained to the parliament which was then in session that the king proposed to set forth on crusade in two years' time.[3] Though they thought the date too early, the magnates did not oppose the plan and from Paris came word of Philip's ardent desire to accompany his cousin. The projected crusade was indeed largely a French enterprise. Its official preacher was Pierre Roger, archbishop of Rouen (afterwards Pope Clement VI), and John XXII had designated Philip VI as its leader. But there is no reason to doubt the sincerity of Edward's crusading zeal in the early years of his reign. It is after the outbreak of the Scottish war that a hint of bargaining begins to appear in the English dispatches: Edward will go on crusade if Philip will first restore the disputed territories in Gascony and will promise neutrality in Scotland. So late as 1334, Edward renewed before parliament his promise to go on crusade; but

[1] J. Barnes, *History of King Edward the Third* (1688), p. 90. The crusading exploits of Chaucer's knight (*Canterbury Tales* (Prologue), ll. 43–67) are a case in point.
[2] Stavegni presented his *Du Conquest de la Terre Sainte* to Edward III, Sanudo offered his *Secreta Fidelium Crucis* to Philip VI. See S. Runciman, *A History of the Crusades*, iii (1954), 440–1. [3] *Rot. Parl.* ii. 64–65.

this time no dates were mentioned and, as the Scottish war began to go badly and the expense of waging it piled up, his departure for the east began to seem increasingly remote, despite the fact that anxiety to prevent a Franco-Scottish alliance led him to keep the crusade in the forefront of his diplomatic parleys. When Armenian envoys came again to London in December 1335, Edward informed them of his desire to go to the help of their king, 'so soon as certain cares and hindrances shall be removed'.[1] Philip, meanwhile, had been proceeding actively with his preparations and nothing would have suited Edward better than to see these come to fruition. A crusading tenth had been levied from the French clergy and in April 1334 Jean de Chepoy was appointed 'capitaine et gouverneur des galies que nous envoioms en ayde des crestiens contre les mescreans Turs'.[2] A year later the ships were assembled at Marseilles awaiting the signal to depart. The peace of Europe might yet have been preserved had Benedict XII been willing to let Philip go. But this pope was of a more cautious disposition than his predecessor; the financing of the crusade and the mounting evidence of French interest in Scotland were causing him anxiety; above all, he dreaded the possible repercussions of a Christian failure in the east. When Philip declared himself ready to set out in August 1336, the pope allowed his qualms to become known; in a long letter to the French king he explained his reasons for postponing the whole project *sine die*. The effect of this decision on Philip was calamitous. Deprived at a stroke of the fruits of his efforts and of his dreams of heroic adventure, he diverted his crusading fleet to the Channel ports and prepared to intervene actively on behalf of the Scots. Though Benedict was still able to restrain him, it was these French activities in the Channel which, as has been seen, gave Edward III his pretext for telling his subjects that war with France was now inevitable. The crusading scheme had come near to precipitating the very conflict it was designed to avert, and events were to show that the pope had played his cards badly. Yet Benedict never remitted his efforts for peace. He sent his nuncios to arbitrate between England and Scotland, he tried to restrain Philip from intervention on behalf of David II, he struggled to prevent an alliance between Lewis IV and either England or France; he begged Philip not to break with Edward, he implored Edward to get rid of Robert of Artois;

[1] Déprez, p. 120.   [2] Ibid., p. 107.

and, in the end, he persuaded Philip to delay the confiscation of Gascony. In the opinion of some French historians the cumulative effect of these manœuvres was to serve the interests of England. It has been argued that by calling off the crusade the pope made Philip look foolish, by restraining him from active intervention in Scotland he deprived him of his best chance of success, by persuading him to delay action in Gascony he reduced his chances of acquiring the duchy and, above all, by frustrating his tentative approaches to Lewis IV he drove the emperor into the arms of England. Such opinions rest on a series of unverifiable hypotheses and however plausible they may appear, they do not tell the whole story. Some credit must be given to the genuinely pacific policy of the Avignon popes. Doubtless their motives, like most motives, were mixed; but at least they spared no effort to preserve the peace. It was their misfortune rather than their fault that they lacked the prestige which might have enabled them to intervene effectively. Benedict XII was, indeed, able to exercise some influence over Philip VI; but probably none knew better than the pope himself how little his counsels of peace were likely to avail in face of the chivalric ambitions of Edward III.

Edward was in his twenty-fifth year when Philip declared Gascony confiscate and from boyhood his favourite pastime had been playing at war. In this respect he was wholly and, it may be, deliberately conventional; he can hardly have been unmindful of the consequences of his father's tastes for rustic and plebeian pursuits. Together with his mother, Mortimer, and the gallant John of Hainault, the young king had witnessed a number of magnificent tournaments and after he had achieved independence, he and his friends continued to indulge in military sports. Typical of such exercises was the great tournament given by William Montagu in Cheapside, when the king and his knights paraded the streets of London in Tartar dress and afterwards entered the lists to challenge all comers. Spiced with real danger and attended with much publicity, performances like this tended to cast a glamour over war; inevitably, they fostered the desire to win yet greater renown on a larger stage. Had the crusading scheme borne fruit, Edward's militant energies might have found their outlet in the east; but the chance of adventure which offered in Scotland was too good to be missed. The young king plunged into a war which laid the foundations of his

reputation as a soldier and further inflamed his knightly zeal.
There are no good grounds for supposing that the Hundred Years
War was the result of long-term planning by Edward III; but it is
inconceivable that the prospect of a trial of strength with France
was distasteful to him. Froissart[1] preserves a story of the appearance of forty young 'bachelors' with the English embassy at
Valenciennes, each of whom had one eye covered with a piece of
silk because, so it was said, they had made a vow among the
ladies of their own country that they would not see with more
than one eye until they had done some deeds of arms in France.
Such tales carry more conviction than the formal protests of
pacific intentions which were being made simultaneously by the
ambassadors. They breathe the atmosphere of a court in which
all young men of birth and spirit desired to take their part in 'the
honourable enterprises and noble adventures in arms' which
for them spelt war. Nor was the appeal only to the spirit of
adventure; for medieval warfare could bring more tangible
rewards and, in pleasing contrast with the rugged hills and
barren moors of Scotland, the fair cities and rich fields of France
lay open to attack. The lure of adventure and the hope of gain,
a mingling of high spirits, greed, and courage drew Edward and
his subjects into war; and the fact that the captains and the
kings enjoyed it was not the least important reason why the war
was fought. Concern for 'rights', we may infer, came second.
The claim to the French Crown had been tacitly abandoned in
1331 and was not resuscitated until Edward had determined on
the war. If the taunts and insinuations of Robert of Artois had
any influence, it was only because they were poured into a
willing ear.[2] Edward's dubious 'rights' in Scotland were not
asserted until they could be made the pretext for a war of conquest; and negotiation on the subject of his 'rights' in Gascony
might well have dragged on for another twenty years. Apart
from the will to wage it, no war can be said to be inevitable;
and, though it is true that our understanding of Edward's motives
must remain imperfect, the facts on record place it beyond
doubt that, at least from the summer of 1336, he and his friends
while they talked of peace, were making ready for battle.

[1] ii. 376.
[2] The story embodied in the *Vow of the Heron* (*Polit. Poems*, i. 1–25) is almost
certainly apocryphal but it may afford interesting evidence of contemporary opinion
(see Déprez, pp. 225–6). B. J. Whiting believes, however, that the whole poem is
a bitter burlesque of a knightly scene (*Speculum*, xx (1945), 261–78).

V

THE HUNDRED YEARS WAR TO 1396

LIKE the second world war of the twentieth century, the Hundred Years War gathered momentum slowly. Gascony was declared confiscate in May 1337 and in October Edward laid his claim to the French Crown; but there was no organized campaigning for another two years. Both kings were engaged in feverish preparations, raising troops, ships, and supplies, cementing alliances, devising schemes of invasion.[1] Edward, with his queen and children, took up quarters at Antwerp; but by the late summer of 1339 he was in a difficult position. He was short of money and supplies, his allies were growing restive and, while French privateers were active in the Channel, Philip was organizing the defences of his northern coasts. Though the season was late, decisive action seemed necessary and in September Edward moved into the Cambrésis, where he ravaged the countryside but failed to take Cambrai itself. The count of Hainault had developed scruples about fighting his uncle on French soil; but the duke of Brabant and the elector of Brandenburg supported Edward who pushed on into the Vermandois, plundering the country round Laon and Soissons. By mid-October he was in the Thiérache, near St. Quentin, while Philip lay not far away at Buironfosse. Moved by one of the rash impulses to which the heroes of chivalry were sometimes prone, the French king at this point suggested a formal encounter (hardly to be distinguished from a tournament), in a field to be chosen by Edward where there was neither marsh, wood, nor river. Nothing could have been more to Edward's liking; he was thirsting for action and his adversary had offered it under ideal conditions. But Philip's second thoughts were wiser and, vociferating their contempt for his cowardice, the English found themselves forced to withdraw to Brussels. A fruitless military demonstration had been carried out at considerable cost against an enemy whose forces still remained intact. But though Edward was disappointed he was not wholly discouraged. There were grounds for hope in the changed situation in Flanders, where a wealthy burgess, James

[1] An order to drive piles into the Thames as a precaution against invasion was issued to the authorities of London in Oct. 1338. *Cal. Letter-Book F.*, p. 28.

van Arteveldt, had made himself master of Ghent, secured accord with Bruges and Ypres, and forced the count to take refuge in France. By the end of the year van Arteveldt had concluded an alliance with Edward III who agreed to raise the embargo on wool exports, to move the staple from Antwerp to Bruges, to pay the Flemings £140,000 for defence purposes, and to furnish them with ships and arms in the event of a French attack. The Flemings, for their part, promised military aid and recognition of Edward's title to the Crown of France. At Ghent, in January 1340, he formally assumed the title to be borne by his descendants for five centuries and received the oaths of his new subjects. He quartered the arms of France, had the fleur-de-lis engraved on his seal, and furnished himself with a red and blue surcoat sewn with leopards and lilies. Such trappings doubtless helped to restore his self-esteem and it was a solid achievement to have converted Flanders from a French into an English dependency; but Edward's financial position was far from satisfactory and when he returned to England to hasten the raising of supplies, his creditors insisted that the queen and her children should be left behind as hostages for the payment of his debts.

An essential preliminary to the projected campaign of 1340 was to put an end to the dangerous activities of French privateers in the Channel and to clear it for the English convoys. Sir Walter Manny's successful raid on Cadsand in November 1337 had been repaid by the French with heavy interest. Portsmouth and Southampton were raided in 1338 and French ships had been seen in the mouth of the Thames; off Middelburg, they had seized the great cog *Christopher* and four other vessels; there were attacks on Dover and Folkestone in 1339. By the following summer a great French fleet had anchored in the Zwin, then a considerable river, at the head of which stood the town of Sluys. Commanded by the admirals Quiéret and Béhuchet, assisted by the Genoese Barbanera who had furnished a contingent of galleys, this fleet was intended to protect the coast against invasion and to cover the route to Bruges.[1] Barbanera perceived the danger to the French ships of remaining crowded together in a constricted space, but Béhuchet refused to listen to him and in the end the Genoese followed his own counsel and escaped to the open sea. Meanwhile, the much smaller English fleet was

---

[1] For details of the French fleet see C. de la Roncière, 'Quatrième guerre navale entre la France et l'Angleterre', *Revue maritime*, cxxxvi (1898), 282–7.

approaching the Flemish coast and from his cog, the *Thomas*,
Edward III discerned 'so great a number of ships that their
masts seemed to be like a great wood.' His own ships were carefully placed, 'the greatest before, well furnished with archers,
and ever between two ships of archers he had one ship with
men-at-arms'.[1] The task of the combatants was made no easier
by the presence on board of a great number of high-born ladies
travelling to join the queen; but 300 men-at-arms and 500
archers were set aside for their protection and, on Midsummer
Day, after waiting till wind and tide were in their favour and
the sun in the faces of the enemy, the English attacked,
steering full sail into the harbour mouth and crashing into the
French at their moorings. Since the fleets were virtually floating
armies and naval tactics did not differ essentially from military,
the methods which had been used at Halidon, and were to be
used again at Crécy, served Edward well at Sluys. The battle,
moult felenesse et moult orible as it seemed to Froissart, was a victory for the English combination of archers with men-at-arms.
Blinded and decimated by the deadly hail of arrows from the
English long-bows and their movements cramped by the site
they had chosen, the French found themselves at the mercy of
the men-at-arms; and, though they fought long and gallantly,
the next morning saw both their admirals killed and their great
fleet almost totally destroyed.[2]

Sluys was a notable victory and after it Edward III's reputation soared still higher. Its most important practical consequence
was that it cleared the way for the English invasion vessels and
gave them command of the Channel for the next few years. But
command of the Channel was merely a preliminary. The conquest of the French army had still to be achieved and Edward
lacked the means for this greater task. The victory at sea was
followed by a land campaign no more effective than that of
1339. Robert of Artois was sent to attack St. Omer while Edward himself invested Tournai, but neither town capitulated.
This time it was Edward who proposed a chivalric exercise to
settle the quarrel; but when Philip insisted that the kingdom of
the vanquished should be the reward of the victor, he dropped

[1] Froissart, iii. 194–206.
[2] Edward III's letter to his son written after the victory (Avesbury, pp. 312–14)
has been described as 'the earliest English naval despatch', W. L. Clowes, *The
Royal Navy*, i (1897), 256.

the challenge. Meanwhile, van Arteveldt was proving incon-
veniently obtrusive, the Flemings and Brabançons were quar-
relling in the English camp, the troops were clamouring for
their pay, and the wool consignments dropped behind schedule.
In short, the allies were tired of the war; and when the dowager
countess of Hainault, who was the French king's sister and the
English queen's mother, emerged from her convent to suggest
a truce, Edward found he had no choice but to agree. A truce
to last till midsummer was arranged at Espléchin on 25 September 1340, on the basis of maintenance of the territorial
status quo and free trade; and both kings pledged themselves not
to seek papal or ecclesiastical mediation. Edward withdrew to
Ghent and at the end of November he returned to England.

Thus ended the first phase of the war. Its only substantial
achievements were the Flemish alliance and the destruction of
the French fleet. The frontiers of France still remained intact
and Edward's other costly alliances had brought him little advantage. His aim had probably been to frustrate Philip's designs on Gascony by diverting the war into northern France;
but he had found himself helpless against an enemy who concentrated on defensive tactics and refused to be brought to
battle. Fighting far from his base and dependent on long lines
of communication, Edward could not afford protracted sieges
and piecemeal reduction of hostile territory. Time was always
his enemy; the best hope of victory lay in the quick, decisive
action which the enemy refused to risk. Subsequent events were
to show that Philip VI had learned little from the experience of
these early campaigns. Yet all unwittingly he was pursuing the
type of strategy which, in the hands of Charles V and Duguesclin
a generation later, was to prove the salvation of France.

In the years following the truce of Espléchin, Edward III's
difficulties mounted. He was deeply in debt; the French succeeded in weaning the emperor Lewis IV from the English
alliance, thereby depriving Edward of the imperial vicariate,
his sole source of authority over the princes; David II returned
to Scotland (June 1341) and a renewal of Border raids followed;
next year Benedict XII was succeeded at Avignon by the abler
and even more determinedly pacific Clement VI. The French
were gaining ground in Gascony, while in Flanders the townsmen had been excommunicated as rebels against their count,

and van Arteveldt's insolence and arrogance were making him
many enemies. Thus, when a succession war in Brittany offered
him the opportunity of a new continental alliance, Edward III
was quick to seize it. The claimants to the duchy of Brittany
were John de Montfort, younger brother of duke John III, who
died in April 1341, and Charles of Blois who had married the
daughter of an elder brother.[1] Both requested Philip VI to receive their homage and both were cited to his court; but before
the case had been heard, de Montfort had visited Edward at
Windsor and had obtained the earldom of Richmond and a promise of military aid in return for his recognition of Edward's
title to the Crown of France.[2] Philip was aware of the bargain
and his acceptance of the claim of Charles of Blois was therefore
almost inevitable. Before the end of the year his eldest son, duke
John of Normandy, had led an army into Brittany, occupied
Nantes, and captured de Montfort whose cause was maintained
henceforward by his wife, the redoubtable duchess Joan. The
confused and protracted war in Brittany had little direct bearing on the main Anglo-French conflict; but it introduced a new
casus belli between Edward and Philip; it constituted a dangerous threat to the north of France while opening up a new
sphere of influence on French soil to Edward; it immobilized
substantial English forces in the garrisons of the towns and castles
held by the Montfortian party; and a war famous for its deeds
of prowess, in the course of which Bertrand Duguesclin first
emerges into history, afforded ideal conditions for the chivalric
exploits of captains like Sir Walter Manny and Sir Thomas Dagworth. Edward III himself took an army to Brittany in the
autumn of 1342 and overran the duchy without meeting serious
opposition until he approached Vannes and found himself
within striking distance of a great French force. A decisive engagement might have been fought here; but the season was
late, the weather was bad, and the ubiquitous papal nuncios
were at hand, pleading for a cessation of hostilities. After the
truce of Malestroit (January 1343) Edward returned to England
and renewed his preparations for the greater conflict.

[1] It will be observed that in this succession dispute, Edward III supported the
nearest male heir, Philip VI, the candidate whose claim derived from a female.
[2] The earldom of Richmond had been an appanage of the dukes of Brittany
since the twelfth century. John de Montfort held the earldom for less than a year.
It was transferred to John of Gaunt in Sept. 1342 and it is unlikely that de Montfort
was ever entitled an English earl (*Complete Peerage*, x. 820, n. *e*).

Though we have little direct evidence as to his plans, he may be presumed to have had a twofold objective. In the south-west the French advance in Gascony must be halted and the English position retrieved; and in the north-east the languishing Flemish alliance must be resuscitated and the Flemings galvanized into activity. Earl Henry of Derby and Sir Walter Manny, two of the ablest among the younger commanders, were sent to Gascony where they achieved a rapid succession of victories. Bergerac, La Réole, Montségur, Aiguillon, and even the northern outpost of Angoulême, all fell into their hands. After the winter truce, however, a strong force under John of Normandy advanced from the Loire, recovered Angoulême and, in April 1346, laid siege to Aiguillon, causing Derby to send home an earnest appeal for reinforcements. Meanwhile, Edward III had decided on personal intervention in Flanders. In July 1345 he interviewed van Arteveldt on board ship in the harbour of Sluys and subsequently declared himself well satisfied with the result, despite the fact that, before he left the Flemish coast, van Arteveldt, after returning to Ghent, had been murdered there by his enemies.[1] But the dependence of the cloth towns on England was in no way affected by the death of their leader and, though the majority of his allies in the Low Countries had fallen away from him, Edward was determined to maintain his contacts in Flanders. In June 1346 he named one of his kinsmen, Hugh Hastings, as his captain and lieutenant there and put him in command of the Flemish troops destined to fight the French. A few days later Hastings was empowered to settle any differences that might arise between the English and the Flemings. At home, naval, military, and financial preparations were reaching their climax and by the beginning of July an army of perhaps some 15,000 men was at Portsmouth ready to embark.[2] Its destination was unknown, and if Edward's design was to keep the enemy guessing, he played his cards well. Uncertainty as to his intentions prevented the French from concentrating their fleet at a fixed point, as at

[1] Edward informed his sheriffs on 3 Aug. '. . . dictam terram Flandrie, laudetur Deus, stabilivimus, ita quod nunquam fuit in fidelitate nostra magis firma' (*Foedera*, iii. 55).

[2] For discussion of the controversial question of the numbers engaged in the campaign of 1346, see F. Lot, *L'Art militaire et les armées* (1946), i. 344–8, and A. H. Burne, *The Crecy War* (1955), pp. 166–8. For a good account of the campaign see J. Viard, 'La Campagne de juillet–août 1346 et la bataille de Crécy', *Moyen Âge*, 2e sér. xxvii (1926), 1–84.

Sluys in 1340, and forced them to keep it dispersed along an extended coastline. Gascony, where Derby was still holding out in Aiguillon, was a likely objective and many modern historians support the widely held contemporary opinion that it was only a last minute change of plan, the result of contrary winds and a false start, which prevented Edward from directing his forces to the Bay of Biscay. Froissart represents this change of plan as the result of advice given by Geoffrey of Harcourt, a French émigré who had been at the English court since 1344. Harcourt is said to have appealed to the profit motive which always operated so powerfully in the wars of chivalry:

Sir, the country of Normandy is one of the most plenteous countries in the world . . . if you will land there, there is none that shall resist you: the people of Normandy have not been used to war . . . you shall find great towns that have not been walled, whereby your men shall have such gain that they shall be the better twenty years after.[1]

Harcourt was lord of St. Sauveur-le-Vicomte and was closely familiar with the Cotentin and western Normandy; the part he played in the subsequent campaign seems to support Froissart's version, though the advice may well have been given at an earlier stage. Edward's concern for the Flemish alliance and his commission to Hastings in June suggest that the possibility of a landing in northern France may have been already in his mind; he may have hoped that such a move would divert French troops from Gascony. But whatever the motives which prompted him, the decision to land in Normandy and to march eastward to effect a junction with the advancing Flemings was a bold one. Edward risked interception by Philip's much greater army and, since the masters of the ships were always eager to return home once the troops had landed, loss of his line of retreat, encirclement, and annihilation. But he was a superb soldier, he had had time to lay his plans, and it must be presumed that he knew what he was about.

After landing at St. Vaast-de-la-Hogue on 12 July, plundering his way through the orchards and cornfields of Normandy and sacking Caen with merciless brutality, Edward reached the Seine to find the bridges broken. News of his landing seems to have been slow in reaching Philip but the French king was in Rouen by the beginning of August. It was vital for him to

[1] Froissart, iv. 381.

defend the Norman capital, capture of which would have given Edward command of the lower Seine and allowed him to effect a junction with the Flemings. The two armies manœuvred for some days on the opposite sides of the river, Edward venturing up as far as Poissy, almost in the suburbs of Paris. Fearing that the English might evade him altogether, Philip then re-entered the city. Edward had deceived him into supposing that he meant to by-pass Paris to the south-west and make for Gascony; in fact, his carpenters were busy repairing the bridge at Poissy and he was over the Seine and well away to the east before Philip grasped the truth about his movements and could wheel round to pursue him. Luck still favoured Edward when he reached the Somme. Here too, the bridges were down, but a peasant showed him a ford with a sandy-bottom—Blanche Tache—and he had his army across just before the tide rose and held up the pursuing French, thereby giving him a precious respite in which to dispose his forces. Supplies of food, wine, and shoes were running dangerously low and the men were tired; the Flemings, who had advanced into Artois, had been repulsed before Béthune; the English were faced with the certain prospect of a major engagement against a much larger and better-provisioned army. But Edward could hardly have found a better defensive position. At the back of his army was the wood of Crécy-en-Ponthieu, in front, the Valley of the Clerks. His forces were disposed on a height above the little river Maie, divided into three corps, the right under the Black Prince and the left under the earl of Northampton, while the king himself commanded the reserve, a short way behind the centre of the line.

Philip VI had already learned from his scouts that the English were at Crécy. Some of his generals, knowing that the enemy could go no farther, advised delay to rest the troops; but the French knights' impetuosity swept counsels of prudence aside. Had they waited for the morning light and attacked when the men were fresh, the famished and exhausted English might have been hard put to it to resist an overwhelming force. As it was, the Genoese crossbowmen advanced against the setting sun; they ran out of arrows and were forced to retreat into the advancing French cavalry who rode them down. Philip soon lost all control of events; but the English, under Edward's direction, remained calm and co-operated well. Successive French attacks broke against their tenacity and the incessant rain of their arrows. In an indescribable mêlée,[1] the battle continued until the late August nightfall and was ended only by the coming of darkness, when the blind king of Bohemia, the count of Flanders, the count of Blois, and many other French nobles, the flower of the aristocracy, lay dead on the field. Only King Philip and a small party succeeded in escaping to Amiens where they rallied the remnants of their army. When dawn broke on 27 August, the whole countryside was blanketed in fog. No better tribute to Edward's capacities as a general can be found than his stern prohibition of pursuit of the scattered French troops. He held his men together while he ordered reconnaissance of the casualties; and on 30 August he resumed his march north-eastwards, through Étaples and Wissant, and on 4 September encamped below the walls of Calais.

Crécy made a hero of Edward III. The battle had given him his chance to use against the chivalry of France the tactics which had proved successful against the rougher patriots of Scotland. Under his skilled direction the combination of archers with dismounted men-at-arms once again brought victory. Edward's judgement and self-control contrasted admirably with the impetuous folly of his adversary; they proved to the world that as a leader of men he was incomparable. The historian must add, however, that Crécy was fought in a fortunate hour, that Edward III was assisted, not only by his own genius and the blunders of the French, but also by the trend of the times. The day of the feudal horseman was nearly over, for the longbow had shown itself capable of neutralizing the danger from cavalry charges; and the science of artillery was still in its infancy.[2] At this particular period in the history of warfare, the scales were so heavily weighted in favour of the defence that to take offensive action against well-supported archers was to court almost certain disaster. This was an unpalatable truth for the heroes of chivalry to digest and many years were to pass before the implications of the battle of Crécy were fully understood.

[1] Literally indescribable. Though we possess many chroniclers' versions of the action at Crécy, detailed reconstruction of its tactics is impossible as M. Lot admits (p. 343, n. 1). No eyewitness was in a position to comprehend the whole scene of battle and the chroniclers' accounts tend to impose artificial order on actual confusion.

[2] There is no reason to suppose that artillery played any significant part at Crécy. A few cannon may have been fired with the idea of frightening the horses, but such weapons were still reserved almost exclusively for siege warfare. For a discussion of this question see A. H. Burne, *The Crécy War*, pp. 192–203.

Calais was Edward's next objective. He needed a port of embarkation and Calais was a prize well worth winning. Possession
of a bridgehead across the straits of Dover would free him
from dependence on the Flemish ports and on the long sea-
route to Gascony; Calais would provide him with a base for
operations on French territory and against French raiders in
the Channel and with an entrepôt for English goods. But it was
not to be had for the asking. Though a small town, it had strong
natural defences in its sand-dunes and marshes; its ramparts
were solid; and its garrison was under the able and resolute
command of Jean de Vienne. To take it by assault looked impossible and Edward determined on a blockade. He and his
army settled down in the fields outside the town and built themselves wooden huts as protection against the coming winter.
The queen and many ladies soon joined the besiegers and in
this aptly named Villeneuve-le-hardi, the Flemings held markets on Wednesdays and Saturdays. Though stirring events were
taking place elsewhere, Edward refused to be ousted from his
position. In October he learned that the Scots were again on the
move. An attempted diversion in the north was not altogether
unexpected and when he left for France Edward had wisely
refrained from calling out levies from the counties north of Trent
and had put Percy, Neville, and the archbishop of York in
charge of the Border defences. It was in response to a summons
from his French ally that David II now crossed the Border with
a large army, seized Liddell, and plunged into the palatinate of
Durham. At Neville's Cross, close to Durham city, he was intercepted by the northern barons, his schiltroms were surrounded
and cut to pieces by the English archers, the famous Black Rood
of Scotland fell into the hands of the English and was hung as a
trophy in Durham cathedral, and, greatest triumph of all, King
David himself was taken. A year later he was joined in the
Tower by Charles of Blois, the French candidate for Brittany,
whom Sir Thomas Dagworth defeated and captured at La
Roche-Derrien in June 1347. In Gascony, too, English fortunes
were in the ascendant, for the siege of Aiguillon had been raised
and the duke of Normandy's army withdrawn across the Loire.
Edward III remained, however, before Calais and, though he
received the papal nuncios engaged on their accustomed mission, he refu:ed to entertain peace proposals which did not admit
his title to the French Crown. For some months the besieged

were able to obtain intermittent supplies by sea; but once Edward had succeeded in blocking the harbour mouth, their only hope was an army of liberation. Philip VI never showed himself more inept. The siege had lasted nearly a year before he appeared with his host on the cliff at Sangatte and then he judged the English position impregnable and withdrew. Next day, Jean de Vienne appeared on the battlements and offered to treat with Edward III.[1] There is no reason to regard the great story of the burghers of Calais with scepticism; but his reluctant generosity to them marked the limits of Edward's clemency.[2] On 4 August 1347 he entered the town and proceeded to evacuate almost all the inhabitants, in order to people it with the English colonists whose descendants were to hold it for another two hundred years.

For nearly a decade after the fall of Calais, the war languished. Though Brittany was still the scene of deeds of prowess and there were skirmishings on the Gascon border and at sea, a Spanish fleet which had been raiding English shipping being destroyed off Winchelsea in 1350 in the action known as Les Espagnols-sur-Mer, elsewhere a series of truces served to maintain an uneasy peace. The combatants were suffering from warweariness and the Black Death came as a dislocation of society and a harsh reminder of the mutability of human fortunes. At one time it seemed as though the new pope, Innocent VI (1352–62), might reap the fruits of his predecessors' labours and persuade the belligerents to agree to a final peace.[3] But John II, who in 1350, succeeded Philip VI on the throne of France, was a man of chivalric ambitions and little judgement; and Edward III was still far from satisfied. He had added Crécy to his laurels and Calais to his possessions and the English position in Gascony had been stabilized. But he had lost control in Flanders where the new count, Louis de Mâle, had secured recognition by the towns; and, above all, the kingdom of France remained unconquered and its Crown had still to be won. Moreover, the prince of Wales had grown to manhood, a son after his father's own heart, and he was not to be denied his share of the glories

[1] The siege is well described by J. Viard, 'Le Siège de Calais', *Moyen Âge*, 2ᵉ sér. xxx (1929), 129–89.   [2] Froissart, v. 198–216.
[3] For the terms of the draft treaty of Guines (Apr. 1354) see F. Bock, 'Some new documents illustrating the early years of the Hundred Years War (1353–1356)', *Bull. John Rylands Library*, xv (1931), 60–99.

and profits of war. It is unlikely, therefore, that Edward III had any serious intention of coming to terms with France. His aim was to renew the war, if and when opportunity offered, and he lent an attentive ear to the suggestion of Charles the Bad, king of Navarre, like himself a descendant of the house of Capet in the female line, that they should divide the kingdom of the Valois between them. Charles the Bad's possession of extensive territories in the neighbourhood of Paris was an undoubted asset; but he proved a less satisfactory ally than Robert of Artois. He engaged in intrigues with both sides and his project for a fresh English invasion based on Normandy came to nothing. Edward did indeed take an army to Calais in October 1355; but he was back in England by the end of November. Meanwhile, the Black Prince, who had been appointed lieutenant in Gascony, went there with a small army in September 1355. He spent the autumn in congenial fashion, firing the suburbs of the ancient cities of Narbonne and Carcassonne and ravaging the Mediterranean provinces; and, though he found no French army to give him battle, the raid could be judged successful in so far as it humiliated the enemy, depleted their resources, cemented the loyalty of the Gascons to Edward III, and greatly enriched the raiders.[1] Next year the prince decided to go farther afield, possibly to try to effect a junction with Henry of Lancaster in Brittany, whence together they might hope to launch a campaign of devastation in central and north-west France. The prince reached the Loire without difficulty early in September; and having already achieved a successful raid, he decided to withdraw to Bordeaux. King John had been mobilizing his army at Chartres but the English had begun their march southward before he could intercept them. The two commanders seem to have been uncertain of each other's whereabouts until, on 17 September, the Black Prince, who was marching across country to Poitiers, collided with a French reconnaissance party at La Chabotrie. Two French counts were taken prisoner, but the English did not break off their march till the next day, when they encountered the main French army in the neighbourhood of Maupertuis on the little river Miosson, where there was a fork in the road from Poitiers to Bordeaux. As it was Sunday, the cardinal nuncios were able to secure a brief truce; but on Monday, 19 September, the battle was joined.

[1] H. J. Hewitt, *The Black Prince's Expedition of 1355–57* (1958), p. 71.

Whether the engagement was of the prince's own seeking or whether, as many French historians believe, his attempt to break away was frustrated by King John's brilliant flanking manœuvre which cut off his line of retreat, cannot be decided with finality, but it seems probable that there was a division of opinion in the English army, some favouring withdrawal to Bordeaux with the spoils already accumulated, others eager to risk a battle. The English were, almost certainly, outnumbered by their opponents, but the wooded country with its vine-clad hills was favourable to the defence and the prince had the good sense to follow the example of his father's tactics at Crécy.[1] He placed his archers in front and his knights, dismounted, in the rear; this arrangement, coupled with the astounding blunders of the French, gave him the victory. King John began by sending a small cavalry force, the flower of his army, under the constable and the two marshals of France, to break through a hedge which divided him from the English: they ran straight into the fire of the archers and were cut to pieces. For a decade before Poitiers the French had been making tentative experiments with the new tactics;[2] and King John now dismounted the remainder of his knights, not perceiving that, unless combined with the archers which he lacked, such a move could be only to his disadvantage. The first French division, led by the dauphin, fought valiantly but was overwhelmed; the second, under Orleans, broke and fled; there remained the third under the personal command of the king. Heedless of his duty to avoid capture at all costs and bemused with dreams of 'vaillance', King John rushed into the fight and into the hands of the enemy.[3] The battle was over and the Black Prince retired to Bordeaux with his plunder and his prisoners. A two years' truce was arranged and the prince, with elaborate courtesies, escorted his royal captive to London.

Had King John not been taken, prestige, plunder, and a great

---

[1] According to M. Lot's calculations, the English forces at Poitiers numbered c. 6,000, the French between 5,000 and 6,000 (pp. 359–64); but nearly all contemporary writers assert that the French force was much larger than the Anglo-Gascon; see Burne, pp. 312–16.

[2] T. F. Tout, 'Some neglected Fights between Crécy and Poitiers', Eng. Hist. Rev. xx (1905), 726–30.

[3] The topography and tactics of the battle of Poitiers are hardly less obscure than those of Bannockburn and nearly as controversial. For a soldier's view see Burne, Eng. Hist. Rev. liii (1938), 21–52, and Crécy War, pp. 275–321, and for a criticism of this view, V. H. Galbraith, Eng. Hist. Rev. liv (1939), 473–5.

haul of ransoms would have represented the sum of the profits of Poitiers. But the capture of the king of France was the greatest English triumph yet secured and a supreme disaster for the French. It left the nineteen-year-old dauphin to face the intrigues of Charles the Bad, the internal chaos resulting from the ravages of the free companies, the communal movement in Paris led by Étienne Marcel, and the peasant rising known as the Jacquerie; and it seems to have convinced Edward III that the time had come for a supreme effort to attain his ends and finish the war. At best, he might hope to secure the French Crown; at worst, to establish himself as sovereign in a greatly enlarged Gascony. His capacities as a general had never shown to better advantage than in the great winter campaign of 1359–60;[1] but all did not go according to plan. He was driven back from before the walls of Rheims and failed to gain effective control of Burgundy; his army was assailed by a tempest; he lost the support of Navarre; and French privateers were once more active in the Channel. His hopes of the Crown vanished; and he could not now insist on the outrageous terms, amounting to confiscation of two-thirds of France, suggested a year before in London. None the less, his possession of King John's person meant that he could negotiate from a position of overwhelming strength.

Peace talks began at the hamlet of Brétigny, near Chartres, on 1 May 1360. A week later a truce was concluded to last until Michaelmas 1361, and the Black Prince and the dauphin accepted a draft treaty to be confirmed later by the sovereigns of both nations. Under its terms Edward III was to be allowed full sovereignty, not only of Gascony but also of the debatable lands on its borders, including the strategically vital county of Poitou,[2] some of which had been in French hands for over a century, and, in northern France, of the counties of Montreuil, Ponthieu, and Guines, the lordship of Marck and the town and environs of Calais. King John's ransom was fixed at 3,000,000 gold crowns (£500,000), the first instalment of 600,000 to be paid at Calais within four months and hostages to be given as security for the payment of the remainder in six annual instalments

---

[1] Burne, pp. 322–53.
[2] Poitou (with Poitiers), Thouars, Belleville, Agenais (with Agen), Périgord (with Périgueux), Quercy (with Cahors), Limousin (with Limoges), and the county of Gaure.

of 400,000. The king was to be released when the first instalment had been paid, the hostages delivered, and a number of strongholds, including La Rochelle, the centre of the salt trade, handed over to the English. For his part, Edward III undertook to renounce his claim to the Crown of France and to all other parts of the realm[1] and to restore the fortresses in his possession which belonged to the French.

Though its terms were so heavy they might have been worse, and the treaty of Brétigny was welcomed in France as marking the end of an intolerable war. It was ratified by the two kings at Calais in October, but with a significant difference. The 'renunciations clauses', under which the ceded lands and the claim to the French Crown were to be formally renounced at a time and place to be arranged, were removed from the text of the treaty itself and embodied in a separate charter, with the additional proviso that the renunciations should be made not later than 1 November 1361.[2] The ultimate effect of this modification was to nullify the most important provisions of the treaty of Brétigny. By making the renunciations conditional on the transfer of the ceded lands and by fixing a limiting date for their cession, those who drafted the Calais treaty had virtually ensured that the renunciations would never be made at all. The transfer of territories was a lengthy and complicated business and both French and English must have been aware that it could not be concluded within twelve months. When the time-limit expired, Edward III made a formal *démarche* but took no further action. He did not resume the title of king of France which he had dropped on 24 October 1360; but, on the other hand, he continued to act as though he were sovereign in the ceded lands. This non-execution of the renunciations clauses has commonly been regarded as a triumph for French diplomacy;[3] but evidence brought to light by M. Chaplais[4] suggests that the responsibility lay with Edward who may have felt that he had conceded too much and was glad to be left with a loophole for

[1] Professor Le Patourel has pointed out (*Rev. historique de droit français et étranger*, 4e sér., t. xxx (1953), pp. 317–18) that the cession of Normandy in particular entailed a real sacrifice for Edward III.

[2] For the text of the treaty, with the omitted clauses, see E. Cosneau, *Les Grands Traités de la Guerre de Cent Ans* (1889), pp. 39–68, 173–4.

[3] e.g. by MM. Petit-Dutaillis and Collier, 'La Diplomatie française et le traité de Brétigny', *Moyen Âge*, 2e sér. i (1897), 1–35.

[4] 'Some Documents regarding the Fulfilment and Interpretation of the Treaty of Brétigny', *Camden Miscellany*, xix (1952).

resuming the war, should he at any time wish to do so. Meanwhile however, he showed himself conciliatory. He agreed to accept 400,000 crowns, the most that the French could raise, as the first instalment of the ransom; King John was released and the hostages, who included three of his sons, were handed over. The unfortunate king was not to enjoy his liberty for long. While he made his way to Avignon to discuss crusading schemes with Urban V, the hostages began to find their captivity irksome. On their own initiative, but pledging the word of their king, they made a treaty with Edward, guaranteeing immediate payment of a further 200,000 crowns for the ransom and the issue of letters of renunciation, in return for which they were brought to Calais to await release. King John expressed himself willing to ratify these arrangements but when his second son, the duke of Anjou, broke his parole and took to flight, he held that honour demanded his return to London. Here he died in April 1364, bequeathing his problems to the astute and capable dauphin who succeeded him as Charles V.[1]

For Edward III the years following the treaty of Calais proved disappointing; for he was beginning to lose ground on three important fronts. The capture of David II at Neville's Cross had not procured the submission of Scotland. Although Edward Balliol had been persuaded to make over his property and rights to the King of England, the nationalist cause was maintained by Robert the Steward and, after a fruitless attempt at terrorization—the lamentable 'Burnt Candlemas' of 1356— Edward had reluctantly offered the Scots a ten years' truce and the liberation of their king, in return for a ransom of 100,000 crowns payable by instalments. David II, whom long residence in the Tower seems to have anglicized, soon found himself at odds with his own people. He was childless and he disliked the Steward who was his heir-presumptive; in 1364 he suggested to Edward that the English and Scottish Crowns should be united and the ransom remitted. Needless to say, the Scottish parliament made short work of this proposal and when David died in 1371, Robert II, the first of the Stewart kings, lost little time in concluding with Charles V the alliance which is commonly regarded as the formal inauguration of the Franco-Scottish

[1] The details relating to the payment of King John's ransom have been worked out by Dorothy Broome in *Camden Miscellany*, xiv (1926); see also Delachenal, *Charles V*, ii. 325–31.

League. In Flanders, Edward III suffered an even more cruel disappointment. Louis de Nevers had fallen at Crécy and his son, Louis de Mâle, who was bent on recovering his position as count, was alive to his dependence on the cloth towns and consequently not unfriendly to England. The establishment of the wool staple at Calais (1363), through which passed the wool cargoes destined for Bruges, was a concession to Flemish interests and about the same time the count opened negotiations for a marriage between his only child Margaret, heiress, not only to Flanders but also to the counties of Burgundy and Artois, and Edward III's fifth son, Edmund Langley, earl of Cambridge. Charles V was determined at all costs to prevent this dangerous alliance. He persuaded Urban V to refuse the necessary dispensation and himself entered into a bargaining contest with Louis which bore fruit in 1369, when Margaret became the wife of the French king's brother, Philip the Bold, duke of Burgundy.[1] France had to pay dearly in the long run for the enhancement of Burgundian power which resulted from this marriage; but at the time, it represented a severe set-back for Edward III and the end of that English domination in Flanders which had opened the way to earlier victories. In Brittany, too, Edward had lost ground. Under the treaty of Calais, the suzerainty of the duchy remained with the king of France and, though Charles of Blois was killed at Auray in 1364 and the succession thereby assured to the anglophil John de Montfort, son of the claimant of 1341, Charles V was well satisfied with a settlement which assured him of the homage of the duke and entailed the removal of English mercenaries from the duchy.

These were discouraging reverses for Edward III; but the most dangerous threat to the maintenance of peace arose elsewhere. In July 1362 Edward transferred Gascony to his eldest son, to be held of himself as sovereign. The Black Prince forthwith established a luxurious court at Bordeaux where, according to his fervent admirer the Chandos herald, 'there abode all nobleness, all joy and jollity, largesse, gentleness and honour, and all his subjects and all his men loved him right dearly.'[2]

[1] F. Quicke, Les Pays-Bas à la veille de la période bourguignonne, 1356–84 (1947), pp. 73–174. The duchy of Burgundy (Lower Burgundy), which was always a fief of the Crown of France, is to be distinguished from the free county (Upper Burgundy) which was a fief of the Empire until it fell to the French dukes of Burgundy in 1384.

[2] Life of the Black Prince by the Herald of Sir John Chandos, ed. Mildred K. Pope and Eleanor C. Lodge (1910), p. 148.

The more truculent members of the Gascon aristocracy may have been in some measure conciliated by this lavish hospitality; and it is to the prince's credit that he did not confine himself to feasting and jousting but also made serious efforts to maintain justice and order, to increase the number of officials, and to strengthen the central government. His mistake lay in bestowing the most lucrative posts, not on Gascons, but on his own followers and friends; coupled with incessant demands for money and with three notorious *fouages*, or hearth-taxes, this policy soon cost him his initial popularity. Discontent within the duchy did not deter him, however, from plunging into war south of the Pyrenees on behalf of Peter the Cruel, whose half-brother, Henry of Trastamara, assisted by Duguesclin and an army of *routiers*, had driven him from the throne of Castile. This was a less reckless adventure than has sometimes been supposed; for the prize at stake was the Castilian navy, or at least its exclusion from the war on the French side.[1] At the head of a mixed army of English, Gascons, Navarrese, and Castilians, the prince advanced into Spain and on 3 April 1367, at Nájera, near Pampeluna, won another famous victory which enabled him to restore Peter to his throne. The prince's abilities as a commander had revealed themselves once more; but his political ineptitude was apparent in his refusal to surrender to Peter those of his political opponents whom he had taken prisoner; for within two years, these men had murdered the king and opened the way for the usurpation of Trastamara and the reassertion of French control over the Castilian fleet. Meanwhile, since the reimbursement of expenses promised by Peter was not forthcoming, the prince found himself driven to return to his duchy and to try to raise his men's wages from the already overtaxed Gascons. A new *fouage* provoked widespread resentment and two of the greatest of the Gascon nobles, the lords of Armagnac and Albret, refused to allow its levy in their domains and, after a formal appeal to Edward III, went to Paris to lay their grievances before the king of France. Charles V hesitated. To receive this appeal would be tantamount to open repudiation of the treaty of Calais and, while secretly encouraging further appeals, he sought the advice of eminent jurists from the universities of Bologna, Montpellier, Orleans, and Toulouse. Their verdict

[1] P. E. Russell, *The English Intervention in Spain and Portugal in the Time of Edward III and Richard II* (1955), pp. 84–107.

was favourable and on 3 December 1368 the French king showed his hand in a letter declaring that he was legally entitled to receive the appeals. Edward III took up the challenge. On 3 June 1369 he resumed the title of king of France and on 30 November Charles V pronounced his French lands confiscate. The war was thus renewed.

The final chapter of its history in the fourteenth century reads as a dreary tale of ineptitude and failure. Edward III was ageing fast; the Black Prince returned home in 1371, his health broken and his reputation tarnished by the brutal siege of Limoges. In Gascony a series of hand-to-hand operations soon reduced the duchy to misery and the French were able to take advantage of the general confusion and of the incapacity of the English leaders to reoccupy most of the ceded lands. Duguesclin overran Brittany and led a successful raid on Poitiers; while the two great *chevauchées* through France led by Sir Robert Knollys and John of Gaunt yielded plunder but no military advantage. Under the guidance of Charles V the French had at last digested the lessons of Crécy and Poitiers. Apart from some harassing of the enemy's rear, they now refused offensive action, barricading themselves in their castles and walled towns which it was beyond the power of the English to reduce piecemeal. Off La Rochelle, in 1372, the earl of Pembroke's fleet was disastrously defeated by the Castilian galleys, the full significance of Charles V's alliance with Henry of Trastamara being thereby effectively demonstrated. When, after protracted negotiations, a two years' truce was concluded at Bruges in 1375,[1] only Calais and a strip of coast from Bordeaux to Bayonne were still firmly in the hands of Edward III, who died 'hauing seene all his great gettings, purchased with so much expence, trauaille and blood-shed, rent cleane from him'.[2] The accession of Richard II coincided with a renewal of French and Castilian raids on England. Rye and Hastings were burnt in 1377 and there were assaults on the Isle of Wight and on the Yarmouth herring-fisheries; it was now the turn of the English to suffer the terrors of enemy invasion. A year later came the schism in the papacy which seemed to presage indefinite prolongation of the war.

[1] E. Perroy, 'The Anglo-French Negotiations at Bruges 1374–77', *Camden Miscellany*, xix (1952).
[2] Samuel Daniel, *The Collection of the History of England* (1626) (*Complete Works*, ed. Grosart, v. 290).

Hitherto, the Avignon popes had laboured untiringly and by
no means wholly unsuccessfully in the interests of peace between
England and France. But the schism involved them in what was
virtually a war of succession and, like John de Montfort and
Charles of Blois, or Peter the Cruel and Henry of Trastamara,
Urban VI at Rome and Clement VII at Avignon found themselves in the position of rival claimants seeking for allies; and, as
neither could hope to crush his rival without military aid, both
were forced to abandon their proper task of mediation and to
encourage war, denouncing their opponents as 'schismatici ac
excommunicati et haeretici, cum quibus participare et tractare
non licet'.[1] When France and Scotland declared for Clement
VII, English support for Urban VI was assured. It was under
papal influence that Richard II's councillors rejected French
proposals for the young king's marriage to a Valois princess in
favour of an alliance with the sister of the emperor Wenceslas.
The disastrous 'crusade' of the bishop of Norwich in Flanders in
1383 was promoted by Urban as a move against the Clementists.[2]
When Anglo-French negotiations did at last begin in earnest,
the question of the schism proved a serious stumbling-block and
it was only by abandoning the 'English pope' and acquiescing
in the 'way of cession', sponsored by the French but unwelcome
to his own people, that Richard II was able to come to terms
with Charles VI.[3]

Meanwhile, the war had begun to peter out for want of
coherent direction on either side. When Charles V died in 1380,
minors ruled, and over-mighty uncles made trouble, in both
countries. Philip of Burgundy was able to use French troops to
secure his position in Flanders and, by 1386, England was
threatened with mass invasion. Thomas of Woodstock led another abortive *chevauchée* to France in 1380 and continued to
maintain a bellicose attitude in face of the pacific tendencies of
his nephew's friends. But soon after he and his supporters had
seized power in 1388, the disaster to English arms at Otterburne
forced them to accept a three years' truce with France. John of
Gaunt missed most of his opportunities in Spain and after his
return home, worked consistently for peace. When Richard II

[1] Higden, ix. 253.   [2] See below, pp. 429-33.
[3] E. Perroy, *L'Angleterre et le grand schisme d'occident, 1377-1399* (1933), pp. 352 ff.
By the 'way of cession' was understood the voluntary or compulsory withdrawal of
both popes, so as to make possible the election of an agreed candidate.

was left a widower in 1394, the way was open for the marriage alliance which seemed likely to prove its best guarantee. The reluctance of both kings to modify their demands on one another made negotiations difficult and a permanent treaty impossible: Richard still called himself king of France and refused to give up Calais; the French still claimed sovereignty in Gascony and stood by their alliance with Scotland. None the less, a truce for twenty-eight years was at last achieved; and when, in March 1396, Richard II was married by proxy to the seven-year-old daughter of Charles VI, the war had been brought to an end for the rest of the reign.

The bewildering cross-currents of the Hundred Years War in this century present the student with many problems and there is room for wide divergence of opinion on some fundamental questions. What, for instance, was Edward III's real objective? Did he seriously envisage conquest of the whole of France, the complete surrender of the French monarchy and people, and his own elevation to the throne of the Capets? Or, recognizing from the start that his dynastic claim was unrealistic, did he put it forward mainly in order to rid himself of the obligation to do homage for Gascony and to afford some kind of legal cover to the rebels in Flanders and, later, in Brittany? No direct evidence is available to enable us to answer these questions with finality. Those who contend that his real objective was the Crown of France have to meet the awkward facts of his voluntary liege homage to Philip VI in 1331 and of his readiness to relinquish the title, not only at Brétigny and Calais but in the two draft treaties drawn up in London in 1358 and 1359, when most of the cards were in his hands. There can be no doubt, however, that Edward wished his claim to be taken seriously by his own subjects; for if he were the rightful king of France, then his war could be depicted as essentially defensive; and the success of such propaganda measures may be read in the pages of the chronicles and of the parliament rolls. As the war proceeded, English successes in Gascony, Normandy, Brittany, Flanders, and the Calais region gave him effective control in large areas formerly dependent on, or forming part of, France and some scholars believe that Edward hoped to take advantage of the deep and dangerous divisions within the French kingdom to

consolidate his claim.[1] Between Crécy and Poitiers, as has already
been noted, he refused to entertain peace proposals which did
not include recognition of his title; but his refusal may have been
merely a pretext for maintaining the war, and it remains doubtful, to say the least, whether discontent with the Valois kings
would ever have reached the point of persuading the inhabitants,
even of the outlying provinces, to accept the king of England
as a permanent substitute. It may be surmised, perhaps, that
though the French Crown represented the summit of Edward's
hopes he had little expectation of attaining it, and that, in the
sphere of practical politics, his principal objectives were sovereignty in a greatly extended Gascony, destruction of the
Franco-Scottish alliance, retention of Calais as an English
possession, and maintenance of his influence in Flanders.

Edward III's capacities as a general have also been the subject of debate and the commonly expressed opinion that, though
a brilliant tactician, he was a poor strategist would appear to
need some revision. Few would deny his gifts as a leader of men.
'Next to God', wrote his seventeenth-century biographer,
Joshua Barnes, 'he reposed his chief Confidence in the Valour of
his own Subjects.'[2] This confidence was not misplaced; for the
discipline and morale of the English armies were magnificent
and without parallel in contemporary Europe and the king
who knew how to control and inspire the ranks also knew how
to choose his officers. We hear nothing of jealousy or contention
among them; their loyalty to Edward persisted to the end. The
principal commanders who served him in his first campaigns—
Henry earl of Derby, the earls of Warwick, Suffolk, and Northampton, Sir Reginald Cobham, and Sir Walter Manny—
were with him in his last.[3] The longbow was the discovery of Edward I; but it was Edward III who developed and perfected
the combinations of archers with men-at-arms which were to
shatter the famous cavalry of France. These tactics were less the
result of brilliant improvisation on the battlefield than of careful planning and rigorous training at home, directed by a man
who must have been able to take long views. Edward, we may
note, never neglected home defences when he took his armies
abroad; the victory at Neville's Cross owed much to his fore-

---

[1] See J. de Patourel, 'Edward III and the Kingdom of France', *History* xliii
(1958), 173-89.
[2] *The History of King Edward III*, p. 112.   [3] Burne, pp. 354-5.

sight. In the opening years of the French war his strategy was intelligible and, on the face of it, intelligent. English numerical inferiority must be offset by a series of alliances both inside and outside the borders of France and an English base in the north-east must balance that already existing in Gascony. The alliances proved, on the whole, unrewarding. But by 1346, when most of them had collapsed, a new field of military activity had been opened up in Flanders and, when the English landed in Normandy, the French who were already heavily engaged in Gascony and Brittany found themselves faced with a third front. The Anglo-Flemish alliance and the English occupation of Calais meant that from 1347 until settlement was reached at Brétigny, they were never free of this menace, though it was neutralized to some extent by difficulties of communication between the widely separated English armies. Hence, Edward's last campaign in 1359–60 took the form of a concerted assault on the heart of France which, though it failed to win him the Crown, enabled him to secure a victorious treaty. If the results of the treaty proved disappointing, this was partly because by this date the ravages of the free companies had greatly enhanced the difficulties of establishing effective control anywhere in France, and partly because Edward had never fully appreciated the importance of sea-power. It is here that his strategy appears most open to criticism. The victory at Sluys was not followed by any attempt to build up the nucleus of a navy which might have converted temporary into permanent control of the Narrow Seas; and there is little indication that Edward appreciated the significance, from the naval standpoint, of the Black Prince's expedition to Castile. This was indeed a serious weakness. Yet, taken as a whole, what we know of his strategy suggests that he lacked neither a capacity for large-scale planning nor the ability to learn from his mistakes which had stood him in such good stead in his dealings with his people at home.

Hardest of all to assess is the balance of profit and loss to the nation at the close of a great struggle which had consumed the energy and treasure of two generations. The terms of the truce of 1396 suggest that the net political gains were small. Except for Calais, nearly all Edward III's conquests had been lost; the Franco-Scottish alliance had not been broken; the French were once more in control in Flanders; and the question of homage for Gascony had been shelved rather than solved. The economic picture is not so clear. The wine trade with Gascony had suffered lasting damage as a direct result of the war; but the decline in the wool trade had been balanced, or nearly so, by the expansion of the cloth trade; and it has not yet been conclusively demonstrated that the people as a whole were less prosperous in 1396 than in 1337.[1] The most lasting and significant consequences of the war should be sought, perhaps, in the sphere of national psychology. Edward III's subjects put up opposition to many of his financial demands and they were fully alive to the decadence of his latter years; but few of them judged him a failure. Pride and satisfaction in the prestige which his victories had brought them endeared the king to his people and far outweighed their resentment at his exorbitant demands on their purses. In the crudely patriotic verses of Laurence Minot, in monastic chronicles, popular histories and parliamentary speeches, the same note is heard. Edward is 'oure cumly king', 'the famous and fortunate warrior', under him, 'the realm of England has been nobly amended, honoured, and enriched to a degree never seen in the time of any other king'. He and his eldest son—'the most valiant prince of this world throughout its compass that ever was since the days of Julius Caesar or Arthur'—had made the name of England known 'in heathennesse and Barbary' and had given her the reputation of a great fighting power.[2] 'When the noble Edward first gained England in his youth', wrote Jean le Bel, 'nobody thought much of the English, nobody spoke of their prowess or courage. . . . Now, in the time of the noble Edward, who has often put them to the test, they are the finest and most daring warriors known to man.'[3] For the victories were the victories, not only of the king and the aristocracy, but of the nation. The debt of the knightly classes to the skill and endurance of the common men who wielded the longbows and manned the ships was plain for all to see. War might be the sport of kings, but it was a sport from which the burgess, the yeoman, and the peasant were not shut out. National pride was no new phenomenon in the fourteenth century; but Sluys and Crécy, Poitiers and Nájera sowed the seeds of that confidence in the invincibility of

[1] See below, pp. 347–8.
[2] *Polit. Poems*, i. 66; Walsingham, i. 327; *Rot. Parl.* ii. 362; Chandos Herald, p. 135; the *Brut* (E.E.T.S.), ii. 334.
[3] *Chroniques*, i. 155–6.

England which was to sustain the spirit of her people through the darkest hours of many future wars.[1]

Less happily, the victories and, still more, the reverses bred a francophobia which died hard in England. 'It is absurd', exclaimed an English chronicler in the last decade of the century, 'that the king of England should do homage and fealty to the king of France for Aquitaine . . . since the king of France may thereby put all Englishmen under his feet.'[2] If, as has been said, the Hundred Years War was a family quarrel, it was a quarrel which finally disrupted the family. The rivalry of Valois and Plantagenet dissolved the sense of kinship which in some degree had softened Anglo-French animosities in the days of St. Louis and Henry III, or even of Charles IV and Edward II. When Richard II attempted to revive it, his uncle of Gloucester made short work of the suggestion that he should seek assistance from the king of France. To do so, he said, would mean inevitable destruction: for the king of France, 'capitalis inimicus vester est et regni vestri adversarius permaximus'.[3] Thus had he come to be regarded by 1386. Richard II's French marriage was widely unpopular in England; in his admiration for French civilization and the French court, he was out of touch with the passions and prejudices of his own people, for one of whom, William Langland, the devil himself took shape as 'a proude pryker of Fraunce'.[4] France was now the national enemy; and desire to secure her humiliation was to persist as a dominant factor in English foreign policy until after the end of the Middle Ages.

---

[1] Froissart (viii. 302) credits the Sire de Clisson with the words, 'Englès sont si grant d'euls-meismes et ont eu pour yaus tant de belles journées, que il leur est avis que il ne poeent perdre'.

[2] Higden, ix. 282.   [3] Knighton, ii. 218.

[4] Piers Plowman, B. ix. 8.

VI

EDWARD III AND ARCHBISHOP
STRATFORD (1330-43)

OPTIMISM was in the air when, at the close of the year 1330, Edward III entered on the long period of his personal rule. To the author of the tract *De Speculo Regis* his release from the tutelage of his mother looked like the delivery of the Israelites out of the house of bondage.[1] The king was in his nineteenth year, strong, brave, and handsome, husband to an amiable princess and already the father of an infant son. By a deed of romantic daring he had rid the Church and the nobility of an odious régime and his popularity with both seemed assured. The magnanimity of temperament which was to stand him in good stead throughout his life was already evident. Mortimer once disposed of, the king allowed no victimization of his adherents; not only Queen Isabella, but other leading partisans of the late earl of March were tolerated, even promoted. Adam Orleton, one of the *alumpni Iezebelae*, was entrusted with an embassy to France, and within three years he had crowned his ambitious career by acquisition of the rich see of Winchester. Henry Burghersh of Lincoln had been removed from the chancery in November, but he was in office again as treasurer by 1334 and soon became deeply implicated in the diplomatic schemes of Edward III. Other prelates of the Mortimer faction, such as Hotham of Ely and the queen-mother's favourite clerk, Wyville of Salisbury, were left in undisputed possession of their sees; and Airmyn of Norwich became treasurer three months after Mortimer's fall. Oliver Ingham was forgiven and restored to his post of seneschal of Aquitaine; Bartholomew Burghersh, nephew to the bishop of Lincoln, after a brief withdrawal, became one of the king's closest confidants and was subsequently entrusted with the important post of master of the household of the prince of Wales. Side by side with such men worked those who had assisted the king in his *coup d'état*. His most intimate personal friend, William Montagu, was rewarded with a large

---

[1] *De Speculo Regis Edwardi*, ed. J. Moisant (1891), p. 123.

share of the forfeited Mortimer estates. Summoned to parliament from 1331 as Lord Montagu, he was created earl of Salisbury in 1337; his brother, Simon Montagu, became successively
bishop of Worcester and Ely. Richard Bury, who had been in
the king's service since his childhood,[1] remained keeper of the
privy seal until 1334 when he became chancellor; by this date
he was already bishop of Durham. Two younger men who had
been concerned in the capture of March at Nottingham were
given posts of importance. William Clinton became justice of
Chester, constable of Dover, and warden of the Cinque Ports;
Robert Ufford became keeper of the forests south of Trent and
subsequently steward of the household. Both were raised to the
rank of earl in 1337, Clinton taking the title of Huntingdon,
Ufford of Suffolk.[2] Such honours signalized the king's proper
gratitude to his friends; but the older families were not forgotten. An act of royal grace restored young Richard Fitzalan
to all the lands and titles of Arundel; later, he was given the
justiciarship of north Wales and the custody of Carnarvon castle.
Thomas Beauchamp, earl of Warwick, was made hereditary
sheriff of Warwick and Leicester; his younger brothers John and
Giles Beauchamp, were influential members of the royal household. Among the new earls of 1337 were William Bohun, a
cadet of the house of Hereford who took the title of Northampton, and Hugh Audley, who attained the much coveted earldom
of Gloucester. Most important of all for Edward III were the
good relations now established between the Lancastrians and
the Crown. Earl Henry's eyesight was failing and after Mortimer's overthrow he retired from public life; but his son, Henry
of Grosmont, a friend of the king's, was created earl of Derby in
1337. John Stratford, bishop of Winchester, who had acted as
Lancaster's spokesman in the Salisbury parliament of 1328, became chancellor in November 1330 and, in 1333, archbishop of
Canterbury. His brother, Robert Stratford, a scholar of some
standing, became archdeacon of Canterbury, chancellor of the
university of Oxford, and, in 1337, bishop of Chichester. Ralph
Stratford, probably their nephew, was given the see of London
in March 1340; Henry Stratford, another kinsman, became an

---

[1] It is likely that Bury had acted for a short while as Edward's tutor. See N.
Denholm-Young, 'Richard de Bury, 1287–1345', *Trans. Royal Hist. Soc.* 4th ser. xx
(1937), 139–40.
[2] Both had already received individual writs of summons to parliament, Clinton
from Sept. 1330, Ufford from Jan. 1332.

important chancery clerk. William Trussell, the Lancastrian official who had played a leading part in the revolution of 1326–7 and in the rebellion of 1328–9, was restored to the office of escheator south of Trent. Thus, the thirties saw the gradual obliteration of the groupings and alignments of the previous decade. Edward III and his friends 'led their young lives in pleasant fashion',[1] engaging in the warlike sports which delighted them, redeeming the national honour in Scotland, and laying their plans for still greater exploits in France. There was still much disorder in the country and much grumbling at the misdeeds of purveyors and other minor officials; but the monarchy was re-establishing itself in the favour of the people, and the ghost of the Ordainers appeared to be laid.

The perennial problems of the medieval ruler—maintenance of an efficient administration and of an adequate flow of supplies—inevitably grew more pressing in time of war. But so long as war was confined to Scotland, administrative difficulties did not prove insuperable. The departments of state and of the household could follow the king to the north and the executive thus remain accessible to his personal direction. From the time of Dupplin Moor until war with France became imminent (1332–7), York was, in effect, the administrative capital of the country. The king and his household lodged in the convent of the Grey Friars, the chancery in the abbey church of St. Mary's, the exchequer and common bench in York castle, other departments in houses built or requisitioned for the purpose. For the greater part of this period, the king's friend and principal counsellor, John Stratford,[2] acted as chancellor and, with amenable officers in other key positions, the governmental machine ran smoothly enough.[3] But even a minor war rendered the financial problem more acute. The ordinary revenue did not normally exceed £30,000 a year and, like his two predecessors, Edward III found continuous borrowing inescapable. The customs afforded the most important regular security for repayment of

---

[1] Scalacronica, p. 158.

[2] 'qui dicti domini regis patricius solebat quasi ab omnibus nominari', Avesbury, p. 324.

[3] J. Stratford was chancellor until Mar. 1337, except for a period of nine months (Sept. 1334–June 1335), when his place was taken by R. Bury. The treasurers were Melton (1330–1), Airmyn (1331–2), R. Aylesdon (1332–4), R. Bury (Feb. to Aug. 1334), H. Burghersh (1334–7). The keepers of the privy seal were Bury (1330–4), Aylesdon (Mar.–Apr. 1334), R. Tawton (1334–5), and W. Zouche (1335–7).

royal debts; but they had to be supplemented at intervals by grants of direct taxation, both ecclesiastical and lay, which would enable the king to discharge his outstanding liabilities. Other devices for raising money might be used;[1] but for the credit system to function properly it was essential for the king to attract powerful lenders to his service and at the same time to keep on terms with parliament and convocation.

Edward III made a good start. Loans from the Bardi and others were regular but not excessive, averaging between some £12,000 and £20,000 a year, with an interest charge of some £4,000. Parliament met frequently, sometimes at York, and taxation was kept within bounds, though attempts to increase the wool custom aroused protests. The king collected an aid from the clergy for the marriage of his sister in 1333, but parliamentary opposition led him to revoke an order for tallage issued the previous year. Tenths and fifteenths were granted in the autumn parliaments of 1333 and 1334; in May 1335 a grant of hobelars (light horsemen) and archers was redeemed for a money payment. From the beginning of 1336, however, pressure of direct taxation increased. A grant of a tenth and fifteenth in the March parliament was followed by another in the great council at Nottingham in September and by two tenths from the clergy. A great council at Westminster in September 1337 gave a tenth and fifteenth for three years;[2] and the convocations of Canterbury and York did likewise. The king was thereby assured, in principle, of an annual income of some £57,000 from subsidies for the period 1337–40. But much more was needed; for eight of the leading allies in the Netherlands had been promised payment of sums totalling £124,000 before the end of 1337, the cost of the projected expedition to the continent had still to be met, and much of the anticipated yield of the first year's subsidies had already been assigned. An increase in the wool tax might offer a partial solution; and an assembly of merchants meeting simultaneously with the great council at Nottingham in

---

[1] One was the exaction of fines for special favours; e.g. in 1334 the county of Kent offered 1,000 marks to be freed from an eyre. The abbot of Bayham refused to contribute on the grounds that the money was not granted in parliament and that he had never assented to the grant. J. G. Edwards, 'Taxation and Consent in the Court of Common Pleas', *Eng. Hist. Rev.* lvii (1942), 473–81.

[2] This grant cancelled those made earlier in the year by individual counties under strong pressure from royal commissioners. J. F. Willard, 'Edward III's Negotiations for a Grant in 1337', ibid. xxi (1906), 727–31.

1336 had granted such an increase in the form of a subsidy of
20s. and a loan of 20s. on the sack, whereby the king might hope
to raise at the utmost another £70,000 a year. Even this was
inadequate, however, and Edward was driven to seek some
other means of exploiting wool, 'the sovereign treasure of the
realm', to his own advantage. By the following summer he had
struck a bargain with William de la Pole of Hull and Reginald
Conduit of London, elected representatives of the principal
wool contractors.[1] In return for a monopoly of export which
would cut out their rivals, the foreign buyers, the merchants
agreed to buy and export 30,000 sacks of wool for the king's use,
the growers to allow them six months' credit for half the amount
owing, twelve months for the other half, the price of all but the
inferior wools being fixed at a fair minimum. There was plenty
of wool in the country; for a strict embargo on its export to the
Netherlands had been maintained since August 1336, thereby
creating a scarcity abroad by which the merchants might hope
to profit, since they would be in a good position to dictate the
price of the wool they sold when the embargo was lifted. Half
of this profit was to go to the king, the merchants receiving
security in the shape of assignments on the wool customs; and
by this means the king expected to raise ready cash to the
amount of £200,000. But the 'ruthless and unscrupulous financiers' with whom the bargain had been made had little in common with the humbler members of the new Wool Company,
many of whom resorted to smuggling and other dishonest practices; and soon after the first consignment of perhaps 10,000
sacks had reached Dordrecht, differences of opinion between the
merchants and the representatives of the government began to
make themselves felt. A conference at Gertruidenberg in December 1337 resulted in a deadlock; and in February the king began
to take over the whole of the wool lying at Dordrecht, offering
the owners notes of acknowledgement to the value of £65,000
(representing the good price of £10 per sack), which authorized
them to recoup themselves through remissions of duty payable
on future exports. Edward's action was less arbitrary and less
sudden than has generally been supposed. A second agreement

---

[1] A revised view of the monopolistic arrangements and of the reasons for their
breakdown has been made necessary by the researches of Dr. E. B. Fryde who has
kindly allowed me access to his doctoral thesis, 'Edward III's War Finance 1337-
41' (Oxford, 1947). Some of Dr. Fryde's conclusions are summarized in *History*,
xxxvii (1952), 8–24.

made with the contract merchants before the first wool-fleet had
sailed had allowed government representatives free disposal of
its cargoes; and the taking over of the wool occupied at least
four months. It is undeniable, however, that the king showed
much bad faith in the execution of his agreements with the merchants which, if he had observed them, were bound to diminish
his revenue from the customs. Periodic embargoes on exports
prevented the merchants from making good their claims; and
many of the owners of the confiscated wool sacks were driven to
sell their 'Dordrecht bonds' to syndicates of financiers specially
empowered by Edward III to buy them up at a discount.[1] The
resultant divergence of interest between these financiers and the
ordinary merchants ended in the destruction of the former and
the disappearance of the extra-parliamentary 'estate of merchants'. For the king himself, the collapse of his plan to split the
profits of the wool trade with a syndicate proved disastrous.
His subsequent bankruptcy, as Stratford was to remind him,
was largely the result of it. From the parliament of February
1338 he received some kind of authorization for pre-emption of
half the wool in the kingdom (estimated at 20,000 sacks), on the
understanding that the remaining half should be left at the free
disposal of his subjects; and on the security of this new grant he
arranged with the Bardi and the Peruzzi for substantial loans.
But the proceeds came in very slowly, and it was in vain that the
king urged speedy payment of the subsidies and tried to persuade boroughs and townships to compound in advance for the
second and third years. When he arrived at Antwerp in July, it
was to discover that of the 20,000 sacks, not more than 2,500 had
been shipped.[2]

Meantime, it had been necessary to meet the administrative
situation created by the prospect of the king's removal overseas.
Shortly before his departure, two of the principal offices of state
changed hands. Richard Bentworth, bishop of London, formerly keeper of the privy seal, became chancellor and William
Kilsby became keeper of the privy seal. Both were household
clerks of long experience, as was also John Charnels who com-

[1] G. O. Sayles, 'The English Company of 1343', *Speculum*, vi (1931), 177–205.
It was said later that only one to two shillings in the £ had been obtained by the
owners, *Rot. Parl.* ii. 169, 170.
[2] G. Unwin, *Finance and Trade under Edward III* (1918), pp. 184–99; Dorothy
Hughes, *The Early Years of Edward III* (1915), pp. 26–39.

monly acted as deputy for the treasurer, Robert Wodehouse, From Walton-on-the-Naze, where he was about to embark. Edward issued a series of ordinances embodying new administrative arrangements for a nation at war. The salient point of his plan was the vesting of effective control of the executive in the privy seal, henceforward to be regarded as the sole instrument for notification of the personal wishes of the sovereign and as an instrument for control of both chancery and exchequer. The Walton Ordinances envisage the necessity of dividing the council in wartime, but they make it clear that all payments by the home government must ultimately have privy seal warrant; moreover, the chamberlains of the exchequer are to account annually to a small committee of audit consisting of a bishop, a banneret, and a clerk, under the supervision of a 'wise, sufficient and knowledgeable man' appointed by the king. The discretion of the exchequer is still further limited by a series of regulations which reflect the financial stringencies of the hour. All exemptions from taxation are cancelled, respite of debts to the Crown and their payment by instalments are disallowed, while at the same time the king repudiates all obligations incurred by his predecessors, pending settlement of those that have arisen since his accession. An ordinance transferring the duty of electing sheriffs and other officers from the exchequer to the shires and boroughs, who were to answer 'at their peril' for the men they chose, suggests a desire to force the local communities to take responsibility for their own administration.[1]

The motive force behind the Walton Ordinances and the promotion of Bentworth and Kilsby was the desire for greater speed and efficiency, particularly in the raising of supplies for the war. Some personal animosities may have played their part but, if so, it was a minor one. Edward III was seeking neither to eliminate the magnates from his counsels nor to follow the example of his father in surrounding himself with courtiers and favourites of inferior rank. His mind was on the war, on the satisfaction of his expensive allies, and on the payment of his soldiers' wages. Without emergency powers vested in officials under his direct control, there seemed to be little hope that he could realize his projects abroad. That the assumption of such powers must inevitably entail attacks on vested interests and

[1] The Ordinances have been analysed in detail by Tout, *Chapters*, iii. 69-79; for the text, see ibid., pp. 143-50.

customary rights, and was therefore likely to frustrate its own
purpose, was not obvious to Edward in the summer of 1338, preoccupied as he was with his financial anxieties. He was already
heavily in debt and as his diplomatic schemes matured his debts
mounted; within three months of his arrival in Brabant he had
borrowed another £100,000.[1] He was looking hourly for delivery
of the wool consignments by which alone his necessities could
substantially be relieved; and he was hoping to anticipate the
direct taxes already granted and to obtain others. The Walton
Ordinances were proof of his determination to secure administrative efficiency. The eight-year-old duke of Cornwall was
appointed nominal regent; he was to be lodged in the Tower
and attended by the earls of Arundel and Huntingdon and by
Lord Ralph Neville. The exchequer was brought back from
York to Westminster where William Zouche succeeded Wodehouse as treasurer;[2] and the common bench soon followed.
Chancellor Bentworth, furnished with a special seal of absence,
remained at home; custody of the great seal was entrusted to
the keeper of the privy seal, William Kilsby, who went abroad
with the king. Already in the Low Countries were the king's
friends, the earls of Salisbury and Northampton, the former
chief justice, Geoffrey Scrope, the bishops of Lincoln and Durham, and the archbishop of Canterbury. Cadets of noble houses
like Reginald Cobham and the Beauchamp brothers, the gallant Hainaulter, Walter Manny, and a number of household
knights including the steward, John Darcy, the chamberlain,
Henry Ferrars, and the surveyor of the chamber, John Moleyns,
reinforced the king's council overseas. As a gesture of confidence
in his allies, Edward appointed his wife's brother-in-law, the
margrave of Juliers, 'privy and very special sovereign secretary'
of the council, naming as his particular associates the bishop of
Lincoln, the earl of Salisbury, Kilsby, and Scrope, and promising to abide by their advice.[3]

It was almost inevitable that a state of cross-purposes should
arise between the king's advisers in the Netherlands and the
government at home. The concern of the former was solely with

---

[1] E. B. Fryde, 'Dismissal of Robert de Wodehouse from the Office of Treasurer,
December 1338', *Eng. Hist. Rev.* lxvii (1952), 75.
[2] Two letters from Wodehouse to Sir John Moleyns (ibid., pp. 77–78) show that
Wodehouse was probably the first victim of the king's dissatisfaction with the home
government.
[3] Tout, *Chapters*, iii. 99–100. (William V of Juliers married Jeanne of Hainault.)

the French war; the latter had to take account of Scotland and of routine administrative expenses. In an age of personal monarchy the absence of the king almost always resulted in some slackening of effort, for the royal authority was difficult to exercise vicariously. Thus, despite the Walton Ordinances, there was little speeding-up of exchequer procedure. On the contrary, resentment at the annulment of exemptions from taxation and of permission to pay debts by instalments found open expression within a fortnight of the king's departure, when, at the council of Northampton, several of the magnates declared that such changes ought not to be made without the consent of the magnates, and that in parliament.[1] Arbitrary revocation of many assignments on revenue agitated the home government, as did the directive sent from Antwerp ordering suspension of the annual fees of all officials possessed of other sources of income.[2] This order was ignored and when the king pressed for its execution he was informed that his officers were threatening mass resignation. Attempts to gather in the wool met with little more success. A staple was organized at Antwerp under William de la Pole, but much of the wool which was received there was damaged and the king was driven to yet more reckless borrowing. Orders to the council at home to explore other possibilities —the probable yield of a scutage, of an aid for knighting the king's son, or of a tallage of the royal demesne—evoked no response, doubtless because it was beyond the capacity of the ministers to answer such questions; a stream of angry letters from Antwerp produced only deferential but evasive replies from Westminster. Edward seems finally to have reached the sensible conclusion that improvement was not to be looked for until he put a strong man at the helm at home and allowed him a measure of independence. On Michaelmas Day 1339 the archbishop of Canterbury was accordingly appointed 'principal councillor to Edward, duke of Cornwall, Keeper of the Realm', and, together with the bishop of Durham and William de la Pole, was empowered to proceed to England and lay the king's necessities before a full parliament.[3]

Edward III's choice of John Stratford for this position of trust proves that there was as yet no hint of a rift between them; but

---

[1] J. F. Baldwin, *The King's Council* (1913), p. 478.
[2] *Cal. Close Rolls 1337-39*, p. 467.
[3] *Cal. Pat. Rolls 1338-40*, p. 394; *Foedera*, ii. 1091.

it is just possible that the archbishop was already beginning to
be troubled by scruples as to the propriety of the king's methods,
if not of his ultimate aims. It was now nearly two years since the
cardinal legates of St. Praxede and St. Maria de Aquiro had
summoned Stratford and his suffragans to the house of the London Carmelites in order to inspect bulls directed against all who
should break the peace of Christendom. In September 1338 the
archbishop had received a personal letter from the pope, urging
him to use his influence with Edward III to persuade him to
drop the imperial alliance; and in the following summer he and
the bishop of Durham had been with the cardinals on a mission
to the French court. A new watchfulness for the interests of the
clergy is apparent in the letters which he issued to the bishops
of the province of Canterbury, warning them that the liberties
of the Church were in danger as the chancellor was attempting
unlawful taxation of ecclesiastical persons.[1] But if doubts were
forming in Stratford's mind, their only effect at this stage was to
stiffen his resolve to protect his clergy from exploitation. He was
not prepared to challenge the king directly, still less to sacrifice
his career as a statesman to the pacifist demands of the papacy.
Thus, when parliament met at Westminster in October he took
the leading part in expounding the king's needs to the assembled
lords and commons.[2] Edward's long sojourn abroad without
engaging the enemy had been due, the emissaries explained, to
lack of funds; but he was now in the neighbourhood of St. Quentin with a great army and the prospects for victory were good,
provided that he could be furnished with supplies to enable him
to discharge his obligations to his allies and to meet the expenses
of the campaign. There was unanimous agreement that he must
be helped, and a tithe for two years, modelled on that customarily
paid to the Church,[3] was suggested. But before making a firm
offer the commons demanded a dissolution to allow time for
consultation with the local communities; and they and the
lords combined to seek abolition of the *maltolte* on wool, the
commons pointing out that it had been levied without their consent. There were further demands for a general pardon for past

    [1] *Reg. R. de Salop* (Somerset Record Society, ix, 1895), pp. 357–8.
    [2] *Rot. Parl.* ii. 103–6.
    [3] 'en la manere quele il les donent a Seinte Eglise.' The king's debts were esti-
mated at £300,000, or more. Professor Prince (*Speculum*, xix (1944), 147) described
the campaign of 1338–9 as the most expensive English expedition in the whole
medieval period.

offences and for exemption from aids, prises, and old debts, in short, for substantial modification of the Walton Ordinances. When the new parliament assembled in January 1340 the magnates confirmed the grant of the tithe, i.e. of the tenth sheaf, fleece, and lamb from the demesne lands; but the commons, after long debate, offered instead a grant of 30,000 sacks of wool, subject to certain conditions which, since they 'touched so nearly the estate of our lord the king', the council decided to send to Antwerp for his personal consideration. As an immediate levy of some kind was essential, in order to equip a fleet for the defence of the realm against the threatened French invasion, the commons, under pressure, agreed to an unconditional first instalment of 2,500 sacks. So meagre a grant was of little avail; and even before news of it reached him Edward had taken a way out of his immediate difficulties by borrowing what he wanted from Pole on the security of the customs. This was not the act of a prudent financier, but the position was desperate. The campaign of the Thiérache had achieved nothing; and, though the long-wished-for Flemish alliance had been secured and Edward had had himself proclaimed king of France in the market-place at Ghent, this served only to increase his commitments in the Low Countries.[1] He now had little choice but to return home and try to extract from parliament a tax, or taxes, which would serve as security for further loans. His credit had sunk so low that he was obliged to leave the queen, his children, and the earls of Derby and Salisbury behind him, as hostages for the payment of his debts.

Edward landed at Harwich on 20 February 1340 and forthwith issued writs for a parliament to meet on 29 March. When it assembled, an unnamed speaker expounded the cause of summons to the magnates *en especialte* and to the commons *en generalte*.[2] Unless a large aid were forthcoming, the king would be dishonoured for ever, his lands and his allies would be lost, and he must return to the Netherlands and remain there as a prisoner until he had paid his debts. This poignant appeal and the presence of the king himself had the desired effect. Within a couple of days lords and commons together had agreed to grant the ninth sheaf, fleece, and lamb from their lands and a ninth of the goods of citizens and burgesses, for the years 1340 and 1341;

---

[1] e.g. he contracted to pay £60,000 to Bruges, Ghent, and Ypres by Pentecost 1340.   [2] *Rot. Parl.* ii. 112.

from those classes which would not be liable for the ninth, a fifteenth of the real value of their goods would be taken. But the schedule of the previous parliament had not been forgotten and the grants were made conditional on favourable answers to petitions, now to be laid before the king and his council; these Edward referred to a committee under the presidency of the archbishop, which included, besides prelates, lay peers, and officials, twelve knights and six representatives of the towns. From their deliberations there emerged four statutes covering a wide range of fiscal and administrative matters, among them, abolition of the *maltolte* and of all unparliamentary aids or charges (a clause which gave tallage its death-blow), pardons for certain debts and fines, appointment of sheriffs in the exchequer (a further modification of the Walton Ordinances), limitation of purveyance and cessation of the obsolete procedure of 'presentment of Englishry'.[1] In the third statute it was laid down that the king's assumption of the title of king of France should never entail the subjection of Englishmen to the French Crown; in the fourth (conceded at their request) the clergy were protected against abuses of royal rights, especially during vacancies.[2] As the price of these substantial concessions, the estates forthwith allowed modification of the statutes to the extent of authorizing the collection of the *maltolte* for a further term of fourteen months. Small committees, on most of which Stratford was prominent, were set up to confer with groups of merchants and to take measures for defence; and those who had received money or goods on the king's behalf were asked to present their accounts.

The recorded proceedings of these three parliaments, of October 1339 and of January and March 1340, afford clear evidence of general uneasiness. The king had gained his main objective in the form of generous grants; but his financial predicament, his unprecedentedly reckless borrowing, was making the lords wary and stimulating the commons to unwonted political activity. Edward had been forced to modify the Walton Ordinances; the determination of the tax-paying population to protect itself against all forms of arbitrary exaction was unmistakable; and, when a demand for a loan of £20,000 from the

---

[1] Obsolete because it had become impossible to distinguish an Englishman from a Norman. The effect of this statute was to relieve the hundred of the murder fine, originally imposed when a Norman was found slain within its boundaries by an unknown hand.

[2] *Statutes* 14 Edw. III, st. 1–4.

City of London yielded only an offer of 5,000 marks, pushed up under pressure to £5,000, it could be seen that the nation was in dangerous mood.[1] But, however much Edward may have resented his enforced concessions, there was little he could do, and he was still content to put his trust in Stratford to whom he restored the great seal on 28 April. As chancellor and president of the council, Stratford was now the minister chiefly responsible for ensuring, on the one hand, that the recent statutes were observed and, on the other, that the king received his supplies; while, as archbishop, he could not escape the obligation to protect his clergy, whose grant of a biennial tenth in the convocation of February 1340 had been accompanied by the condition that they should not be burdened simultaneously with any further contribution. It was ominous enough that, on 18 April, the king should have found it necessary to write to Benedict XII repudiating scandalous rumours that he was an oppressor of churchmen.[2] Meantime, however, the cumbrous machinery of tax-collection was set in motion. Receivers and supervisors in every shire were appointed to assess the value of the ninth sheaf, fleece, and lamb which were to be collected and sold locally, and of the ninth of movables in cities and boroughs. Robert Sadington, chief baron of the exchequer, replaced Zouche as treasurer, and two receivers of the ninth were appointed for the districts north and south of Trent. A tax on corn and wool could not be gathered in till after the harvest and the shearing; but the king naturally did not hesitate to make large assignments on its potential receipts. On the eve of Edward's departure, Stratford's apprehensions got the better of him and he seems to have wished to throw in his hand. Fearful of the consequences, chiefly, no doubt, of the financial consequences of the loss of the English fleet, he tried to dissuade the king from sailing and, when Edward refused to listen to him, resigned the seal and his membership of the council on the plea of ill health.[3] But Edward still had no wish to quarrel with his archbishop. He accepted his resignation, granted him certain immunities, put his brother, Robert Stratford, in his place as chancellor and, after boldly ordering his receiver to pay war-wages to the amount of £6,000, sailed out of Orwell to find the French.

[1] Cal. Letter-Book F, pp. 45-46.
[2] E. Déprez, Les Préliminaires de la guerre de cent ans, p. 304, n. 4.
[3] Avesbury, p. 311.

The English victory at Sluys proved some of the late chancellor's fears to have been groundless; but it in no way alleviated Edward's financial predicament and a situation similar to that which he had returned home to remedy soon reasserted itself. Once again, the king found himself virtually a prisoner in a foreign town, waiting for supplies which did not arrive. Though Stratford was no longer chancellor, he continued to direct the regency and, subject to his care for the interests of the clergy,[1] it is likely that he did his best. But he was struggling against impossible odds. Attempts were made to stimulate the collectors of the ninth to a sense of the urgency of their task though it was already becoming clear that the sheaves, fleeces, and lambs could not be sold at the high prices desired by the government; and in the parliament of July 1340 there was discussion as to the best means of meeting this difficulty. It was far from encouraging to discover that, despite the magnitude of his existing debts, the king now had fresh enterprises in mind. Arundel, Gloucester, and William Trussell, who appeared with royal letters describing the victory at Sluys, announced his intention to lay siege to Tournai and once more stressed his urgent need of funds. Mollified to some extent by the news of the victory and recognizing that (since Stratford and the councillors were unwilling to lower the minimum price) the proceeds of the ninth could not be realized immediately, the estates allowed the king a loan of 20,000 sacks of wool, on the understanding that the owners would be compensated from the proceeds of the subsidy; the failure of the similar project of February 1338 seems to have aroused no qualms. Arrangements were worked out with the merchants, and the king was told that the profits of this grant would be in his hands as soon as possible, the money to be paid into the wardrobe at Bruges on various dates during the late summer and autumn. But when nothing had been received by the middle of September and the rich prize of Tournai had slipped from his grasp, Edward had no choice but to agree to the unwelcome truce of Espléchin. His allies of Brabant and Hainault had withdrawn and his Flemish creditors were pressing him hard; in a letter written from Ghent on 31 October, the king hinted plainly that if all those responsible had done their

---

[1] In July the council under Stratford's presidency issued orders suspending the collection of the ninth from all men of religion not summoned to the last parliament and therefore not consenting to it. *Cal. Close Rolls 1339–41*, p. 613.

duty he might have won greater honour abroad.[1] The ministers were in a dilemma. The Bardi and the Peruzzi, to whom the subsidy from eight counties had been assigned, were waiting to lay hands on it, for the £125,000 which they had advanced to the king had exhausted them; the proceeds from eleven other counties had been assigned to other creditors; only eight remained for the king; and meanwhile the tax-collectors had to face 'conspiracies and false covins of certain wicked barretors'.[2] According to one source, the ministers warned the king that, rather than make further contributions, the people would revolt;[3] the sheriff of Essex had orders to deal with those of the county who 'refuse to pay the ninth . . . but resist the vendors and assessors in the levying and sale thereof to the utmost of their power'.[4] The wool grant met with no better fortune. By the end of October it was known that in twelve counties nothing had been raised; alien merchants were said to be combining to get control of most of the wool of the realm and attempts to investigate the conduct of customs officers proved useless.[5] Baulked of his victories, harassed by his debts, exasperated by his officials, and (to a degree which still remains somewhat inexplicable) with his mind poisoned against Stratford personally, Edward determined to return home without notice, to remove those officials whom he believed to be betraying their trust, and to make a supreme effort to rehabilitate his finances.

The true causes of the crisis which had thus been reached were not altogether as they appeared to the king. He himself was a principal offender, for he had exhausted his credit, the policy of subsidizing selfish and lukewarm allies was beyond the capacity of the nation to sustain and was in no way worth the effort it entailed, and Edward had been unwise enough to embark on a fresh campaign, lacking the means to pay for it and seemingly ignoring the difficulties and delays that must inevitably attend his ministers' efforts to realize the taxes which had been granted. Yet the dilemma in which he found himself was real enough. To abandon his continental enterprises would have cost him a loss of prestige both at home and abroad which, despite his victories in Scotland and at Sluys, he could not yet afford. It was vital to retain the friendship of the Flemish towns; and he was

[1] Déprez, pp. 355–6.   [2] *Rot. Parl.* ii. 117.
[3] *French Chronicle of London*, p. 83.   [4] *Cal. Close Rolls 1339–41*, p. 536.
[5] Hughes, p. 95.

not wholly unjust in holding his officials to blame for the difficulties which confronted him. There is no reason to suppose that Stratford or any of his colleagues were traitors; but both energy and intelligence had been lacking in their handling of the royal finances. Their refusal to lower the price of the sheaves, lambs, and fleeces had deprived the king of the anticipated profits of the ninth; their adoption of the wool levy in July 1340 had been a bad blunder; and from the end of that month they had shown a tendency to let matters drift.[1] The exasperation of those who had the king's ear at Ghent is understandable. War-captains like Salisbury, Manny, and their followers, had been cheated of victory and they, together with Kilsby who held Stratford responsible for his failure to secure the archbishopric of York, may have been the prime movers against the minister. But this cannot be the whole story. Stratford must have made other enemies while he was abroad. A worldly prelate's criticisms of lax morals at court were unlikely to be well received;[2] and there was inevitable jealousy of his influence with the king and, perhaps, suspicion that his exchanges with the cardinals had tempered his zeal for the war. Whatever the cause, there can be no doubt that at Ghent hostility to the archbishop was little short of murderous;[3] and the London chronicler suggests that at home the courtiers had an ally in the council (possibly, Sadington or Parvyng, neither of whom was a victim of the king's subsequent purge), who advised Edward's speedy return.[4] Edward himself had been converted to these views. His affection for his old mentor had turned to bitter hatred and on 18 November he dispatched a confidential mission to Avignon with instructions to inform the pope that his faith in Stratford had been misplaced, that the archbishop had deliberately withheld supplies in order to achieve his ruin and even his death, and—a mysterious allusion which may possibly afford the real reason for the king's change of front—that he had done his best to deprive him of the affection of the queen.[5] Much that Edward wrote was patently unjust; but rage and frustration had clouded his mind and the stage was now set for a trial of strength.

[1] These points have been fully developed by Dr. Fryde in the dissertation referred to above, p. 156, n. 1.

[2] The Canterbury chronicler ascribes Stratford's influence with the king partly to the need for 'effoeminatorum in Regno Angliae dominationem . . . reprimendam', Birchington, p. 20. [3] Ibid.

[4] *French Chronicle*, p. 83. [5] Déprez, p. 352.

Edward's dramatic arrival at the water-front of the Tower
on St. Andrew's night (30 November) was followed by a
drastic, albeit selective, purge of the administration.[1] While
the archbishop prudently withdrew to Canterbury, the mayor of
London was ordered to arrest the principal ministers. Robert
Stratford and Roger Northburgh both lost their posts, only their
episcopal orders saving them from imprisonment. The privy
seal remained in the hands of Kilsby; but the new chancellor
and treasurer, Bourchier and Parvyng, were laymen and lawyers,
for the king had declared his intention to have as his ministers
men justiciable in his courts.[2] The constable of the Tower, Sir
Nicholas Beche, who was found absent from his post, Sir John
Stonor, chief justice of the king's bench, four judges of the common bench, a number of lesser officials and some prominent
merchants, including William and John de la Pole and Reginald
Conduit, all suffered arrest.[3] Writs of *oyer et terminer* were issued
for the trial of the ministers and judges, general commissions of
trailbaston[4] covered misconduct of lesser officials and of other
persons throughout the land. London alone received different
treatment. An attempt to extend the commission of trailbaston
to the city was resisted as a violation of liberties 'allowed beyond
the memory of man' and a writ for a general eyre at the Tower was
substituted.[5] Great severity was exercised almost everywhere.
Judge Willoughby protested vigorously against a trial deriving
from irregular bills and popular clamour; but in the end, perhaps because he knew himself to be not altogether blameless,
he declared that 'he would not plead with his liege lord' and

[1] The principal contemporary sources for the ensuing conflict are the *French
Chronicle* of London, the *History of the Archbishops of Canterbury*, wrongly ascribed to
Birchington (*Anglia Sacra*, i. 19–41), Murimuth, Avesbury, Hemingburgh, and the
further correspondence printed in *Foedera* ii. 1143 ff. For critical discussions see
Hughes, pp. 100–81; G. T. Lapsley, *Eng. Hist. Rev.* xxx (1915), 6–18, 193–215;
Tout, *Chapters*, iii. 118–42; Maude Clarke, *Fourteenth Century Studies*, pp. 128–30,
242–57; B. Wilkinson, *Eng. Hist. Rev.* xlvi (1931), 177–93; and *Const. Hist.* ii.
176–87.

[2] *French Chronicle*, p. 90. The new appointments were dated 14 and 15 Dec. but
Bourchier and Parvyng are described as chancellor and treasurer in the commission
of *oyer et terminer* dated 10 Dec. Bourchier was the first lay chancellor. Parvyng replaced Stonor as chief justice of the king's bench but early in January this office
passed to William Scot.

[3] I learn from Dr. Fryde that Pole, who had raised £100,000 for the king, was
refusing further loans.

[4] 'Trailbaston' was a nickname for any commission with unusual powers, especially to impose heavy financial penalties.

[5] *Cal. Letter-Book F*, pp. 59–60.

was thrust into the Tower.[1] Maladministration was alleged
against officials high and low, ecclesiastical and lay, from judges,
chancery clerks, and archdeacons, down to customers, arrayers,
and bailiffs, special stress being laid on smuggling offences, maintenance, and assault.[2] Six escheators and twelve sheriffs were
deprived in the opening months of 1341 and writs sent out for the
election of new coroners in every shire. Had the inquiries been confined to offending officials, they might have commanded some
popular sympathy; but the commissions of trailbaston covered
all breaches of the peace and an attempt was made simultaneously to enforce payment of taxes long overdue. On
26 January orders were issued for an assessment by inquest in
every parish of the true value of the ninth. In London, though
the general eyre was finally bought off for a fine of 500 marks,
the presence of the judges at the Tower gave rise to violence.[3]
Such was the harshness of the commissioners, says Murimuth,
that guilty and innocent alike had to buy themselves freedom
from imprisonment; the customary processes of purgation were
not allowed.

Meanwhile, the king and the archbishop were engaged in a
protracted battle of words. On 4 December Edward sent Sir
Nicholas Cantilupe to Canterbury with a request to Stratford to
come to London, prepared to go abroad as a hostage for the
debts due to the merchants of Louvain. Stratford asked for time
and it was not until 29 December, the feast of St. Thomas the
Martyr, that he made his first public declaration. Choosing the
pulpit of the cathedral as his rostrum, he treated his congregation to a 'long and various discourse' in English. Criticisms of his
failings as a bishop were forestalled by self-accusation; he blamed
himself for his long preoccupation with secular affairs, for his
share in oppressing the clergy and community of England and
for his payment of tithes to the king; he was not unfamiliar with
the history of the martyr of Canterbury.[4] After promising

---

[1] *Year Book*, 14 and 15 Edw. III, ed. L. O. Pike (Rolls Ser. 1889), p. xxvi; *French
Chronicle*, p. 87. Cf. T. F. T. Plucknett, 'The Origin of Impeachment', *Trans. Royal
Hist. Soc.*, 4th ser. xxiv (1942), 64–68, and Hughes, pp. 83–84.

[2] *French Chronicle*, pp. 88–89.

[3] *Cal. Letter-Book F*, p. 92; Murimuth, pp. 118–19.

[4] If, as seems probable, the sermon published by W. D. Macray in *Eng. Hist.
Rev.* viii (1893), 87–91, was preached on this occasion, Stratford made some points
likely to come home to Edward III. He suggested that the loss of Henry II's overseas dominions had been a just judgement on his treatment of Becket and hazarded
a connexion between the English defeat at Bannockburn—'et hoc a gente quam pro

amends for the future, the archbishop pronounced sentence of excommunication on all, the royal family alone excepted, who should violate the Great Charter, the peace of the land, or the liberties of the Church; the bishops were ordered to see to the publication of the sentence in all parish churches.[1] Stratford must already have drafted the long reproachful letter to the king which is dated 1 January; but Edward can hardly have received it when he sent Lord Stafford to Canterbury to ask for the articles of the excommunication and for Stratford's attendance. As both requests were refused, the king waited some weeks before trying again. On 26 January he sent Cantilupe with a further summons and a safe-conduct—a move unlikely to inspire confidence since the sheriff of Kent had simultaneously been ordered to produce the archbishop to answer for his contempt of the king's commands. If Stratford was aiming at the role of a second Becket, Edward, it seems, was not unwilling to play the part of Henry II.[2] So far the king had written nothing, but Stratford now followed up his letter of 1 January with others addressed to the chancellor, to the king, and his council, and to the bishops.[3] Edward's supporters had meantime been preparing his case. On 10 February he issued writs prohibiting the bishops from publishing ecclesiastical censures against the tax-collectors and, on the same day, he sent to sixteen bishops, and eleven abbots and priors, the lengthy justification of his actions to which Stratford gave the name of 'the infamous tract' (*libellus famosus*).[4] A week later Kilsby, *principalis incentor discordiae*, appeared at Canterbury in the company of some merchants from Brabant.[5] Writing to the earl of Huntingdon, a good friend of his house, the prior of Christ Church told of the efforts of these merchants to obtain an interview with Stratford and of how, after he had dined at the prior's expense, Kilsby made public proclamation at the priory gates, summoning Stratford to Brabant as security for certain debts.[6] Nothing came of these

nichilo habebant, scil. a Scotis'—and the injuries done by secular princes to the Church.

[1] *Reg. Grandisson*, ii. 933.

[2] In 1164 Henry II had deliberately insulted Becket by citing him to a council through a writ directed to the sheriff of Kent, FitzStephen, *Vita Sancti Thomae*, p. 51.

[3] Hemingburgh, ii. 367-75.

[4] *Foedera*, ii. 1147-8; Hemingburgh, ii. 380-8; Avesbury, pp. 330-6; Birchington, pp. 23-27.

[5] Ibid., pp. 21-22.   [6] *Lit. Cant.* ii. 226-30.

manœuvres; and on Ash Wednesday (21 February) the archbishop, who seems to have been taken into some kind of protec-
tive custody for a time, again preached in his cathedral. After
the sermon, he caused the *libellus* to be read aloud and answered
its charges in order.[1]

His need of supplies now forced Edward to concede one of
Stratford's main points and on 3 March, he issued writs for a
parliament to meet on 23 April. Three days later he directed
the sheriff of Kent to announce that he did not propose to take
the ninth irregularly and that the statutes of 1340 would be observed. But these were the limits of his concessions, for at the
same time he protested strongly against the action of the bishop
of Exeter in threatening the judges of trailbaston in Devon with
ecclesiastical censures[2] and on 14 March he laid his case before
the pope. Meanwhile, Stratford had issued on 10 March what
appears to have been a circular letter to the bishops, denouncing
the *libellus* and explaining that fear of death had been his
reason for refusing to attend the king's council.[3] He had been
preparing a full answer to the charges in the *libellus* and this was
now issued, as he himself explains, for propaganda purposes—to
prevent a further decline in his reputation before parliament
should meet.[4] Edward's reply, in the form of a letter to the
bishops dated 31 March, had a similar end in view.[5]

Shorn of their rhetoric and personal recriminations, these
letters and tracts are of very great interest. No other political
crisis in medieval England is documented in just this way, by a
series of manifestoes issuing from the headquarters of the rival
parties and comprising a full statement of their respective positions as they wished these to be generally understood. Whoever
composed the royalist propaganda, it was certainly not Edward
himself. Orleton of Winchester denied responsibility for the
libellus, but its abusive tone suggests either he or Kilsby was probably the author.[6] Stratford, on the other hand, probably conducted his own defence. He had not been trained in the schools
for nothing and his superior skill in argument is manifest
throughout. None understood better than he the advantage of

[1] Birchington, p. 23.   [2] *Foedera*, ii. 1151-2.
[3] Wilkins, *Concilia*, ii. 672-3.   [4] Birchington, pp. 27-36.
[5] Ibid., pp. 36-38.
[6] Ibid., p. 39. Cf. Avesbury, p. 330, 'Litera quam dominus Adam Wyntoniensis
episcopus, praefato domino archiepiscopo semper infestus, ad quorumdam ipsius
archiepiscopi aemulorum instantiam, prout dicebatur a pluribus, fabricavit'.

raising large issues of principle as framework for his defence against particular charges of incompetence or worse. Subsequent events suggest that he may not have been altogether convinced by his own arguments and that in no part of the whole correspondence was he more sincere than in his reiterated protests of affection for the king. But he knew how to construct an argument and he understood clearly to which sections of public opinion it was important to appeal. Indeed, as has been pointed out, the enormous importance attached by both sides to public opinion, 'that ultimate political force', is among the most impressive aspects of the whole controversy.[1]

Stratford found no difficulty in returning plausible replies to the specific charges laid against him in the *libellus*. It was not he but Philip of Valois who was responsible for the outbreak of the war and he was not to blame for the expensive alliances which were its necessary consequence. If supplies had not been forthcoming, this was partly because certain merchants had broken their covenants, not because of any failures on his part. Nor was he responsible for the squandering of such taxes as had been collected; it was with the king's full assent that the ninth for the first year had been assigned to his creditors. This was true enough; but Stratford glosses over his own failure to reorganize the ninth when it became clear that it was not fulfilling expectations. To the charge of having encouraged the king to make prodigal grants, he replied that the only donations he could recall were to the newly created earls and these he believed to have been to the king's honour and advantage; the only excessive favours he remembered were the remissions of debt which had the assent of the whole council and without which the ninth would not have been conceded. As to his alleged contempt of the king's citations, only fear of death at the hands of the councillors—'those tyrants who now dominate the land'—had kept him away from court. As a sympathetic contemporary put it, he was willing enough to come to the king's parliament, but not to councils.[2] To the charge that he was not a good pastor he had already pleaded guilty, but it was hardly for the king to reproach him with having procured subsidies which damaged the clergy.

Few of these points were readily answerable; but Stratford was much too astute to confine himself to defensive action. With ingenuity and skill he sought to broaden the basis of his position so as to engage the sympathies of a wider public. The attempts of the tax-collectors to exact the ninth from the clergy who had already paid the tenth, were indefensible on any grounds; but it was not enough for Stratford to proclaim himself the protector of the clergy from lay extortions, the champion of clerical immunities, the heir of St. Thomas. Such an attitude might satisfy the bishops; but it was unlikely to make much appeal to any section of the laity at a time when the political activities of the papacy were serving to foment anti-clericalism, particularly among the military classes.[1] It was with the wisdom of the serpent that in his sentence of excommunication the archbishop coupled violators of the Great Charter with offenders against the clergy, that he protested against arbitrary arrests, not only of clerks, but also of free laymen; for the Charter was the subject's main defence against encroachments on his rights at common law and the assertion that the royal power was threatening new oppressions, that the people were being subjected to arbitrary exactions and reduced to poverty, was well calculated to evoke sympathetic response from the taxpayers and from those who, as Murimuth complains, were being denied the customary processes of justice.

It was less clear that such arguments would appeal to the lay magnates, most of whom were in full sympathy with the king's military and diplomatic objectives and, hence, with his financial predicament. But the magnates, as Stratford well knew, could hardly question the justice of his claim as a peer of the realm to judgement by his peers; and they were unlikely to dispute his contention that, as a peer holding office under the Crown, he was answerable for the conduct of his office to the peers as well as to the king. To deny the archbishop the right to answer the specific charges against him before the magnates in parliament, to condemn him, a major peer of the land, unheard and unconvicted, would be, as he said, to the manifest prejudice of all the peers and a most dangerous precedent.[2] By raising such issues

[1] In his sermon for St. Thomas's Day, Stratford himself referred to the decline of the saint's reputation: 'Hae pro tanto dixi diffusius de causa martirii Thomae, quia multi solent detrahere loquendo, scribendo, et publice praedicando' (*Eng. Hist. Rev.* viii. 88). None the less, annual offerings at the shrine averaged over £400 in the first half of the fourteenth century, R. A. L. Smith, *Canterbury Cathedral Priory* (1943), p. 12.

[2] 'omnium Parium Regni terrae Angliae praejudicium manifestum ac perniciosum exemplum', Birchington, p. 28.

and by associating with them the explosive subject of treason
(albeit only to deny the right of any secular authority to judge
him on this charge), Stratford was using the arguments most
likely to win him aristocratic support. For many of the lords,
and not least those whose ancestors had been victimized, were
still mindful of the arbitrary sentences and forfeitures of 1320-1;
and the archbishop drove home the lesson with his insinuations
that they were threatened with a return to those evil times, that
the young king was beginning to follow his father's example in
submitting to the influence of a closed circle of clerks and
courtiers and excluding the 'great and wise of the land' from his
confidence.

Thus, before parliament met, Stratford had provided each of
the estates with a cause to champion. For the clergy, the issues
were arbitrary arrests of clerks, lay encroachments on their
courts, and the levy of taxes to which they had not consented.
For the peers, the issues were the privileges of their order and
their right to share with the king responsibility for the appointment and conduct of officials. For the knights, Londoners, and
other townsmen, the issues were defence of the common law and
of the Charter and maintenance of their franchises and customary rights. The archbishop himself must be credited with
some genuine concern for the interests of the clergy; but his subsequent conduct strongly suggests that for the rest he cared little
and that his main object in stirring up so many hornets' nests
was to clear his own reputation and to effect a reconciliation
with the king.

The immediate outcome of the parliament of April–May 1341
was a paper victory for the estates on every issue.[1] The clergy
obtained a statute guaranteeing them against arbitrary arrest,
intrusions of laymen and officials on their fiefs, encroachments
on their jurisdiction and the payment of the ninth by those
clergy who, not having been summoned to parliament, had not
consented to it. The peers won from the king the concession that
none of their order, whether ministers or not, should be arraigned
and brought to judgement on any charge laid against them by
the king, save in parliament and before the peers. It was further

---

[1] *Statutes* 14 Edw. III, st. 1; *Rot. Parl.* ii. 131. For reconstruction of the course of
this parliament, the official roll (ibid., pp. 126-34) has to be supplemented by the
narratives in Birchington and the French Chronicle. The chronology of its first ten
days has been outlined by Professor B. Wilkinson in *Eng. Hist. Rev.* xlvi. 132-9.

conceded that breaches of the Charter should be reported in parliament and redressed by the peers of the land, whatever the status of the offender; that officers of state and of the household should be appointed and sworn in parliament, and commissioners be there appointed to audit accounts of all monies received since the beginning of the war. The commons had combined with the lords to press for these last three points; and of particular importance to them were the further clauses of the statute, ensuring maintenance of the privileges of the city of London and of all other communities; repeal of the commissions of trailbaston and of the eyre; and maintenance in all points of the statutes of 1340. Urgent need of supplies constrained the king to yield; the commutation of the ninth for 30,000 sacks of wool was his reward. But he did not yield without a struggle and, before parliament was dissolved, the chancellor, treasurer, and some of the justices had made formal protest that they withheld their assent from any clauses of the statute which were contrary to the laws and customs of the realm which they had sworn to maintain.[1]

Though his own passage had been stormy, Stratford, too, had good reason to be satisfied. Monday, 23 April, had been the day assigned for the opening of parliament; but it was not until the Thursday following that the magnates had assembled in sufficient numbers for the business to begin. Accompanied by his kinsmen, the bishops of London and Chichester, and by 'a great multitude of clerks and squires', the archbishop presented himself at Westminster on Tuesday. The chamberlain and steward (Darcy and Stafford) directed him to go to the exchequer to answer certain charges against him and, *ad placendum regi*, he concurred; but presently he returned and took his accustomed seat in the Painted Chamber where he remained until the chancellor adjourned the meeting. Next day he again sat with such of the bishops as were present and took occasion to expound to them the reasons for his attendance.[2] On Thursday he went again to the exchequer and laid certain information relating to

---

[1] It seems unlikely that the judges on this occasion were concerned to uphold customary as against statute law. Their declaration was probably made with an eye to the king's certain wrath if they should fail him.

[2] Birchington, p. 38. The reasons he gave were: (1) The honour, rights, and liberties of the English Church. (2) The public good. (3) The honour of the king and the good of the queen. (4) To purge himself *in pleno parliamento*. (5) Because he had the royal writ of summons.

the charges against him before the king's councillors. On Friday, the magnates now being assembled, he appeared again and was again ordered to the exchequer; but this time he refused to go and insisted on his right to join his fellow-bishops in the Painted Chamber forthwith. The king was not present; but Orleton, Bourchier, and Darcy, alleging that they spoke on his behalf, urged the archbishop to submit;[1] and when he and his kinsmen reappeared on Saturday, they found their way barred by some of the household knights who addressed them in most unseemly language. But it is likely that by this time the lay peers were beginning to be uneasy; and when Stratford appealed to two of the king's most intimate friends, the earls of Salisbury and Northampton, he received a sympathetic response. The famous protest of Arundel and his uncle Warenne (to both of whom the rights of peers were of particular interest) was probably made at the same time. Those who should be foremost in parliament, said Warenne, were being excluded and others, not peers of the land, were admitted. When Kilsby, Darcy, and the rest had left the chamber, Arundel reinforced his uncle's plea: 'Sire, let the archbishop come before you and if he can defend himself, well: if not, we will settle his fate.'[2] Edward yielded to their plea; and later in the day he and Stratford met face to face for the first time since they had parted on board the cog *Thomas* some ten months before. For what took place at their meeting we have only the evidence of Birchington who tells us that terms of re-conciliation were discussed;[3] but it may be surmised that they struck a bargain, which enabled the archbishop to spend his Sunday at Lambeth *in pace* while Kilsby and Darcy were trying vainly to stir up the Londoners against him. Stratford can hardly have been unaware that his fellow-peers were engaged in draft-ing the bill relating to their privileges which was to be presented to the king within the next few days; he must have known that

---

[1] Birchington, p. 39. It was on this occasion that Orleton denied authorship of the *libellus famosus*.

[2] *French Chronicle*, p. 90. Professor Wilkinson (*Eng. Hist. Rev.* xlvi) argues that, since parliament was not in session on 28 Apr., the estates having separated to dis-cuss the king's request for supplies, the earls' plea must refer to the exclusion of Stratford, not from parliament but from the king's council. But, as the author him-self suggests (p. 182), such a distinction would probably have seemed meaningless to Warenne himself and, in any event, to admit Stratford's claim to sit among the bishops in council would have been tantamount to admitting his claim to sit in parliament.

[3] Birchington, p. 40: *tractatu pacis habito*.

the game was now in his hands. But it is likely that he was no
more anxious than the king to have the whole sorry story of administrative incompetence thrashed out in public. Thus, while
continuing on the Tuesday and Wednesday to repeat his formal
claim to clear himself *in pleno parliamento* of the charges against
him (which had been presented in the form of thirty-two articles), he had tacitly abandoned it. Lords, clergy, and commons
went forward with their respective bills which, as has been
seen, conceded all the issues of principle raised by Stratford; he
himself made public submission and, having been granted the
right to clear himself as he had asked, allowed his case to be
shelved.[1] On 12 May a committee consisting of two bishops
(Durham and Salisbury) and four earls (Northampton, Salisbury, Arundel, and Warwick) was empowered to hear the archbishop's answers and, if these seemed sufficient, to hold him
excused, the record to remain with the keeper of the privy seal.
Nothing more was heard of it until the parliament of 1343, when
the king directed that the record of the arraignment be annulled
as untrue and contrary to reason and that the documents be
brought into parliament and there destroyed.[2]

Judgement of Stratford as an opposition leader or as a prophet of 'Lancastrian' political theory, must take account of
the sequel to the crisis. His reconciliation with Edward seems
to have been complete and in no way affected by the king's
arbitrary action in repealing the statutes of 1341. On 1 October
of that year, letters close to the sheriffs notified them that the
king, after consultation with his magnates and other counsellors,
had resolved to repeal the 'pretended statutes' of the last parliament, with which he had never agreed and to which he had consented only in order to obtain supplies. They had not proceeded
from his spontaneous goodwill and were now annulled as contrary to English law and the royal prerogative; only those parts
of them which were consonant with law and reason were to be
re-enacted on the advice of the judges and those learned in
the law.[3] No parliament was then in session and none was

---

[1] Professor Plucknett (*Trans. Royal Hist. Soc.*, 4th ser. xxiv. 68–69) points out that
Stratford admitted the propriety of his arraignment in parliament on the basis of
articles which 'notoriously defamed' him and that the procedure against him therefore 'shows a tendency' in the direction of impeachment.

[2] *Rot. Parl.* ii. 139.

[3] *Foedera*, ii. 1177. As Professor Sayles and Mr. Richardson have pointed out
(*Law Quarterly Review*, l. 552) there were precedents for Edward's action. In 1333

summoned until the spring of 1343 when, despite repeated and spirited protest by the commons, the lords acquiesced in the repeal. The ease with which Edward thus recovered the ground he had lost doubtless owed much to the reluctance of both king and magnates to renew the quarrel at a time when good prospects of military action were opening up in France; but it must also have owed something to the complaisance of the archbishop. Less than three weeks after Edward had issued his letters to the sheriffs, he and Stratford were publicly reconciled and henceforward the latter seems to have resumed his role of principal confidant and counsellor.[1] It was Stratford to whom the king turned for advice in his negotiations with the papacy; it was Stratford and his episcopal kinsmen who rallied the bishops in the convocation of 1344 to grant tenths to the king;[2] it was Stratford who presided over the council during the crucial Crécy campaign; it was Stratford who read aloud at Paul's Cross the French plan, discovered in the archives of Caen, for the invasion of England; and it was Stratford to whom the king wrote describing his exchanges with the French during the siege of Calais.[3] If the archbishop had ever sympathized with papal censures of the war, such scruples had long since been repressed; if he had ever regarded Edward as an oppressor of the clergy, such animosities had long been forgotten; if he had ever sought to lay down limits to the royal prerogative, he had now learned wisdom. Perhaps Warenne felt that the cause of the peers had been betrayed, for when Stratford excommunicated the bishop of Norwich, who had resisted his visitation, the earl welcomed the offender into his domains.[4] But it would be unjust to dismiss Stratford as either a humbug or a time-server; for he had never sought a quarrel with the king and had done no worse than defend himself to the best of his very considerable ability when charged with offences to most of which he could fairly plead not guilty. By choice, Stratford was always the king's friend; and Edward, who was never vindictive, probably came to recognize that, if his chief counsellor had blundered, his intentions had been good; and, once the storm had subsided, the king was willing to let bygones be bygones. The reconciliation was

with the consent of the council and the merchants he had revoked a recent parliamentary ordinance; and in 1337 he had suspended the application to Cheshire of a statute passed two years earlier.   [1] Murimuth, p. 122.
   [2] Ibid., pp. 125-6, 157-62, 176-7; *Rot. Parl.* ii. 146.
   [3] Avesbury, pp. 391-3.   [4] Murimuth, p. 147.

creditable to both; but it makes it hard for the historian to por-tray Stratford as a convinced champion of the aristocracy against the Crown, or of common right against the royal prerogative. If the service which he paid to these causes had been much more than lip-service the history of his remaining years must surely have been other than it was.

The voluminous correspondence and tangled history of this crisis of 1340–1 have proved puzzling to students and there has been little agreement as to its true significance. In origin, the crisis was purely administrative, deriving from Edward III's financial embarrassments and his determination to search out and punish those officials whom he believed to be responsible for the failure of his sources of supply. It was Stratford's in-genuity which achieved a diversion from administrative to con-stitutional issues, and in the course of the ensuing conflict its original cause tended to be lost sight of. The question of respon-sibility for the king's shortage of funds was never really answered, though relief was afforded, not only by the wool grant of 1341, but also by the reorganization of the ninth which, when the minimum price was removed, yielded a sum about 70 per cent. higher than the average tenth and fifteenth.[1] Few, if any, of the constitutional issues raised by Stratford were new and the net constitutional gains appear to have been small. The attempt to hold ministers of the Crown answerable to parliament and to insist on the maintenance of charters, franchises, and customary rights, takes us back to the days of the Ordainers; and such claims to restrict his control over his servants and to limit his dis-cretionary power were bound to be resisted by Edward III, as they had been resisted by his father. To translate such notions into common practice was, indeed, hardly possible; for it was only at periods of acute political tension that medieval civil servants could be made accountable to an occasional and un-professional body like parliament; and the vesting of some dis-cretionary power in the executive is a necessary condition of waging a great war. The ease with which adjustment was reached on these matters, after Edward's repudiation of the statutes of 1341, may serve to remind us how small was the area of necessary conflict between the medieval king and his barons. An able and vigorous ruler who had the good sense to

[1] Dr. Fryde's calculation.

go through the motions of consulting his councils and parliaments on matters of high policy, to choose reasonably efficient and personally inoffensive ministers and to keep his favours to them within the limits of discretion, usually found the lords ready enough to leave him to manage his own business. Baronial clamour to control the executive arose only when a king like Edward II acted as a fool, or like Richard II, as a tyrant, or when, like Edward III in his declining years, he allowed his hold on the reins of government to slacken. Even in matters that touched them most nearly, questions of their right to a fair trial, questions of the nature of treason and its penalties, no further conflict arose, once the lords recognized that in Edward III they had a king whom they could respect and need not fear. Edward's extravagance and unwisdom had opened the way for Stratford to raise dangerous issues; but the king never repeated the mistakes of his early years and for a generation after the crisis, these issues lay dormant. Edward was a popular and successful king and his subjects, above all the magnates to whom he was bound by so many ties of interest and affinity, lacked the kind of incentive which had moved their forebears under Edward II and was to move their descendants under Richard II, to seek to impose limits upon the royal prerogative.

None the less, it remains true that these opening years of the French war had afforded striking demonstration of the dependence of the king on parliament. Much could be had by borrowing; but the only acceptable security for the raising of a series of large loans was the parliamentary subsidies and the wool grants. And, within parliament, it was always the commons who had to be wooed. Criticisms of the king's financial policy, opposition to his demands, attempts to make grants of supply dependent on redress of grievances, came almost always from them. The part played by the commons in the crisis of 1340-1 stands out in sharp contrast to their passive role in 1310-11; and it was the commons alone who, in 1343, contested, albeit unsuccessfully, the right of the king to repudiate parliamentary acts. It is not to be supposed that all the commons spoke with one voice, nor that many of them did not share the magnates' enthusiasm for Edward III and his wars. But as a body representative of the middle-class taxpayers and traders, they could never afford, as could some of the magnates, to relax their vigilance; and unprecedented opportunities lay open to them to force concessions from a martial king, whose ability to fight depended ultimately on their willingness to pay. Thirty years were to elapse after 1341 without a major parliamentary crisis; but the dependence of the king on the knights and burgesses ensured that these should not be years of political or constitutional stagnation.

VII

PARLIAMENT, LAW, AND JUSTICE

No very precise answer can be given to the question, 'What is parliament?', for even in the fourth decade of the century a flavour of colloquialism still clings to the word, and such attempts as have been made to establish distinctions between similar assemblies, variously described as *parliamentum, tractatus, colloquium, consilium*, fail to carry conviction.[1] None the less, the main outlines of what was to become the historic structure of the national assembly are already clearly perceptible at the death of Edward II. To all the forty-eight parliaments of Edward III there were summoned lords spiritual and temporal, certain councillors and officials, proctors for the clergy, and representatives of shires, cities, and boroughs. The lords and some of the councillors receive individual writs of summons, the lower clergy and the commons are summoned indirectly. Parliament may continue to sit after the commons have been dismissed; but the days when it meets without them are over.[2] Surviving records of parliament for the early years of Edward III are disjointed, miscellaneous, and unorganized; the four rolls of proceedings (as distinct from petitions) which are all that remain to us for the period 1327–39, seem to have been the work of an enterprising chancery clerk, Henry Edenstowe, or Edwinstowe, the first known clerk of the parliament. But after 1339 the records improve—possibly in consequence of their transference from the treasury to the chancery a few years earlier[3]—and, though some of them, even for the reign of Richard II, are still disappointingly meagre, there are only five gaps in a continuous series from 1339 to the end of the century.[4]

[1] See T. F. T. Plucknett, 'Parliament', *The English Government at Work, 1327–36*, ed. Willard and Morris, i (1940), 82–90.

[2] The last parliament to which the commons were (probably) not summoned was that of midsummer 1325, *Interim Report of the Committee on House of Commons Personnel and Politics* (1932), p. 70.

[3] See V. H. Galbraith, 'The Tower as an Exchequer Record Office in the reign of Edward II', *Essays . . . presented to T. F. Tout*, pp. 231–47.

[4] No rolls are extant for the parliaments of 1357, 1358, 1360, 1361, and Sept. 1388.

In addition to the rolls and the miscellaneous notes on parliament provided by the chroniclers, we have to take account of the mysterious document known as the *Modus Tenendi Parliamentum*. On the face of it, this is a tract, or short treatise, belonging to a familiar medieval genre (other examples are the *modus faciendi duellum* and the *modus omnium cartarum*), a description of procedure intended to serve as a guide to those responsible for the conduct of a parliament (or of a duel or for the drafting of charters). But closer investigation soon makes it clear that the *Modus Tenendi Parliamentum* does not altogether conform to this conventional pattern. It is true that it includes a description of parliamentary procedure which is asserted to have been in use at the time of the Norman Conquest, and that the whole document reveals the hand of an expert who knew all about such matters as the technique of enrolment, the method of summons, the duties of clerks. But the author uses this allegedly historical exposition as cover for a radical, if not revolutionary, political theory and what he tells us, for instance, of the balance of power between the different parliamentary estates does not square with what is known of any parliament in this period. This discrepancy in itself poses a serious problem for the investigator; and it is a further complication that two distinct versions of the *Modus* survive—an English version, the earliest extant manuscript of which was probably written between 1390 and 1400 and an Irish version dating, in the earliest known manuscript, from 1419. A formidable consensus of learned opinion favours the priority of the English version, regards the Irish as a later adaptation, and places the composition of the original treatise in the reign of Edward II, somewhere between 1316 and 1324.[1] Against this, Mr. H. G. Richardson and Professor Sayles believe it to be 'infallibly true' that the Irish version preceded the English and that it could not have been written earlier than the reign of Richard II.[2] No argument in support of their view has yet been published, however, and as matters now stand the weight of the evidence must be said to favour the earlier date, the onus of proof resting on those who would question it. Yet an element of doubt remains; and the student must hesitate to take

---

[1] W. A. Morris, 'The Date of the "Modus Tenendi Parliamentum"', *Eng. Hist. Rev.* xlix (1934), 407–22; Maude Clarke, *Medieval Representation and Consent* (1936); V. H. Galbraith, 'The *Modus Tenendi Parliamentum*', *Journal of Warburg and Courtauld Institutes,* xvi (1953), 81–99.

[2] *The Irish Parliament in the Middle Ages* (1952), pp. 137–8.

the *Modus* as a guide to understanding of the medieval parliament. Its propagandist intention is too obvious; it tells us, not what parliament is, but what the anonymous author, or authors, think it ought to be. It may throw some light on the political programme of some of the rebels of 1321–2; but where it appears to be in conformity with contemporary practice it tells us little or nothing that could not be gleaned from other sources. As a source of information about the structure, procedure, and functions of the fourteenth-century parliament, it is suspect and of minor importance.[1]

First in dignity among those summoned to parliament come the lords spiritual, bishops, abbots, and priors. The archbishop of Canterbury sits on the king's right hand, next to him the archbishop of York, then the bishops of London, Winchester, and Salisbury, in that order, then the remaining sixteen bishops, in order of seniority of consecration. Below the bishops sit the abbots and priors headed by the prior of St. John of Jerusalem, next to whom sits the abbot of St. Albans, then the abbot of Glastonbury. Despite much heated contemporary argument, it seems clear that no perfectly consistent principle underlay the summons of the higher clergy. The ecclesiastical contention that only those who held of the king *per baroniam* ought to be summoned was never urged on behalf of the six bishops who did not so hold;[2] and the fact that, during the vacancy of a see, it was the vicar-general and the guardian of the spiritualities (both of whom were ecclesiastical officers) who represented the absent bishop in parliament, suggests that it was primarily by reason of their spiritual position that the bishops attended. On the other hand, the royal contention that tenure-in-chief was irrelevant, and that the king might summon whom he would to his councils and parliaments, had to be abandoned under pressure of events. Edward I had summoned very large numbers of abbots and priors—seventy to the so-called Model Parliament of 1295, forty-eight to the Carlisle parliament of 1307; and this practice was maintained in the early years of Edward II, though with some reduction in the number of Premonstratensians and

---

[1] For the text see Clarke, *Medieval Representation*, pp. 374–92.

[2] Rochester, Carlisle, and the four Welsh bishops. So late as 1388 Archbishop Courtenay made the unhistorical claim that it belonged to prelates *holding of the king by barony* to be present in the king's parliaments as peers of the realm, *Rot. Parl.* iii. 236–7.

Cistercians. After 1322, however, there was a marked decline. To the parliament of February 1324 there were summoned only twenty-two black monk abbots, two black monk priors, six Austin canons, one Cistercian, and the prior of St. John, thirty-two in all. The years that followed saw variations, but by 1364 the list has become stereotyped and consists of the prior of St. John of Jerusalem, twenty-three black monk abbots, the prior of Coventry, and two abbots of the Austin canons, twenty-seven in all.[1] Tenants in frankalmoign, as well as those who did not hold directly of the king, had been allowed to drop out; the development of convocation as a tax-granting assembly had reduced the king's interest in their attendance. Even so, there were exceptions and omissions which make it impossible to hold that tenure by barony, and that alone, would determine the obligation of attendance for the head of any particular house. A tradition that certain abbots and priors should be summoned was gradually established, and resulted in the emergence of a class of ecclesiastical 'barons by writ', by no means necessarily coincident with the class of ecclesiastical 'barons by tenure'.[2]

A similar process was at work in regard to the lords temporal who, sitting on the king's left hand, constituted the second of the parliamentary groups, or estates. They included magnates of different ranks, dukes, earls, barons, and others. The title of duke, hitherto borne only by the king himself in his capacity of duke of Aquitaine, was, in 1337, bestowed by Edward III on his eldest son whom he created duke of Cornwall, possibly in anticipation of his own assumption of the title of king of France.[3] Henry of Lancaster received the title of duke in 1351, Lionel of Antwerp and John of Gaunt became respectively dukes of Clarence and Lancaster in 1362, their younger brothers became dukes of York and Gloucester in 1385. Richard II also conferred the title of duke upon Robert de Vere in 1386 and on five other magnates in 1397; all of these could claim some sort of kinship with the king, by blood or marriage. The title of marquis (a rarity before the sixteenth century) first appears in 1385 when de Vere became marquis of Dublin; it was renewed in 1397, for the benefit of John Beaufort who became marquis of Dorset. The only other territorial titles were the earldoms which varied in

[1] M. D. Knowles, *The Religious Orders in England*, ii (1955), 302-4.
[2] Helena M. Chew, *The English Ecclesiastical Tenants-in-Chief and Knight Service* (1932), p. 179.   [3] *Complete Peerage*, viii, App. A, p. 721.

number, according to the vicissitudes of family fortunes and national politics. There were nine earls in 1307; when Edward III became king there were only six; six more were added in 1337; in the latter years of Edward III the earls numbered about fourteen; ten were summoned to the last parliament of Richard II. The dukes and earls were the natural leaders of the baronage, but it is only in the crises of Edward II's reign that the earls appear to act as a politically self-conscious group. The growth of the concept of peerage under Edward's two successors tended to minimize the importance of particular classes within the peerage. As with the ecclesiastical barons, so with the lay magnates below the rank of earl, no consistent principle of selection seems to have been applied and their number varied greatly. Under Edward III, men of substance were summoned to parliament by reason of their wealth, territorial influence, or individual importance; the choice of the persons to be summoned still lay within the discretion of the Crown. It was not until after Richard II had initiated the practice of creating peers by letters patent, without reference to the lands they held, that the list of those customarily summoned began to harden into rigidity.[1] The kings of the fourteenth century felt themselves free to vary the writs at will. Thus, Edward II summoned ninety barons in 1321, fifty-two in 1322; Edward III summoned sixty-three in 1334, thirty in 1348, fifty-six in 1351, thirty-eight in 1376. Richard II summoned forty-eight in 1384, forty-five in 1388, thirty-eight in 1397. Moreover, a son could be summoned together with his father, as were Richard and Gilbert Talbot between 1330 and 1344; or a son might attend, as did John Cobham in 1333, on behalf of his father.[2] Bannerets, whose titles were of purely military significance, are found among the lords; others might make their way into the ranks through marriage, or deeds of prowess. The whole system under Edward III was fluid and, as such, was well adapted to the political climate of the period, when the general harmony prevailing between king and magnates deprived both of any motive for seeking more precise definition. Claims for privilege of peers, particularly in judicial processes, were likely to produce de-

---

[1] The first barony by writ was created for Beauchamp of Kidderminster in 1387, *Complete Peerage*, vii, App. A, p. 763.
[2] I owe these examples to Dr. G. A. Holmes's unpublished thesis, 'The English Nobility in the Reign of Edward III' (Cambridge, 1950).

finition, in the long run; but a carefully delimited hereditary peerage, carrying with it an exclusive right to a summons to parliament was not in being before 1399.[1]

A third and smaller group always found in parliament comprised officials and administrators, some of whom were summoned by individual writs. The two chief justices were always summoned, other judges occasionally; so were the serjeants-at-law, whose business was to act as legal advisers to the Crown. One or two barons of the exchequer might be summoned; and the king's principal ministers, the chancellor, treasurer, keeper of the privy seal, chamberlain, and steward of the household, would normally be there. These *curiales* were the hard core of parliament and probably responsible for the direction of most of its business; but the magnates were intermittently suspicious of royal officials and, as in 1341, resentment at their presence might flare into open protest. A special place was accorded to the chancellor; but a wise ruler kept his other officials in the background and allowed them to do their work behind the scenes.[2]

Three other groups—clerical proctors, knights, and burgesses —were summoned, not by name, but indirectly. The *praemunientes* clause in the writs addressed to the bishops instructed them to ensure the presence of the dean, or prior, of the cathedral and of the archdeacons and to warn the chapter to elect one proctor, the other clergy of the diocese to elect two. But there was opposition to the summons of the lower clergy to a secular assembly and by the middle of the century they had ceased to be of any significance in parliament.[3] The lay commons were summoned by means of writs addressed to the sheriffs of all the English counties (except Chester and Durham), directing them to procure the election of two knights from every shire, two citizens from every city, and two burgesses from every borough, from among those most discreet and able for work, and to cause them to come to parliament with full power to do and consent to whatever might be ordained by common counsel.

Though the writs demand the election of knights as representatives of the shires, few sheriffs took, or were expected to take, this literally. It is rare indeed to find a sheriff troubling himself to produce the kind of excuse offered by the sheriff of

---

[1] Tout, *Chapters*, iii. 138, n. 2.

[2] H. G. Richardson and G. O. Sayles, 'The King's Ministers in Parliament 1272–1377', *Eng. Hist. Rev.* xlvii (1932), 377–97.   [3] Below, pp. 287–8.

Northumberland in 1360, that there were no knights in his county except Walter of Tynedale and he was feeble and useless.[1] From an early date it seems to have been accepted that men of substance in the shire, whether knights or not, were competent to represent it; and occasional specific demands—as in 1353, 1355, and 1371—for knights 'girt with the sword and having the honour of knighthood', arose from the exigencies of war and were not notably effective. Hardly less ineffective were the attempts, culminating in an ordinance of 1372, to keep out lawyers, who were believed to devote their energies to furthering the interests of private clients. The same ordinance excluded the sheriffs, and this regulation, springing as it did from a reasonable objection to their removal from their proper sphere of office, seems to have been better observed.[2] Orders to elect representatives for the counties were rarely disobeyed; but all too little is known of the conduct of the elections themselves. The fairly numerous occasions before 1372 when the sheriff was allowed to return himself may suggest apathy on the part of the electors; but the sources afford no positive evidence either of reluctance or of enthusiasm to elect or to be elected. At the conclusion of each parliament writs were issued to the knights to enable them to secure payment of their wages and expenses; and the regular enrolment of these writs by the chancery clerks may be accepted as proof that the persons elected for the counties very seldom failed to attend. Moreover, it was by no means uncommon for the same individual to be returned to more than one parliament or to represent different communities (sometimes a shire, sometimes a borough) in different parliaments.[3] Viewed as a whole, the parliamentary knights of the fourteenth century were a heterogeneous body of men, ranging from the belted knight who might be the younger son of an earl, to the prosperous yeoman whose ancestor might have been a serf. Many were, or had been, retainers of magnates, both ecclesiastical and lay; some were the sons of former members; many had trading interests; as the century progressed, an increasing number had been in the wars; probably the majority had served

---

[1] 'languidus et impotens ad laborandum', Prynne, *Brief Register of Parliamentary Writs*, iii. 166–7.

[2] Kathleen Wood-Legh, 'Sheriffs, Lawyers, and Belted Knights in the Parliaments of Edward III', *Eng. Hist. Rev.* xlvi (1931), 372–88.

[3] Kathleen Wood-Legh, 'The Knights' Attendance in the Parliaments of Edward III', ibid. xlvii (1932), 398–413.

their counties in some other capacity, as sheriffs or coroners, as keepers, or justices of the peace, as tax-collectors, or commissioners of array. Not all were of unimpeachable character and it was by no means uncommon for a parliamentary knight to have been in the courts on a charge of felony;[1] but they brought into parliament a wide range of administrative experience and knowledge of local conditions and, overlapping at one end with the nobility and, at the other, with the bourgeoisie, their presence there prevented alignment into rigid social or professional groups and made possible their own consolidation with the burgesses into a single *collegium*, or house of commons.[2]

Defective returns make it impossible to state with any degree of precision how many towns elected representatives to parliament in the fourteenth century. After the reign of Edward I the numbers certainly decline; but the average for the last twenty years of Edward III is much higher than for the reign of Edward II and that for the reign of Richard II is higher still. We may not be far wrong if we postulate an average of seventy for Edward II, seventy-five for Edward III, eighty-three for Richard II. Though procedure was never uniform, elections usually took place in the towns themselves and seem often to have been managed by a few of the wealthier burgesses; but there is no evidence to suggest that the ordinary townsman felt himself in any way defrauded by this practice. Some of those elected may have failed to attend, but the sources do not bear out the suggestion that absenteeism was widespread and persistent.[3] Attempts by the Crown to impose qualifications for election are rare and seem to reflect little beyond a desire to secure the presence of discreet and substantial persons. Like the knights, the citizens and burgesses form a heterogeneous group, including weighty capitalists, members of knightly families, and small traders and shopkeepers. Many of the burgesses sat in several parliaments; and, though it is likely that the smaller men were willing to accept direction from above, the wealthy merchants, whose co-operation was indispensable to Edward III, were probably as influential (if not more influential) there as any of the knights.

[1] See the examples cited by Professor Plucknett, pp. 102–3.
[2] Walsingham refers to the commons in the Good Parliament of 1376 as *collegium*, *Chronicon Angliae*, p. 70.
[3] I have discussed this question in my *Parliamentary Representation of the English Boroughs during the Middle Ages* (1932), pp. 66–81.

A customary order of proceedings in parliament is already establishing itself in the early years of Edward III. The writs of summons normally allowed forty days' notice, but lay magnates and prelates caused great difficulty by arriving unpunctually, or not at all, and the formal opening of parliament often had to be postponed for several days. In December 1332 it was found necessary to prorogue parliament for a month because most of the prelates and many of the lay lords had failed to appear at York; in 1344 the king heartily thanked those few who had arrived in time.[1] As a prelude to the session, public proclamation forbade the bearing of arms by those assembled for the parliament and threatened with imprisonment children or others who played at 'barre' or similar games, knocked people's hats off, or in other ways hindered those attending the parliament from going about their business.[2] Within the parliament chamber itself, proceedings were opened by the delivery of one or more formal orations. Sometimes the chancellor, sometimes a chief justice or an archbishop, declared the cause of summons. Petitions were then presented, these being addressed, not to parliament, but to the king and his council. The appointment of receivers of petitions (all of them senior chancery clerks) and of triers or auditors (a mixed body of magnates and judges) is first recorded in 1333.[3] Next came the deliberative stage, the assembly commonly dividing into groups for this purpose; but grouping arrangements were still very flexible and in the parliament of January 1333, for example, we find six prelates, two earls, and four barons in one group, the remaining lay magnates and the clerical proctors in a second, and the knights and burgesses in a third. Only after 1341, when the clerical proctors are tending to disappear, does it become normal for knights and burgesses to confer together and apart; lay and ecclesiastical magnates still frequently confer in separate groups. The normal way for the commons to communicate with the king or the lords was by means of a small deputation appointed *ad hoc*; the election of an undoubted Speaker is not traceable before 1376.[4] In these

---

[1] *Rot. Parl.* ii. 67, 146. Only 40 out of 129 magnates summoned were present at York in Dec. 1332. See J. S. Roskell, 'The Problem of the Attendance of the Lords in Medieval Parliaments', *Bull. Inst. Hist. Res.* xxix (1956), 165, 167.

[2] *Rot. Parl.* ii. 64. 'Barre' was probably the game known as 'prisoner's base' or 'chevy' (see *New English Dictionary*).   [3] *Rot. Parl.* ii. 68.

[4] J. S. Roskell, 'The Medieval Speakers for the Commons in Parliament', *Bull. Inst. Hist. Res.* xxiii (1950), 33 and n. 1.

informal gatherings, reports were drawn up and important decisions reached; they are distinct from the public sessions, *in pleno parliamento*, where the formal orations were pronounced and formal offers of supply or suggestions for action were made.[1] Sessions were usually short, sometimes lasting no more than a week or ten days, and not all those summoned necessarily remained for the whole session. Before their departure, knights and burgesses would be given leave to apply for the writs *de expensis* which authorized them to claim wages at the rate of four shillings a day for the former, two shillings for the latter, payable for each day of attendance at parliament and for the journey to and fro.

Parliamentary procedure was determined by parliamentary functions, deliberative, taxative, judicial, and legislative. The very ancient tradition whereby great councils, or parliaments, afforded opportunity for the king to discuss affairs of state with his barons is being maintained and, by the time of Edward III, sometimes extended to include other groups. In 1331 and 1332, for example, Edward sought the advice of his magnates on Gascony, Scotland, Ireland, and the projected crusade; in 1339 both lords and commons were consulted on questions of defence; in 1343 on coinage and the price of wool.[2] Social, commercial, military, and diplomatic questions were all laid before lords and commons, as were those touching the papacy and the Church; in short, parliament was used by Edward III as a clearing-house for discussion of all important affairs of state. He did not always follow the advice proferred; but the practice of seeking it was useful to him, partly because parliamentary clamour could serve as a diplomatic lever, notably in negotiations with the papacy, and partly because lords and commons could fairly be asked to face the financial consequences of following their advice.

A tradition of direct parliamentary taxation was well established by 1327 and the subsidies on movable property were always taken with the consent of the knights and burgesses. The rate of these levies was still variable, but the greater concentration of wealth in the towns, and the liability of both boroughs and demesne lands to tallage, led to their being taxed normally at a higher rate than the rest of the country: a tenth from the boroughs and demesne lands and a fifteenth from the rest was becoming the rule. In 1340, as has been seen, a ninth was offered;

---

[1] Plucknett, p. 108. [2] *Rot. Parl.* ii. 60–61, 62, 103, 196–8.

but after this date all the taxes on movables took the form of tenths and fifteenths. A standard valuation, comparable to the clerical *Taxatio* of 1291, was achieved in 1334 as a result of persistent complaints of corrupt practices, a tenth and fifteenth being valued at the sum of £38,170 and the contribution of each shire, borough, and township assessed proportionately.[1] Standardization had the effect of turning the tax on movables into something resembling a land tax and the simplification thus obtained was not without its disadvantages. The standard valuation was far below the real value of a tenth and fifteenth and it was too rigid; for, though adjustment to meet changing circumstances was possible in theory, in practice this was apt to prove difficult. Later experiments with novel forms of direct taxation, such as the tax on parishes proposed in 1371 and the poll-taxes which followed it, were the natural result of dissatisfaction with subsidies based on assessments which had never been realistic and were fast becoming obsolete.

In the meantime Edward III looked chiefly to the profits of the wool trade for the additional funds necessary for the financing of his wars. He was legally entitled to the *antiqua custuma* of half a mark on the sack of wool, or three hundred woolfells, payable as an export duty by both native and foreign merchants, and to the *nova custuma* of forty pence payable by foreign merchants only. As the revenue yielded by these duties was not very substantial, Edward III, like his grandfather before him, was bent on securing additional subsidies. Under the *Confirmatio Cartarum* of 1297 increases in the wool custom required some kind of consent; but the consent of the merchants and a few magnates, even of the merchants alone, seems to have been accepted as sufficient warrant for moderate additional levies. In January 1333 an additional tax of half a mark was taken with

---

[1] J. F. Willard, 'The Taxes on Movables in the Reign of Edward III', *Eng. Hist. Rev.* xxx (1915), 69–74. A new and more accurate transcript of the tax assessment of 1334 will be found in W. G. Hoskins, *Devonshire Studies* (1952), pp. 215–16. The valuation affords interesting evidence of the relative prosperity of different parts of the country. Norfolk (with an acreage of *c.* 1,315,000) heads the list with a valuation of £3,485. Kent (*c.* 976,000), Gloucestershire (*c.* 804,000), Wiltshire (*c.* 861,000), and the Lindsey division of Lincolnshire (*c.* 973,000) were assessed at sums ranging from £1,500 to £2,000. Ten counties were assessed at less than £500—Cornwall (*c.* 868,000 acres), Derby (*c.* 648,000), Huntingdon (*c.* 234,000), Hereford (*c.* 539,000), Lancashire (*c.* 1,200,000), Middlesex (*c.* 149,000), Northumberland (*c.* 1,292,000), Cumberland (*c.* 973,000), Rutland (*c.* 97,200), and Westmorland (*c.* 505,000). London's assessment was £733, Bristol's £220, York's £162.

the consent of 'prelates and magnates'; and though the ten-shilling levy, imposed later in the year by king and council on the advice of the merchants, gave rise to complaints in parliament, these were said to have been made on the grounds that the tax damaged the people, not that it had been taken without parliamentary consent.[1] In 1339 the lords demanded merely the abolition of the *maltolte* of 40s. granted by the great council at Nottingham in 1336 and the restoration of the *antiqua custuma*; it was the commons who in this year first raised the issue of consent. The *maltolte* had been imposed, they said, 'saunz Assent de la Commune ou des Graundz'; and they preferred to offer instead the profits from the sale of a fixed quantity of wool.[2] Their protest marks the opening of a protracted struggle for control of the wool-tax which was to culminate in the parliamentary victory of 1362.[3]

In the sphere of justice, parliament remained what it had always been, the highest court in the land; and expansion of its judicial functions is implicit in the claims of the peers to act as a special parliamentary tribunal competent to judge members of their own order. Before Edward III was dead they had accepted responsibility for the judgement of others, not of their order, but accused by the commons; the lords in parliament had acted as judges in the first clear case of impeachment (1376). But trials of peers and impeachments—state trials—were exceptional cases; and the routine judicial functions of parliament were tending, on the whole, to contract.[4] Professor Plucknett has cited examples of parliament acting as a court of error, as a court for the discussion of difficult points which had arisen in the lower courts, and as a court adjudicating on a large miscellany of administrative matters; but, as he rightly insists, it is difficult, in these contexts, to distinguish parliament from council; and many decisions of the council, taken while parliament was in session, may have been described as acts of parliament. So far as justice is concerned, the council, with its hard core of experts—judges and others—dominates parliament; and the council does not cease to exercise its judicial functions when parliament has ceased to sit.[5] The development of parliament as a deliberative,

---

[1] *Cal. Fine Rolls*, iv. 342, 365; *Cal. Close Rolls 1333–37*, p. 257.
[2] *Rot. Parl.* ii. 105.   [3] See below, pp. 222–3.
[4] See J. G. Edwards, ' "Justice" in Early English Parliaments', *Bulletin Just. Hist. Res.* xxvii (1954), 35–53.   [5] Plucknett, pp. 109–13.

taxative, and legislative assembly, the tendency, if not to exclude officials, at least to keep them in the background, implies that it grew correspondingly less professional. Adjudication on the type of civil case which came before parliament, demanded some degree of *expertise*; it was in council, chancery, and exchequer that the experts were to be found and it was to these bodies that litigants looked for redress. The classic distinction between the French *parlement*—a body of professional lawyers—and the English parliament—a body of amateur politicians—is already clearly foreshadowed in the reign of Edward III.

While the council thus absorbed many of parliament's judicial activities, parliament itself was concerned increasingly with legislation. In so far as it is remedial, legislation is, of course, closely linked with justice and even before the accession of Edward III, it was beginning to be founded on the *petitions des communes*, or *communes petitions*, which were submitted to parliament in its judicial capacity. We know little of the procedure lying behind these petitions; but, since many of them embody grievances of purely local or sectional interest, it seems certain that not all the petitions can have originated with the commons as a whole; and it is very unlikely that they were systematically sifted or examined by the knights and burgesses. The *commune petition* is a roll of requests which the commons have agreed, or been persuaded, in some sense to sponsor; it is distinct from the *singulere* petitions presented by individuals.[1] In the first parliament of Edward III the commons presented a long roll of forty-one articles, sixteen of which were embodied in a statute; and, although a statute was not the inevitable consequence of the king's acceptance of a *commune petition*, seven such have survived for the first decade of the reign, all systematically enrolled and followed, on the roll, by a writ directing their publication. Their subject-matter is very various, touching many aspects of private and public life; but it has to be recognized that they may bear no very close resemblance to the original petition on which they purport to be based. Clauses in the petition were sometimes overlooked or forgotten; sometimes, the statute contained provisions which were not in the petition and had the effect of invalidating it. Even if a statute were formally enrolled, it

[1] Doris Rayner, 'The Forms and Machinery of the "Commune Petition" in the Fourteenth Century', *Eng. Hist. Rev.* lvi (1941), 198–233; 549–70.

might still remain a dead letter, for lack of any provision for its execution; or it might be superseded by the will of the king, or, as in 1341, repealed as contrary to his prerogative. The stage of presenting draft statutes, or bills, had not yet been reached; and the parliament rolls reveal something of the prolonged and never wholly successful struggle of the commons against abuses and evasions. In 1341 they complained that the answers to their petitions were not so satisfactory as reason would demand; in 1344 and again in 1362 they asked that petitions might be examined and redress ordered 'before the end of the parliament'; in 1348 that the petitions might be answered and endorsed in parliament and that they might be shown the endorsement; and again that the full answers given might remain in force without being changed.[1] And, even if the statute were drawn up in a form satisfactory to the petitioners, there still remained the difficulty of securing its enforcement. The frequency with which the statute limiting purveyance was re-enacted, the history of Edward III's handling of the anti-papal legislation,[2] the numerous petitions for observance of statutes, the protests against pardons granted for breaches of statute law, all tell the same tale. The legislative activity of parliament is increasing and initiative lies more and more with the commons, though some of their petitions may have been inspired from without, by magnates or councillors; but the execution of the law is an administrative problem and, as such, lies largely outside parliamentary control. Moreover, legislation did not necessarily originate with a *commune petition*; like the Statute of Staples of 1354, it might originate in an ordinance of the king and his council. The distinction between statute and ordinance establishes itself very slowly; in the early years of the reign, they appear to be synonymous terms. But by 1354 when, at the request of the commons, the Ordinance of the Staple, approved in the great council of the previous year, was re-enacted in parliament as a statute *pur greindre feremete*,[3] then a distinction between the ordinance as a temporary administrative measure and the statute as a permanent addition to the law, seems to have been accepted in principle, though in practice it was not to be consistently observed.[4]

[1] *Rot. Parl.* ii. 130, 149, 165, 203, 272.   [2] Below, pp. 281–2.
[3] *Rot. Parl.* ii. 253.
[4] H. G. Richardson and G. O. Sayles, 'The Early Statutes', *Law Quarterly Review*, l (1934), 201–23, 540–71.

This great development of statute law, characteristic of the fourteenth century, owed much to the initiative of the commons in parliament; but it has also to be remembered that it was the almost inevitable outcome of the expanding scope of the law and of the accumulation of business in the king's courts. Earlier kings, like Henry II, Henry III, and even Edward I, had modified the customary law merely by issuing instructions to the justices; but this practice could not be maintained indefinitely and the written statute is the outcome of the obvious need to put changes in the body of the law into writing, in a form accessible to the public, through proclamation in the county court.[1] Multiplication of statutes gave rise, however, to many problems; the mere fact that it was written and proclaimed did not make a law simple. 'This is what the law says, but what does the law mean?' was the practical question facing judges, counsel, and litigants daily in the courts. Many of Edward III's statutes, based as they were on common petitions and hastily issued in order to expedite grants of supply, were very loosely constructed and difficult to master, even by lawyers, who indeed often misquoted them.[2] In short, statutes needed interpretation. This need had been less urgent in the days of Edward I when it was commonly the same men—the judges—who both drafted and applied the statutes. 'Do not gloss the statute', said the great Chief Justice Hengham on one occasion, 'for we know better than you: we made it.' But as the courts moved farther away from council and parliament and as parliament itself lost much of its judicial character, such guidance was no longer available; and by the early years of Edward III, the judges often had to try to infer the meaning of a statute from the text alone. The process whereby they thus assumed responsibility for the interpretation of statutes was a gradual one, for the tendency to regard interpretation as a form of fraud, died hard. So late as 1336, the final clause of a statute directs that it shall be observed without any kind of evasion or contrivance, 'ou par interpretacion des paroles.[3] But by the end of the reign, though the

[1] For much of what follows I am indebted to Professor T. F. T. Plucknett's two studies, *Statutes and their Interpretation in the First Half of the Fourteenth Century* (1922) and *Concise History of the Common Law* (5th ed. 1956).

[2] Chaucer's man of law is thought to have been Thomas Pynchbek, admitted serjeant in 1376 and later chief baron of the exchequer and a justice of the common pleas; but it is hard to believe, even of so eminent a lawyer that 'every statut koude he pleyn by rote' (*Canterbury Tales* (Prologue), l. 327).

[3] *Statutes* 10 Edw. III, st. 3.

judges still occasionally refer doubtful points to parliament or council, they are not far from the modern position of accepting statutes as the commands of an authority external to themselves whose will has to be deduced from the written text.

Interpretation was not the only problem arising from the multiplication of statutes. Statutes were still something of a novelty and this 'novel ley' had to be reconciled with the unwritten common law practised in the courts. In the early years of the fourteenth century some judges seem to have taken the view that litigants were free to choose between the remedies offered under common and under statute law. A large measure of discretion was allowed to the bench and, on the whole, it seems to have been used in the interests of justice. But the growth of statute law caused judges to look more and more to the guidance of the written text, an attitude epitomized in a judicial dictum of 1346: 'We cannot do this in the absence of any statute.'[1] The reluctance of some judges to exercise their discretion brought in its train a loss of flexibility, a hardening of procedure in the common-law courts, which the statutes were powerless to remedy. Statutes were, indeed, often concerned with law reform. They could abolish, or attempt to abolish, such archaisms as the murder fine or presentment of Englishry (1340), or the use of unintelligible Norman-French in the courts (1362),[2] they could define, clarify, and amend the existing law, but they could never supply the need, recognized from ancient times, for some residue of discretionary power, in the Crown or its officers, able to mitigate the rigours of the customary or written law. Thus, it is no accident that the same century which saw the acceptance of statute law as a normal and necessary form of legislation, should also have seen the beginnings of a system of equity administered in courts other than the courts of common law. Moreover, the common law, even when supplemented by statutes, could not of itself supply all the needs of an increasingly complex society. Its chief concern had always been with real property and it was not

---

[1] *Gisors* v. *Anon. Year-Book, 19 Edw. III* (Rolls Ser.), p. 12.

[2] *Statutes*, 14 Edw. III, st. 1, c. 4; 36 Edw. III, st. 1, c. 15. Though a litigant who appeared in person could avail himself of the act of 1362, it proved impossible to substitute English for Norman-French as the language of the law, French legal terms having become *vocabula artis*. See W. S. Holdsworth, *A History of English Law* (3rd ed. 1923), ii. 477–82. Examples of the growing precision of the legal language are discussed by J. P. Collas and T. F. T. Plucknett in their Introduction to the *Year-Books, 12 Edw. II* (Selden Soc. lxx, 1951), pp. xii–lxiv. See also G. E. Woodbine, 'The Language of the English Law', *Speculum* xviii (1943), 395–436.

well adapted to the needs of litigants who were neither landowners nor tenants; it was slow to admit the need for evidence other than the evidence of written records; the juries empanelled in its courts were neither impartial nor incorruptible; and its procedure was intolerably slow. Persons for whom its actions afforded no remedy, persons who went in fear of their neighbours, persons in need of speedy or extraordinary relief, turned almost inevitably to the king as the ultimate source of justice and sought to obtain their ends by way of petition.

It was the king's council which supplied the deficiencies of the common-law courts and entertained petitions from subjects of all types. Such petitions were commonly presented in parliament and, if the case were of sufficient importance, might occasionally be heard there. When Blanche Lady Wake brought her suit against the bishop of Ely into the parliament of 1355, the king intervened in person with the words, 'Jeo prenk la querele en ma main.'[1] But the bulk of petitions came before the council, the small group of officials, clerks, and advisers attendant on the king, who exercised the residuum of justice and equity inseparable from the royal person. Since the council also from time to time took upon itself to deal with offenders too powerful to be reached by the ordinary courts, it soon found itself overwhelmed by a multiplicity of business.[2] Some delegation was clearly necessary and as the century advanced, the custom grew of delegating the judicial business of the council to the chancellor, the official most constantly in attendance at court, who was already furnished with a staff of clerks expert in the technique of issuing both judicial and administrative writs. Parliament itself encouraged this development by expressly assigning such subjects as misdemeanours of officials (1347), or of purveyors (1362), or of offences under the Statute of Praemunire (1353) to the jurisdiction of the chancellor and council.[3] In the council, the chancellor was the presiding officer and it was left largely to his discretion to decide how many or which of the ordinary members should be summoned to deal with judicial business. Even before the beginning of Edward III's reign, petitioners occasionally addressed themselves 'to the chancellor and council', and such forms become increasingly common; but

[1] *Rot. Parl.* ii. 277.
[2] For examples see *Select Cases before the King's Council*, ed. I. S. Leadam and J. F. Baldwin (Selden Soc. xxxv, 1918).
[3] *Statutes* 20 Edw. III, c. 6; 36 Edw. III, c. 9; 27 Edw. III, c. 1.

there are also numerous examples of petitions addressed to the chancellor alone and, by the time of Richard II, the frequency of these seems to indicate a new stage of development in the separation of the chancellor's court from the council.[1] Such petitions were still greatly outnumbered by those addressed to the council; but before the close of the century, the evolution of chancery as a court of equity had begun.

Even if ecclesiastical, private, and peculiar jurisdictions are left out of account, the elaboration of the judicial system under Edward III remains impressive. At its apex were king, council, and parliament; below them were the great central courts—the exchequer which had long entertained cases arising from the audit of the revenue and which, in 1357, developed a new off-shoot, the court of exchequer chamber, statutorily erected to hear errors from the court of exchequer;[2] the court of king's bench, a separate institution but still closely allied to the council, handling cases supposed to be of special concern to the king and correcting errors in the common pleas; and the court of common pleas, or common bench, hearing such cases between subject and subject as came within its purview and correcting errors in local courts.[3] From time to time (though, by an accepted convention, not more than once in seven years), specially commissioned royal justices visited the counties to hold comprehensive inquiries (*ad omnia placita*) in extraordinary sessions of the county courts. The unpopularity of these general eyres is attested by the frequent petitions for their abolition and they had fallen out of use by the end of the reign.[4] But it had long been the custom for the possessory assizes to be taken locally by commissioners who need not be justices;[5] and under the system

[1] J. F. Baldwin, *The King's Council*, p. 249: *Select Cases in Chancery*, ed. W. P. Baildon (Selden Soc. x (1896), 1–50).   [2] *Statutes* 31 Edw. III, st. 1, c. 12.

[3] References to 'the bench' in the Middle Ages are always to this court. The king's bench and the common pleas shared certain actions, such as conspiracy and trespass, and there was some rivalry between them. See Margaret Hastings, *The Court of Common Pleas in Fifteenth-Century England* (1947), pp. 16–24. The courts of king's bench and common pleas were given seals of their own in 1344. B. Wilkinson, 'The Seals of the Two Benches under Edward III', *Eng. Hist. Rev.* xlii (1927), 397–401.

[4] There are some brief proceedings for a Kentish eyre in 1348; the latest reference to a commission for the eyre is for Kent in 1374. Bertha H. Putnam, *Proceedings before the Justices of the Peace in the Fourteenth and Fifteenth Centuries* (1938), p. xlvi.

[5] Often, like Chaucer's man of law, they were serjeants:

> Justice he was ful often in assise,
> By patente and by pleyn commissioun.
> (*Canterbury Tales*, Prologue, ll. 314–15.)

known as *nisi prius*, certain types of cases initiated in the common pleas could be terminated in the county courts before the justices of *nisi prius* who, because it became customary for them to take the assizes also, were later known as justices of assize. These same justices, under commissions of trailbaston, *oyer et terminer*, or jail delivery, could also entertain criminal cases. Hence arose the judicial circuit on which justices of the superior courts made regular tours of the country and brought these courts into direct contact with the localities.

In the counties, where antiquated procedures such as compurgation and judgement by the suitors were still in use, the judicial—as distinct from the administrative—powers of the sheriff had long been declining. By the opening of Edward III's reign, these powers were normally confined to small pleas of debt and trespass, though, under special writs of *justicies* or *loquelam audias*, the sheriff might be empowered to hear cases of greater significance.[1] Twice a year, on his tourn, he presided over a session of each hundred in his shire which was not in private hands, executed summary jurisdiction over such petty offences as brawls and affrays, and received indictments for pleas of the Crown. He was also responsible for the empanelling of juries and for the arrest and custody of offenders. It was the business of the coroner, whose functions were primarily those of a recorder, to keep rolls of inquisitions into sudden deaths, indictments of felonies, and concealment of treasure-trove. Subordinate officials included the bailiffs whom the sheriffs might authorize to take distraints, make arrests, and summon persons to court; and the constables of hundreds and townships whose duty it was to follow up the hue and cry, make arrests and imprison suspects, and on whom the brunt of enforcement of the labour laws seems to have fallen.[2] To assist in the maintenance of order, commissions were also issued to a number of local knights and gentry who in this century came to be known as justices of the peace.

The evolution of the justice of the peace is a complicated story, reflecting a division of opinion as to the best means or maintaining order in the country. The commons in parliament

---

[1] Such extensions of the sheriff's judicial powers were very unpopular; e.g. the complaint of the commons in 1354 that he was using them for purposes of extortion. *Rot. Parl.* ii. 262.   [2] Putnam, *Proceedings*, p. cxxvi.

and some of the magnates favoured an extension of judicial powers to the local knights and gentry who, for over a century, had been appointed periodically as keepers of the peace; but the government, the judges, and the professional lawyers were suspicious of these local amateurs and preferred to make use of extraordinary commissions of lawyers and magnates (notably of commissions of trailbaston) which could be expected to yield a rich harvest of fines and amercements, an expectation which goes far to explain their unpopularity in the shires. Under Edward II, many commissions *de custodibus pacis* were issued, the main duty of the commissioners being enforcement of the Statute of Winchester of 1285; by 1314 they had power to arrest suspects and by 1316 to inquire into felonies as well as trespasses; in Kent, at least, though without any statutory sanction, they were virtually acting as justices.[1] The act of 1327 was indicative of a re-action, for, while confirming the traditional functions of the keepers, it yielded nothing to the clamour of *la commune* that they should be given power to punish as well as to inquire;[2] and distrust of the keepers is clearly evident in the Statute of North-ampton, the great police measure of the following year.[3] Sir Geoffrey Scrope, then chief justice of the king's bench, was bent upon the restoration of the general eyre as the best and most profitable means of restoring order;[4] the keepers were entrusted merely with enforcement of the new provision against bearing arms, and that only in conjunction with other local officials. An unrecorded controversy in parliament may have led to their powers being increased in the commissions of May 1329 which include authority to determine both felonies and trespasses; but this policy was reversed by statute after the fall of Mortimer in 1330, when the powers of the keepers were once again reduced to insignificance. Power to determine was restored in the com-missions of February 1332, yet dissatisfaction with the keepers is shown by the appointment, a few weeks later, of certain magnates as 'keepers of counties' to whom the keepers of the

[1] Bertha H. Putnam, 'Kent Keepers of the Peace 1316–17', *Kent Archaeological Soc., Records Branch*, xiii (1933), p. xxi; 'The Transformation of the Keepers of the Peace into the Justices of the Peace', *Trans. Royal Hist. Soc.*, 4th ser. xii (1929), 19–48.

[2] *Statutes* 1 Edw. III, st. 2, c. 16.

[3] Ibid. 2 Edw. III, cc. 1–7.

[4] Helen M. Cam, 'The General Eyres of 1329–30', *Eng. Hist. Rev.* xxxix (1924), 241–9.

peace and other local officers were to be answerable. Fluctuations in the powers entrusted to the keepers persisted for more than thirty years, uneasiness as to their capacity being reflected from time to time in the provision that they shall be 'afforced' by one or more professional lawyers. In 1350 and 1351 they were made responsible for enforcement of the new labour laws; but these powers were withdrawn during the period of special labour commissioners (1352–9). They acted as commissioners of array at intervals between 1338 and 1359, when this function was transferred to others.[1] It was not until 1368 that their judicial powers were placed firmly and finally on a statutory basis.[2] Under the act of this year, the commissions to the justices include maintenance of the peace and of the Statutes of Winchester (1285) and Northampton (1328); inquiry into felonies and trespasses, labour laws, weights and measures, forestalling and regrating;[3] determination of felonies and trespasses, a quorum of professional lawyers being necessary for the former; and provision for the appointment of a *custos rotulorum*.[4] The six justices in every county, provided for by an act of 1388, were raised to eight in 1390.[5]

Thus, after long delays and many setbacks, the justices of the peace came to be established as an integral part of the machinery of English justice. To Sir Edward Coke, with a century of Tudor rule behind him, they seemed to be 'such a form of subordinate government for the tranquillity and quiet of the realm, as no part of the Christian world hath the like';[6] but the justices brought no access of tranquillity to fourteenth-century England and it is easy to understand why an experienced judge like Scrope should have viewed the system with apprehension. It

[1] For details see Putnam, *Trans. Royal Hist. Soc.*, loc. cit., and *Proceedings*, pp. xxii ff.
[2] *Statutes* 42 Edw. III, cc. 4, 6. Professor Putnam pointed out that the importance of the earlier act of 1361 has been exaggerated, since the powers of the justices were again curtailed in 1364.
[3] A forestaller was one who intercepted sellers on the way to market and attempted to raise prices artificially: a regrator, one who bought goods with a view to reselling them at a higher price in the same market.
[4] The origin of the Quorum is traced by Professor Putnam to the act of 1344 (*Statutes* 18 Edw. III, c. 2) under which men 'sages et apris de la ley' are to be added to the keepers when they act as justices. The intention probably was that only those members of the commission learned in the law (the Quorum) should exercise judicial functions; but the practice grew of making all the J.P.s members of the Quorum, the necessary legal knowledge being supplied by the clerk who kept the rolls. See Holdsworth, i. 127.　　[5] *Statutes* 12 Ric. II, c. 10; 14 Ric. II, c. 11.
[6] Coke, *4th Inst.*, p. 170.

was economical, for the justices were unpaid; and it may have had its uses in so far as the justice who had sat in parliament should have been able to preside over the local sessions with some understanding of governmental policy and problems, and the parliamentary knight who had been a justice to bring some knowledge of local conditions into the national assembly. But its effective application demanded closer supervision from the centre and better policing of the countryside than any medieval government was able to provide. The evidence at our disposal makes it abundantly plain that fourteenth-century England, for all its multiplicity of courts, statutes, and justices, was not a law-abiding country and that those responsible for the maintenance of order were faced with obstacles beyond their power to surmount.

It did not make for good order in the land that, the clergy apart, almost the whole male population of fighting age should have been not only armed, but trained in the use of arms. Under Edward I's Statute of Winchester (modified in points of detail by the Statutes of Northampton, 1328, and Westminster IV, 1331), laymen between the ages of fifteen and sixty were bound to possess arms, ranging from the helmet, hauberk, and sword of the knight down to the poor man's bow, arrows, and knife. The object of this legislation was, of course, the defence of the realm and the maintenance of the peace. Constables were in command of the forces of township and hundred which made up the *posse comitatus* under the command of the sheriff; and on many occasions and in many places, most notably on the northern border, the *posse* did good service. But to arm the population entailed arming its numerous criminals and accustoming ordinary citizens to the expert handling of weapons, a tendency further encouraged by Edward III's exhortations concerning the practice of archery.[1] Doubtless, the ordinary man needed no official encouragement to carry a knife in his belt when he went abroad; but it was a more serious matter that the able-bodied men of the village, trained to act as a military unit, could without difficulty organize themselves in defiance of the law.[2] It may well have been the experience gained by the country

[1] *Foedera*, iii. 704, 770.
[2] The smallest unit of foot-soldiers in the *posse* was the *vintenary* of twenty men. See A. E. Prince, 'The Army and Navy', *English Government at Work*, i. 355.

folk in the *posse* which made it necessary for the Black Prince himself, with Lancaster and other lords, to take armed forces into Cheshire to suppress the rising of 1353;[1] or which facilitated the rising of Oxfordshire villagers, *agrestes arcitenenses*, with their strange cries of 'Havak, havok, smygt faste, gyf good knok', at the time of the Oxford riots on St. Scholastica's day, 1355;[2] or, indeed, of the men of Kent and Essex in 1381.

Mass movements on this scale were exceptional; but no student of the period can fail to be impressed by the general prevalence of criminal groups and gangs, the *compaignies*, *conspiratours*, *confederatours*, which were the subject of so many parliamentary complaints.[3] The 'draw-latch', 'wastor', or 'roberdesman' might work alone;[4] but it was the roving bands of criminals, many of them highly organized, for whom the sparsely-populated, thickly-wooded countryside afforded such ample cover, who terrorized the ordinary citizen and defied the officers of the law. The activities of the Folvilles, as recently revealed to us, show six brothers of a knightly family (one of them a priest) forming the nucleus of a criminal gang in Leicestershire in the early years of Edward III, a gang which after sixteen years of recorded crime was still untouched by the law.[5] Besides countless less conspicuous crimes, the Folvilles were responsible for the murder of a baron of the exchequer (Roger Bellers) on the high road, near Leicester, and for the capture and holding to ransom of Sir Richard Willoughby, a justice of the king's bench. In one of his indictments, Eustace de Folville is referred to as *capitalis de societate*: and we know that his associates included a village parson, a clerk (described also as *miles*), and the constable of Rockingham castle, Sir Robert de Vere. Crime on this scale among the landed gentry can hardly have been general; but the exploits of the Folvilles do not appear to have been very different from those of Sir Robert Holland and Sir Adam Banaster in Lancashire, or of Sir Gilbert Middleton in Northumberland; and the extraordinary letter sent to a Yorkshire parson, by one

---

[1] H. J. Hewitt, *Medieval Cheshire* (1929), p. 17.

[2] *Collectanea* III, ed. H. Furneaux, *Oxf. Hist. Soc.* (1896), p. 185. The cry *havak* or *havoc* was the signal for general spoliation and pillage. It was strictly forbidden to the ranks in wartime. See *Black Book of the Admiralty* (Rolls Series), i. 286–7, and note 5.

[3] e.g. *Rot. Parl.* ii. 64, 165.　　　　　　　　[4] *Statutes* 5 Edw. III, c. 14.

[5] E. L. G. Stones, 'The Folvilles of Ashby-Folville, Leicestershire, and their Associates in Crime', *Trans. Royal Hist. Soc.*, 5th ser. vii (1957), 117–36.

describing himself as 'Lionel, king of the rout of robbers', from his 'Castle of the North Wind', is unlikely to have been the work of a common footpad.[1] No doubt, as is often alleged, crime and disorder increased as a result of the Black Death and when disbanded soldiers began to return from the wars; criminals were said to be flocking into London after the cessation of the pestilence and the chroniclers refer to an increase of crime, particularly of crimes of sacrilege, in the sixties.[2] But there is abundant evidence to show criminal gangs at work before the war began and those who joined the army included many professional criminals, enticed into service by the offer of a royal pardon for their past misdeeds. Murderers, robbers, and poachers are said to have been responsible for the victory of Halidon Hill (1333); and over 200 holders of royal pardons for crime are to be found among the shire levies serving in Scotland in 1334–5.[3] Of the notorious Folvilles, one, Eustace, went overseas in the retinue of the earl of Northampton in 1337, another, Robert, was summoned, together with several of his associates in crime, to serve in Flanders in 1338. So late as 1390 the commons were still protesting against the issue of such pardons to men convicted of murder, treason, or rape.[4]

Impartially administered justice was not to be looked for in such a world. Officials from the highest to the lowest were corruptible and the people knew it. Even the judges, with their high professional qualifications, salaries, fees, and pensions, were by no means above suspicion. In 1341 John Inge, a justice of the common pleas, pleaded guilty to the charge of having taken money on many occasions from accused persons and litigants while he was acting as justice of assize and jail delivery; Sir John Willoughby was accused 'by clamour of the people' of selling the laws as if they had been oxen or cows.[5] In 1350 Sir William Thorp, chief justice of the king's bench, was deprived and imprisoned for taking bribes. Fifteen years later Sir Henry Green, one of his successors in office, and Sir William Skipwith, chief

[1] *Trans. Royal Hist. Soc.* (1957), pp. 134–5. This case came into the courts in 1336.
[2] *Cal. Letter-Book F*, p. 210; Knighton, ii. 120; John of Reading, p. 178.
[3] J. E. Morris, 'Mounted Infantry in Medieval Warfare', *Trans. Royal Hist. Soc.*, 3rd ser. viii (1914), 93; A. E. Prince, 'The Strength of English Armies in the Reign of Edward III', *Eng. Hist. Rev.* xlvi (1931), 354.
[4] *Rot. Parl.* iii. 268.
[5] Dorothy Hughes, *Early Years of Edward III*, p. 184; *Year Book 14–15 Edward III* (Rolls Series), p. 258.

baron of the exchequer, suffered a like fate on account of what are described as their *enormes infidelitates*.[1] The commons, in 1339, alleged that trailbaston commissions did more harm to the innocent than to the guilty; in 1365 they were protesting against the appointment of commissioners of *oyer et terminer* for life, with a grant of a third of the fines and amercements, an arrangement which encouraged the justices to hold superfluous sessions, to bribe jurors, and to procure false indictments.[2] Attacks on judges during the rising of 1381 may have derived much of their impetus from hatred of the labour laws in particular; but it is hardly to be doubted that they also owed something to widespread popular suspicion of the integrity of even the most highly placed administrators of the law.

The corruptibility of the sheriffs was notorious. Though shorn of most of their judicial powers, they were still indispensable to the smooth running of the machinery of justice. The commission to the justices of the peace was always accompanied by the writ *de intendendo* to the sheriff whose duty it was to summon panels of jurors and, if the persons accused pleaded not guilty and put themselves on the country, to summon a trial jury, usually for a later session. Thus, it was open to the sheriffs to empanel jurors to suit one of the parties, to procure wrongful indictments, and to make false returns. In 1314 the commons of Suffolk petitioned against false indictments preferred by subordinate officials for their masters' advantage; outcries against the sheriffs' habit of compelling people to prefer indictments led, in 1330, to the removal of all the sheriffs and their subordinates throughout the country.[3] In the case of *Ughtred* v. *Musgrave* (1366) the sheriff of Yorkshire, Thomas Musgrave, was accused of malicious arrest, false imprisonment, extortion of money, and an endeavour to entrap complainants into an attempt to abuse the forms of law by arranging for a collusive indictment by a packed jury. Musgrave was said to have seized Ughtred's servant, one Robert Woolman, put him in durance and tortured him to within a little of death until he agreed to lay false accusations of felony and larceny against his master.[4] The number of cases of corruption which found their way on to the assize rolls proves, however, that such offences did not always go unpunished; and we read of a jury

---

[1] Tout, *Chapters*, iii. 259.
[3] Ibid., i. 293; ii. 9, 12.
[2] *Rot. Parl.* ii. 286.
[4] *Select Cases before the King's Council*, pp. lxxxiv–lxxxvii, 56.

empanelled for a trial during the Yorkshire sessions of the peace in 1362 being successfully challenged by the king's attorney on the grounds that the person accused was of the livery of the sheriff (*ad robas vicecomitis*), Marmaduke Constable.[1]

The government was by no means indifferent to the prevailing lawlessness and petitions on the subject seldom failed to elicit a favourable response. The four main methods of defeating the ends of justice—bribery; maintenance (the unlawful upholding of another's suit by word, writing, encouragement, or other act); embracery (the attempt to influence a jury by money, promises, threats, or persuasion); and champerty (maintaining a suit in consideration of receiving a part of the land, damages, or chattels recovered) were all forbidden by statute under Edward III and Richard II. But statutes were of little avail when a session of the commissioners of *oyer et terminer* could be broken up by a knight invading the hall with drawn sword and scizing one of the justices by the throat;[2] when the earl of Devon himself could send a message to a justice of the peace telling him 'that he was false and that he should answer with his body, that he (the earl) knew all the roads by which he must come and go and that he should not escape the hands of the said earl who was sure of him';[3] when a canon of Sempringham and the cellarer of a Cistercian house could be found among those who employed the Folvilles to destroy a rival's water mill; and when a country parson could be dragged from his bed at daybreak and carried away to Sherwood.[4]

The great forests of medieval England must be reckoned among the most formidable obstacles to the enforcement of law throughout the land. They afforded safe refuge to innumerable fugitives from justice; and the forest law itself acted as an inducement to trespassers and poachers. In the fourteenth century, however, its severity was somewhat relaxed; and the long struggle for disafforestment comes to an end in 1327, when, probably as a bid for popularity, Isabella and Mortimer sponsored an act which ordered that the perambulations made under Edward I should be taken as establishing the future boundaries of the forest and that in those counties where the

---

[1] Bertha H. Putnam, *Yorkshire Session of the Peace 1361-64* (Yorks. Arch. Soc. C. 1939), p. 29. [2] *Cal. Pat. Rolls 1354-58*, p. 166.
[3] *Select Cases before the King's Council*, p. 79 (1390). [4] Ibid., p. 47.

bounds had not been ridden, perambulations should be instituted forthwith.[1] This legislation was confirmed in general terms in later years; and though complaints of illicit afforestments by officials continued to be heard, neither Edward III nor Richard II attempted to repudiate the principles laid down in 1327. Large areas of the country, notably in Essex, the north midlands, Yorkshire, the Welsh border shires, and between Southampton Water and the river Kennet, still lay under forest law;[2] and the elaborate regulations laid down by earlier kings for protection of the beasts of the forest (the venison) and of the timber and undergrowth (the vert), with the hierarchy of courts and officials set up to enforce them, remained much as they had been in the early thirteenth century;[3] but clamour against them had become rather less vociferous. Such complaints as were heard in parliament related to abuses of the system, rather than to the system itself, the commons asking that the boundaries established under Edward I should be maintained; that persons dwelling outside them should not be charged with forest offences; that there should be no suborning of witnesses or imprisonment without due indictment; and that the forest officials should not demand contributions to which they had no just claim.[4] Under pressure from parliament, both Edward III and Richard II expressed themselves as desirous of protecting their subjects against abuses of this kind; they ordered inquisitions, issued rebukes to offending officials and reminders of the correct procedure for the punishment of forest offenders;[5] and similar pressure produced certain modifications in the law itself. Edward III gave permission to certain dwellers in the forest to take wood for their houses and fences; and in 1369 he allowed a general indemnity to offenders against the forest law.[6] An act of

[1] *Statutes* 1 Edw. III, st. 2, c. 1.

[2] The exact area of the forest is very difficult to determine, however, partly because the bounds laid down as a result of the perambulations were not always observed, partly because identification of local landmarks referred to in the records (particularly in the more inaccessible regions of England) has become almost impossible. See Nellie Neilson, 'The Forests', *English Government at Work*, i. 397 ff. A map indicative of the afforested areas in the decade 1327–36 will be found at the end of Miss Neilson's essay.

[3] See A. L. Poole, *Domesday Book to Magna Carta* (vol. iii of the present *History*), pp. 29–35.

[4] *Rot. Parl.* ii. 239, 311, 335; iii. 164.

[5] *Cal. Close Rolls 1343–46*, p. 257; *Statutes* 25 Edw. III, st. 5, c. 7.

[6] C. Petit-Dutaillis, *Studies Supplementary to Stubbs's Constitutional History*, ii (1915), 242.

1390, restricting hunting-rights to the wealthier classes, was provoked by complaints from the gentry of an increase in poaching;[1]
but this act is to be regarded as a foretaste of the later game-laws
rather than as an intensification of the medieval forest code. Even
in the twelfth century, as Dr. Poole has reminded us, life in the
forest had its cheerful side; by the fourteenth, the forest law is no
longer a national grievance and it is significant that its abolition
finds no place among the demands of the rebels in 1381, Wat
Tyler contenting himself with a claim for free hunting and fishing.[2] Sloth, in *Piers Plowman*, already knew 'rymes of Robyn
Hode'; the forest, which so often harbours crime and tragedy,
may also harbour romance.[3]

[1] *Statutes* 13 Ric. II, st. 1, c. 13; *Rot. Parl.* iii. 273. Poachers were said to be particularly active on feast-days, 'a temps qe bones Cristiens . . . sont as Esglises'.
[2] Knighton, ii. 137.
[3] *Passus*, C viii. 11. Dr. Poole seems to favour identification of this elusive character with a Robert (or Robin) Hood, *fugitivus*, known in Yorkshire in 1230.
Dr. E. K. Chambers (*English Literature at the Close of the Middle Ages*, 1945, p. 130)
thought it likely that the story originated with a Robin Hood who was in prison in
1354, awaiting trial for trespasses of vert and venison in Rockingham forest; but
this hypothesis seems to allow very little time for the growth of the legend.

VIII

EDINGTON AND WYKEHAM
(1344–71)

THE most important reason for the lightening of the political atmosphere after the crisis of 1340–1 was the turn for the better taken by English fortunes abroad. In the Low Countries, the defection of many of his allies relieved Edward III of some of his most burdensome commitments; and the strategy of large continental coalitions was wisely abandoned. The opening of the succession war in Brittany in 1342 offered prospects so alluring that even Sir Robert Bourchier chose to exchange custody of the great seal for the pleasures of active service and drew £400 in wages for his company of 100 archers and seventy men-at-arms. Disappointed in his hopes of the see of York, the notorious Kilsby followed his example.[1] Crécy, Neville's Cross, Derby's brilliant campaign in Gascony, the capture of Calais, all served to make heroes of Edward III and his captains, to enrich the fighting-men and to swell the pride of parliament and the nation. Experience had taught the king, however, that successful foreign policy demanded sound domestic administration; and the easing of tension in these years owed much to the men whom he put in key positions at home.

On his return from the Netherlands at the end of 1340, Edward III had declared his intention of having as his ministers laymen who would be justiciable in his courts; but lay ministers were costly to the state. Bishops had their town houses and the emoluments of their sees and of other ecclesiastical preferments; the average layman was less well endowed. Parvyng, it is true, had a house of his own in Aldermanbury, but both Bourchier and Sadington had to rent episcopal lodgings and all three had to be provided with allowances over and above the normal chancellor's fee.[2] Thus, it was not long before Edward reverted to the practice of his ancestors in appointing clerical ministers;

---

[1] Tout, *Chapters*, iii. 157, 162.

[2] Bourchier lived in the house of the bishop of Worcester, Sadington in that of the bishop of Lichfield (Chester's Inn). Parvyng and Sadington each had allowances of £200 a year, Bourchier had £600.

and, throughout the period when he and his generals were winning their most spectacular victories abroad, responsibility for the direction of affairs at home rested with churchmen. Not only the chancellors, but also the treasurers and the keepers of the privy seal were clerks. For four years after Sadington's resignation in 1345, the chancellor was John Offord, dean of Lincoln, a trusted household clerk who had gained useful diplomatic experience with Edward in the Netherlands between 1338 and 1340. When Stratford died in 1348, Clement VI provided Offord to Canterbury, but the chancellor died of the plague before he could be consecrated. His successor at the chancery was John Thoresby, bishop of Worcester, who, in 1352, became archbishop of York. Thoresby was the professional civil servant *par excellence*. As a clerk of the chancery, he had been 'constantly attendant on the king's business' so early as 1333; twelve years later he was keeper of the privy seal and he had acted as keeper of the great seal during the illness of Parvyng. When, at his own request, he was relieved of office in 1356, the new chancellor was an official of a similar type, William Edington, bishop of Winchester, a former keeper of the wardrobe and treasurer of the exchequer. After him, in 1363, the chancery passed to the bishop of Ely, a Benedictine monk named Simon Langham, who had not found his position as abbot of Westminster incompatible with tenure of the treasurer's office. Langham became archbishop of Canterbury in 1366 and was succeeded as chancellor by the keeper of the privy seal, William of Wykeham, who was simultaneously elevated to the see of Winchester. Episcopal treasurers in this period included, besides Edington and Langham, John Sheppey, bishop of Rochester (1356–60), John Barnet, bishop of Bath and Wells, and, from 1366, of Ely (1363–9), and Thomas Brantingham, bishop of Exeter (1369–71); while, of the nine keepers of the privy seal between 1343 and 1371, six were, or became bishops.[1]

Capable, devoted, and generally discreet, his ecclesiastical ministers served Edward well. They were neither servile nor forgetful of the immunities of their order;[2] but there was no

---

[1] Thoresby, Wykeham, Hatfield of Durham, Islip of Canterbury, Northburgh of London, and Buckingham of Lincoln.

[2] In 1355 Edward administered a sharp rebuke to the chancellor and treasurer (Thoresby and Edington) for their failure to carry out his orders to confiscate the temporalities of the bishop of Ely, suggesting that they would not have hesitated if the offender had been a great lay peer. B. Wilkinson, 'A Letter of Edward III to his

recurrence of the kind of friction that had proved so disruptive in the early years of the war. The ministers were content to serve not, indeed, for no reward, but without claiming the kind of intimacy and special privilege which would have rendered them obnoxious to magnates and people. To meet the situation created by the king's absences abroad, an administrative pattern established itself which was followed with little deviation. A child regent—the duke of Cornwall in 1342-3, Lionel of Clarence in 1346-7, Thomas of Woodstock in 1359-60—was provided with a privy seal which, in the keeping of officials like Simon Islip or John Buckingham, became, in effect, a privy seal of absence. To advise the regent, the chancellor, furnished with a great seal of absence, and a selected number of councillors remained at home. Abroad with the king went the keeper of the privy seal, with the great seal in his custody, the wardrobe and household officers, and a large number of clerks. War conditions did not allow of exchanges of acrimonious letters or violent clashes of policy such as had been almost inevitable when the king and his ministers were lodged in a foreign capital surrounded by watchful and suspicious allies, creditors, and mischief-makers. The officials of 1346-7 and of 1359-60 were on active service. In 1346 contact between the government at home and the government abroad was entirely suspended from the time of the king's departure from Caen until his arrival before Calais;[1] and in 1359-60 it became difficult to maintain communication between the two branches of government while his officials accompanied Edward III from Calais to Rheims, from Rheims to Burgundy, from Burgundy to the gates of Paris, and from Paris to Chartres and Brétigny.

William Edington is the outstanding minister of the most active period of the war. A man of obscure origin who may have studied at Oxford,[2] he owed his first appointment as king's clerk to his patron, Bishop Orleton of Salisbury, and for five years (c. 1335-40) he served the king in this capacity. Collector of the

Chancellor and Treasurer', *Eng. Hist. Rev.* xlii (1927), 248-51. It was later pointed out, however, that the confiscation was a breach of the statute of 1340. See H. G. Richardson and G. O. Sayles, 'The Early Statutes', *Law Quarterly Review*, l (1934), 554.
  [1] Tout (*Chapters*, iii. 170) points out that there must have been a large number of chancery clerks in the Calais camp to write and enrol the numerous writs issued. This roll, known as the *Rotulus Normanniae*, is calendared in *Cal. Pat. Rolls 1345-48*, pp. 473-577; a continuation, commonly called the *Calais Roll*, is summarized ibid., pp. 518-70.   [2] See A. B. Emden, *Biog. Register of the Univ. of Oxford*, i. 629.

ninth south of Trent 1340-1, it was his elevation to the important post of keeper of the wardrobe, 1341-44, which assured his future. He was treasurer of the exchequer, 1345-56, chancellor, 1356-63; and bishop of Winchester from 1346 until his death twenty years later. A good friend, we are told, to the commons whom he protected from royal extortions,[1] and assuredly a good friend to the king, Edington has left only a faint impress on the pages of history; something of the anonymity of the model civil servant still clings to him. Yet it is likely that the successful financing of the war in the years of victory owed more to him than to any other single man. Before his retirement in 1363 he had seen Edward through all his great campaigns and it was fitting that he should have been a witness to the treaty of Calais which ended the first phase of the Hundred Years War. He was a munificent benefactor to his cathedral and to his native village of Edington in Wiltshire where he built a splendid collegiate church; and it was in order to devote himself to the interests of his diocese that he retired from the royal service some three years before his death. His refusal of Edward's offer of the primacy says much for his wisdom and disinterestedness.

When Edington became chancellor in 1356 he found himself at the head of an elaborately organized bureaucracy.[2] The chancery was lodged at Westminster in the *hospicium* where the chancellor provided a table for the clerks, twelve of the first and twelve of the second grade, who led a semi-collegiate existence with well-established traditions and a strongly-marked *esprit de corps*. With its two subordinate offices, the office of the rolls at the Tower, where classified records of all the writs issued were compiled and stored, and the office of the hanaper which collected the fees for the writs and delivered them to the exchequer, the chancery had already acquired the status of a semi-independent administrative department. Many of its duties were of a formal character, routine administrative processes initiated by the great seal, such as the issue of writs of course initiating actions at common law, or of writs of summons to parliament. But, though the chancery enjoyed a measure of independence, the independence of the chancellor (unless the king were a weakling

---

[1] John of Reading, p. 113.
[2] See Sir H. Maxwell-Lyte, *Historical Notes on the Use of the Great Seal in England* (1926) and B. Wilkinson, *The Chancery under Edward III* (1929).

or a minor) was strictly limited. As head of the secretariat it was his duty to carry out the commands of the king, whose personal servant he was. He held the great seal at the king's discretion and it could be removed from him at any time. None the less, he was a key man in the state. An individual chancellor might become significant politically because of his personal influence with the king; this was probably true of Edington and certainly of Wykeham. But in general he owed his importance to his position as the leading member of the king's council and this was inevitably enhanced under Edward III whose policy, after 1341, was to govern with the advice of his council and to conciliate the baronage, a policy which depended for its smooth working on chancellors like Edington and Thoresby.

The practice, established by the Walton Ordinances, of entrusting custody of the great seal to the keeper of the privy seal when he went abroad with the king, inevitably enhanced the importance of the privy seal; and the office of keeper proved capable of attracting some of the ablest clerks in the country. By the middle of Edward III's reign the days were long past when the privy seal could be regarded merely as a household secretariat. It had established itself as the third of the great offices of state, its keeper discharging the duties which, as Tout has said, would fall in modern times to the secretaries of state for foreign affairs and war.[1] This development, though undoubtedly stimulated by war conditions, was not entirely dependent on them; for administrative expansion at home meant that the chancellor needed to be relieved of some of his routine responsibilities and the keeper of the privy seal thus came to act as a kind of second chancellor. The privy seal was commonly used as the instrument of the council to give executive force to its decisions and the normal way of 'moving' the great seal was by a warrant under the privy seal, the so-called 'chancery warrants'. Similar warrants to the exchequer authorized disbursements from its funds. Chancellor, treasurer, and keeper were now recognized as the three principal ministers of the Crown; and, though the keeper was always inferior in dignity to the other two, suspicion of the activities of the seal, such as had been voiced under Edward II, manifests itself under his grandson only in relation to privy seal interferences with the processes of the common law. The seal is no longer regarded as

[1] *Chapters*, v. 55-56.

an instrument of the king's personal wishes. For this he has recourse to the *signum secretum* or signet.[1]

The keynote of Edington's policy during his twelve years as treasurer is probably to be found in his desire to restore the financial supremacy of the exchequer over all other government departments; such, at least, would seem to be the implications of the administrative changes that took place during his period of office. This was a policy which could hardly have been pursued except under the stress of a great war; for it entailed abandonment by the Crown of its efforts to divert certain revenues into the household departments, a reversal of the aims pursued, for example, by the younger Despenser. In time of war it was a yielding to necessity. Efficiency demanded that all the king's resources should so far as possible be canalized and the great campaigns could not conceivably be financed from the revenues of the household. They could be financed only from the exchequer. Into this the proceeds of taxation were paid and from it alone could the king hope to raise the large sums needed as security for loans and for paying his soldiers' wages and the multifarious expenses which the war entailed. Whether or not he recognized its full implications, Edington's policy was tantamount to recognition that in the last resort the nation must pay for the war and that, if Edward was to wage it successfully, he must keep on terms with his parliaments and be prepared even to suffer some diminution of his prerogative as the price of the supplies he could not afford to forego. It was not to be expected, however, that a young king should relinquish the traditional policy of his forebears overnight; and Edington found that he had to work slowly and wait on the passage of events.

Reservation of lands to the chamber had been abandoned in 1327, in the full tide of reaction against Despenser; but it was resumed in 1333 and certain other sources of income were added in the course of the following twenty years.[2] Thus, the forfeited lands of rebels and contrariants were commonly reserved to the chamber; these were not numerous under Edward III, but the chamber profited by the forfeitures of the earl of Desmond in 1346 and of Chief Justice Thorp in 1351, as well as of minor malefactors. The lands of the alien priories which the king took into his hands on the outbreak of war were likewise reserved and

---

[1] See Maxwell-Lyte, *Use of the Great Seal*, pp. 101–17.
[2] See Tout, *Chapters*, iv. 228–311.

so were the lands of royal wards; while, so late as October 1349, the king ordered that escheated lands should be removed from exchequer control and reserved to the chamber. The chamber acquired a new seal—the griffin seal—for the affairs of its estates.[1] Yet by this date, the subordination of the chamber to the exchequer was already becoming unmistakable. Chamber activity under Edward III had reached its peak about 1340, at a time when the king's financial embarrassments were threatening to overwhelm him; the process of subordination began after the fall of Calais with the transfer of the chamber records to the custody of the exchequer and the decision that the officers of the exchequer should henceforth audit the chamber accounts. It reached completion with a royal writ of 20 January 1356, ordering that all lands, tenements, and other things previously reserved to the chamber should now be reunited to the exchequer.[2]

Inevitably, the chamber under Edward III exercised many of the same functions as under Edward II. Its resources were used for the purchase of plate and jewels (custody of which was the responsibility of the chamber officers), for the royal alms, for personal gifts from the king to his friends and servants, for certain charges for building operations, and, of course, for the wages, livery, and rewards of the chamber staff. But, whereas under Edward II the accounts tell of little except these things, under Edward III they are subsidiary to the king's 'secret business', that is to say, to the provision of men and ships for the war. For Edward III subordinated even his personal extravagances to his military ambitions and in this lies the most likely explanation of his willingness to allow the merging of chamber lands in the exchequer. There had been no parliamentary criticism of chamber activities—a fact which speaks volumes for the discretion of the ministers—and it was certainly not parliament which had constrained him; the change must rather be ascribed to the king's tardy recognition that a self-supporting court office could not function satisfactorily as a war department. Edington, who was shortly to leave the exchequer for the chancery, may well have persuaded him of the wisdom of a decision which neither the king nor his people were to have reason to regret.

[1] This seal was in use by 1335. See Maxwell-Lyte, *Use of the Great Seal*, pp. 109-10.
[2] For the text of this writ, see Tout, *Chapters*, iv. 305, n. 2.

The subordination of the king's wardrobe to the exchequer presented fewer problems.[1] At home, it was mainly a court office concerned with the expenditure of the royal household. Abroad, it achieved a new importance under Edward III through its function as a war treasury, acting as paymaster of the major expeditions when the king personally was in command. The Scottish expedition of 1341–2 and the Breton expedition of 1342–3, which was largely under Edington's direction, were both financed by the wardrobe; and the preservation of two detailed accounts for the Crécy-Calais campaign of 1346–7 makes it possible to discover something of the part it played in these memorable years. Walter Wetwang's Book of Receipts, covering the whole period of his office as keeper, from April 1344 until November 1347, gives precise details of the wardrobe's income, though only a bald summary of its expenses;[2] and William Retford's Kitchen Journal, a great roll of ninety-five membranes which covers most of the same period, affords particulars of this branch of wardrobe expenditure.[3] Its isolation from the home government during such periods of active service meant that the wardrobe had to be temporarily self-sufficient; but it was never in any sense a rival to the exchequer, from which it received the bulk of its income and which kept firm control of its activites. It was not with the wardrobe but with the exchequer, where the relevant records were kept, that the increasingly important military commanders computed individually.[4] Wetwang's account bears witness to the advantages resulting from exchequer control. Under Norwell as keeper (1338–40) the issues of the wardrobe had reached a figure of £410,292, whereas under Wetwang they dropped to £242,162; the successful campaign of 1346–7 cost the king less than the fiascos in the Netherlands at the beginning of the war. An attempt under keeper Cusance (July 1349 to February 1350) to provide the wardrobe with an independent income by assigning to it the lands of certain royal wards, was soon abandoned and the complete dependence of the wardrobe reasserted. Its last great effort as a war treasury came during the campaign of 1359–60; in the latter years of Edward III, when the king no

[1] Ibid. iv. 110–68.
[2] Part of this roll is summarized in *Crecy and Calais*, ed. G. Wrottesley (Wm. Salt. Arch. Soc. Collns. xviii. ii, 1897), pp. 191–219.   [3] E. A. 390/11.
[4] A. E. Prince, 'Payment of Army Wages in Edward III's Reign', *Speculum*, xix (1944), 137–60.

longer took command of his armies, it lost its importance even in this capacity.

The king's wardrobe, or wardrobe of the household, has to be distinguished from the great and privy wardrobes, both of which served mainly as storehouses. The king's craftsmen, tailor, armourer, and pavilioner were attached to the great wardrobe and during the war it functioned as an army clothing and military stores department, often accompanying the king beyond the seas. At home, it was peripatetic until 1361 when it settled in the house near Baynard castle which it was to occupy for another 300 years. By an ordinance of 1324, Stapledon had freed the great wardrobe from dependence on the king's wardrobe, instructing its officers to account directly to the exchequer; but, perhaps because it was of less importance, the great wardrobe was never so securely under the control of the exchequer as was either the chamber or the wardrobe of the household. The privy wardrobe was a later development, a product of wartime conditions, when it was found convenient to separate arms and armour from other stores and to house them in the Tower of London. Here were accumulated large stocks of bows and arrows, horse-trappings, pikes, lances, standards, tools, and raw materials for the manufacture and repair of weapons. We hear also of gunpowder being manufactured at the Tower in 1346-7, when not less than 2,683 lb. of saltpetre and 1,662 lb. of quick sulphur were provided by the great wardrobe for the purpose. There were guns and cannon balls in the Tower at the same period; but it was not until the last years of the reign that the privy wardrobe began to concern itself with firearms on a large scale, providing twenty-nine iron guns for the expedition of 1372 and equipping the new fortress of Queenborough with modern artillery.[1]

These administrative arrangements underlay the great victories in France and did much to make them possible. Formalization of the privy seal and restoration of the financial supremacy of the exchequer were indicative of a swing away from a curialist type of government towards one that would bear public scrutiny. It was not the least of Edington's achievements that he succeeded in keeping on good terms with the magnates while pursuing a

[1] T. F. Tout, 'Firearms in England in the Fourteenth Century', *Eng. Hist. Rev.* xxvi (1911), 666-702. Cf. *Chapters*, iv. 470-2.

clear-cut policy of administrative reorganization; and it is reasonable to allow him some of the credit for successful government handling of parliament in the years that separated the crisis of 1341 from his resignation of the chancellorship in 1363. Then, as always, taxation was unpopular and there was some overt and doubtless much subterranean grumbling; but by and large, the commons gave Edward III what he wanted—supplies to equip his armies, to pay the wages of his soldiers and sailors, and to afford security for his borrowings. Their complaisance appears the more remarkable when we remember that 1348 saw the first visitation of the Black Death and that the period as a whole is one of contracting population and falling agricultural prices. Yet despite all adverse conditions, fundamental goodwill persisted between Edward III and his parliaments. They were behind him, partly, no doubt, because, as lords and commons admitted in 1346, he had given them value for their money;[1] but such opposition as there was could well have become more formidable had it not been for the skill and address shown by the king and his ministers in handling it.

Edward's policy was to present the war as a joint-stock enterprise, undertaken for the defence of the realm and of his legitimate claim to the Crown of France, and to carry the estates along with him by keeping them informed of what was happening abroad and seeking their advice on the conduct of his diplomacy. Thus, in the parliament of 1343, Bartholomew Burghersh, acting on the king's instructions, reviewed the course of the war to date explaining that the truce of Malestroit had been agreed to at the request of the pope but that, since the war had been undertaken on the advice of the lords and commons, the king did not wish to conclude a final peace without their concurrence. Should he send an embassy to Avignon to present his case? In the next parliament (June 1344) Chancellor Sadington reported that the Bretons had violated the truce and that the Scots were supporting the French and asked for advice from lords and commons. The victory at Crécy was reported in detail to the parliament of 1346 and the draft scheme, discovered in the archives at Caen, for a French invasion of England, read aloud to lords and commons. In January 1352 Chief Justice Shareshill, speaking for

[1] '. . . all thanked God for the victory that He had granted to their liege lord . . . and said that all the money they had given him had been well spent'. *Rot. Parl.* ii. 159.

the king, reminded parliament of the causes of the war and dilated on King John's breaches of the truce and his overtures towards the Scots. The estates were notified of the draft treaty of Guines (April 1354) and asked if they wished the king to try for a perpetual peace (to which they made *entierement et uniement*, their famous reply, 'Yes! Yes!'); and the treaty of Calais was solemnly ratified in parliament.[1] In order to persuade parliament that money was still needed, despite the conclusion of peace, there seems even to have been an attempt, in 1363, to present a national balance-sheet embodying particulars of all expenses accountable through the exchequer.[2]

It is clear from some of their responses that the lords and more particularly the commons appreciated the king's confidence, though they were by no means blind to the financial implications of tendering their advice. Edward and his ministers showed their wisdom in seldom pressing too hard; they knew when to conciliate, when to flatter, when to withdraw. Protests against the king's attempt in 1345 to introduce a system of compulsory military service for all landowners without warrant from parliament were met with the assurance that this was an emergency measure which would establish no precedent.[3] In January 1348 the commons again showed themselves awkward; they declined to advise the king on the conduct of his war; and their grant of a triple subsidy in March was accompanied by a remonstrance that they could bear little more taxation.[4] With the subsidy in his pocket and the Black Death to serve as a pretext, Edward could take them at their word. He summoned no further parliament for three years and when the estates assembled in 1351, he put their needs and their sacrifices in the forefront of the picture. Parliament had been summoned, said the chancellor, because the king wished to redress all the grievances of his people; he thanked magnates and commons for the great love which they had always shown him, for the large aids and subsidies which they had granted him, and for all that they had

---

[1] *Rot. Parl.* ii. 136, 147, 158, 237, 262; Walsingham, i. 294.

[2] T. T. Tout and Dorothy Broome, 'A National Balance Sheet for 1362-3', *Eng. Hist. Rev.* xxxix (1924), 404-19.

[3] *Cal. Pat. Rolls 1343-45*, p. 427; Murimuth, pp. 192-3, 198; *Rot. Parl.* ii. 160. In 1351 it was established by statute that, except for those holding by military service, no man should be constrained to find men-at-arms, hobelars, or archers, unless by assent of parliament. *Statutes* 25 Edw. III, st. 5, c. 8.

[4] *Rot. Parl.* ii. 164, 200-1.

suffered, both in body and goods, for the maintenance of the war and the defence of the realm; he wished, in return, to do all in his power for their 'ease, comfort and favour'.[1] Again, in 1355 when the breakdown of the peace negotiations with France might have been expected to cause unfavourable reactions among the taxpayers, Edward was careful to send Sir Walter Manny, one of the most distinguished of the war heroes, into parliament to explain what had occurred.[2] The steady flow of taxes into the exchequer during these years affords evidence of the success of such manœuvres. A single subsidy (a tenth and fifteenth) was granted in 1357 and 1360, subsidies for two years in 1344 and 1346 and for three in 1348 and 1352; a forty-shilling export duty (a *maltolte*) on wool for three years in 1343, for two in 1351, for six in 1355, and for three in 1362; an aid for the knighting of the Black Prince in 1346 and, in 1347, a new cloth duty and the duties on wine and merchandise which came to be known as tunnage and poundage. The commons maintained fairly consistently the attitude they had adopted in the parliament of 1343. Let the king, they had then said, try for peace, if possible; if it proved not to be possible, they would maintain his quarrel with all their power.[3] The demands of the royal purveyors were provocative of the kind of bitter resentment which finds expression in the tract *De Speculo Regis*; but there seems to have been remarkably little criticism of the court, either inside or outside parliament, and the labours of Edward's best years were triumphantly vindicated in the apparently spontaneous tribute offered by the commons in 1363:

Sire, the commons thank their liege lord so far as they know and can, for the graces, pardons and good will shown to them . . . and they beg that it may please him to continue them, as to his lieges who from their hearts entirely thank God who has given them such a lord and governor, who has delivered them from servitude to other lands and from the charges sustained by them in times past.[4]

But there is another side to the record of Edward III's financial dealings with his parliaments. His wooing of the commons depended for its success largely on his readiness to concede their demands and to abandon most of his extra-parliamentary resources; behind the grants of subsidies and *maltoltes* lies the story of a long and obstinate struggle for redress of grievances

[1] Ibid. ii. 225-6.
[2] Ibid. ii. 264.
[3] Ibid. ii. 136.
[4] Ibid. ii. 276.

and for control of the purse-strings. In their grants of subsidies, the commons showed an increasing tendency to seek to impose specific conditions. A premature attempt, in 1344, to secure appropriation of supplies by insisting that the money should be spent on the war alone and that the aid from the counties north of Trent should be allocated to the defence of the north, was not followed up;[1] but a favourable answer to petitions was almost always made the condition of a grant and further demands were often added. For example, in 1346, Edward had to promise that the subsidy should be taken in cash only and remitted in part if the war were over before the end of the second year. In 1348 he had to undertake not to turn the subsidy into wool and to appoint a committee to hear petitions still in arrears; in 1352 he had to promise that the profits of the Statute of Labourers should be set against the tax.[2] It was by no means easy to hold the king to his promises; but at least they were there on record for future generations to read.

The question of the wool-tax was more complicated. The demand for abolition of the *maltolte* made in 1339 was reiterated in 1340 and conceded in the statute of that year, though with the important proviso that, if the king were unable to keep his promise, any fresh imposition must be sanctioned by magnates and commons in full parliament. In the meantime, they had already sanctioned a *maltolte* for a further fourteen months: but the king did not hold himself bound by these time-limits. Abolition of the *maltolte* was again demanded in 1343 and there were strong protests against its imposition, on the plea of military necessity, by a council of prelates and merchants.[3] It did not prove possible to prevent the king from taking it, but public indignation found expression in outbreaks of smuggling and in the parliament of January 1348 there were renewed demands for restoration of the old duty.[4] In 1351 the commons complained that the tax which the merchants had granted fell on the whole people and asked that the king should not in future levy this tax, except *en plein parlement*, and that any such grant made outside parliament should be held as void.[5] Concession of this point enabled Edward to extract a further *maltolte* for two years; but he was not thereby deterred from summoning, in 1353, a novel form of council to which the sheriffs were directed to send one

[1] *Rot. Parl.* ii. 149.   [2] Ibid. ii. 159–60, 200–1, 238.
[3] Ibid. ii. 140.   [4] Ibid. ii. 168.   [5] Ibid. ii. 229.

knight from each shire and the mayor and bailiffs of forty-three selected towns. This assembly sanctioned a *maltolte* for a further three years; but, since the members insisted that what had been done should be recited at the next parliament and entered on its roll, so that it might be clearly recorded as if enacted *par commune parlement*, it was obvious that the king had little to gain by such tactics.[1] The injunction was duly carried out and no more councils of this type were called. The commons had been converted to the necessity of the tax, but they were now well on their way towards securing exclusive control of it. The conclusion of peace in 1360 put them in a better position to enforce their demands and two years later the king at last conceded that no subsidy or other charge be laid on wool or merchandise without consent of parliament.[2] This was a notable victory: and though further *maltoltes* were taken, that of 1369 even reaching the unprecedented figure of 43s. 4d., for the rest of the reign there were no more wool-taxes without parliamentary consent.

For explanation of this victory we have to look to the history of Edward III's credit operations in the years succeeding the crisis of 1341. The Italian houses of Bardi and Peruzzi had been the principal victims of the king's reckless dealings in the Netherlands. He treated them badly but responsibility for their ruin cannot be laid solely to his charge, for the war itself had been a major calamity for the foreign bankers, who found it impossible to keep on terms with all the combatants. After 1342, when their loans virtually ceased, they were of no further use to Edward and their unpopularity in the country made it safe for him to repudiate the bulk of his debts. The Peruzzi, to whom the king owed some £77,000 (inclusive of interest), went bankrupt in 1343 and shortly after this date they disappear from the records. The Bardi, to whom the king's debt was at least £103,000, collapsed in 1346. Some of them suffered imprisonment and were released only on condition of renouncing all claims to payment of interest.[3] As substitutes for the Italians, Edward turned perforce to the native financiers, some of whom,

---

[1] Unwin, *Finance and Trade*, p. 230; *Rot. Parl.* ii. 253.   [2] Ibid., p. 271.

[3] They were reorganized in a small way in 1357 and thereafter held a modest position as bankers into the reign of Richard II. It is to the credit of the monarchy that the debt to them was not wholly repudiated, considerable sums being paid to them between 1346 and 1391, when Walter Bardi finally quit-claimed all debts due to the company. See Alice Beardwood, *Alien Merchants in England 1350 to 1377* (1931), pp. 3-7.

as has been seen, had already done him good service. A few of
these, like William de la Pole, were very wealthy men; but the
majority lacked large capital assets and they had to be provided
by the king with the securities necessary to enable them to raise
substantial sums. Thus, the English company of 1343, organized
by de la Pole (though, on account of recent scandals his name
was kept out of it), consisted of thirty-three merchants who
were granted the customs and subsidies for three months together with a monopoly of export which would enable them to
control the price of wool. In return for these privileges, the
company agreed to pay the king 10,000 marks a year and 1,000
marks every four weeks.[1] On the eve of the Crécy campaign in
May 1346, Edward entered into agreement with another company headed by Walter Chiriton and Thomas Swanland, who
were granted the farm of the customs for two years in return for
a guarantee of £50,000 per annum and an immediate advance
of £4,000; since the estimated yield of the customs was £60,000
a year, Chiriton, Swanland and Co. might look for substantial
profit. But the strain of financing the long siege of Calais had
made it impossible for them to fulfil their contracts even before
the drastic interruption of trade consequent on the Black Death;
and in April 1349 they went bankrupt.[2] A group acting as their
sureties kept their heads above water for a little longer; but the
whole episode was discouraging to potential lenders and after
1353 the king found that he had to forego regular borrowing
from large financiers. Such a change of policy was inevitable
for other reasons, the most important of which was the growing divergence of interest between the relatively small number
of wealthy monopolists and the mass of the lesser merchants.
Hostility to the monopolists helped to promote the suspension
of the Bruges staple and the establishment of home staples in
1353, a measure which was intended to prevent monopolies by
taking the export trade entirely out of the hands of native merchants and transferring it to aliens. It was the voice of these
anti-monopolists which made itself heard in parliament and
it was they who inspired the struggle for such parliamentary
control of the wool-tax as would prevent the king from negotiat-

[1] G. O. Sayles, 'The "English Company" of 1343 and a Merchant's Oath',
Speculum, vi (1931), 177-205.
[2] See the forthcoming article by Dr. E. B. Fryde in *Trans. Royal Hist. Soc.* (I am
much indebted to Dr. Fryde for his kindness in sending me a proof of his paper.)

ing the *maltolte* with groups of merchants outside parliament. Fortunately for Edward III, he now had other resources to fall back on, so that the need for large-scale borrowing was in some degree reduced. London was becoming the centre of national credit and its citizens, whose proximity to the exchequer increased their chances of repayment, were often ready to advance moderate sums to help the king in his chronic predicament of shortage of ready cash.[1] Moreover, Edington's policy of bringing the chamber into dependence on the exchequer now began to pay dividends. Between 1355 and 1377 the chamber received from the exchequer a comfortable annual income of not less than 10,000 marks and in the years following the treaty of Calais it was in a position to offer loans, the sum of which amounted to over £31,000 in the period 1362–70.[2] Officials, councillors, and bishops could also be appealed to for small loans. The wool trade soon recovered from the effects of the Black Death; exports were high for most of the decade 1350–60 and the profits of the customs rose accordingly. Peace with France brought some relaxation of financial strain, while the victors continued to reap the harvest of war. In the years following the treaty of Calais, Edward III received at least £268,000 on account of the ransoms of the kings of France and Scotland and the duke of Burgundy; so that it was not without some reason that, after the renewal of hostilities, the commons suggested that the war ought to pay for itself.[3] None the less, the disappearance of a class of professional financiers and the enforced abandonment of systematic borrowing by the Crown were to prove a serious handicap to the government in the early years of Richard II when, for other reasons, the financial position had already deteriorated.

It was in the sixties that William of Wykeham replaced Edington as the right-hand man of Edward III. Wykeham was a man of humble origin whose rapid rise to power affords a striking example of the career open to talent offered by the medieval Church. It was probably Edington who introduced him to the service of Edward III, perhaps when the king came into Hampshire after the siege of Calais in 1347. Not long after

[1] Unwin, pp. 237–43.　　　　　　　[2] Tout, *Chapters*, iv. 314–27.

[3] '. . . aiant consideration as grandes sommes d'or q'ont este apportez deinz le Roialme des Ranceons des Roys de France et d'Escoce & d'autres Prisoners & pays, q'amonte a une tres-grande somme'. *Rot. Parl.* ii. 323 (1376).

this, Wykeham removed to court and in the fifties, as king's clerk, he was engaged on a miscellany of business, keeping the king's dogs and selling his horses at Windsor, reassessing rents there, carrying a bundle of memoranda down to Wiltshire for the justices of the forest.[1] Edington continued to watch his interests and it was at the treasurer's request that he received the appointment which was to prove crucial for his future, that of surveyor of the king's works in Windsor castle, where his duties were to engage workmen and materials, pay wages, and imprison contrariants.[2] Edward III cherished a particular affection for his birthplace; and its enlargement to serve both as a royal palace and as home for the new Order of the Garter took first place among his building schemes. In Wykeham he found a collaborator after his own heart. John of Malvern says unkindly that this astute man of lowly origin had discovered the best means of winning the king's favour;[3] but if Wykeham's zeal for building was not disinterested, at least there is no denying that it was lifelong: and it is noteworthy that the plan of the new buildings at Windsor closely resembled those later adopted for Winchester and New Colleges—hall and chapel, divided by a partition wall, formed a continuous range with courts or quadrangles lying alongside. Certainly, it was his achievement at Windsor which established Wykeham in the favour of Edward III and assured his future. In 1362, he became joint keeper of the forest south of Trent. The records make it clear that between 1360 and 1365 Edward was becoming increasingly dependent on Wykeham's advice in such matters as ecclesiastical preferments, grants of lands and privileges, and official appointments. Letters patent issued in 1363 describe 'the king's clerk, William of Wykeham' as 'his secretary who stays by his side in constant attendance on his service'.[4] Edward rewarded him so lavishly that he soon became the most notorious pluralist in England;[5] and the truth of Froissart's suggestion that he was in such high

---

[1] *Issues of the Exchequer*, i. 163; *Cal. Fine Rolls*, vii. 28, 91; *Cal. Close Rolls 1354-60*, p. 306.   [2] *Cal. Pat. Rolls 1354-58*, p. 364.

[3] *Polychronicon*, viii. 360. According to a well-known if not very well-authenticated tradition, Wykeham marked the completion of his work at Windsor by inscribing the words 'Hoc fecit Wykeham' on one of the inner walls. When the king took this presumption amiss, he explained that the words meant, not 'Wykeham made this', but 'This made Wykeham'. R. Lowth, *Life of Wykeham* (1777), p. 22.

[4] *Cal. Pat. Rolls 1361-64*, p. 444.

[5] In 1366 Wykeham held the archdeaconry of Lincoln, eleven canonries or prebends, and a Cornish rectory.

favour at court that 'all things were done by him and without him nothing was done' is reflected in the papal correspondence which shows an increasing tendency to seek his intervention with the king.[1] Keeper of the privy seal in 1364, Wykeham was elected to succeed Edington in the see of Winchester in 1366 and in 1367 was appointed chancellor.

His rapid rise to favour provoked understandable resentment. The pope, says the Westminster monk, John of Reading, driven by fear rather than love and passing over many more worthy persons, provided Wykeham to the see of Winchester;[2] and Wyclif makes bitter allusion to a bench of bishops which included a clerk 'wise in building castles'. It was natural that in some ecclesiastical circles Wykeham should be regarded as an unscrupulous climber; but he was no Gaveston. He served the king faithfully and although he enjoyed the royal favour and confidence to a degree which may suggest a decline of Edward's own initiative there was never a hint of unbecoming intimacy. Wykeham's instincts were conservative and, like Edington, he knew how to keep on terms with the nobility while controlling the inner circle of the royal household. If, as Tout suggests, he was, in a sense, rebuilding a court party, he did this with the utmost circumspection.[3] Even while keeper of the privy seal he enjoyed an authority surpassing that of the chancellor, Langham, or the treasurer, Barnet; but he did not make use of this authority to introduce departmental novelties, encourage hangers-on of his own, or provoke the baronage. Such conservatism was unusual in royal favourites of humble birth. Wykeham's may have owed something to the example of Edington, though Edington never stood so high in the royal favour; it may also, perhaps, have owed something to his own long-term ambitions. We cannot tell at what date he conceived the notion of using his wealth in the service of religion and education; but he under-took the reconstruction of the chapel and cloister of St. Martin le Grand soon after he became its dean in 1360 and his purchases of land for New College began in 1369. Beyond question an able administrator, Wykeham's career in the sixties affords little evidence that he possessed the qualities of a statesman. He was an unpopular chancellor mainly because, in a period of nominal peace, he did little to alleviate the burden of national taxation

[1] Froissart, iv. 205. *Cal. Pap. Reg.* (Letters), iii. 2, 12, 15, 26, 92.
[2] John of Reading, p. 178.   [3] *Chapters*, iii. 239.

or to put the royal finances on a sound basis. The weightiest
charge brought against him at his trial in 1376 was that, although

the said bishop had the administration of all the royal revenues . . .
the greater part was not used for the benefit of the king and the kingdom. And when peace had lasted ten years and the second war began, nothing was found in the king's treasury but he was in such great
poverty . . . that he had to burden the clergy and the commons with
subsidies and loans.[1]

The obstacles confronting him were admittedly formidable and
the incurable extravagance of the king, the Black Prince, and
the other captains was not to be offset by such measures as the
abortive sumptuary legislation of 1363. But it was inevitable
that as the king's chief minister he should be held responsible
for the failures and disasters of the period and that he should fall
victim to the rising tide of anti-clericalism, manifesting itself in
the revolt of the military aristocracy against the predominance
of churchmen in the government. Yet those who, in the parliament of 1371, demanded the removal of the bishops of Winchester
and Exeter and the substitution of a lay chancellor and treasurer,
made no attack on Wykeham personally. He was discredited by
his dismissal from office; but he remained on the best of terms
with the war leaders and was not in any sense disgraced.

Ireland under Edward III

In view of their many preoccupations elsewhere, it is no matter
for surprise that the ministers of Edward III should have failed
to arrest the decline of the Anglo-Norman colony in Ireland or
that the king himself should have found it impossible to go
there.[2] After the Bruce invasion, the Gaelic revival went on
apace and when the young earl of Ulster and Connaught,
William de Burgh, was assassinated at the ford of Carrickfergus
in 1333 by some of his Irish cousins, the earldom of Connaught
reverted to two of his kinsmen who established what were
virtually native principalities, while in Ulster the murder was
followed by a notable expansion of the power of the O'Donnells
and O'Neills, with the result that the northern earldom, once

---

[1] *Anon. Chron.*, p. 96.
[2] Edward planned a visit to Ireland in the summer of 1332; but the English
parliament decided it would be too dangerous to allow him to leave the country in
view of the possibility of a Scottish invasion. *Foedera*, ii. 828; *Rot. Parl.* ii. 66.

the bulwark of the English colony, was to all intents and purposes destroyed.[1] Under the domination of the Gaelic princes,
the eastern and north-eastern coasts of Ulster were thrown open
to the gallowglasses and the Gaels of western Scotland were free
to join hands with their Irish brethren. There was little hope of
recovering either Ulster or Connaught for the murdered earl's
infant daughter, Elizabeth de Burgh, later to become the wife
of Edward III's third son, Prince Lionel of Clarence. The
prospects would have been brighter if the home government
could have relied on the co-operation of the Anglo-Irish magnates who, having won their lands from the Irish, were always
opposed to plans for their legal enfranchisement or to any recognition of the independent rule of Irish princes within their own
territories. But by the time of Edward III, most of these Anglo-
Norman families had been settled in Ireland for over a century;
the majority were Irish-born and many had Irish mothers. They
were, in a sense, Irishmen, and the culture of the Gael held a
strong attraction for them, isolated as they were from English
society. John de Bermingham, earl of Louth, was accompanied
everywhere by his Irish harper; native music and poetry were
listened to eagerly in the stone castles of magnates like Desmond
and Kildare. Moreover, native custom influenced them in other
and more dangerous ways. They adopted, for instance, the
system of *coign and livery*, the compulsory billeting of their military followers on private persons, which was to scandalize the
Tudor rulers of Ireland. They also adopted the custom whereby
females could neither inherit land nor transmit a claim to it.
It was this sympathy with Gaelic culture and tendency to
adopt Gaelic customs, when these seemed to favour their interests, that provoked the English government to accuse the Anglo-
Irish of 'degeneracy'. They came to be regarded as traitors to
their own race attempting to overset the well-established
customs of the civilized world. Yet they themselves had no
wish to repudiate the authority of the English Crown or
their own status as its highly-privileged tenants. For all their
'Irishry' they were still in a precarious position and what they
wanted was, not to set up a government independent of the
English Crown, but to control the government in accordance
with their own ideas. They were proud men; and while they
would bow to the authority of the king in person, they were

[1] Orpen, *Ireland under the Normans*, iv. 245–9.

little disposed to respect justiciars, many of whom were of much humbler birth than themselves. The outcries against officials of mean birth which echo through the English history of the period

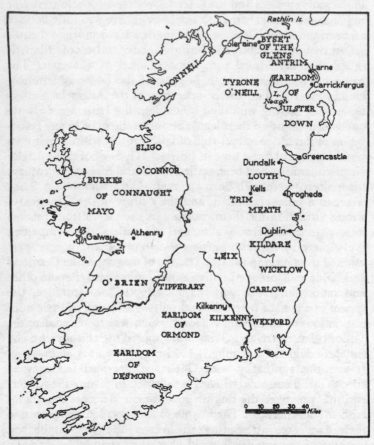

FIG. 6. Fourteenth-Century Ireland.

were much more vociferous in Ireland, where the magnates accused the justiciars of blundering incompetence and argued that they themselves, as Englishmen born in Ireland, were much better fitted to represent the authority of the English Crown. Yet they were even less capable than their English counterparts of maintaining a solid front or of reaching agreement on a common policy; and this, to some extent, reduced the threat to the Dublin government from their intransigence.

The justiciars, for their part, enjoyed certain solid advantages. They were the legal representatives of the English Crown in Ireland and as such they controlled the revenues and the garrisons of the royal towns and castles. From time to time, though not nearly often enough, a justiciar was allowed to bring to Ireland a professional army raised in England and to use it in support of his authority. The justiciar had the right to call parliaments to which he might expound government policy and appeal for the aid and co-operation of the colonists; and he could count on the support of the towns whose interests lay in the maintenance of order, the cessation of civil war and the promotion of commerce. Even Edward Bruce had made no headway against Dublin. Yet the series of insignificant knights whom Edward III appointed to the justiciarship of Ireland were given little opportunity to exploit these assets. After Sir Anthony Lucy, Sir John Darcy, and Sir Ralph Dufford had all embroiled themselves with the magnates, Edward, in 1346, conferred the office of justiciar on a member of one of the great Anglo-Irish families, Sir Walter de Bermingham; but he proved incapable of checking the Gaelic revival. His successor, an English knight named Sir Thomas Rokeby (1349–54), did better; but after his death the colony was ruled for seven years by two of its greatest barons, the earls of Desmond and Kildare. Edward III was not prepared, however, to allow control of the Dublin government to fall permanently into the hands of the feudatories. Their appointments to office were in the nature of bribes and were always for short terms; the vacillation and confusion which resulted were enough in themselves to wreck the state. The problem was immensely difficult and so long as the English government continued to regard it as secondary there was little hope of even a partial solution. Thus, it was a step in the right direction when, shortly after the conclusion of peace with France, Edward III appointed Lionel of Clarence as his lieutenant in Ireland.

Clarence, who in 1361 was twenty-two years of age, was himself the greatest of the absentee Irish landlords, being earl of Ulster and Connaught in right of his wife, Elizabeth de Burgh. He landed in Dublin on 15 September, accompanied by a force of 1,500 men under the command of the earl of Stafford; and, except for a visit of eight months to England in 1364, he remained in Ireland for five years—by far the longest stay yet made by any member of the royal family. During this period he conducted a series of campaigns designed to bring the magnates under control and to secure his hold on his own earldoms. But his military successes were few and their results insignificant; he had not enough troops and the absentee lords in England showed little desire to come to his assistance. A seneschal and constable were maintained at Greencastle in south Ulster, but north of Carrickfergus, Clarence's control of his earldom was purely nominal and in Connaught he achieved even less. Nor was he successful in rehabilitating the colony as a whole. The revenues from Ireland fell steeply during his tenure of office; feuds and jealousies between the English-born tenants and officers of the Crown and the Irish-born feudatories were as bitter as ever; and the prince found that he could rely only on the loyalty of the towns and of the so-called 'obedient shires'.[1] His departure from Ireland in November 1366 was followed by an Anglo-Irish re-action signalized by the appointment of the Geraldine earl of Desmond as justiciar. Yet Clarence's administration had one memorable achievement to its credit; it secured the passing of the Statutes of Kilkenny (1366) which were to dominate English policy in Ireland from the fourteenth to the sixteenth century.[2]

It is sometimes suggested that these famous statutes are to be read as a declaration of war against the Irish; on the contrary they are to be read as a confession of defeat. The wide lands which have already slipped from English control are virtually abandoned; the statutes represent a determined effort to save the 'obedient shires' from the same fate and to keep them pure English. It is tacitly admitted that the original attempt to colonize the whole of Ireland and to bring it under English control has broken down completely. Henceforward, the English colony is to be limited to the district that was coming to be known as the English Pale and 'Irish enemies' becomes the official designation of the native Irish living beyond its borders. They are excluded from ecclesiastical office; the king's lieges are to have nothing to do with them; they are not to parley with them, nor to marry them, nor to sell them horses or armour. But the concern of the

[1] The shires and liberties of Louth, Meath, Trim, Dublin, Kildare, Carlow, Kilkenny, Wexford, Waterford, and Tipperary.
[2] Berry, *Early Statutes of Ireland*, i. 431–69.

statutes is less with the 'mere' Irish than with the descendants of the English settlers, and their principal intention is to arrest the process of 'degeneracy' in the areas of English influence. Recourse to Brehon law is forbidden; Englishmen may not entertain Irish minstrels, story-tellers, or rhymers; all Englishmen and all Irishmen dwelling *inter anglicos* must use English surnames and the English language and follow English customs; Englishmen are to forsake Irish sports, such as hurling and quoits, and are to learn the use of the bow and of 'other gentle games which pertain to arms'.

Needless to say, the process of 'degeneracy' could not be arrested by act of parliament. The Burkes continued to rule Galway and Mayo in the style of native chieftains; in the north, around Coleraine, the Norman de Mandevilles changed their name to MacQuillan and dominated a great part of north Antrim; in south Down, another English family, the Savages, retained their English surname but became almost completely Irish in language and custom. Even within the Pale itself, there was no perceptible change. Irish surnames and the Irish language continued to gain ground and the third earl of Desmond, who succeeded Clarence as lieutenant, was known to his contemporaries as 'Gerald the Poet' because he composed verses in Irish and 'excelled all the English and many of the Irish in knowledge of the Irish language, poetry and history'.[1] The English government even had to sanction breaches of its own law, as when, in 1376, it gave permission to 'Donal O'Moghane, Irish minstrel' to dwell among the English and be in their houses, and when the earl of Desmond was allowed to have his son fostered with Conor O'Brien.[2] Special licences had to be granted to Irishmen to hold benefices, for as the English Pale contracted it began to be impossible to find Englishmen willing to take ecclesiastical office among the Irish where, indeed, their ignorance of the Irish language would have been an insuperable handicap. Grants of pardon to abbots and priors for admitting Irishmen to their houses appear frequently on the rolls. The importance of the Statutes of Kilkenny lies, not in their application but in their long-lasting influence on English policy in Ireland. Between 1366 and 1495 they were many times confirmed and they remained on the statute-book till 1613 when their repeal was secured by Sir Arthur Chichester.

[1] *Annals of the Four Masters*, iii. 761. [2] Curtis, *Medieval Ireland*, p. 234, n. 2.

IX

WAR AND CHIVALRY

BY the end of the twelfth century the feudal levy, as an effective fighting force, was already out of date. Designed to supply the officers of the army, the quotas of knights which made up the *servitia debita* appeared at the muster as incoherent units, hard to combine either with one another or with the levies from the shires and towns which constituted the rank and file of the host. The inadequacy of the whole system and the need for paid professional soldiers—other than foreign mercenaries who were odious to the people—may have been recognized at a much earlier date than used to be supposed; it is certain that the native *solidarius* was not a new phenomenon in the time of Edward I.[1] Yet, whatever may have been achieved before his day, Edward seems to have been the first king to offer pay systematically to all ranks of the army except the very highest. He understood that the key to organization of the inchoate host was the systematic use of pay; and his genius for military organization did much to make possible the spectacular triumphs achieved by his grandson on the battlefields of France and Scotland in the fourteenth century. The two generations which preceded the outbreak of the Hundred Years War saw widespread and general acceptance of the principle of payment for military service, until the earl himself became a stipendiary and even kings and princes did not scruple to draw their wages. The change was complete when Edward Balliol, titular king of Scots, drew 30s. a day in peacetime and 50s. in wartime and when the Black Prince took his pay at the daily rate of 20s.[2] Though the old tenurial obligations were still maintained in principle, their enforcement became increasingly difficult and was seldom attempted. The feudal host was called out for the inglorious Stanhope Park campaign of 1327, but ten years later the scutages due were still in arrears and the king had to aban-

[1] J. O. Prestwich, 'War and Finance in the Anglo-Norman State', *Trans. Royal Hist. Soc.* 5th ser. iv (1954), 26; A. L. Poole, *Obligations of Society* (1946), p. 52; Powicke, *The Thirteenth Century*, pp. 542–59.

[2] A. E. Prince, 'The Army and Navy', *English Government at Work*, i. 336, notes 5 and 7.

don the attempt to collect them. Later summonses to selected tenants-in-chief (for Ireland in 1332 and for Scotland in 1333) seem to have yielded little result. Only once again before the close of the century were writs of summons issued for a general feudal levy. Richard II's ministers arranged that the host should be called out for the Scottish expedition of 1385—perhaps in the hope of enhancing his prestige while making formal affirmation of his royal rights, perhaps in order to compel the tenants-in-chief to act as contractors for the recruitment of men-at-arms and archers. But so far from being a purely feudal levy, the force which accompanied Richard across the Border was a contract army of the normal type, in which all the leaders, from the duke of Lancaster downwards, drew their pay; and in the next parliament to meet after the campaign the king explicitly renounced his claim to any scutages which might be exigible in respect of it.[1]

In the upper ranks of the army, pay was guaranteed by contracts, oral or written. Edward I's contracts with his barons were probably oral, and formal indentures to which the king was party do not seem to have become normal until the early stages of the Hundred Years War, though some written sub-contracts have survived from the close of the thirteenth century.[2] The indenture with Maurice de Berkeley to stay with the king for life, supplying fourteen men-at-arms in time of war,[3] and the bargains struck with the princes and magnates of the Low Countries for service in Scotland and Flanders, prepared the way for the typical indentures of the French war; those drawn up in August 1341 between the king and certain captains for the impending campaign in Brittany show that by this date the system was well established. The captain—Edward Montagu will serve as an example—agrees to serve in Brittany for forty days with six knights, twenty men-at-arms (*homines ad arma*), twelve armed men (*homines armati*), and twelve archers. Wages for this period, totalling £76 are met by assignments on the wool subsidy in Suffolk.[4] Another type of indenture which becomes increasingly

---

[1] *Rot. Parl.* iii. 213. See N. B. Lewis, 'The Last Medieval Summons of the English Feudal Levy, 13 June 1385', *Eng. Hist. Rev.* lxxiii (1958), 1–26.

[2] Dr. G. A. Holmes points out (*Estates of the Higher Nobility*, p. 81) that the beginning of the well-documented indenture system coincides with the establishment of exchequer control over war finance. See above, pp. 215–16.

[3] *Cal. Pat. Rolls 1327–30*, p. 530.

[4] *Foedera*, ii. 1173; *Cal. Close Rolls 1340–43*, pp. 260–1.

common as the war progresses, is for a specific military under-taking, for example, when Henry Husee in 1347 assumes cus-tody of the Isle of Wight for six months, with forty men-at-arms and sixty archers at the king's wages.[1] When, however, the con-tractor was a person of consequence assigned to high command, the conditions attaching to his service are specified in much greater detail. Thus, the indenture drawn up in 1345 between the king and the earl of Northampton lays it down that the earl is to have supreme command in Brittany; that the king will be responsible for expenses incurred as a result of military under-takings; that the earl shall draw his wages and 'regard' for three months before his departure and quarterly thereafter; that compensation shall be paid for lost horses;[2] that the king will find transport in both directions; that he will send help if the earl's forces are besieged; and that the earl and his men shall be entitled to all prisoners valued at less than £500.[3] From both the administrative and the military standpoints, the superiority of the indenture system to older methods was obvious. It was al-most always short-term; it freed the military tenants from the burden of scutage and the exchequer from the labour of collect-ing it:[4] the problems arising from fragmentation of knights' fees could be ignored; it made even the greatest of the magnates directly dependent on the king or his appointed representative, while safeguarding all captains against serious loss and offering them the chance of successful profit; and it substituted discipline and a proper subordination of commands for the unruly indi-vidualism of the feudal musters. The nucleus of the forces which accompanied the more important commanders to the wars would probably be members of their households and life-retainers; others—the majority—would have been enlisted under temporary sub-contracts; but for the rank and file of an expedi-tionary force it was still necessary to rely on the ancient obliga-tion of the shires and boroughs to raise men for national defence.

To satisfy the requirements of a great war a principle of selec-tion had to be applied to the *posse comitatus*; and the method

[1] *Foedera*, iii. 114.

[2] This was often a heavy liability, e.g. £6,656 was paid as compensation for lost horses between July 1338 and May 1340. See A. E. Prince, 'Payment of Army Wages in Edward III's Reign', *Speculum*, xix (1944), 150.

[3] *Foedera*, iii. 37.

[4] See Helena M. Chew, 'Scutage in the Fourteenth Century', *Eng. Hist. Rev.* xxxviii (1923), 19–41.

most commonly employed was that of the commission of array, whereby commissioners appointed by the Crown surveyed the able-bodied men between the ages of sixteen and sixty in each hundred, township, and liberty within the shire and, under the authorization of the Statute of Winchester of 1285, selected the best of them to serve at the king's wages, the cost of initial equipment being borne by the localities.[1] The commissioners themselves were normally men of the knightly class, sometimes members of the king's household, experienced staff officers who knew what they were looking for and may be presumed to have chosen well. Their efforts were assisted by the proclamations issued at the beginning of each campaign, offering inducements to all and sundry to serve in the wars. Pardons for criminal offences, the prospect of good wages, and of a share in the incidental profits of war attracted many to active service. The troops raised under commissions of array were conscripts in principle and compulsion may have had to be applied for service in Scotland; but there was no lack of eager volunteers for the campaigns in France with their infinitely more attractive prospects of material reward. Parliamentary petitions on the subject of the commissions of array complained, not of the obligation to serve, but of the expenses incurred by the local communities; for the campaign of 1359, Edward III had more volunteers than he could use.[2]

The troops raised in these ways, by indenture, array, or voluntary enlistment, fall into fairly distinct categories. The commanding officers were the leading members of the aristocracy together with those captains of lower rank who had won renown by deeds of prowess. All were knights; but a graduated scale of payments reveals a hierarchy within the order of knighthood. The titles of duke and earl still retained something of their original military significance, recognized by special rates of pay, 13s. 4d. a day for a duke, half a mark to 8s. for an earl. Below them came the bannerets, paid normally at the rate of 4s. a day and chosen from the body of the knights by reason of their

---

[1] Under a statute of 1344 it was laid down that all troops going out of England were to be at the king's wages from the day they left their own counties. *Statutes* 18 Edw. III, st. 2, c. 7.

[2] e.g. *Rot. Parl.* ii. 149, 160. See A. E. Prince, 'The Indenture System under Edward III', *Essays in Honour of James Tait*, ed. J. G. Edwards, V. H. Galbraith, and E. F. Jacob (1933), p. 290; M. R. Powicke, 'Edward II and Military Obligation', *Speculum*, xxxi (1956), 92–119.

military skill and their capacity to sustain the expenses consequent on conversion of the knight's pennon into a rectangular
banner. Dukes and earls, like the Black Prince, Henry of Lancaster, Bohun of Northampton, and Beauchamp of Warwick,
combined with the king to direct the general strategy of the
great campaigns; to the bannerets fell the tasks of commanding
retinues in the field, arraying contingents, garrisoning castles,
and acting as general staff officers. Some of these tasks might
be shared by knights whose lack of resources forbade them to
aspire to higher rank. Even so distinguished a commander as
Sir John Chandos did not attain the status of banneret until late
in his career when he had obtained estates sufficient to support
it.[1] Thus, the knights bachelor who ranked next to the bannerets
and were normally paid at the rate of 2s. a day, might include
men of standing and experience as well as youthful scions of
noble families or minor tenants who had been distrained into
knighthood.[2] All were heavily armed; for it was their arms,
armour, and the weight of their war-horses which distinguished
the knightly, or 'armigerous' classes in the field from those of
inferior status and put knighthood out of the reach of poor men.
Armour of the period, which was in process of transition from
mail to plate (the tendency being to use the former mainly as
auxiliary protection), was elaborate and costly. A memorial
brass of Edward II's time shows the knight in a quilted tunic, or
gambeson (stuffed, probably, with wool or rags), over which he
wears a short-sleeved *hauberk*, or mail tunic and, over this, in
succession a *habergeon*, of some kind of soft material, a *haketon* of
light armour, the long surcoat known as a *cyclas* and a trans-
verse sword-belt. On his head the knight wears a fluted *bascinet*,
or burnished helmet, his arms, shoulders, and legs being protected by plate.[3] By the time of Crécy, the *cyclas* has vanished
and is being replaced by the tight short surcoat, or *jupon*, worn
over a metal breast-plate and with a *baldric*, or horizontal belt
across the hips. In battle, the knight's head was protected by
a heavy plate helm, closed with a *vizor*. The development of

[1] J. E. Morris, *The Welsh Wars of King Edward I* (1901), p. 72.
[2] Orders for distraint of knighthood were issued periodically by Edward III.
In July 1346 one Roger Normand received formal pardon for his failure to take up
knighthood on the understanding that he would do so before the end of the campaign. (*Foedera*, iii. 86.) Mass knightings were common on the eve of great campaigns.
[3] Brass of Sir John de Creke at Westley Waterless, Cambs., c. 1325 (reproduced
in *Medieval England*, ed. H. W. C. Davis (1924), p. 176).

this heavier type of armour led to the disappearance of the shield which had become superfluous; the knight's armorial bearings were transferred to his *jupon*.[1] The only weapon worn by the knight of Edward II's day was his sword, though he might carry a lance on horseback; but by the middle of the century there appears the short narrow-bladed dagger known as a *misericord* (used for slipping between the plates of armour or through the vizor-holes), and henceforth this becomes an essential part of knightly equipment. The care and cleaning of arms and armour were entrusted to pages who, during an action, also took charge of the great *destriers* and other war-horses, three or four of which always accompanied the knight. Bearing such high-sounding names as *Bayard Dieu*, *Morel de Francia*, and *Bauzan de Burgh*,[2] these expensive animals, the best of which might cost well over £100, were carefully registered and valued; for, though the knights now normally fought on foot, strong horses were indispensable for the great *chevauchées* and for the pursuit of a broken enemy. Edward III had agents buying war-horses for him so far afield as Sicily and Spain.[3]

The category of 'men-at-arms' is difficult to define precisely. Sometimes the term is used to denote all combatants other than archers, sometimes it refers to a special class of troopers, otherwise known as esquires, who ranked below the knights, were paid at the rate of one shilling a day, and were often accompanied by one mounted archer apiece. Such men-at-arms might be aspirants to knighthood, or merely members of the shire and borough levies. But Edward Montagu's indenture of 1341, distinguishing *homines ad arma* from *homines armati* suggests that the former may have been light horsemen and the latter foot.[4] Such light horsemen may have resembled the *hobelars*, whose name derived from the *hobyns*, or ponies of Irish origin, which they

[1] The earliest known brass of a knight not wearing a shield is that of Sir John de Wantone, Wimbush, Essex (1347); the latest example with shields is that of two effigies at Dorchester, c. 1370 (*Medieval England*, ed. A. L. Poole (1958) i. 324). See also S. J. Herben, 'Arms and Armour in Chaucer', *Speculum*, xii (1937), 477, n. 3.

[2] *Black Prince's Register*, iv. 67. Bayard, Morel, and Bauzan denoted the horse's colour, to which the place of origin was often added.

[3] *Cal. Pat. Rolls 1334-38*, pp. 52, 166. A grey charger bought for the king in 1331 cost £120; £200 was paid for the charger ridden by Richard II at his coronation (*Issues of the Exchequer*, pp. 141, 206). The author of the tract *De Speculo Regis Edwardi III* (ed. Moisant, p. 142), urges the king to consider the great expense of war-horses and not to waste money on them.

[4] Above, p. 235.

rode. Hobelars are found, both in the retinues of the magnates and in the shire levies and, like the foot-archers, they played an important part in the Scottish campaigns, before their eclipse by the mounted archers in the early stages of the Hundred Years War. It was in Scotland that the English learned the value of light horsemen for reconnoitering and harrassing the enemy and for rapid movement in difficult country. The hobelar's armour normally consisted of a haketon, a bascinet, a vizor, and iron gauntlets; his arms were a sword, a knife, and a lance, and he was paid 6d. a day. The foot-archer, paid at the rate of 2d. or 3d., was usually drawn from the *posse* or enlisted as a volunteer, though some magnates had foot-archers in their retinues, notably in the Low Countries in 1338–9.[1] His weapons were a knife, a short sword, and a bow with one or two sheaves of arrows and he wore a haketon and a guard (*bracer*) for his arm to catch the string when the bow was loosed. But as the war in France proceeded, both hobelars and foot-archers fade out of the picture, their place being taken by the horse-archers whose skill made such an important contribution to the victories of Edward III and his generals.

Mounted archers, nearly a thousand strong, first appear in the Scottish campaigns of 1334–5, 200 of them, drawn mainly from Cheshire, in the king's bodyguard, the rest in the retinues of the magnates where they seem to have been roughly equal in numbers to the esquires, or men-at-arms. In the Hundred Years War, their normal rate of pay was 6d. a day, their characteristic weapon the Welsh longbow. The value of the horse-archer lay in his ability to move quickly and dismount to shoot; he could not, of course, shoot from the saddle, since a solid foundation was needed for the pull. As was pointed out by J. E. Morris, the longbows used in the Welsh wars were probably no more effective than the old crossbows.[2] But they were capable of much greater development and this was ensured when, from the time of Edward I, archery became the national sport *par excellence*. As in other sports, proficiency was attainable only after long training; the archer had to learn the use of a delicate instrument which a clumsy man could break and a weak man fail to control. Yet it was a skill well within the reach of the

---

[1] A. E. Prince, 'The Strength of English Armies in the Reign of Edward III', *Eng. Hist. Rev.* xlvi (1931), 361.
[2] *Welsh Wars*, p. 100.

ordinary civilian, for it required neither mechanism nor professional drill; all that was needed was steady practice at the town
or village butts. Generations of such practice in the conservation
of energy and the correct use of weight and strength went to
produce the skilled archer, who stood sideways to the enemy, so
that the acts of aiming and loading were practically one and
both hand and eye were brought into play. A good longbow-
man could shoot ten or twelve arrows a minute, as against the
crossbowman's two;[1] and it was this rapid hail of arrows hurt-
ling around the ears of an advancing enemy or, as at Crécy,
maddening his horses which made the longbow so deadly a
weapon. By the time of Poitiers, the great six-foot bows of yew,
maple, or oak, were capable of penetrating chain mail and their
maximum range may not have been far short of 400 yards. But
there were limits to the effectiveness even of the longbow, for it
was essentially a defensive weapon for use against advancing
enemy cavalry and the archers were to some extent dependent
on a suitable terrain. If the enemy failed to advance, or the site of
the battle were ill-chosen, there was little they could do.[2]

The lowest ranking troops in an Edwardian army were the
foot-soldiers, paid at the rate of 2d. a day, among whom the
Welsh were prominent. Though despised by the courtly chroniclers for their unchivalric methods of fighting, the green and
white uniformed Welsh, with their knives and daggers, did
useful service. To their contingents were attached such super-
numeraries as chaplains, standard-bearers, and interpreters.
Finally, every army needed its technicians and workmen.
Miners from the Forest of Dean were useful in siege operations
and forty of them were summoned to join the expeditions of
1346 and 1359; eight carpenters and eight cementarii were raised
in East Anglia in 1354 for the repair of the castle of Guines.[3]
Smiths, armourers, and pavilioners all had their functions, and
minstrels, often attached to the royal household, provided entertainment and helped to sustain morale. A new German dance
lately introduced by Sir John Chandos, beguiled the troops
aboard off Winchelsea on a summer's day in 1350 while they
were waiting for the advent of the Spanish ships.[4]

Commissariat needs were met by various devices. The nature

[1] Lot, L'art militaire, i. 314–15.
[2] P. Pieri, Il Rinascimento e la crisi militare italiana (1952), pp. 225–6.
[3] Foedera, iii. 78, 279, 417.   [4] Froissart, v. 260.

of medieval warfare made it essential for an invading army to try to live off the enemy's country; erratic movement and difficulties of communication would have frustrated any attempt to supply an expeditionary force from the home base over a long period. But supplies had to be collected at the outset of a campaign and to meet certain emergencies, when it was usual to issue orders forbidding the export of corn and instructing purveyors and merchants to collect victuals and other stores and to disburse them as directed. Thus, in March 1333 when there was war in Scotland, merchants of sixteen counties were instructed to take corn and other victuals to the northern ports and sell them there.[1] In March 1338 the government issued orders for the purveyance of 2,000 quarters of wheat in London and the home counties for shipment to Aquitaine, 'for the sustenance of those whom the king is sending over for the defence of the realm'; and William Dunstable was ordered to arrange for corn, meat, salt, and beer to be purveyed for the king's own expedition.[2] The lieutenant of Brittany, finding his supplies running low in the winter of 1343–4, sent William Wariner of Southampton with orders to buy up food supplies at home for the use of his English troops; London was ordered to send victuals to Calais just before its fall;[3] and Froissart tells us that, for the winter campaign of 1359, Edward III, who had heard rumours of famine in France, took elaborate commissariat measures. When the great army issued forth from Calais at the beginning of November, it was accompanied by 6,000 chariots laden with provisions and other stores, including handmills, ovens, and fishing-boats 'made of leather subtly wrought', as well as hawks, hounds, and falcons.[4]

Military effort on this scale seems to presuppose effective transport arrangements and mastery of the Narrow Seas; but such command of the Channel as the English enjoyed was always precarious. The paramountcy at sea of the Mediterranean nations had not yet been challenged seriously; and in a society organized for land warfare, the part played by the fleet was inevitably subsidiary and dependent.[5] If Edward III cherished any vision of naval development, his other commitments pre-

---

[1] *Foedera*, ii. 855.   [2] *Cal. Pat. Rolls 1338–40*, p. 84; *Foedera*, ii. 1021; iii. 3.
[3] *Cal. Letter-Book F*, p. 166.   [4] Froissart, vi. 220–3.
[5] N. H. Nicolas, *A History of the Royal Navy* (1847), ii. 159; M. Oppenheim, *The Administration of the Royal Navy* (1896), p. 6.

vented him from realizing it.[1] The nucleus of what was to become the Royal Navy may be seen in the king's own ships which included two or three 'cogs' (broadly-built cargo ships with blunt prow and stern), and perhaps as many galleys and barges, all brightly painted and ornamented with heraldic and other devices. But it does not seem that Edward added greatly to the number of the vessels he inherited from his grandfather, nor that he did much to encourage ship-building in general; no doubt it was cheaper to commandeer merchant vessels. A permanent fleet might have developed from the ancient *servitium debitum* of the Cinque Ports which had to supply fifty-seven ships, with a crew of about twenty-four apiece, for a short period each year; some such notion seems to have been in the mind of Edward I who made them 'the core of a naval command under a single captain, or admiral'.[2] But the creeks which had served the Ports were silting up and their protecting headlands were being eroded and, under Edward III, the full *servitium* was seldom forthcoming;[3] the tendency here, as in the army, to substitute paid for customary service went far to destroy the unique position of the Ports. In 1335 the wages of the crews of thirty ships supplied by Winchelsea, Romney, Hythe, Dover, and Sandwich, were made chargeable to the exchequer; in 1339 the exchequer, 'of special grace', paid half the wages of the sailors; and in 1344, when the king claimed eight of their greater ships from the Ports, he made it plain that he would pay the wages.[4]

Even if the old customary services had been maintained, the Cinque Ports (with or without the assistance of the royal ships) were altogether incapable of meeting the needs of Edward III. The only way of transporting great armies across the Channel was to conscript large numbers of merchant and fishing vessels for the king's service; and the great majority of the ships which carried the knights, men-at-arms, and archers to France were of this type. At the opening of the war (April 1338) the king complained that his passage overseas was being delayed for

---

[1] P. E. Russell, *English Intervention in Spain and Portugal*, pp. 228–9.

[2] Powicke, *The Thirteenth Century*, p. 655.

[3] J. A. Williamson, 'Geographical History of the Cinque Ports', *History*, xi (1926), 97.

[4] A. E. Prince, 'The Army and Navy', p. 385. Montagu Burrows, *The Cinque Ports* (1888), p. 138; *Foedera*, iii. 10. (Wages were usually at the rate of 6*d*. a day for the master and the constable (or boatswain). The owners were paid 3*s*. 4*d*. per three months for each tun of shipping.)

want of ships and the admirals were ordered to seize all they could find and send them to Yarmouth. In February 1341 Edward III wrote to the mayor and bailiffs of twenty-eight seaports to prepare ships for his service; in April he ordered that 100 fishing boats and small vessels should be assembled for his passage to France. Three years later (June 1344) the admirals were instructed to assemble at Portsmouth and send to the Solent all ships capable of transporting horses and men. In the spring before Crécy, all ships of ten tuns and upwards were impressed; ships in the port of London were detained for the king's service in February 1347; and in 1355 officials were ordered to seize and take to certain specified ports all the ships they could find at sea or in harbour.[1] Measures like these entailed considerable hardship for the owners and merchants whose business connexions were often interrupted by peremptory orders to unload their cargoes at short notice.[2] A parliamentary petition of 1371 reveals some of the disabilities to which they were subject. Ships were arrested long before they were needed, the owners receiving no compensation in the interval and the sailors no wages; superfluous regulations hampered the merchants and drove many seamen to seek employment on land; and the practice of impressing the masters and the best men of the merchant service for the king's ships was most damaging in its effects.[3] By this date, the English fleet was in decline and only a few months later it was decisively defeated by the Spaniards off La Rochelle. Yet even in the days of the great campaigns, there had been frequent complaints of evasion and desertion and it was no easy task to secure adequate transport for the return of expeditionary forces serving in France.[4]

Consistent and effective control of the Channel could not be secured without a permanent navy and not even the victories off Sluys and Winchelsea served to free English merchantmen from the danger of attack by Genoese, French, or Spanish squadrons; the disaster off La Rochelle showed the incapacity of armed

[1] *Foedera*, ii. 1027, 1150, 1156; iii. 15, 78, 299; *Cal. Letter-Book F*, p. 159. For lists of the ships requisitioned or manned in 1355 for the Black Prince's expedition to Bordeaux, see H. J. Hewitt, *The Black Prince's Expedition of 1355-7*, pp. 40-42.

[2] e.g. in 1348 and 1360. *Foedera*, iii. 174, 478-9.   [3] *Rot. Parl.* ii. 307.

[4] e.g. in 1343 orders were issued for the arrest of the masters and mariners of seven ships which, *cum multis aliis*, had withdrawn from the port of Brest after landing the king and his army (*Foedera*, ii. 1226). In 1346 Edward and the army were left stranded in Normandy because the fleet disappeared. Morris, *Welsh Wars*, p. 108.

merchant ships to resist the skilful tactics of the Castilian galleys. None the less, in the period between the battle of Sluys and the treaty of Calais, Edward III undoubtedly had the upper hand at sea. It is no coincidence that this period saw the emergence of the court of admiralty whose function was to keep the king's peace upon the seas and, in particular, to deal with piracy claims made by or against foreign sovereigns. The office of admiral, which appears first at the end of the thirteenth century, was by this date well established. Two admirals chosen from the knightly class,[1] the one controlling the coast northwards from the Thames, the other from the Thames westwards, were appointed regularly in time of war, on a temporary basis, continuity being supplied by the clerks of the king's ships, who arrested ships in the ports and prepared them for war, supervised the choice of men to be impressed, and arranged for supplies of provisions and naval stores. The original functions of the admirals were mainly disciplinary and administrative; they were allowed rights of impressment in the maritime regions and disciplinary powers over the sailors and other members of the fleet. The court of admiralty developed from the extension of their powers, between 1340 and 1357, to include jurisdiction in maritime cases which did not fall within the purview of the courts of common law. Even if Edward III's proud title of 'King of the Sea'[2] derived mainly from his personal achievement in the two naval victories of his reign, the emergence of such a court affords some evidence that, at least for a few years, the claim to the sovereignty of the seas was not altogether baseless.[3]

For men on the make the attractions of war were powerful. In an age when 4d. an acre was judged a fair price for arable land and a ploughman's yearly wage might be no more than 12 or 13s., even the Welsh footman was not badly off with his

[1] e.g. Sir Walter Manny and Sir Bartholomew Burghersh were appointed admirals in Aug. 1337.

[2] Rot. Parl. ii. 311. On the golden noble, first introduced in 1344, Edward III is depicted standing in a ship, with sword and shield.

[3] For the origins of the court of admiralty see R. G. Marsden, Select Pleas in the Court of Admiralty (Selden Soc. 1898), pp. xii–xlv. Under Richard II the jurisdiction of the court was statutorily defined (Statutes 13 Ric. II, c. 5 and 15 Ric. II, c. 3) in consequence of conflicts between its franchises and those of some of the seaports. Like the other prerogative courts, it acquired an evil reputation in the last years of the reign when cases withdrawn to it from the common-law courts were sometimes sent on appeal to special commissions appointed ad hoc. See Tout, Chapters, iii. 468; iv. 45.

2d. a day, and the hobelars, mounted archers, and men-at-arms were very well paid, all the more so because in addition to their regular wages, serving soldiers of all ranks might expect the 'regard', or bonus, payable quarterly at the rate of 100 marks for thirty men. But pay was often in arrears, and, though undoubtedly useful, was by no means the strongest material inducement to service in the field. The man who was looking for really substantial enrichment would pin his hopes to the plundering of the enemy which was an essential element in all the wars of chivalry; and he would be unlucky indeed if he returned from one of the great *chevauchées* through France with nothing in his pocket. For, 'by reason of these hot Wars many poor and mean Fellows arrived to great riches, as Fortune favour'd that side they served';[1] even the ill-starred adventure of the bishop of Norwich yielded loot to the value of over £36 to a Nottingham tailor and his friend.[2] Walsingham tells us that after the fall of Calais, it seemed to the English as though a new sun had arisen 'because of the abundance of peace, the plenitude of goods and the glory of the victories'. There was scarcely a woman in the land who was not decked out in some of the spoils of Caen, Calais, or other towns abroad.[3] Though a few commanders made occasional efforts to spare the churches (according to Froissart twenty men were hanged in Normandy for burning an abbey, and Derby tried to prevent church-burning in Poitou),[4] pillaging the enemy, non-combatants included, was accepted as the soldier's right. From the Black Prince, who robbed King John of his jewels, down to the common soldiers, who found in Normandy such abundance of booty that good furred gowns were despised, and those who, in Gascony, thought only gold, silver, and feathers for arrows worth collecting, all threw themselves with zest into the business of plundering the French, which, indeed, was one of the major objectives of the war. Long and hard experience had taught the lowland Scots to frustrate an enemy bent on pillage; but so little did the people of Normandy know of war that when they were cruelly invaded they had no notion of how to save their stuff.[5]

---

[1] Barnes, *Edward III*, p. 446.
[2] *Records of the Borough of Nottingham*, i. 230–3.   [3] *Hist. Ang.* i. 272.
[4] Froissart, iv. 30, v. 114. There can be no doubt, however, that both churches and houses of religion suffered severely in these regions, see H. Denifle, *La Guerre de cent ans et la désolation des églises, monastères et hôpitaux en France* (1899), i. 24–49.
[5] Froissart, iv. 388–405.

Ransoms were the most valuable form of plunder; for all men, other than the *ribaudaille*, had their price. A burgess taken in war might be ransomed for axes, swords, coats, doublets, or hose; knights and esquires for money or horses; a poor gentleman with nothing to give might buy his freedom by a term of service with his captor.[1] At the top of the scale, ransoms were enormous and their value mounted as the war went on. The ransoms of King John of France and David II of Scotland, fixed, the one at £500,000, the other at 100,000 marks, were exceptional, of course; but ransoms running into four figures were not uncommonly demanded for important prisoners. In 1347 Sir Thomas Dagworth was offered £4,900 for Charles of Blois, the French candidate for the duchy of Brittany; Sir John Harleston's share in the ransom of a French knight taken in Normandy amounted to £1,583. 6s. 8d.[2] Moreover, ransoms were marketable commodities. The Black Prince's sale of some of the prisoners taken at Poitiers to Edward III for £20,000 was an unusually large-scale deal; but many other bargains of the same type were struck.[3] Edward paid Sir John Wingfield £1,666. 13s. 4d. for his prisoner, the Sire Daubigny; Sir Thomas Cheyne, who captured Duguesclin at Nájera, sold him to the king for £1,483. 6s. 4d.; Sir Thomas Holland sold the count of Eu, whom he had taken at Caen, for 20,000 marks; and Robert Clinton engaged in a complicated transaction with the king over the sale of a quarter of the ransom of the archbishop of Le Mans whom he had been lucky enough to capture, receiving in exchange £700 in cash and a life-interest in some Irish manors.[4]

As time went on, distribution of the profits of a wealthy prisoner gradually became a matter of careful regulation. Very ancient custom put all prisoners of war at the disposal of the Crown; but some modification of this royal monopoly was clearly inevitable and by the end of Edward III's reign it had become an accepted convention that each soldier should pay over one-third of his spoils of war to his captain who, in his turn, paid over to the Crown a third of his soldiers' profits together with a third of his own.[5] This rule did not apply,

---

[1] Ibid. vi. 98–99.   [2] *Issues of the Exchequer*, pp. 153, 217.

[3] H. J. Hewitt, *The Black Prince's Expedition of 1355–57*, pp. 152–65.

[4] Ibid., pp. 167, 177; Russell, *English Intervention*, p. 106, n. 3; *Foedera*, iii. 126, 399.

[5] D. Hay, 'The Division of the Spoils of War in Fourteenth-Century England', *Trans. Royal Hist. Soc.*, 4th ser. iv (1954), 91–109. Cf. E. Perroy, 'L'Affaire du Comte de Denia', *Mélanges . . . Louis Halphen* (1951), pp. 573–80.

however, to royal prisoners or to commanders-in-chief of hostile armies whom the king always reserved to himself, though the actual captor would be handsomely rewarded for his services. John Coupland, who took David II at Neville's Cross, was granted an annuity of £500 and the status of a banneret; and both claimants to the honour of having captured King John at Poitiers received satisfaction.[1] But enforcement of the rules and conventions relating to ransoms was not always easy. It was hard to keep track of what went on in the north parts and the Scottish prisoners were a source of constant anxiety to Edward III whose numerous orders forbidding their ransoming without his leave and instructing the captors to bring them to London were freely disregarded. Rumours that an archer named John Ballard had captured the archdeacon of Paris, deserted from the army before Calais, smuggled his prisoner into Colchester abbey, and finally sold him in London for £50, provoked a stream of angry letters threatening the archer, the abbot, and other supposed accomplices with severe penalties.[2] It is likely that, from the king downwards, most of the captors had to be content with less than the large sums originally postulated. The law of arms permitted servants of important prisoners and sometimes the prisoners themselves, to return home to seek their ransom money; but, though the servants of the count of Eu were negotiating his ransom in 1350, he himself was still in custody eight years later.[3] The affair of the count of Denia, which was responsible for a notorious sacrilege case during the minority of Richard II, seems to prove that many ransoms were rated much too high. Nevertheless, the spoils of war were on a scale which made it possible for the commons in the Good Parliament of 1376 to suggest that, if the king had better councillors and servants, he need levy no subsidies because the war should pay for itself.[4] Well might the duke of Gloucester complain to Richard II that his peace policy was disheartening to the poor knights, squires, and archers of England whose comforts and stations in society depended upon war.[5]

[1] *Issues of the Exchequer*, p. 180. The Black Prince is said to have silenced this dispute with the words, 'I am so great a lord as to make you all rich.' Froissart, v. 455.
[2] *Foedera*, iii. 94–98, 100–1.   [3] Ibid., pp. 191, 393.   [4] *Rot. Parl.* ii. 323.
[5] D. Hay (quoting Froissart, xiv. 314). Many years earlier, the Sire d'Albret had reproached the Black Prince for retaining superfluous knights and thereby preventing them from taking their profit elsewhere, in Prussia, Constantinople, or Jerusalem, 'as all knights and squires do, who wish to advance themselves'. Froissart, vii. 145.

Lively hope of financial gain was much; but medieval warfare held deeper and subtler allurements. Again and again, in the pages of Froissart, and for all his awareness of its horrors, we catch echoes of what can only be described as the joy of war, its challenge to professional skill, its appeal to the spirit of adventure and to the desire for glory, the aesthetic, almost spiritual satisfaction which it might afford. Knights were trained in the use of arms; the skill acquired in the lists showed to best advantage on the battlefield. The Black Prince was not the only one who 'in the lusty flower of his youth' was 'never weary nor full satisfied of war ... but ever intended to achieve high deeds of arms'; and most of Froissart's contemporaries would have endorsed his judgement that the companions of Edward III 'in his battles and happy fortunate adventures . . . ought right well to be taken and reputed for valiant and worthy of renown'.[1] Men not of the highest social rank, like Sir Robert Knollys or Sir Thomas Dagworth, a gallant foreigner, like Sir Walter Manny, could win European renown by their prowess in the field; alike to the common criminal and to the greatest in the land, active service offered opportunities of rehabilitation and pardon. Men sensitive to beauty must have shared something of Froissart's delight in the pageantry of war, in 'the fresh, shining armour, the banners waving in the wind, the companies in good order, riding a soft pace', in the sight of a famous knight, like Sir John Chandos, 'with his banner before him and his company about him, with his coat of arms on him, great and large, beaten with his arms'.[2] The ideals which the code of chivalry set before every knight—'trouthe and honour, fredom and courteisie'—were such as to evoke response from many noble hearts. For, despite the intrusion of much that was far from Christian into late medieval notions of chivalry, a volume of religious and semi-religious sentiment still surrounded the concept of knighthood; and a layman of profound personal piety like Henry of Lancaster, felt no scruple in associating himself with the militant enterprises of Edward III. A poet might sometimes sigh for the lost Age of Gold when,

No trompes for the werres folk ne knewe,
Ne toures heye and walles rounde or square

. . . .

Unforged was the hauberk and the plate,[3]

[1] Ibid. vii. 122, ii. 6.   [2] Ibid. vi. 210, vii. 455.   [3] Chaucer, *The Former Age*, ll. 23–24, 49.

Wyclif might allude to 'the sin of the realm in invading the kingdom of France';[1] there were ecclesiastics and, no doubt, some others for whom war between Christians was an abomination; and all men, from the king downwards, paid lip-service to the blessings of peace. But in lay society, by virtue of a tradition inherited from antiquity and destined to endure for many centuries, the profession of arms was supremely honourable; and, despite reiterated papal protests against the Anglo-French conflict, every one of Edward III's expeditions was blessed by the native Church. Feudalism had changed its character, but society was still organized for war; its aristocrats were its generals, its generals its aristocrats. Not merely ransoms and booty, but prestige, adventure, and romance were the rewards of the fighting men. 'Every man', said Dr. Johnson, 'thinks meanly of himself for not having been a soldier, or not having been at sea.'[2] Edward III would have agreed with him. He understood better than his grandson, and better than some of his modern critics, how little a policy of peace, retrenchment, and reform was likely to appeal to the knightly classes among his subjects.

It was the king's understanding of these classes who held the balance of power in the state, his readiness to meet them on equal terms which ensured his popularity among them. Thatching and ditching were not for him. A series of splendid tournaments, given at the charges of the king or one of his friends and honoured with his presence, ministered to the knightly passion for warlike exercise and lavish entertainment. For the tournament was no longer what it had been, a country sport, a gathering in the open for trial of horses and arms.[3] Fought in the lists (an enclosed area not unlike a modern football ground), fourteenth-century tournaments, over which the constable and the marshal presided, were governed by elaborate formal regulations. A series of single combats, or joustings, replaced the mock battles of an earlier age and were made the occasion of much fantastic pageantry, affording opportunity for the display of individual

---

[1] *De Ecclesia*, ed. J. Loserth (1886), p. 427.
[2] Boswell's *Life of Johnson*, ed. Birkbeck Hill, revised L. F. Powell, iii. 265.
[3] Powicke, *King Henry III and the Lord Edward* (1947), i. 21. Cf. N. Denholm-Young, 'The Tournament in the Thirteenth Century', *Essays presented to Sir Maurice Powicke* (1948), pp. 240–68. Tournaments and joustings were strictly forbidden without royal licence, but the number of proclamations to this effect points to the difficulty of enforcing them, e.g. *Foedera*, ii. 808, 815.

prowess, for hazardous encounters[1] and the winning of knightly reputations, and forging between the king and the aristocracy the powerful bond of pleasures shared. Edward's delight in them was undoubtedly spontaneous and, at least in his younger days, may have owed little to conscious recognition of their value as a specific against political disorders. But the crisis of 1341 stood as a warning against complacency; and an element of propaganda is unmistakable in what was, perhaps, the most brilliant inspiration of the Age of Chivalry, the foundation of the Order of the Garter.

The loss of the early statutes of the Order and the imperfection of the early records have opened the way to much speculation as to its origins.[2] It may have been the example of a voluntary association of knights at Lincoln which first suggested to Edward III that he himself should found a knightly Order;[3] and it seems to have been at the conclusion of a great tournament held at Windsor in 1344 that the king took solemn oath that within a certain time he would follow in the footsteps of King Arthur and create a Round Table for his knights. The precise form which the Order should take had not yet been determined; but some barons and knights were sworn forthwith as companions of the Round Table and provision was made for an annual feast at Whitsuntide. Froissart, who places the foundation of the Order of the Garter in 1344, is almost certainly conflating two distinct occasions;[4] for it was not until the king returned to England wearing the laurels of Crécy and Calais that the Order was formally inaugurated, perhaps on St. George's Day, 1348. The symbol of the blue garter had first suggested itself at a ball held at Calais to celebrate the fall of the town. Recent research gives credence to the tale (told by Polydore Vergil and Selden, but scorned by Ashmole), of how the young countess of Salisbury, with whom the king was at this

[1] Among those who lost their lives in tournaments were John, son of Lord Beaumont (1342), and the king's friend, William Montagu, earl of Salisbury (1344).

[2] The best history of the Order is still that of Elias Ashmole, published in 1715. See also J. Anstis, *The Register of the Most Noble Order of the Garter* (1724); G. F. Beltz, *Memorials of the Order of the Garter* (1841); N. H. Nicolas, *History of the Orders of Knighthood*, 2 vols. (1841–2); and W. A. Shaw, *Knights of England* (1906). The early statutes survive only in a fifteenth-century transcript.

[3] John Duke of Normandy projected a similar idea about the same time, but the Order of the Star did not emerge until 1351.

[4] Froissart, iv. 203–6.

time in love, dropped her garter as she danced and of how he stooped and bound the blue ribbon round his knee, rebuking the jests of the bystanders with words that were to become famous in the history of England—'honi soit qui mal y pense'. The garter, the king declared, would soon be most highly honoured; and it was not long before he set about fulfilling his oath of 1344.[1] Garters and robes were issued to a dozen knights for the victory tournaments in England and for the initiation of his Order Edward provided a state banquet in the great tower of Windsor castle, then in process of reconstruction by the master mason, William de Ramseye. Within the castle, the collegiate chapel of St. Edward was enlarged and rededicated as the chapel of St. Edward and St. George; and, although in the event few seem to have benefited, it was part of the original intention that the endowments of the college should provide for the needs of twenty-six impoverished warriors.[2]

The Order of the Garter was a brotherhood, 'a Society, Fellowship, College of Knights', and within it all were equal, 'to represent how they ought to be united in all Chances and various Turns of Fortune; co-partners both in Peace and War, assistant to one another in all serious and dangerous Exploits: and thro' the whole Course of their Lives to shew Fidelity and Friendliness one towards another'. Its basis was deliberately exclusive, for it consisted of only twenty-six knights, including the sovereign, between whom and the knights the blue garter signified 'a lasting bond of Friendship and Honour'.[3] By means of this ingenious device, Edward harnessed the idealism of chivalry to his cause and linked to himself under an obligation of honour some of the greatest names in the land. Edward, the Black Prince, Henry, earl of Derby, Thomas Beauchamp, earl of Warwick, Ralph (later earl) Stafford, Roger Mortimer, earl of March, William Montagu, earl of Salisbury, Sir Bartholomew Burghersh, Sir Hugh Courtenay, Sir James Audley, Sir John Chandos, and the loyal Gascon, Sir Jean de Grailly, Captal de Buch, were all among the Founder Knights. As stalls fell vacant, there was no lack of other famous men to fill them—the king's

---

[1] Margaret Galway, 'Joan of Kent and the Order of the Garter', *Univ. of Birmingham Hist. Journal*, i (1947–48), 13–50. Miss Galway has identified the countess as the celebrated beauty, Joan of Kent, later the wife of the Black Prince.

[2] W. St. John Hope, *Windsor Castle* (1913), ii. 531–3; E. H. Fellowes, *The Military Knights of Windsor* (1944), p. xvii.

[3] Ashmole, pp. 124–6.

younger sons, Lionel of Clarence and John of Gaunt; the earls of Pembroke, Suffolk, and Northampton; Sir Walter Manny and Sir Thomas Felton; and friendly foreigners, like the duke of Brittany and Sir Robert of Namur.[1] Its exclusiveness and the opportunities which it afforded for personal contact with the sovereign soon caused membership of the Order to be regarded as a high distinction, and that not only in England. Edward sent his heralds to publish it in France and Scotland, in Burgundy and the Empire, in Hainault and Brabant. The Order soon had its own herald, the Windsor herald; and the soaring military reputations of the king and his eldest son reflected glory upon the brotherhood which they had made peculiarly their own.

The Order of the Garter was exclusive, but the king himself was not; and there is no reason to suppose that enthusiasm for his wars was confined to the knightly classes. The archer, too, was master of a warlike craft and the profits of war were open to all. Good service in the wars might even win a serf his freedom;[2] and the commissioners of array did not find it necessary to fill the ranks with unwilling conscripts.[3] There must have been many old soldiers with the rebels of 1381 and it is hard to believe that Gaunt's unpopularity owed nothing to his military failures, or that Wat Tyler's men, who flew the banner of St. George, were averse from war as such.[4] Pacifist sentiments find no place in the tirades of John Ball or among the grievances presented to Richard II at Mile End and Smithfield. There were periods of war-weariness; there was the inevitable grumbling at war taxation, purveyance, and dislocation of shipping; in the sixties and seventies there may even have been some revulsion from the excesses of the Black Prince and others among the captains. But the conclusion is irresistible that, on the whole, the war was popular with the laity of all classes. It was less for peace that men clamoured than for good leadership, enterprise, and victory.

[1] For a complete list, see E. H. Fellowes, *The Knights of the Garter 1349-1939* (1940).

[2] e.g. *Black Prince's Register*, iv. 206: 'Manumission of Reynold, son of Alexander Rede of Weybridge, the prince's bondman of the manor of Byfleet, and all his issue, as a reward for his good service in Gascony.'

[3] A roll of *c.* 1355 shows that more than half the fighting-men of seventeen Derbyshire manors were then actually at the wars. *Vict. County Hist. Derby*, ii. 168.

[4] It may be more than coincidence that the most successful peasant rising in England should have been in the period of the longbow. See S. Andrzejewski, *Military Organisation and Society* (1954), p. 117.

Thus did the winter of baronial discontent give place to glorious summer. All the leading families were behind Edward III in his great adventure. Emblazoned on his roll of honour were names which in the past had spelt turbulence and treason—Lancaster, Mortimer, Arundel, Bohun, Beauchamp, and Warenne. Pride of place among the captains was accorded to Henry of Grosmont, active representative of the house of Lancaster from 1330 and for a generation thereafter the close companion of Edward III in nearly all his major endeavours. Already a hero of the Scottish wars and of the naval victory of Sluys when, in 1345, he became earl of Lancaster, Henry crowned his military reputation by a series of brilliant victories in Gascony and Poitou before joining the king at the siege of Calais. In 1349 he is found raiding in the neighbourhood of Toulouse, in 1350 at the sea-fight off Winchelsea; and although, in the summer following, he allowed himself the interlude of a crusade in Prussia and Poland, he was back in time to represent Edward in his negotiations with Charles of Navarre and Innocent VI. Lancaster saw further active service in Brittany and Normandy in 1356–7, accompanied Edward on his last campaign in 1359, and was acting as principal negotiator of the peace treaties the year before he died. His services were generously rewarded by a king never unmindful of the debts he owed his friends. Grants of the castle and town of Bergerac, which Henry himself had captured and from the profits of which he built the great palace of the Savoy, of the life shrievalty of Staffordshire and of the earldom of Lincoln culminated in his elevation in 1351 to the rank of duke and of his county of Lancaster into a palatinate.[1] The title (hitherto held only by the prince of Wales as duke of Cornwall) was purely honorary and carried with it no specific rights or privileges: but the grant of the palatinate allowed the duke virtually royal powers within the county of Lancaster.[2] Edward III may have had some notion of creating a bulwark against the Scots in north-western England, to balance the palatinate of Durham in the north-east; but there can be little doubt that the grant was inspired, first and foremost, by his desire to honour his friend and companion-in-arms. The king had no reason to fear the consequences of allowing quasi-royal powers to devolve upon a subject, for the grant was for

[1] The title of earl of Moray was bestowed on Lancaster by David II in 1349.
[2] R. Somerville, *History of the Duchy of Lancaster*, i. 40–45.

life only and, as it happened, duke Henry had no son. His great inheritance would be divided between his two daughters; and it is likely that Edward was already determined on the marriage between Blanche of Lancaster and John of Gaunt which would secure the reversion of at least a part of it to one of his own children.

Even more impressive as evidence of the changed political climate of the new reign is the volte-face of the house of Mortimer. With the ignominious end of the first earl of March in 1330, it must have seemed to many that the fortunes of his line had finally collapsed. But Edward III was seldom disposed to visit the sins of guilty fathers upon their children and the traitor's son, Edmund Mortimer, was allowed to retain a block of territory in central Wales. When Edmund died within the year, leaving an infant heir, the estates were temporarily dispersed; but the interests of the minor, Roger Mortimer, were not forgotten. In 1342, while still well under age, he was allowed the castle of Radnor with other lands in Wales; next year, he received Wigmore, the cradle of his race. Young Mortimer's chance to prove himself came on the eve of the Crécy campaign when, in company with the Black Prince (who was his exact contemporary) and many other sons of noble houses, he was admitted to the honour of knighthood. He rose to the occasion and was rewarded with admission to the select circle of the Garter and with seizin of his father's lands. Mortimer was summoned to parliament as a baron from 1348; and the last vestige of stigma attaching to his name was removed when, six years later, he secured parliamentary reversal of the judgement on his grandfather, together with restoration of the title of earl of March and all the family estates. Many of these had by this date fallen into other hands; but restoration was effected by a series of arbitrary decisions, thinly veiled in judicial forms and given in the king's bench, to which Edward did not hesitate to send a privy seal letter whenever any contestant of Mortimer's claim looked like winning his case. Salisbury was forced to relinquish the valuable marcher lordship of Denbigh and Arundel was allowed to retain Chirk, only as a result of a private bargain, cemented by a marriage alliance between his family and Mortimer's.[1] For the remaining years of his short life, March was closely associated with the king. As a member of the council, he

[1] See G. A. Holmes, *Estates of the Higher Nobility*, pp. 15–17.

was with Edward in Picardy in 1355 and he rode in the van of the host during the great campaign of 1359. When he died, a year before Henry of Lancaster, his obsequies were solemnly celebrated, as became a knight of the Garter, in the royal chapel at Windsor; and his infant heir, destined to marry the king's granddaughter, became a royal ward.

Richard Fitzalan, earl of Arundel and Surrey, shared his friend's good fortune and lived much longer. His father had been one of the victims of 1326 and it was as a gesture of compensation for the disruption of the family property that Fitzalan (commonly known as 'Copped Hat') had been allowed the Mortimer lordship of Chirk in 1331. He too secured reversal of the judgement on his ancestor in the parliament of 1354, by which date he had already succeeded to most of the southern estates of his uncle Warenne who died in 1347. After adventuring with Balliol in Scotland, Arundel, in his capacity of admiral of the western fleet, had commanded a squadron at Sluys; together with Northampton, he had led the left division at Crécy; and he had been with the king in the fight off Winchelsea. He was lord of extensive lands and, though, curiously enough, he was never a knight of the Garter, he may be presumed to have done well out of the war, for the king, the Black Prince, and John of Gaunt were all indebted to him for a series of very substantial loans and he died possessed, in cash or bullion, of the enormous sum of £60,000.[1] In the two Bohun brothers, Humphrey earl of Hereford and William earl of Northampton, grandsons of Edward I in the female line, Edward III found no less loyal adherents. As constable of England from 1338, Northampton played a leading part in nearly all the great campaigns. When he died in 1360 his son Humphrey, whose wife was Arundel's daughter, inherited all the wide lands of the Bohuns in Essex and the Welsh marches. Thomas Beauchamp, earl of Warwick, John de Vere, earl of Oxford, Robert Ufford, earl of Suffolk, and Ralph, Lord Stafford (for whom an earldom was created in 1351) were active in almost every major engagement between 1340 and 1360; and Lawrence Hastings, earl of Pembroke, contrived to see active service at Sluys and Tournai, in Brittany and Gascony, and before the walls of Calais, in the course of a life of under thirty years.

Treason became almost inconceivable to a generation which, after 1341, knew nothing of obnoxious favourites and saw no state trials. And, since Edward III had finally repudiated his own obligations as a vassal, the atmosphere was propitious for a settlement of this explosive subject on lines acceptable to all parties. The great statute of 1352 defined high treason under six heads: (1) compassing or imagining the death of the king, the queen, or their eldest son: (2) violating the queen, the eldest unmarried daughter of the king, or the wife of his eldest son: (3) levying war against the king or adhering to, or comforting his enemies; (4) counterfeiting the royal seals or the royal money; (5) importing counterfeit money into the realm; (6) killing the chancellor, the treasurer or any of the judges. Moreover, it was specifically declared that to ride armed against other persons, to slay, rob or imprison them, should in future be adjudged, not treason, but felony or trespass, 'according to the laws of the land of old time used.'[1] The primary object of the statute was probably legal, rather than political, to establish a clear distinction between high and petty treason and so to settle the rules about forfeitures. As such, it was welcome to both king and lords, not least, perhaps, to those who remembered the dark days of Edward II; and, though events were to show that the definitions of the statute had been too narrowly drawn, the cry of treason was never raised again under Edward III, not even in the turbulent parliament of 1376.[2]

Satisfied with abundance of military adventure and relieved of the nightmare of treason, the magnates were ready to leave Edward III to manage his own business while they themselves looked to their personal, family, and territorial interests. They were better equipped for these tasks than has sometimes been supposed, not only because of the experience they had gained in the royal service but also because by this date most of them were at least literate and some were men of good education.[3] Henry of Lancaster was the author of a devotional treatise, the

[1] Statutes 25 Edw. III, st. 5.
[2] See S. B. Chrimes, 'Richard II's Questions to the Judges, 1387', Law Quarterly Rev. lxxii (1956), 365–90.
[3] I am much indebted to the courtesy of Mr. K. B. McFarlane in allowing me access to the manuscript of his Ford Lectures (1953), 'The English Nobility 1290–1536', which greatly illuminate a difficult subject.

Livre de Seyntz Medicines;[1] John Montagu, earl of Salisbury, was something of a poet, the Berkeleys were responsible for Trevisa's translation of the *Polychronicon*, Guy Beauchamp, heir to the earl of Warwick, bequeathed forty-two books to Bordesley abbey, Thomas of Woodstock possessed a considerable library. Literary tastes and aptitudes may have been unusual; but there can be no question as to the general capacity of the landed classes to organize their private affairs and, despite frequent absences abroad, to exercise careful supervision over their servants and officials. Lordship, like monarchy, was still intensely personal and when he went to the wars a lord seldom failed to do as the king did and to designate someone—his wife, a son, or a near kinsman—to exercise the seignorial authority, assisted by the group of friends and retainers, officials, and lawyers which constituted the normal baronial council, appointed to advise the lord on all questions relating to the exploitation of his resources. Surviving records of the great estates of the Black Prince and of John of Gaunt reveal an elaborate hierarchy of officials and complex administrative arrangements, closely resembling those of the king himself. Within the Black Prince's highly diversified appanage, bailiffs and farmers of manors were answerable, directly or indirectly, to the local financial officers—the chamberlains of Chester, or of north or south Wales, or the receiver of the duchy of Cornwall—whose accounts were submitted to the prince's auditors, themselves answerable to the receiver-general, or keeper of his exchequer at Westminster. The exchequer was the core of this increasingly centralized system. Here, the prince's council normally sat and pleas were entertained. Though the prince had no great seal before 1362, the keeper of his privy seal is sometimes styled his chancellor; and he had a chamber, as the source of his privy purse, and a wardrobe to supply his domestic needs in peacetime and his military needs in time of war. The clerks who organized his affairs were often king's clerks borrowed from the departments of state—a fact which goes far to explain the analogy between the two systems—and one of these, Peter Lacy, was for a time simultaneously keeper of the king's privy seal and receiver-general to the prince. Within his English appanage the prince was active, despite his many commitments overseas. He himself frequently authorized the use of his seal; and, though he

[1] Ed. J. Arnould, Anglo-Norman Text Society (1940).

seems never to have visited Wales, he was well known in Chester and Cornwall.[1] The organization of the Lancastrian estates under Duke Henry and John of Gaunt was hardly less intricate and was also modelled on the royal administration. Above the minor local officials there were local receivers and local auditors; at the centre, there was the receiver-general (the chief financial and accounting officer of the duchy), there were also the wardrober and treasurer who supplied the duke's personal needs and paid the wages of his household staff, there were stewards and secretaries, and, at least from the time of the first duke, there was a chancellor.[2] But the numerous warrants issued by Gaunt to this official show how close was the personal supervision which he himself exercised over the affairs of his widely scattered lands. We have all too little evidence as to the organization of other lay estates; but it seems that any magnate of consequence was normally assisted by a council; and we know that Thomas II, Lord Berkeley, kept a watchful eye on every detail relating to the profits of his manors.[3] Under Edward III, the profits of landowning were beginning to contract, at least in some areas, but the wages and profits of war relieved the landlord from exclusive dependence on his farms and a shrinkage of agricultural yields might be offset by inheritance or purchase of additional properties. In general, it was not failure of resources or financial mismanagement which brought baronial houses to disaster in this century; it was political miscalculation or failure of heirs.

Failure of heirs constituted by far the most serious threat to baronial stability. Families like the Mortimers or the Beauchamps, which maintained an unbroken line of male descent from the thirteenth to the fifteenth century, are the exceptions; and such advantages as continuity brought to the Mortimers were largely outweighed by short lives and long minorities.[4] The case of Edward III himself, a father of seven sons whose patrimony had passed to a childless grandson within a generation of his own death, affords only the most conspicuous example of what was happening to many noble families. In the fourteenth century alone, thirteen earls—Gilbert de Clare III of

---

[1] Margaret Sharp, 'The Household of the Black Prince', Tout, *Chapters*, v. 289–400.
[2] Somerville, i. 71–133.   [3] *Lives of the Berkeleys*, i. 156.
[4] The first earl of March was the only one of his line to reach his fortieth year.

Gloucester, Robert de Vere III and Robert de Vere IV of Oxford, John de Warenne II of Surrey, William Ufford of Suffolk, John Bohun of Hereford and Essex, John, earl of Kent, William Clinton of Huntingdon, Thomas and William, earls of Stafford, William Montagu II of Salisbury, and Aymer de Valence and John Hastings II, earls of Pembroke—left no surviving legitimate issue. Lionel, duke of Clarence, Henry, duke of Lancaster, Henry Lacy, earl of Lincoln, and Humphrey Bohun V of Hereford, Essex, and Northampton, left only daughters. When a landowner was childless, he might arrange in his will for his lands, or some of them, to be sold for the benefit of a religious foundation or at a low rate to his friends or kinsmen; or he might sell certain manors, or the reversion of them, for ready cash in his own lifetime. Dispersal of the inheritance was not, however, a necessary consequence of the childlessness of a lord. If the heir-presumptive were a nephew or a brother, this was usually sufficient incentive for trying to hold the family estates together. Robert de Vere III of Oxford, for instance, when his only son died without issue, obtained licence to entail his estates on his nephew, John de Vere.[1] If the heir were a daughter (or daughters), the lands and titles would normally pass to their husbands, and fathers of wealthy heiresses could take their pick among likely suitors. The daughters of Henry of Lincoln, Henry of Lancaster, and Humphrey V of Hereford all married princes of the blood. But transmission of land to sons-in-law inevitably meant the concentration of ever greater estates in the hands of fewer landlords; and this was the prevailing tendency throughout the later Middle Ages when the number of magnates was decreasing and their wealth and influence growing. Recruits to their ranks were few, for Edward III was sparing of new creations outside his immediate family circle and many of those which derived from the ill-considered impulses of Edward II or Richard II failed to take root. There was little to compensate for the toll taken of the nobility by plague and disease, battles and tournaments, political misadventure, and the normal hazards of medieval daily life.

But for the man of property, great or small, who had children the prospects were brighter than ever before. The legal practice of the fourteenth century was allowing the landowner a new freedom of action in disposing of his land; and it was increasingly

[1] *Cal. Pat. Rolls 1327–30,* p. 500.

evident that, like the obligation of military service, feudal rights and obligations in regard to land were fast becoming anachronisms. Landlords, who were also tenants, were seeking new freedoms—freedom to prefer a younger son, or even a daughter, to the heir at law; freedom to endow a wife (or a husband) at the expense of the heirs; freedom to make their own arrangements for the wardship and marriages of their children; freedom, in short, to treat their land as absolute property, under the Crown. Various devices at common law opened the way for them to achieve these ends. The purpose of the statute *De Donis*, designed to safeguard the rights of the heir at law, was defeated when it became possible for a man to entail land upon himself with reversion to a younger son. By endowing his wife with a jointure, that is, with a joint tenancy of some of his lands or rents, with remainder to the survivor of the marriage, a husband might ensure that his wife lived in comfort to the end of her days, at whatever cost to his ultimate heirs. When John, Lord Segrave, married Margaret Brotherton, the countess Marshal, he took the unusual step of giving her a jointure of his whole property, to which his heir at law was John Mowbray of Axholme. Segrave died in 1353 but his widow outlived not only her second husband, but also Mowbray and his wife, dying at a ripe old age in 1399. It was not until fourteen years later that Mowbray's great-grandson obtained possession of the Brotherton and Segrave lands.[1] The system of jointures goes far to explain the prominence of dowagers in this period. But of far more widespread application was the system of uses or trusts, which, since it had the effect of making it possible to devise land by will, amounted to a revolution in the law of real property. Under this system a man might alienate some of his lands to a group of his friends, relatives, or retainers, with instructions to hold them to his use while he lived and to dispose of them after his death, in accordance with his will. Direct instructions about real property become common in wills after 1380, the main object usually being to provide for younger children.[2] The system of uses made it possible for a tenant to keep land permanently out of the hands of his overlord, to avoid the feudal incidents and, if he so desired, to partition his property among his issue. In so far as it was resorted to by tenants-in-chief it was evidently detrimental to the interests of the Crown and such enfeoffments

[1] Example cited by Mr. McFarlane. [2] Holmes, p. 55.

needed royal licence. Either because he failed to perceive its full implications or, as is more likely, because he thought no price too high to pay for the goodwill of his magnates, Edward III licensed them freely; and the system did not, in fact, result in the fragmentation of many estates. Most landlords preferred to keep their property intact, to seek rich marriages or a career in the Church for their younger sons and to endow their daughters with marriage-portions in cash. But, however applied, the system was enhancing the power of the magnates and diminishing that of the Crown in regard to land; it came near to reversal of the policy pursued by Edward I.

The personal prestige of landowners of all ranks owed much to the size of their households, the numbers of their dependents, and the standard of display and hospitality which they could afford to maintain. The indentured retinue, already well established in the reign of Edward II,[1] was hardly less valuable in peace than in war; and though military necessity was the main reason for its development, this was by no means the only reason for its growing popularity. Since the statute *Quia Emptores* (1295) had put a stop to sub-infeudation, it became necessary to devise a new type of contract to attract dependents who were not necessarily soldiers. In addition to the nucleus of permanent military retainers whom he took with him to the wars, a lord would normally retain a number of officials—chaplains, lawyers, minstrels, even cooks—on a life basis. John of Gaunt retained Brother William Appleton, 'phisicien et surgein' for life, at a wage of 40 marks in peace-time and 80 marks in war-time, with allowances for servants and horses.[2] Except in great households like his, however, life-retainers were not very numerous, for they constituted a burden on the estate and it was cheaper to offer fees and liveries on a temporary basis to neighbours, tenants, or others in search of a patron. It was against these temporary retainers, potentially lawless and apt to be left in search of employment when their contracts expired, that the statute of 1390 was directed, making it illegal for anyone below the rank of banneret to retain others and for anyone below the rank of esquire to be retained, and prohibiting the wearing of liveries, except at the wars or in the lord's service.[3] Household attendants and life retainers were accepted as necessary and the

[1] Holmes, pp. 81–82.   [2] *John of Gaunt's Register 1372–76*, ii. 335 (no. 836).
[3] *Statutes* 13 Ric. II, c. 3. See N. B. Lewis, 'The Organisation of Indentured

fourteenth century saw no attack on all retainers as such, nor any attempt at statutory limitation of the size of noble households. Next to mutual protection, the main purpose of the life retinue was undoubtedly display. On any occasion   such as a meeting of parliament—when a lord wished to show to advantage in the world, he would expect to be accompanied by a respectable complement of liveried retainers; and even a plain knight would normally travel with a retinue of at least half a dozen servants; to go abroad unattended would be to lose prestige. And the system had other advantages. It enabled the man of property to help his remote kinsmen and others whom he might wish to patronize; and to the retainers themselves it offered a pleasing prospect of wages, security, and probably a pension. Some even found it possible to be simultaneously retained for life by more than one lord.[1] So far as the system affected social stability in general, it seems likely that the contrast between the classic age of feudalism when a lord's tenants were bound to him by homage, and the 'bastard feudalism' of the later Middle Ages has been overdrawn. It was to the mutual advantage of lords and their followers in any period to be serviceable one to another; and although, as has been seen, the fourteenth century affords many examples of disorder and perversion of the law among the landed gentry, maintenance, in the sense of gangs of armed men seeking to overawe the courts in the interests of a patron, had not yet become widespread. None the less, it is evident that under a weak ruler the new system, like the old, might easily get out of hand.

Splendid living and lavish hospitality were expected of great lords in an age which admired ostentation; and the fourteenth-century magnates fulfilled expectation on a scale which does not suggest straitened resources, or (since few of them seem to have been seriously embarrassed by debt) reckless improvidence. Surviving wills give us some idea of the splendid furnishings of the great houses and their chapels, of the plate, jewels, and fantastically elaborate clothing upon which thousands of pounds were spent. Many magnates were zealous builders—at Berkeley, we may see an example of a fine new hall added to an ancient fortress and at Bodiam a new-fashioned castle built entirely in this

Retinues in Fourteenth-Century England', *Trans. Royal Hist. Soc.*, 4th ser. xxvii (1945), 29–39.

[1] K. B. McFarlane, 'Bastard Feudalism', *Bull. Inst. Hist. Res.* xx (1947), 173–5.

century. Many allocated part of their wealth to religious or educational foundations, as did Richard II of Arundel at Oriel College, Oxford, where he founded a chapel, or Humphrey III of Hereford and Essex, who made it possible for the Austin friars of London to rebuild their church, or the great ladies of Pembroke and Clare, foundresses of the Cambridge colleges which bear their names.

Taken as a whole, the knightly classes, the chivalry, constituted a distinctive but by no means closed society, the armigerous families including men of widely diverse ancestry and fortunes. It was not easy for a man of humble origin to gain entry to the select circle of the earls; the age affords no parallel to the sensational rise of the de la Poles. But into the ranks of the barons and knights there was always a steady influx of aspirants to gentility including, not only the war heroes and those who had earned the gratitude of a powerful patron, but also sons of successful lawyers, like the Scropes[1] and the Cobhams, kinsmen of influential bishops, like Burnell or Melton, prosperous traders like John Pulteney or John Philipot. Contemporary complaints of a shortage of knights suggest that intake may not have been adequate to balance losses and that the numbers of the gentry, as well as of the higher nobility, may have been declining in relation to the total population. None the less, the chivalry remained the dominant social caste, subscribers to a common code of behaviour, linked to one another and distinguished from those below them by the coat-armour which they cherished as proof of noble or gentle birth.[2] Armorial bearings were military in origin, serving the practical purpose of identification on the battlefield;[3] but handed down from father to son, they had become the hallmarks of social distinction; and the experts in such matters were the heralds. Rapid multiplication of heralds in this century reflects their growing importance. King's heralds are not heard of before the reign of Edward I and then only in connexion with tournaments; but under Edward III there

---

[1] See E. L. G. Stones, 'Sir Geoffrey le Scrope (*c.* 1285–1340), Chief Justice of the King's Bench', *Eng. Hist. Rev.* lxix (1954), 1–17.

[2] By heraldic tradition, perfect *noblesse* or gentility was acquired only after inherited arms had been borne for four generations; but the man whose right to a coat of arms had been officially recognized had at least set his foot on the lowest rung of the social ladder.

[3]   'But by hir cote-armures and by hir gere.
        The heraudes knewe hem best . . .' *Canterbury Tales*, ll. 1016–17.

appear the 'kings of arms'—Windsor for the Order of the Garter, Norroy for the regions north of Trent, Surroy for the regions south of it, Clarenceux for Lionel of Clarence, as well as many others in noble households, or attendant on the great captains. Hereford herald served the Bohuns; and it was the herald of Sir John Chandos who composed the metrical life of the Black Prince. Froissart, indeed, tells his readers that he looked to the heralds as the best purveyors of material for his history. But, though empowered to adjudicate on claims to arms, it was not within the competence of the heralds to pronounce between rival claims to the same coat. This was a matter for the king and his two principal military officers, the constable and the marshal, on whom rested responsibility for the maintenance of discipline in the army and for the enforcement of martial law. Under the presidency of these officers there emerged the court of chivalry, concerned with personal disputes between gentlemen, touching their honour.[1] The *cause célèbre* of *Scrope* v. *Grosvenor* over the right to bear the arms, azure a bend or, dragged on in this court for nearly five years: and the case of *Lovell* v. *Morley* was almost as protracted.[2] Like all the prerogative courts, the court of the constable and marshal was guided, in the main, by civil lawyers and already in 1379, it was being seen as a potential threat to the common law. Statutory definition of its functions eleven years later was a direct consequence of parliamentary complaints.[3] None the less, the subservient parliament of 1398 agreed to remit the dispute between Hereford and Norfolk to the court of chivalry.

In so far as he was armigerous, a landowner, and a parent, Edward III's problems were not essentially different from those of his barons and knights. His dispute with Philip of Valois was, on the face of it, a quarrel about armorial bearings, to be settled by battle like other such quarrels. Hasty condemnation of the appanages which Edward created for his sons ignores the fact

[1] See A. R. Wagner, 'Court of Chivalry', *Chambers's Encyclopaedia* iv (1955), and 'Heraldry', *Medieval England* i (1958), 338–81. It is now thought likely that this court originated in a delegation by the Crown to the constable and marshal in 1347 or 1348.

[2] N. H. Nicolas, *The Scrope-Grosvenor Roll* (1832). The case of *Lovell* v. *Morley* is still unpublished.

[3] *Rot. Parl.* iii. 65; *Statutes* 13 Ric. II, st. 1, c. 2. John Barnet, Official of the court of Canterbury, and Robert Weston, D.C.L. were among the judges in the Scrope-Grosvenor case.

that he—the first English king since Henry II with a large
family to provide for—was merely doing for his children what
his magnates were doing for theirs, what it was taken for granted
that he must do. Paternal duty demanded that he should help
his children to maintain their proper station in life, that he
should promote the solidarity of his family and extend its influence where possible. Since he was a king, the scope of his
activities was wider than that of his barons and it was natural
for him to seek marriage alliances outside as well as inside England; but it is inconceivable that he should have sought to disinherit or impoverish his sons in the interests of a future which
no one then living could foresee. What brought Edward's house
to disaster was not the so-called 'appanage policy' but two un-
predictable calamities—the premature death of his eldest son
and the tragic incapacity of his grandson. His family policy was
eminently successful in his own lifetime; and his children had
reason to be grateful to a father whose military preoccupations
had never rendered him neglectful of their proper interests.

Provision for the eldest son was a matter of no great difficulty.
The earldom of Chester, already earmarked as the appanage of
the heir-apparent, was conferred on Prince Edward in 1333,
before he was three years old; and it was taken for granted that
in due course he would be prince of Wales. Edward III's decision to allow him the earldom of Cornwall was another matter;
for this dignity was normally reserved for a younger son. It may
have been the death in infancy of the king's second son, William,
which prompted him to add Cornwall to his heir's considerable
estates.[1] The Black Prince's marriage proved something of a
disappointment. Schemes to marry him to a princess of Portugal
and to a daughter of the duke of Brabant came to nothing; and
he had reached the advanced age of thirty when he made a love-
match with his father's first cousin, the beautiful countess Joan
of Kent. As an heiress with royal blood in her veins, Joan was
sufficiently eligible; but she had been twice married already,
one of her former husbands (the earl of Salisbury) was still alive,
and the Black Prince was godfather to one of her children. Thus,
the couple had to go through a form of penance before the necessary papal dispensation was obtained. Since he was already so

[1] In Mar. 1337 the earldom of Cornwall was converted into a duchy annexed in
perpetuity to the eldest son of the king of England. Edward became prince of Wales
in 1343.

well endowed, the prince did not need a very wealthy wife; and
his prospects were still further improved when, in 1362, the year
after his marriage, his father granted him the principality of
Aquitaine.

The king's third son, who was born at Antwerp in 1338 and
given the romantic name of Lionel (perhaps in honour of the
Lion of Brabant), could look for little from the paternal estates.
An important marriage had to be arranged for him and before
he was four years old, Lionel had been married to Elizabeth de
Burgh, granddaughter of the Lady Elizabeth of Clare and only
child of William de Burgh, earl of Ulster. His marriage gave
Lionel the honour of Clare, from which he derived his title *de
Clarentia*, as well as nominal control of the earldom of Ulster and
the de Burgh lands in Connaught. But fortune-hunters stood
little chance of success in Ireland and, after the death of his wife,
Lionel turned his thoughts elsewhere. Bernabo Visconti of
Milan was willing to pay heavily for the prestige of a connexion
with the English royal house: his niece, Violante, brought a portion of 2,000,000 gold florins as well as towns and castles in
Piedmont, when she became Clarence's second wife at a splendid ceremony in Milan in June 1368. Wild rumours current in
England that Lionel was about to become king of Italy, or even
emperor, were nonsensical of course; though in the tangled state
of contemporary Italian politics, it was not impossible that
Clarence, had he lived, might have carved out some kind of
Mediterranean principality for himself. But he survived his marriage by no more than a few months and died, leaving behind
him as his sole heir, his daughter Philippa, whom he had already
married to Edmund Mortimer, earl of March.

The king's fourth son John, born at Ghent in 1340, also had
to be provided for by marriage. For him, his father obtained, in
1359, the hand of Blanche, younger of the two coheiresses of
Henry, duke of Lancaster; and the death, in quick succession, of
his father-in-law and his wife's sister soon brought Gaunt the
whole of this great inheritance, comprising the duchy of Lancaster, the earldoms of Leicester, Lincoln, and Derby and the
high stewardship of England. Blanche (in whose memory
Chaucer composed the exquisite *Book of the Duchess*) died young;
and for his second wife, Lancaster, like his brother Lionel, looked
to the south. His marriage, in 1371, to Constance, elder daughter and coheiress of Peter the Cruel, gave him a claim to the

throne of Castile and Leon, the style of which he assumed forth-with. Edward III's fifth son, Edmund, born at King's Langley in 1341 and created earl of Cambridge in 1362, was married to Isabella, younger of the two Castilian princesses. The marriage of the seventh son, Thomas, born at Woodstock in 1355 (a sixth son died in infancy) brought him something more substantial than castles in Spain. For him, Edward III secured the hand of the elder coheiress of Humphrey Bohun, earl of Hereford, Essex, and Northampton.

Like other parents, Edward had also to provide for his daughters. The eldest, Isabella, became the wife of Euguerrand de Coucy, an important French prisoner of war who, it was hoped, might settle in England; the king admitted him to the Order of the Garter and gave him the earldom of Bedford. But de Coucy returned to France a few weeks after the old king's death and his wife, who had accompanied him, was sent home again with their only child, the unfortunate Philippa de Coucy, later the wife of Robert de Vere.[1] Edward's second and third daughters died in childhood; and the two youngest—Mary, who became the wife of John IV, duke of Brittany, and Margaret, who married John Hastings, earl of Pembroke—did not long survive their weddings. None the less, the king had good reason to be satisfied with a family policy which had secured to his dynasty so many great lordships in England and Ireland, opened up to his children the possibility of further acquisitions abroad, and ensured that the natural leaders of the baronage should be princes of the blood.

If Edward perceived no dangers inherent in such a policy, this was for the very good reason that he himself had none to fear. The solidarity of his family remained unbroken; the king's sons were his trusted lieutenants in the French wars, in Gascony, in Ireland. No hint of discord or disobedience escapes any contemporary writer, and many are at pains to emphasize the co-operation of the sons with their father, their deference to his wishes,[2] and the mutual love and admiration of the Black Prince and Gaunt, in particular. All Edward's sons seem to have been in full sympathy with their father's tastes and ambitions and, if only the eldest was his equal as a soldier, none of them was be-

[1] Below, p. 447.
[2] e.g. the Black Prince's insistence, throughout the campaigns of 1355–7, that he was the instrument of his father's policy. H. J. Hewitt, *The Black Prince's Expedition of 1355–57,* p. 143.

hind in love of war and warlike sports. Whatever its long-term consequences, Edward's consolidation of wealth and power in the hands of his family must be reckoned as a stabilizing influence in his own generation and as one of the most important factors enabling him to retain the loyalty of the nation through nearly forty years of foreign war.

Some of the credit for this family solidarity must undoubtedly go to Edward's wife; for the girl who had been brought from Hainault as a pawn in the unsavoury schemes of Isabella and Mortimer, proved herself a paragon among English queens. Her devotion to Edward III seems never to have been seriously affected by his gallantries; the rapid deterioration of his latter years followed hard upon her death in 1369. Philippa it is true, had her weaknesses, the most serious being her inability to make ends meet. She ran deeply into debt, there were complaints in parliament about the excesses of her purveyors and, in 1362, Edward was forced to make himself answerable to her creditors. Yet there is no reason to suppose that she was inordinately extravagant. The source of her difficulties probably lay, partly in the discrepancy between her nominal and her actual income and partly in the corruptibility and inefficiency of many of her servants. The queen's household was always elaborate, cumbersome, and expensive and able administrators were seldom attracted to it.[1] Philippa was a generous patron of two notable Hainaulters—Jean Froissart, who acted for a time as her secretary, and Walter Manny, who came to England as her carver; but she was never guilty of the common failing of foreign queens in allowing her household to become a happy hunting-ground for her compatriots. All contemporaries agree in praising 'this full noble and good woman', who won the love of her husband's people by her readiness to assume the traditional queenly role of intervention on behalf of the oppressed. The histories of Eleanor of Aquitaine and of Isabella of France stood as examples of the power of an evil queen to divide the royal family and the State; the unity of both, under Edward III, owed much to Philippa of Hainault.

Edward III was king of England for over fifty years; only Henry III had reigned longer. Yet for many of his fellow-countrymen he remains a pasteboard figure. The eccentric Edward

[1] Tout, *Chapters*, iv. 174–5, v. 250–9.

of Carnarvon, the unstable Richard of Bordeaux, rebels both against their environment, have stirred the imagination of poets and dramatists and the lively curiosity of later historians. But Edward of Windsor, who conformed to his age as he found it and accepted its standards, has merged into the vanished age of chivalry and now partakes of its unreality. Lauded to excess by his contemporaries, he has been roughly handled by modern historians, most of whom concur in Stubbs's judgement that Edward was no statesman, that his obligations as a king sat lightly on him, that he was ambitious, extravagant, ostentatious, and unscrupulous. Between extremes of praise and censure, the man himself too readily eludes us. We know him to have been courageous and of sanguine temperament, 'not accustomed to be sad', we know that he was magnanimous to his enemies and generous to his friends. We know, too, that his private life was not austere and that his dissipations took heavy toll of his middle years; for he was not yet sixty when he fell into premature senility. He may fairly be charged with vanity, ostentation, and extravagance, if it is remembered that Edward's subjects called these things by different names and thought them proper to a king. Like almost all his predecessors, he was an unscrupulous financier and his word in such matters was not his bond. He may be judged unstatesmanlike, in that he was not a man of profound or original vision; and it is arguable that in his dealings with his magnates and his parliaments he gave away too much. Whether we agree with Stubbs that his obligations as a king sat lightly on him, must depend on our view of what those obligations were; but we may be sure that this is a verdict which would have astonished Edward's own contemporaries. We know all too little of his mind and it remains doubtful whether the policy which he pursued with what looks like remarkable consistency was deliberately contrived in the interests of the monarchy and the nation, or was merely the outcome of a conventional, even short-sighted outlook on his world. Yet some incontrovertible facts remain. Edward III succeeded, where nearly all his predecessors had failed, in winning and holding the loyalty of his people and the affection of his magnates, even in the years of his decline. He accepted the chivalric and militant ambitions of his age and used them, as he used the devotion of his wife and sons, in the service of his dynasty. He raised that dynasty from unexampled depths of degradation to a place of

high renown in western Christendom. His armies won for him
and for themselves a military reputation seldom equalled and
never surpassed at any period of English history, before or since.
He blundered badly in his early years but, after 1341, he chose
his servants well and favoured them discreetly. He avoided
clashes with his parliaments, with the pope, and with the clergy;
and, while maintaining his royal rights to the best of his ability,
he never permitted himself his grandson's folly of openly chal-
lenging the laws and customs of the realm. If under him the
Church, sound government, and trade were all in some degree
sacrificed to the overriding demands of the war, Edward's subjects, for the most part, acquiesced in the necessity; for they saw
him as the pattern of chivalry and the maker of England's fame,
and when he lay upon his death-bed they mourned the passing
of a great English king. It is not altogether easy to share Stubbs's
confidence that they were wrong.

X

THE CHURCH, THE POPE, AND THE KING

A NEW epoch in the history of the western Church opened
inauspiciously when the disturbed condition of Italy led
Pope Clement V (1305–14) to take up his residence at
Avignon. Committed by their predecessors to an expensive
policy of centralization and harrassed incessantly by its demands, the popes of the 'Babylonish Captivity' at Avignon
found themselves confronted with national monarchs whose
aims were very similar to their own. Popes and kings alike cast
covetous eyes on the wealth and patronage of the English
Church; but the days of heroic conflict were over. Their residence at Avignon put the popes at a disadvantage which was
later intensified as a result of the Great Schism (1378); and, for
reasons which will be discussed in this chapter, none of the
fourteenth-century kings of England could afford to alienate
the head of the Church. Rivalry between the two powers thus
remained on the level of bickerings and mutual reproaches,
jockeyings for position, strategic advances and withdrawals.
Highly skilled curialists, first at Avignon and afterwards at
Rome, and their counterparts at Westminster (many of whom
rose to high office in the Church) directed these manœuvres and
ensured the maintenance of working compromise. Papal provisions and the financial exploitation of the clergy were their chief
preoccupations; and these were, indeed, matters of vital importance, helping to determine the relations, not only of the pope
with the king, but also of king and pope with the bishops and
clergy and of the clergy with the laity and thus subtly affecting
the whole character of the late medieval Church.

There is no mistaking the unpopularity of papal provisions
with the laity in parliament, nor the widespread and active resentment at papal encroachments on the revenues and endowments of the English Church. The view was frequently expressed
that the ancestors of both king and magnates had endowed the
native Church to serve the spiritual needs of the English people
and that papal provisions to foreigners and other absentees

implied misuse of these endowments. A well-known petition from the laity in the Carlisle parliament of 1307 underlined this point, with special reference to provisions, tenure of ecclesiastical dignities by cardinals, and the export of English money abroad. In 1309 the barons sent a further protest to Clement V; in 1324 Edward II forbade papal letters touching the appointment of John Stratford to Winchester to enter the country. After the opening of the French war, it was widely believed that the revenues which the pope and other foreigners derived from English churches went to sustain the king's enemies overseas;[1] and in 1343 Edward III voiced the resentment of his people in a strong letter.[2] These non-resident aliens, he reminded the pope, did not know the faces of their flocks and could not speak their language. In consequence, there was a decline of devotion among the people, hospitality was neglected, and the fabric of churches became ruinous; moreover, the rights of patrons were damaged and the treasure of the realm was carried abroad. Further protests, in 1344 and 1348, culminated in the first Statutes of Provisors and Praemunire (1351 and 1353), and papal provisions continued to be attacked in almost all the later parliaments of Edward III. It is not surprising that the older protestant historians, who took these complaints at their face value, should have given the Avignon popes a bad name and regarded the system of provisions and reservations for which they were responsible as an almost unmitigated evil and a principal cause of the Reformation.

Modern opinion has reacted strongly against this standpoint. Generalized complaints of abuses are now regarded with scepticism, the character of the Avignon popes has been largely vindicated and papal provisions have been defended on grounds both of principle and expediency. It has been shown that the pope rarely interfered with lay patronage; and that his frequent interferences with ecclesiastical patronage might—and often did—serve to restrain the tendency of cathedral and collegiate bodies to promote relatives or dependents, without regard to merit or suitability. Papal influence has even been described as 'the only effective check on the dominance of strictly material class interests within the Church.[3] The penniless clerk without powerful friends, the deserving university graduate, often stood

[1] Murimuth, pp. 173–5.   [2] Foedera, ii. 1233–4.
[3] G. Barraclough, Papal Provisions (1935), p. 61.

little chance of promotion except by way of papal provision; and John XXII's declaration that the system of general reservations was designed to avoid simony need not be cynically received. In so far as it threatened private rights, the law of the Church, or high standards of Church government, the system of papal provisions was no more dangerous than any other system. Recent work also tends to stress the point that the struggle over provisions derived much of its impetus from a clash of interests between king and pope, between royal and papal administration.[1] The king wished to use his rights of patronage primarily in the interests of his own servants whose stipends would thereby become a charge upon the Church instead of upon the hardpressed royal exchequer; and the pope wished to make use of canonries and prebends in English secular cathedrals to provide for the cardinals and members of his household. Yet the history of papal provisions in fourteenth-century England cannot be written as the history of a straight fight between king and pope; for, though his reign witnessed the first anti-papal statutes, Edward III himself cherished no radical concepts of the relations between Church and Crown. In this, as in so much else, he was conventional and he never wholly repudiated the system of provisions to which he had always been accustomed. Claims which looked absolute and irreconcilable, in practice were often the subject of compromise. Thus, it was seldom possible for the pope to force unwelcome candidates for bishoprics upon the king whom he was naturally loth to antagonize completely; and, for both Edward III and Richard II, the friendship and co-operation of the cardinals were essential to the promotion of their own schemes, political as well as ecclesiastical. Such friendship and support could best be assured by assisting the cardinals to draw the revenues from their English benefices and by allowing, or even helping them to acquire, fresh promotion in England. And what was true of the king was true also of several of the lay magnates and of the great ecclesiastical patrons, many of whom kept agents at the Curia. Thus, behind the great national protests, even those presented by the king himself, we have to recognize the existence of numerous shifts and compromises

[1] I am greatly indebted to Dr. J. R. L. Highfield for allowing me to make use of his unpublished dissertation, 'The Relations between the Church and the English Crown, 1348-78' (Oxford, 1950), which throws much light on this subject and supplements Anne Deeley, 'Papal Provision and Royal Rights of Patronage in the early Fourteenth Century', *Eng. Hist. Rev.* xliii (1928), 497-527.

which owed almost everything to political and diplomatic, and very little to moral or religious considerations.

Bishoprics ranked first in importance among the types of benefice to which provision might be applied. Under English common law, the right of electing a bishop lay with the chapter of the cathedral, subject to the royal *congé d'élire*, royal confirmation of the temporalities and confirmation of the election by the metropolitan; but the political importance of medieval bishops rendered a system of free elections impracticable.[1] When a vacancy occurred the normal usage was for the king to transmit a nomination to the electors and to the pope, sometimes, if the chapter seemed likely to make difficulties, to the pope alone. Whether the pope accepted the royal candidate or put forward one of his own, the electoral rights of the chapter and the archbishop's right of confirmation would almost certainly be set aside, with the result that elections to bishoprics in the fourteenth century tended to be hardly less of a formality than in the Church of England at the present day. The first half of the century affords some examples of apparently free elections, but very few chapters showed any independence in the face of opposition. In 1313 the election of the learned and virtuous Thomas Cobham to Canterbury was overruled by Clement V, who agreed to promote the king's favourite, Walter Reynolds, bishop of Worcester.[2] When the queen's cousin, Louis de Beaumont, was a candidate for Durham in 1316, the magnates gathered in the church during the election and dared the monks to make any other choice.[3] John XXII forced his chaplain, Rigaud d'Assier, on the reluctant chapter of Winchester in 1319.[4] By supporting rival candidates in 1322, the chapters of Coventry and Lichfield gave the pope the opportunity to reject both, in favour of Roger Northburgh, keeper of the privy seal.[5] John XXII also set aside

---

[1] Elections of abbots and priors who, by this century, were of less public importance, were seldom subject to interference; but a notable exception was the election at Bury St. Edmunds which occasioned a bitter contest between pope, king, and chapter from 1379 to 1384. See *Memorials of St. Edmunds Abbey* (Rolls Series), iii. 113–37 and A. H. Sweet, 'The Apostolic See and the Heads of English Religious Houses', *Speculum*, xxviii (1953), 468–84.

[2] W. E. L. Smith, *Episcopal Appointments and Patronage in the Reign of Edward II* (1938), pp. 17–18.

[3] R. de Graystanes, *Historia Dunelmensis* (Surtees Soc. 1839), p. 98.

[4] Smith, pp. 33–35. (Beaumont and d'Assier were the only foreigners appointed to English sees in this century.)

[5] *Anglia Sacra*, i. 443.

the election by the chapter of Exeter of John Godeleghe, one of the canons, in favour of his own chaplain and former pupil, John Grandisson.[1] Clement VI, in 1345, ordered the chapter of Winchester not to proceed to an election at all, as provision of the see had already been reserved to the pope.[2] By this date papal provision to bishoprics had become the norm; but it was proving increasingly difficult for any pope to thwart the will of the king and his family.[3] Thus, the Black Prince's nominee for Coventry and Lichfield was appointed in 1360, despite the initial support lent by Innocent VI to Archbishop Islip's spirited refusal to confirm an ignorant candidate; and the only satisfaction left to Islip, when the pope yielded to pressure and ordered him to confirm Robert Stretton, was to have it recorded in his register that the new bishop's lack of learning made it necessary for his profession of canonical obedience to be read for him.[4] When Edward III, in 1362, put forward John Bucking-ham for Lincoln, Urban V expressed doubts as to the candidate's capacity to rule so noble and populous a diocese; but after subjecting him to examination, the pope issued a bull of provision in accordance with the king's wishes. The only serious obstacle ever encountered by Edward III over a bishopric was at Winchester, when he was seeking this rich see for William of Wykeham. Elected by a complaisant chapter, Wykeham had to suffer long delay before papal confirmation was granted; and Edward found it necessary to write no fewer than twenty-four letters to cardinals, as well as three to the pope, before obtaining his wish.[5] Such arrangements continued to be effected until the end of the century; but in the latter years of Edward III and, increasingly, under Richard II, episcopal vacancies tended to be filled by translation instead of by direct provision. Translations were acceptable to the papacy because they brought money to the Curia and entailed automatic reservation of the benefices of the last man in the series of those translated; and at time of political crisis they were acceptable to the party in power

---

[1] Grandisson's *Register*, ed. F. C. Hingeston-Randolph (1894), III. vi.

[2] *Cal. Pap. Reg*. iii. 20.

[3] Trilleck of Hereford (1344) seems to have been the last English bishop not provided to his see in this century.

[4] Stretton's *Register*, ed. R. A. Wilson (Wm. Salt Arch. Collns., N.S. viii, 1905), pp. viii–ix. The editor makes the charitable suggestion that his illiteracy was due to defective eyesight. Stretton is known to have been blind by 1381.

[5] J. R. L. Highfield, 'The Promotion of William of Wickham to the See of Winchester', *Jour. Eccles. Hist*. iv (1953), 44.

in England because responsibility for the appointment of episcopal supporters could then be thrown on to the shoulders of
the pope. Urban VI agreed without demur to the episcopal
changes desired by the appellants in 1388 and Boniface IX did
the same for Richard II ten years later. Such translations were
facilitated by the Schism which made it possible to remove
politically undesirable prelates to sees in the jurisdiction of the
anti-pope which they could never hope to occupy. Undoubtedly,
they were a convenient political device; but it would be hard
to find a more notorious example of the extent to which these
popes were prepared to subordinate ecclesiastical to purely
secular interests.

Next in importance to bishoprics came the higher cathedral
offices—the canonries, prebends, and archdeaconries—which
Maitland described as 'the staple commodity of the papal
market'.[1] Under the Avignon popes many of the best of these
benefices were flooded with foreign cardinals and other papal
provisors. Every bishop with such dignities at his disposal found
himself confronted with a long waiting-list of candidates furnished with letters of provision.[2] At Hereford, between 1305 and
1317, five canonries and prebends were filled in this way.
John XXII provided to twenty-one canonries in the diocese of
Exeter and to forty-seven in Bath and Wells. In 1326 Bishop
Mortival of Salisbury protested to the same pope that, out of
fifty dignities to which he had the original right of collation,
eight had been filled by Clement V and twenty by his successor
and that hardly more than three of these persons had ever been
seen in Salisbury.[3] More than half the chapter of York and about
a quarter of the chapter of Lincoln were foreigners, in the middle
years of the century. Under Roger Northburgh, a devoted
servant of the papacy as well as of the Crown, the deanery of
Lichfield was twice filled by foreigners who never appeared
there.[4] Some of the provisors were royal servants provided by
the pope at the king's request. Robert Baldock was given a

[1] *Roman Canon Law in the Church of England* (1898), p. 67, n. 2.
[2] e.g. Hethe of Rochester in 1327 gave the number of papal provisions which had
to be met, as grounds for his refusal to institute a royal clerk to a benefice. Hethe's
Register, ed. C. Johnson (Cant. and York Ser. xlviii), p. 383.
[3] E. K. Lyle, *Office of an English Bishop* (1903), pp. 15–16. Dr. Kathleen Edwards
(*English Secular Cathedrals in the Middle Ages* (1949), p. 85) shows that in 1320–1,
fourteen out of twenty-nine non-resident canons at Salisbury were foreigners.
[4] Northburgh's *Register* (Wm. Salt Arch. Collns. i, 1880), p. 242.

canonry and the reservation of a prebend in Salisbury in deference to the wishes of Edward II; Queen Isabella's secretary,
Robert Wyville, was made a canon of Lichfield in 1327. Sometimes, and increasingly in the second half of the century, the
will of the king prevailed against the pope. Edward III succeeded in thwarting the execution of the provision of cardinal
Elias Talleyrand to the deanery of York. At Hereford, under
Bishop Trilleck (1344–60), only six out of thirty-three papal
mandates are known to have been executed and the episcopate
of Lewis Charlton (1361–9) saw a further decrease.[1] These tendencies were greatly strengthened by the Schism; for by no
means all the benefices held by Clementist cardinals and confiscated by the English government in 1378 were handed over
to Urban VI. Inquiry into the benefices held by foreigners in
1384 revealed a marked decline in their numbers.[2] No trace of
papal influence is to be found at Hereford under Courtenay and
Gilbert (1370–89); and at Salisbury in the late nineties for the
first time in the century, all four principal dignitaries of the
cathedral were residents with English names.[3]

   John XXII admitted very few limits to the papal claim to
provide to lesser benefices in ecclesiastical patronage; Benedict
XII followed his example; and Clement VI went farther still,
beginning his pontificate with a lavish offer to grant benefices
to all poor clerks who should come to Avignon to claim them
within two months of his coronation. Though many of the grants
doubtless remained inoperative, the number of this pope's
reservations was altogether unprecedented.[4] Innocent VI was
more moderate in the matter of provisions, possibly because he
was in some degree influenced by the Statute of 1351; but with
the accession of Urban V came a renewed crisis which led to the
re-enactment of the statute, though this seems to have had little
effect on the number of provisions. Urban did, however, tell Sir
Hugh Calveley, who had petitioned him on behalf of three of
his nephews, that he had decided to promote no more candidates of tender age, as the Church had been too heavily burdened with such appointments.[5] By the end of Edward III's
reign the difficulties of papal provisors were certainly on the

---

[1] J. T. Driver, 'The Papacy and the Diocese of Hereford, 1307–77, *Ch. Qu. Rev.*
cxlv (1947), 31–47.
[2] E. Perroy, *L'Angleterre et le schisme*, pp. 284–5.
[3] Edwards, *English Secular Cathedrals*, pp. 85–86.
[4] *Cal. Pap. Reg.* III. vi.   [5] Ibid. iv. 17.

increase and the voices of protest were growing louder. Gregory XI made some temporary concessions in 1375 and the Schism forced his successors to moderate their claims; but complaints about provisions continued to be heard, in parliament and elsewhere. None the less, they were still found useful, not least by the universities, desirous of exercising their influence on behalf of their own members. Well-sponsored collective petitions to the pope stood a much better chance of success than individual requests and both Oxford and Cambridge were in the habit of compiling rolls of their masters for presentation at the Curia.[1] Other candidates for benefices might find a place on the roll of some powerful magnate.[2] If his qualifications were judged satisfactory and his plea accepted, the candidate would be given a papal letter of grace providing him to a particular benefice. Such letters did not, however, ensure possession of the benefice named. A group of executors was charged with the duty of inquiring into the legal aspect of the grace and, as rival claimants could be heard, the pope's action was not wholly authoritarian. What his letter did was to set the machinery in motion. Armed with it, the candidate would approach the bishop, or other patron, wherever he could find him. An entry in one of the Hereford registers shows a certain John of Lugwardine appearing before Bishop Orleton in Port Meadow, outside Oxford, and presenting his papal letter of grace for a benefice in the bishop's collation worth twenty marks.[3]

Pluralism and non-residence were inevitable concomitants of the medieval tendency to regard office in the Church primarily as a *beneficium*;[4] but it is a mistake to suppose that these evils became intensified as a result of papal provisions. On the contrary, the Avignon popes made strenuous efforts to secure the acceptance, at least in principle, of the canon forbidding any individual to hold more than a single benefice with cure of

---

[1] E. F. Jacob, 'Petitions for Benefices from English Universities during the Great Schism', *Trans. Royal Hist. Soc.*, 4th ser. xxvii (1945), 41–59; 'On the Promotion of English University Clerks during the later Middle Ages', *Jour. Eccles. Hist.* i (1950), 172–86. (Wyclif was included in the roll presented by Oxford to Urban V in Nov. 1362.)

[2] e.g. fourteen papal mandates for benefices were issued to Bishop Trilleck of Hereford as a result of petitions from the Black Prince, the earl of Hereford, and other magnates.

[3] 'in prato inter Godestowe et Oxoniam'. Orleton's *Register*, ed. A. T. Bannister (Cant. and York Series v), p. 75.

[4] Barraclough, pp. 71 ff.

souls. Three papal constitutions—the *Si plures* of Clement V, the *Execrabilis* of John XXII, and the *Consueta* of Urban V—were all directed against pluralities; and, although it is true that dispensations were freely granted in return for a fee payable at the Curia, papal legislation did put an end to the indiscriminate habit of collecting benefices without dispensation which had prevailed under lay patronage in the thirteenth century. A pluralist on the scale of William of Wykeham, who held benefices to the annual value of £873. 6s. 8d., is altogether exceptional in our period. The majority of fourteenth-century pluralists were civil servants, promoted at the king's request; but their total income from their benefices seldom ran into three figures.[1] And, though there was much non-residence in secular cathedrals, this, as the work of Dr. Kathleen Edwards has demonstrated, was less than has generally been supposed. The worst effect of the whole system was that it tended to deepen the gulf between the well-to-do pluralists who were drawing their incomes from the Church and the mass of the stipendiary clergy who were doing its work.[2]

The famous anti-papal statutes of the fourteenth century have to be interpreted in the light of what is now known of the system of provisions in working. The first Statute of Provisors (1351) did little more than give statutory authority to repressive measures which had already been used by the king and his officers. Under the new act, they could now freely imprison papal provisors and hold them until they had paid compensation and undertaken not to appeal to the Curia. It is to be noted, however, that the statute offers no protection against the king. If the pope makes provision to a higher ecclesiastical dignity or interferes with the rights of an ecclesiastical patron, the king will expel the intruder and will himself fill the vacancy.[3] The first Statute of Praemunire (1353) introduced no new principle. It was framed with the object of bringing to order those who opposed the decisions of the royal courts and, as its title shows, it is directed, not against all appeals to foreign courts, but *contra*

[1] A. Hamilton Thompson, 'Pluralism in the Medieval Church', *Ass. Archit. Socs. Reports*, xxxiii (1915), 35–73; xxxiv (1916), 1–26; xxxv (1919–20), 87–108, 199–242; xxxvi (1921), 1–41. The return for London diocese to the *Consueta* of Urban V is printed as an appendix to Sudbury's *London Register*, ii. 148–82, those for the other dioceses of the Canterbury province in the *Register* of Simon Langham, i. 5–111.

[2] W. A. Pantin, *The English Church in the Fourteenth Century* (1955), p. 98.

[3] *Statutes* 25 Edw. III.

adnullatores judiciorum curiae regis.[1] Edward III had already, in 1343, issued writs of arrest against persons who appealed against the patrons of their benefices or pursued in any court matters which might be prejudicial to the king or the realm. In the preamble to the statute it is asserted that persons have been taking out of the realm cases cognizable in the king's courts, in contempt of the common law; henceforward, no such cases are to be taken out, nor shall judgements given in the king's courts be referred to the decisions of another court. The allusion is clearly to cases touching advowsons which, under English common law, were matters for the royal judges. Specific mention of the papal court does not appear until the second Statute of Praemunire (1365).

Strictly interpreted, these acts could have constituted an effective barrier against both provisions and improper appeals. But, whatever hopes may have been entertained by the commons, it was never Edward III's intention to apply the statutes literally and *in toto*. He ignored them,[2] or used them for purposes of diplomatic blackmail, whenever it suited him, and it was still possible for individual benefice-hunters to buy licences to petition the Curia. There was no attempt to put a stop to all litigation at Avignon on the subject of provisions. None the less, there is evidence[3] that the second half of the century saw some decline, not only in the number of papal provisions, but also in the volume of English business at the Curia; though it remains an open question how far this decline is to be attributed to the statutes. Further protests against provisors in the Good Parliament of 1376 and in the reign of Richard II were, however, normally coupled with demands for their enforcement. The exasperation of the commons found expression in an ordinance of the Cambridge parliament of 1388, threatening confiscation to anyone seeking favours from the pope; and the lollard preachers were active. Urban VI unwisely ignored these danger-signals and when he died, Anglo-papal relations had reached a state of tension unparalleled since the Carlisle parliament of 1307. The

[1] *Statutes* 27 Edw. III, st. 1. See E. B. Graves, 'The Legal Significance of the Statute of Praemunire, 1353', *Anniversary Essays . . . by Students of C. H. Haskins* (1929), pp. 57–80. Professor Graves has shown that the importance of the statute is mainly legal.

[2] e.g. the king's offer of canonries to cardinals whose support he hoped to enlist for the promotion of Wykeham to Winchester. *Jour. Eccles. Hist.* iv. 45.

[3] Analysed by Dr. Highfield in the dissertation referred to above, p. 274.

result was the issue of two further statutes designed to supplement the legislation of the fifties and sixties.

The second Statute of Provisors (1390) went beyond its predecessor in threatening with the most drastic penalties any prelate or clerk accepting a favour from Rome and any person,
ecclesiastical or lay, helping a clerk to obtain such a favour. It
was forbidden to introduce into England any sentence which
the pope might pronounce by way of retaliation and the commons at the same time attempted (though unsuccessfully) to
secure the exclusion of the papal collectors.[1] Boniface IX replied
with the extravagant gesture of annulling all the anti-papal
legislation of the English parliament as contrary to the liberties
of the Church; but the commons refused to withdraw or modify
the statute and in 1393 they reinforced it with a third Statute of
Praemunire, designed to prevent the pope from nullifying the
act of 1390. No citations to Rome, sentences of excommunication, or bulls of provision were to enter the country and the
Curia was prohibited from entertaining any suits relating to
patronage of English benefices.[2] These were brave words; but
both sides recognized the impossibility of maintaining an attitude of complete intransigence and this same parliament empowered Richard II to try to come to terms with Rome.[3] In
the event, the statute remained a dead-letter; for, by the mid-
nineties, the situation of the mid-eighties was beginning to re-
cur. Once again, papal support was needed for the furtherance
of political ambitions; and the history of the closing years of
Richard's reign reflects his strenuous efforts to reach agreement
with the Curia. He made no attempt to fulfil the terms of his
treaty with Charles VI, pledging him to the policy of ending the
Schism by securing the resignation of both popes; for he hoped,
by means of papal provision, to secure a subservient bench of
bishops. Boniface IX, for his part, was no less eager to retain the
support and friendship of England; and long negotiations at
last produced the concordat of November 1398 which, as M.
Perroy points out, served to revolutionize relations between the
Holy See and the English Crown. Richard II agreed that nomi-
nations to bishoprics should be taken out of the hands of the

---

[1] *Statutes* 13 Ric. II, st. 2; *Rot. Parl.* iii. 270.
[2] *Statutes* 16 Ric. II, c. 5. See W. T. Waugh, 'The Great Statute of Praemunire
1393', *Eng. Hist. Rev.* xxxvii (1922), 173–205; and Perroy, *L'Angleterre et le schisme*,
pp. 332–5.
[3] *Rot. Parl.* iii. 301.

chapters and reserved to himself and the pope; that the pope
should have the right to collate to one out of every three of the
greater dignities in cathedral and collegiate churches; and that,
at least until Easter 1400, he should enjoy the right of provision
to all minor benefices on the first occasion when they fell
vacant. Thus, the king had used the liberty allowed him by parliament to sell the pass in the matter of provisions, to abandon
the defensive positions carefully established by his grandfather.
Cynical as was his attitude to the Statutes of Provisors and
Praemunire, Edward III would never have agreed to such
wholesale concessions, even on a temporary basis. It was enough
for Richard II that he judged papal support indispensable to
realization of his dreams of unrestricted power. In this, as in so
many other respects, he showed himself altogether oblivious of,
or indifferent to, the sentiments of his own people.

Provisions, reservations, and their attendant processes were
not the only channels through which the papacy might gain
access to the resources of the English Church. Supplementing
them was a system of papal taxation, the onus of which fell
almost entirely on the clergy. The only forms of papal tax
exacted from the nation as a whole were the ancient hearth-tax,
known as Peter's Pence and long since commuted for a small
annual census of just under £200, payable through the bishops;[1]
and the cess, or tribute of 1,000 marks a year, promised by King
John, in token of England's feudal dependence on the Holy See.
John XXII made a determined, though unsuccessful, attempt
to increase the amount of Peter's Pence and both he and his
predecessor tried to secure more regular payment of the tribute;
but Edward II followed the policy, already traditional, of using
the tribute as a bargaining counter. The papal agent, William
de Testa, reported to Clement V in 1310 that it was twenty
years in arrears. Something was paid in return for a loan from
John XXII to the king in 1317, and a little more in 1319; but
after 1320 Edward II paid nothing at all, and by 1327 another
13,000 marks had been added to the accumulated arrears of the
previous century. Indeed, it was during the reign of Edward II,
as Professor Lunt has shown, that the policy of passive resistance

[1] e.g. the total of Peter's Pence due from the diocese of Worcester in 1302 was
£34. 2s. 7½d.; less than one-third of this sum was paid to the papal collector, the
bishop retaining the balance which he regarded as part of his income. See Rose
Graham, *English Ecclesiastical Studies* (1929), p. 310.

to the tribute reached its climax.[1] Nothing was paid between 1333 and 1365 when, as a *riposte* to the second Statute of Praemunire, Urban V demanded its renewal (with arrears) in order to finance his wars with the Visconti.[2] Knowing very well how such a request would be received there, Edward III referred it to parliament, where the lords gave the blunt answer that neither King John nor any other person could place the realm in subjection without the consent of the magnates, that John's submission had been in violation of his coronation oath, and that, should the papacy attempt to enforce its claim, magnates and commons would unite in resisting it to the uttermost.[3] This was the end of the tribute as a live issue in Anglo-papal relations.[4]

Papal taxation of the clergy took three principal forms— income-taxes (*subsidia caritativa*), annates, and procurations. The subsidies were the least profitable to the pope because, by virtue of a custom already well established, a high proportion of subsidies taken by papal mandate always went to the king, without whose co-operation they could not be raised at all. Three-quarters of the triennial tenth raised by Clement V in 1309 and half of the quadrennial tenth of John XXII in 1329, went into the royal exchequer. Cessation of papal demands for subsidies was thus a natural consequence of the outbreak of the French war, when the difficulty of transferring funds abroad would almost certainly have entailed diversion of an even larger proportion to the use of the king and hence to the maintenance of a war with which the Avignon popes had little sympathy. It was not until after the conclusion of the treaty of Calais in 1360 that a papal subsidy was again demanded and then the object was unusual. Innocent VI had promised a contribution towards the ransom of King John of France and had decided to raise it from the English clergy; quittance was subsequently given them from the exchequer for the sum of £15,000.[5] Since the proceeds of

---

[1] W. E. Lunt, *Financial Relations of the Papacy with England to 1327*, p. 170. Cf. O. Jensen, 'The Denarius Sancti Petri in England', *Trans. Royal Hist. Soc.*, N.S. xv (1901), 171–200.

[2] *Cal. Pap. Reg.* iv. 16.   [3] *Rot. Parl.* ii. 290.

[4] Wyclif's discussion of the subject in the tract *Determinatio* was almost certainly later but seems to have arisen from a purely academic controversy. See H. Workman, *John Wyclif* (1926), i. 231 ff.; M. D. Knowles, *The Religious Orders in England*, ii (1955), 67.

[5] Dorothy M. Broome, 'The Ransom of John II', *Camden Miscellany* xiv (1926), xvii–xviii.

this subsidy went into the king's pocket, he raised no objection
to its levy. But when Gregory XI, twelve years later, demanded
another for his own purposes,[1] the government forbade its collection; and the papal collector, Arnulf Garnier, was forced to
take an oath (afterwards demanded of all such officials) that he
would do nothing prejudicial to the king, the realm, or the
common law, and would export neither money nor plate without the king's leave. The English government was determined
that the clergy should not be taxed by the pope without profit
to the Crown; and it was only in return for his concessions on
provisors and other matters that Gregory was allowed, in 1375,
to collect three-fifths (£10,000) of the sum originally proposed.[2]
Even so, resentment at what the commons regarded as a surrender to the pope found expression in the Good Parliament,
where there were outbursts against the papal collector and demands for a ban on the export of bullion; and the raising of the
subsidy in the dioceses proved difficult.[3] It was becoming increasingly clear that direct taxation of the English clergy for the
sole benefit of the papacy would not be tolerated by the laity.
When Urban VI, in 1388, tried to exact a subsidy in return
for his complaisance in the translation of bishops and, again,
in 1391, when the convocation of Canterbury agreed to allow
Boniface IX a levy of 4d. in the £, the clergy were sharply reminded that the exactions were illegal and that payment would
entail forfeiture of their benefices.[4]

The unsatisfactory yield from subsidies forced the fourteenth-
century papacy to look elsewhere; and it was in this period that
papal annates became a regular charge on the lower clergy.
Earlier popes had made grants of annates to local prelates and
to lay rulers, but it was under Clement V that the demand for
the assessed valuation of a benefice in the year following the
appointment of a new incumbent, developed into a general and
perpetual claim deriving from the pope's right of provision. The

[1] 'for the recovery and preservation of the lands of the Roman Church' (i.e. for
the war against Milan). *Cal. Pap. Reg.* iv. 101.

[2] E. Perroy, 'The Anglo-French Negotiations at Bruges, 1374–77', *Camden
Miscellany*, xix (1952), xii; *L'Angleterre et le schisme*, pp. 29–40.

[3] e.g. the letter of Gregory XI to Wykeham complaining that the subsidy due
from his diocese has not been paid and reminding him that the English clergy are
rich, not having suffered invasion. Wykeham's *Register*, ed. T. F. Kirby, 2 vols.
(Hants Record Soc., 1896, 1899), ii. 244–5.

[4] Perroy, pp. 305–8, 319–20. Boniface IX made further abortive demands for
subsidies in 1394 and 1398. *Cal. Pap. Reg.* iv. 288, 302.

English clergy, 'like good asses', agreed to pay; and the English government, though watchful of its own interests, made no systematic attempt to impede the new development. John XXII's constitution, *Execrabilis*, by depriving many pluralists, created a large number of vacant benefices subject to papal collation, and the activities of the papal collector, Rigaud d'Assier, resulted in a rich haul of annates for the Curia.[1] Throughout the century, annates constituted a principal source of papal income; and from the seventies onwards they were loudly denounced in many parliaments. Wild notions of their value were in the air. A commons' petition of 1376 suggested that they amounted to five times the revenue of the English Crown.[2] Closely associated with them in popular complaints were procurations, a small tax levied on the clergy for the expenses of papal envoys and resident papal collectors, payment of which was normally allowed by the government. The number of writs issued by the bishops for the enforcement of procurations suggests, however, that their collection must have been difficult. Though the amounts levied were not large, they were by no means insignificant; and the unpopularity of the collectors and other papal agents rendered payment for their benefit peculiarly unwelcome to the English clergy.[3]

The fourteenth-century layman's complaint of papal exactions sprang, not from sympathy with the sufferings of the clergy, but from fear lest depleted resources should make it impossible for them to contribute adequately to the needs of the State. Though some obstacles remained to be overcome, the battle for the immunity of the spiritualities from lay taxation had been lost under Edward I; all his three successors were in a position to make large inroads on the wealth of the Church. Edward II, as has been seen, made substantial profit from clerical subsidies raised under papal mandate. Professor Lunt has calculated that 75 per cent. of the total of some £255,000 taken in

---

[1] Lunt, *Financial Relations*, pp. 486–505: 'The First Levy of Papal Annates', *Amer. His. Rev.* xviii (1912), 48–64.   [2] *Rot. Parl.* ii. 337.

[3] Knighton (ii. 2, 58) complains that 4*d*. in the mark was taken from every church in the land for cardinals' procurations in 1338 and double this amount in 1348. In 1340 arrears of procurations in the diocese of Ely amounted to nearly £170. *Ely Diocesan Remembrancer* (1891), p. 532. Cf. *Piers Plowman*, B xix, ll. 410–13.

'I knewe neure cardynal · that he ne cam fro the pope,
And we clerkes, whan they come · for her comunes payeth,
For her pelure and her palfreyes mete · and piloures that hem folweth.'

taxation from the clergy in his reign derived from this source; the remaining 25 per cent. derived from grants made directly by the clergy to the Crown.[1] When, after the outbreak of the French war, papal mandates were no longer forthcoming, the king was thrown back on this kind of direct taxation; and, despite opposition and grumbling, both Edward III and Richard II were able to use it to good effect. A lump sum of £50,000 was granted by the clergy in 1371 and there were one or two poll-taxes and other forms of special levy; but the great majority of clerical taxes took the form of tenths based, like the papal subsidies, on the Valuation of Pope Nicholas IV (1291), as modified for the northern province by the *Nova Taxatio* of 1318. Under the 1291 valuation a clerical tenth was estimated at about £20,000, £16,000 of which was payable by the province of Canterbury, £4,000 by the province of York; in 1318 the depredations of the Scots led to a reduction of the northern contribution by £1,600. Movable property attaching to lands acquired by the Church since 1291, or for any reason not taxed when clerical subsidies were levied, was assessed with the lay subsidies; but certain privileged churches, such as the royal free chapels, by custom or special grant were exempt from contribution.[2] Though the valuation soon became obsolete and the clergy of the southern province complained, in 1356, that their benefices afforded barely half the sums for which they were assessed,[3] no general reassessment was attempted between 1318 and the close of the century.

Direct taxation of the clergy by the Crown raised constitutional issues of some importance. The struggle over the bull *Clericis laicos* of 1296 had made it plain that the Crown would no longer respect the immunity of the clergy from taxation; but such taxes were always, in principle, free gifts conceded by reason of the urgent need of the secular power. In other words, a form of consent was necessary and at the opening of our period, the question that still remained to be answered was where and how this consent should be given. Edward I had hoped to bring

[1] 'Clerical Tenths levied in England by Papal Authority during the Reign of Edward II', *Haskins Anniversary Essays*, pp. 178–9.

[2] These exemptions sometimes proved hard to maintain. Even St. George's chapel, Windsor, was unable to protect its appropriated churches, despite a special grant of exemption from Edward III himself. See A. K. B. Roberts, *St. George's Chapel, Windsor Castle, 1348–1416* (1948), p. 47.

[3] Dorothy B. Weske, *Convocation of the Clergy* (1937), p. 159.

the clergy into parliament, under the *praemunientes* clause, and to tax them there, together with the laity; but at the time of his death it was by no means clear that this policy would succeed. The clergy disliked any form of summons which seemed to imply their obligation to attend a secular court; in particular, they disliked the supplementation of the *praemunientes* clause by a royal letter to the archbishops containing the formula *mandamus . . . venire faciatis*. When Edward II attempted to revive this formula, Archbishop Winchelsey offered strenuous resistance; but it was not until after Stratford's protest in 1341 that the supplementary royal letter was dropped. Edward III never admitted that the clergy were not bound to attend his parliaments and he and his successors continued to summon them under the *praemunientes* clause; but the attempt to secure the presence of a full complement of clerical proctors was abandoned. The clergy preferred a system of separate bargainings with the Crown; and already by 1322 it had become clear that all that could be had from them in parliament were conditional grants, subject to ratification in their own assemblies. When Edward III discovered that subsidies would be granted him in provincial councils summoned by the archbishops at his request, he was ready enough to connive at the absence of the lower clergy from parliament. Henceforward, clerical taxes were granted in these assemblies to which, from the time of archbishop Whittlesey (1368–74), the name *convocacio* was commonly applied.[1]

The clergy did not pay without protest and schedules of *gravamina* frequently accompanied their grants; on more than one occasion, the king was forced to be content with less than the sum demanded. Thus, in 1356, his request for six tenths produced a grant of only one; and in the Canterbury convocation of 1370 and 1371, the proctors offered strong resistance to the claims of the government on their purses. In convocation, as in parliament, Edward III often made use of influential laymen to put his case for a grant. Sir Walter Manny led a group of

---

[1] For a detailed analysis of the constitutional conflict, see Maude Clarke, *Medieval Representation and Consent*, chap. vii. J. Armitage Robinson's valuable essay, 'Convocation of Canterbury: its early history', in *Ch. Qu. Rev.* lxxxi (1915), 81–137, should be read in conjunction with E. Kemp, 'The Origins of the Canterbury Convocation' in *Jour. Eccl. Hist.* iii (1952), 132–43. Canon Kemp draws attention to the election of certain proctors to parliament so late as the sixteenth century. See also Edith Lowry, 'Clerical Proctors in Parliament and Knights of the Shire 1280–1374', *Eng. Hist. Rev.* xlviii (1933), 443–55.

judges and officials into the convocation house in 1356; another
group of judges appeared there in 1370; and in 1371 the whole
assembly was moved over to the Savoy palace to listen to
harangues from the Black Prince and other magnates. Moreover,
it was by no means uncommon for the bishops to take sides with
the government; Whittlesey, in 1370, was supported by his
suffragans in his appeals to the generosity of the proctors.[1] Edward III's policy of seeking ecclesiastical preferment for his
officials and dependents began to yield a rich dividend in his
latter years; and, though a realistic outlook on secular politics
was no bad thing for a bishop, lack of effective leadership was a
fatal weakness in convocation. For more than thirty years after
Stratford's protest, until Courtenay's outburst in 1373 against
taxation without redress of grievances, though there may have
been some passive resistance, especially under Islip, no prelate
seems to have been ready to raise his voice in defence of the
poorer clergy against the king. Lack of solidarity in the clerical
estate, however advantageous to the government, was clearly
damaging to the Church, all the more so as it coincided with the
phase of violent anti-clericalism among the laity which first
found open expression in the parliament of 1371.

The anti-clericalism of the seventies was, in part, the product
of contemporary social and political unrest. Recurrent visitations of the Black Death, trade depression, the renewal of the
war, reverses abroad, the prospect of heavy and sustained taxation, loss of confidence in the leadership of Edward III, jealousy
and mistrust of his ecclesiastical ministers, all contributed to a
lowering of national morale; while provisions, and the reiterated financial demands of the papacy, were serving to stiffen
clerical resistance to lay taxation and to anger and alarm the
lay magnates and the commons in parliament. But the movement against a wealthy Church had deeper roots than these.
Denunciations of ecclesiastical riches were not, of course, new;
what was new was the attempt to find a metaphysical basis for
anti-clericalism and at the same time to translate it into terms of
political action.

The metaphysical argument derived from three main currents
of speculation. Giles of Rome's tract *De Ecclesiastica Potestate*
(1302) had stimulated discussion of the relation of lordship and

---

[1] Wilkins, *Concilia*, iii. 82–84, 91. The bishops were responsible for the raising of
these taxes, but commonly delegated the task of collection to abbots and priors.

grace. Some held with Giles himself, that the visible Church, as the vehicle of divine grace, alone enjoyed lordship as of right; but there were others who contended that the test of authority must be character and that, therefore, the right to it lay, not with the visible Church, but with the 'invisible body of the just'.[1] Secondly, the movement of the Spiritual Franciscans had brought the closely related and perennial problem of Christian poverty into the forefront of academic controversy. Was absolute poverty binding upon the religious? Were richly endowed monastic establishments and a wealthy and worldly minded episcopate consistent with the principles of the Gospel or the practice of the primitive Church? The Spiritual Franciscans had been condemned by the highest authority but their ideas were still being discussed, particularly among other mendicants, and were still giving rise to obstinate questionings. And, thirdly, philosophers were taking note of the radical notions propounded by Marsiglio of Padua and William of Ockham who, between them, had developed a fully secularist and anti-papal theory of the State. A combination of arguments drawn from all three schools of thought could readily afford warrant for attacks on ecclesiastical 'possessioners' and on ecclesiastical endowments. Such a policy was being vigorously advocated by the Oxford doctor, John Wyclif, who was present in the parliament of 1371, when two Austin friars argued the case for mulcting the clergy.[2] It was probably in this same parliament that a lay lord retailed the fable, preserved by Wyclif, of the featherless owl (the Church) whom the other birds (lay benefactors) had provided with feathers (endowments), which they demanded back when threatened by the hawk.[3] Wyclif's academic reputation was already high; in conjunction with his notorious erastianism it made him a valuable ally for Gaunt and others of the laity who were eager to conscript ecclesiastical wealth in the service of the State. A series of treatises emanating from Oxford within the next few years—On *Divine* and *Civil Dominion, On the Church, On the Office of a King, On the Power of the Pope*—soon provided the anti-clericals with all, and more than all, that they needed in the way of theoretical justification.

Simultaneously, anti-clericalism began to manifest itself in

---

[1] Knowles, *Religious Orders*, ii. 62.
[2] The friars' arguments have been printed by V. H. Galbraith, *Eng. Hist. Rev.* xxxiv (1919), 579–82. Significantly enough, the text comes to us from a cartulary of Bury St. Edmunds.   [3] Workman, i. 210–11.

action. A group of lay lords inspired, it seems, by the radical young earl of Pembroke, secured the removal of the bishop of Winchester from the chancery and of the bishop of Exeter from the exchequer; and the commons, in this parliament of 1371, agreed to raise £50,000 for the war, only on condition that the same amount was forthcoming from the clergy. After the dissolution of the Good Parliament of 1376, Gaunt was able to secure forfeiture of the temporalities of William of Wykeham, probably the wealthiest of the bishops, and, though these were subsequently restored, the whole transaction cost Wykeham 10,000 marks. In 1380 the commons clamoured that, since the clergy possessed a third of the land they should pay a third of the £100,000 being demanded in taxation. Further attempts, in 1383 and 1384, to make lay grants conditional on ecclesiastical drew sharp protests from Archbishop Courtenay. It was probably only a few hot-heads who, if Walsingham is to be believed, suggested next year, that the temporalities of the Church should be confiscated;[1] and too much weight need not be attached, either to the anti-clerical sentiments voiced by the rebels of 1381, or to the wilder utterances of the lollards whose extravagant opinions soon began to alienate many of their supporters. Courtenay's courage and determination provided convocation with the leadership which had been lacking and served to avert the worst dangers threatening the Church. None the less, the accepted doctrine of the voluntary basis of ecclesiastical grants to the lay power had been openly challenged; the title of ecclesiastics to their endowments had been questioned; and once again they had been sharply reminded that these derived from the generosity of lay founders and benefactors.[2]

The papacy itself had opened the way for yet another form of lay exploitation of ecclesiastical possessions. Clement V did the Church no service when, at the opening of the century, he allowed Philip IV of France to bully him into dissolving the Order of the Templars. The wealth of the English province was not comparable to that of the French; but the Knights had a considerable number of small houses in this country and, both at the London Temple, their central English house, and at some

---

[1] *Hist. Ang.* ii. 140.
[2] In both the second (1377) and third (1393 or later) versions of his poem, Langland advocated confiscation of Church property: 'Taketh here londes, ye lordes; and leet hem lyve by dymes.' (*Piers Plowman*, B xv. 526; C xviii. 228.)

of the larger provincial establishments, they did useful business
as bankers and as custodians of records, valuables, and money.[1]
There was no spontaneous movement against them in England,
but the prospect of acquiring a share in their goods was naturally not unwelcome to Edward II and his friends. After some
hesitation, Edward complied with the papal request and, in
January 1308, ordered the apprehension of all Templars in England, Scotland, and Ireland. Their property was taken into the
king's hands, while a commission, which included some papal
inquisitors, was appointed for the trials at London and York.
The Grand Preceptor of the English province, William de la
More, steadfastly upheld the innocence of himself and his brethren in the face of appalling charges of heresy, idolatry, and
moral corruption; but the trials, conducted in strict secrecy,
and discredited by the use of torture, gave the inquisitors what
they wanted in the way of evidence.[2] Clement V dissolved the
Order by apostolic authority in April 1312, and decreed the
transference of its property to the Knights of St. John of Jerusalem; there was, of course, no suggestion that the dissolution
was in the interests of the secular power. Yet the goods of the
Templars provided Isabella of France with her dowry when she
became the bride of Edward II; and Roger Wingfield, the royal
clerk who was appointed general keeper of the Templars' English lands, saw to it that his own master did not go unrewarded.
Tout has shown that it was largely from the profits of the former
Templar lands that the revenues of the king's chamber were
replenished in the period between the death of Gaveston and
Edward II's reconciliation with the earls.[3]

[1] Agnes Sandys, 'The Financial and Administrative Importance of the London
Temple in the Thirteenth Century', *Essays presented to T. F. Tout* (1925), pp. 147–
62; C. Perkins, 'The Wealth of the Knights Templars in England', *Amer. Hist. Rev.*
xv (1910), 252–63. A list of the Templars' English houses will be found in Knowles
and Hadcock, *Medieval Religious Houses* (1953), pp. 235–9.

[2] For a good account of the northern trials, see Greenfield's *Register*, ed. W.
Brown and A. Hamilton Thompson (Surtees Soc. 1930–8), pt. v, Introd., pp. xxxiii–
xl. The bishops of the southern province were persuaded to seek permission to use
torture, though the inquisitors subsequently complained that they could find no one
to do the work. The northern prelates showed more spirit, declaring that torture
was unknown in England and asking if they were to import torturers from abroad.
Guisborough, p. 392.

[3] *Chapters*, ii. 316–24. Something of the significance of the dissolution was apparent to Langland. Writing of the cross on the gold noble he predicted

'For coueityse of that crosse · men of holykirke
Shul tourne as Templeres did · the tyme approcheth faste.'

(*Piers Plowman*, B xv, 508–9.)

Suspicion of foreigners and potential enemy agents, rather than a calculated policy of disendowment, probably prompted the movement against the alien priories, begun by Edward I when, between 1295 and 1303, he took into his hands the estates of 'alien religious of the power of the king of France and his allies'. Edward II followed his father's example at the time of the war of St. Sardos in 1324; and Edward III seized alien priories between 1337 and 1360 and again after 1369. The commons' petition in the parliament of 1346 that the king should expel all alien monks and take their estates into his own hands was only the first of many such; and prejudice against them grew with the general growth of anti-clericalism. A violent petition of 1377 is typical of the period in its demands for the expulsion of all aliens, monks included, on the ground that they were spies, and its suggestion that patrons should take over the lands of the monks and pay their profits to the king during the war; and there were similar requests in 1379 and 1380.[1] If, under Richard II, the danger to the aliens was receding, this was only because the difficulties consequent on the Schism and restrictions on the immigration of French monks were making it almost impossible for any of them to gain access to England.

The alien priories fall into two main groups—priories dependent on Cluny, and priories dependent on autonomous Norman houses.[2] The first group comprised some thirty-eight houses and included such important priories as Lewes, Wenlock, Bermondsey, and Montacute, where most of the monks were Englishmen. In the second group were some large priories, like Spalding and Stoke-by-Clare, but the great majority were small cells—manors, usually with the advowson of the parish church attached, to which the mother-house in France would from time to time send two or three monks to act as land-agents and to collect the revenue. Professor Knowles concludes that not more than seventy in this group were at any time fully conventual houses and that, at the end of the fourteenth century, there were probably not more than forty such in being. In principle, the taking of an alien priory into the king's hands meant that a royal *custos* would be appointed to keep an eye on the monks and

[1] *Rot. Parl.* ii. 162, 367, iii. 64, 96.
[2] For an excellent short account of the alien priories, see Marjorie Morgan, 'The Suppression of the Alien Priories', *History*, xxvi (1941–2), 204–12, which is supplemented by Knowles, *Religious Orders*, ii. 157–66.

to prevent spying; and that, after allowance had been made for the bare necessities of the inmates, the proceeds of the house would be paid into the exchequer. It seems, however, that in practice these harsh conditions were seldom fulfilled for long. The larger houses could bargain for better terms and the smaller tended to disappear altogether. Payment of a lump sum and acceptance of responsibility for a heavy annual tribute were usually allowed in lieu of occupation and forfeiture of revenue. But, since the burden of these tributes was apt to prove crushing (500 marks was demanded from Lewes, £120 from Montacute, and £100 from Bermondsey),[1] some of the larger houses began to petition for charters of denization. Such charters cost money and powerful advocacy was needed;[2] but Lewes obtained one in 1351 and by the end of the century most of the larger priories had followed this example. The smaller, unable either to command support or to raise the necessary funds, began to dwindle in number. Many of them were sold; and by the end of Richard II's reign the dispersal of the alien priories, which was to be completed in the following century, had begun in earnest.

The majority of these priories were not sold, however, for the benefit of acquisitive individuals, or even, directly, of the Crown. Deliberate conversion of monastic property to secular uses, if practised on a large scale, would have offended the conscience of the age. It was one thing to allow the king to make his profit from monastic lands that were temporarily in his hands and to charge a high price for charters and other privileges granted to alien religious; it would have been quite another to allow lands set aside for the service of God to lose their dedicated character and become permanently secularized. In 1385 the commons demanded the repeal of any letters patent authorizing the transfer of the property of the alien priories to laymen or secular chaplains;[3] and though there were a few such sales, the bulk of the revenues of the former alien priories continued to be used in the service of religion. Thus, the cells of the abbey of St. Tiron and other alien priories were bought by William of Wykeham and given to Winchester and New College; Thomas Mowbray's Charterhouse at Epworth owed its foundation to

---

[1] Rose Graham, *English Ecclesiastical Studies*, p. 49.
[2] e.g. the charter of denization granted to Eye for £60, 'at the special request of the queen, who is their patron'. *Cal. Pat. Rolls 1381–85*, p. 491.
[3] *Rot. Parl.* iii. 213.

his purchase of the alien priory of Monk's Kirby; and Richard II gave Ecclesfield, once the property of St. Wandrille, to the Coventry Charterhouse which he and Queen Anne had founded.[1] Steventon priory, though first acquired by a layman, Sir Hugh Calveley, passed ultimately to Westminster Abbey.[2] Analogy between the protracted dissolutions of the alien priories in the fourteenth and fifteenth centuries and the cataclysm which overwhelmed the native houses in the sixteenth, should not be pressed too far. We are not yet in the age of Simon Fish. None the less, in the dissolution of religious communities at the bidding of the secular power and not without profit to it, the historian may discern, albeit faintly, the shape of things to come. A less scrupulous generation would know how to turn the medieval precedents to good account. 'For if they do these things in a green tree, what shall be done in the dry?'[3]

Successful administration of the ecclesiastical system and the maintenance of a just balance between the claims of pope and king, depended largely on the bishops and their officers. From Wyclif's time onwards, these 'Caesarean clergy' of the later Middle Ages have never lacked their critics; but the injustice of many of the charges against them is now gaining general recognition. It is important to remember that the complex system which the medieval bishops were called on to administer was not of their making and that they cannot fairly be blamed for failing to try to alter it; nor, indeed, would substantial alteration have been possible, except by revolution of a type for which the age was not prepared. The task of the bishops was to make the system work, to defend their churches against royal and papal usurpation, to execute justice and, as heretics began to multiply, to defend their faith. And if their stature was less than that of their thirteenth-century predecessors and sanctity or intellectual eminence were rare among them, industry, devotion, and integrity were not. Robert Winchelsey, whose reign at Canterbury lasted till 1313, was a lonely giant, survivor from a former

---

[1] Hamilton Thompson, *English Clergy*, pp. 182–4.

[2] Marjorie Morgan, *The English Lands of the Abbey of Bec* (1946), pp. 124–5.

[3] 'The dissolving of these Priories made a dangerous impression on all the rest . . . here was an Act of State for precedent. . . . Use was made hereof beyond the king's [Henry V's] intention . . . whereas now some Courtiers by his bounty tasting on the sweet of Abbey lands, made their breakfasts thereon in the time of Henry the Fift, which increased their appetites to dine on the same in the daies of King Henry the eighth . . .'. Fuller, *Church History of Britain* (1655), p. 304.

age; but none of his successors among the southern primates, save, perhaps, Walter Reynolds (1313–27) was wholly unfitted for the office; and almost all were vigorous, capable, and well-educated men. The greatest scholar among them, Thomas Bradwardine, *doctor profundus*, fell victim to the plague within a week of receiving the temporalities; but both Simon Islip (1349–66) and William Courtenay (1381–96) would have been memorable at any period. Some justice has lately been done to Courtenay;[1] but little as yet to Islip.

Educated at Oxford, a distinguished ecclesiastical lawyer and an expert administrator, Islip had all the necessary qualifications for high office in the Church. He had been vicar-general to the bishop of Lincoln, dean of the court of arches, and keeper of the privy seal. But there was more to him than *curialitas*. Elected to Canterbury at the request of Edward III and afterwards provided to the see by Clement VI, he showed wisdom, courage, and realism in meeting both ecclesiastical and secular claims. Though the ascription to him of the tract *De Speculo Regis*, directed against purveyance, has been shown to be wrong,[2] it is certain that he challenged the Black Prince's claim to Crown dues in the diocese of St. Asaph;[3] and it was when he was archbishop that the clergy refused Edward III's demand for a sexennial tenth. Yet Islip was not prepared to claim benefit of clergy for the bishop of Ely when, in the course of his lawsuit with Blanche, Lady Wake, the bishop was cited before the king's bench; in its dispute with the bishop of Lincoln over the chancellorship he upheld the cause of his old university.[4] It was during his primacy (perhaps, as has been suggested, because both were former keepers of the privy seal) that the unedifying feud between the two archbishops was settled on a basis of reasonable compromise.[5] Islip was a poor man and he came to Canterbury in the worst plague year; his attempts to stabilize his resources by taking heavy procurations and by selling the timber from his estates won him some unpopularity.[6] Yet he could be generous in a good cause. He would have no needless

[1] K. B. McFarlane, *John Wycliffe* (1952), pp. 71–73.
[2] By Professor J. Tait in *Eng. Hist. Rev.* xvi (1901), 110–15.
[3] *Archaeological Journal*, xi (1954), 275.
[4] Rashdall, *Universities*, ed. Powicke and Emden (1936), p. 124.
[5] J. R. L. Highfield, 'The English Hierarchy in the Reign of Edward III', *Trans. Royal Hist. Soc.*, 5th ser. vi (1956), 137.
[6] *Anglia Sacra* (Birchington), i. 46.

expenditure on his funeral and to the monks of Christ Church he bequeathed, not only plate and vestments, but also a thousand of his best ewes to improve the breed of their sheep.[1] At Oxford he was the founder of Canterbury College, a mixed society of monks and seculars designed, like William of Wykeham's, to make good the wastage of educated clerks following the Black Death;[2] and for the humbler clergy of his diocese who could not hope to go to Oxford, he composed a *libellus*, or catechism, to help them in teaching their people.[3] His much criticized ordinance regulating the wages of the lower clergy sprang, not from a proud prelate's indifference to the needs of the poor, but from anxiety lest offers of higher pay should lure priests from their parishes to serve private chantries, as such offers were luring ploughmen from their lords to join the ranks of casual labour. Courtenay, with whom Islip is comparable in stature, was, perhaps, more ultramontane in outlook. He led the prelates in a formal protest against the Statute of Provisors of 1390 and he agreed to offer subsidies to Boniface IX; but he, too, was ready to serve the State, to keep the great seal for Richard II and to sit on the commission of 1386. His king was indebted to him, not only for sage counsel, but also for at least one substantial loan.[4]

Of the other archbishops of Canterbury in this century, only the sickly Whittlesey (1368–74) and Roger Walden (1398–9), who was thrust into the primacy by Richard II, can fairly be described as ecclesiastical nonentities. Simon Meopham (1328–33) was a man of little judgement but he was zealous for the maintenance of discipline and made some attempt to check both royal and papal encroachments.[5] John Stratford (1338–48) was more than an expert civil servant and a zealous upholder of the privilege of peers. He had served as a bishop's official and as dean of the arches before John XXII provided him to Winchester; he was a generous benefactor of his native parish church of

---

[1] *Inventories of Christ Church, Canterbury*, ed. J. Wickham Legg and W. H. St. John Hope (1902), p. 95; *Dict. Nat. Biog.* x. 513.

[2] 'considerantes viros in omni sciencia doctos et expertos in epidimiis preteritis plurimum defecisse'. W. A. Pantin, *Canterbury College, Oxford* (Oxf. Hist. Soc., N.S. viii (1950), iii. 159).

[3] Pantin, *English Church*, p. 212. (This libellus has not survived.)

[4] £1,000 for the Irish expedition of 1394–5. See Steel, *Receipt of the Exchequer* (1954), p. 71.

[5] In 1329 Meopham refused to institute the archbishop of Naples to the church of Maidstone and was cited to the Curia and suspended by John XXII.

Stratford-upon-Avon; and he was a considerable preacher.[1] The pontificate of Simon Langham (1366–8), formerly chancellor and treasurer, and the last monk to occupy the chair of St. Augustine, was too short to leave much impression on the see, which he was forced to resign on accepting Urban V's offer of a cardinalate; but he was unquestionably a man of high character and outstanding ability, a zealot for discipline, as well as a munificent benefactor to his abbey of Westminster.[2] Simon Sudbury (1375–81), papal auditor and former bishop of London, may have lacked force of character but had shown himself a capable and conscientious diocesan;[3] and at Canterbury he initiated the fund for rebuilding the nave and himself made generous contributions to it. Even Thomas Arundel found opportunity during his first brief tenure of the see (1396–7) to make a visitation of his province and to issue new statutes for the court of arches.

In the northern province the four distinguished civil servants who occupied the see of York from 1306–73 were all able and successful archbishops. William Greenfield (1306–15), who came to York when the war with Bruce was at its height, was forced to devote much of his energy to organizing the defence of the Border; but he found time to care for individual victims of Clement V's purge of the Templars,[4] to make visitations in his diocese, and to issue statutes for the guidance of his courts and officials.[5] His two successors, William Melton (1317–40) and William Zouche (1341–52), both took the field in person against the Scots, the former at the disastrous 'Chapter of Myton', the latter at Neville's Cross. Devout and austere in his private life, the wise and wealthy Melton[6] was a man of unbounded generosity throughout the diocese and beyond. The sum of his loans (on which he often received little or no interest) to the king, to the northern barons and knights who found themselves in difficulties through the Scottish wars, to fellow-bishops and to religious houses, is said to have reached over £23,500; but Melton himself kept free from debt and opened the way for his nephew

---

[1] Extracts from his sermons are printed in *Eng. Hist. Rev.* viii (1893), 85–91.
[2] J. Armitage Robinson, 'Simon Langham, Abbot of Westminster', *Ch. Qu. Rev.* lxvi (1908), 339–66.
[3] Sudbury's (London) *Register*, I. vi (ed. R. C. Fowler, Cant. and York Ser. xxxiv).
[4] *Historians of the Church of York* (Rolls Ser.), ii. 414.
[5] Printed in Wilkins's *Concilia*, ii. 409–15.
[6] 'sapiens et abundans divitiis'.

and namesake to found a knightly family.[1] At York minster, he restored the tomb of St. William and gave £700 towards the completion of the nave. William Zouche, whom the chapter, with a rare show of independence, elected in preference to the king's candidate, William Kilsby, was finally provided to the see, at great cost to himself, by Clement VI. Thereafter, he seldom left the north where he showed himself no less active than Islip in seeking remedies for the shortage of clergy arising from the Black Death. John Thoresby (1352–73) is memorable as the benefactor responsible for the eastern bays of the great choir of the minster and as the author of a *libellus* which was translated by his orders into English verse.[2] There is less to be said for the two aristocrats who succeeded him. Though he came of a great northern family, Alexander Neville (1374–88) showed little understanding of the problems peculiar to his diocese and embroiled himself in quarrels with the dean and chapter and with the citizens of York, with the bishop of Durham and the canons of Beverley[3] and Ripon, before his association with the courtiers brought about his downfall.[4] Thomas Arundel (1388–96) gave generously to the minster but was rarely seen in the north.[5]

Only men of proved capacity in the public service, or of noble birth, or both, were likely to achieve the position of primate; but what we know of the episcopal bench as a whole in this century does not suggest that many of the bishops are fairly described as 'devil's proctors for dispersing the flock of Christ' or as 'dumb fools in the realm of hell'.[6] The list includes some distinguished names. There are the two magnificent prelates of Exeter who, between them, rebuilt a large part of the cathedral, Walter Stapledon (1306–26) and John Grandisson (1327–69), the former the founder of Stapledon Hall (later Exeter College) in the university of Oxford, the latter, refounder of St. Mary Ottery as a splendid collegiate church; there is William Bateman of Norwich (1344–55), the eminent canonist who was

[1] L. H. Butler, 'Archbishop Melton, his Neighbours and his Kinsmen, 1317–1340', *Jour. Eccles. Hist.* ii (1951), 54–68.

[2] Pantin, *English Church*, p. 212. This so-called 'Lay Folk's Catechism' has been edited for the Early Eng. Text Soc. (1901) by T. F. Simmons and H. E. Nolloth.

[3] A. F. Leach, 'A Clerical Strike at Beverley Minster in the Fourteenth Century', *Archaeologia*, lv (1896), 1–20.

[4] Below, pp. 448–59.

[5] *Historians of the Church of York*, ii. 425–6.

[6] The words are Wyclif's or one of his followers'. See Workman, ii. 109.

founder of Trinity Hall and second founder of Gonville Hall in the university of Cambridge; there is Thomas Cobham of Worcester (1317–27) who bequeathed his books to Oxford to form the nucleus of its first university library and built the old congregation house, its earliest known public building. There are the well-known bibliophiles, Richard Bury of Durham (1333–45) and William Rede of Chichester (1368–85); there is William of Wykeham, bishop of Winchester from 1367 to 1404, whose vast wealth, if dubiously acquired, was nobly invested; there is Thomas Brinton of Rochester (1373–89), scholar, preacher, and fearless critic of abuses.[1] Against them we have to set notorious careerists, able but often unscrupulous men like Adam Orleton, bishop successively of Hereford (1317–27), Worcester (1327–33), and Winchester (1333–45),[2] or John Droxford of Bath and Wells (1309–29); and the unlearned, like Robert Stretton of Lichfield (1360–85), or Robert Wyville who ruled Salisbury for forty-five years (1330–75) and whom the pope, says Murimuth, could never have appointed if he had first seen him.[3] But only a minority of the bishops fall into any of these categories. Most of them were men of sufficient education and capacity (it has been calculated that about two-thirds were university graduates), resident in their dioceses for a great part of the year and doing their best to preserve discipline and sound doctrine among clergy and laity and to maintain their estates and the fabric of their cathedrals in the face of incessant demands upon their resources. To name only three out of many possible examples, a bishop like the Benedictine, Henry Woodlock of Winchester (1305–16), appears as the type of normal episcopal administrator whose rule in his diocese is said to have been characterized by care, intelligence, and mercy. Though deputed by Winchelsey to crown Edward II, Woodlock, so far as possible, held aloof from affairs of state,[4] In the middle years

---

[1] Knowles, *Religious Orders*, ii. 58–60. Brinton's sermons have been edited by Sister Mary Devlin for the Camden Series (2 vols.) 1954.

[2] A popular verse ran:

'Trinus erat Adam: talem suspendere vadam.
Thomam despexit, Wlstanum non bene rexit.
Swithinum maluit. Cur? Quia plus valuit.'

Anglia Sacra, i. 534.

[3] Murimuth, pp. 60–61.

[4] Hilda Johnstone, 'Henry Woodlock of Winchester and his Register', *Ch. Qu. Rev.* cxl (1945), 154–64. The *Register* has been edited by A. W. Goodman (Cant. and York Ser. xliii and xliv).

of the century, John Gynewell of Lincoln (1347–62) was cease-lessly active in his diocese throughout the worst months of the plague;[1] and towards the end, when the extirpation of heresy had become an important episcopal responsibility, we meet the former papal auditor, John Trefnant of Hereford (1389–1404), litigious but hard-working and the hammer of lollards, like William Swynderby and Walter Brute.[2]

The employment of bishops as diplomats and civil servants necessarily entailed some absenteeism; but the diocese suffered less from this than has sometimes been supposed. Suffragans were readily available from among holders of inaccessible and impoverished Irish sees, or of titular sees *in partibus infidelium*, many of whom were friars provided by the pope for the specific purpose of helping the diocesan bishops. Edington, as treasurer, and Edmund Stafford, as chancellor, were assisted in their re-spective dioceses by the bishops of Cashel and Ardagh; the archbishop of Damascus helped Zouche in York and Hethe in Rochester; and the bishop of Basel acted for Braybroke in Lon-don. Ordinations, confirmations, and consecrations were en-trusted to these suffragans in the absence of the holders of the see.[3] But whether he were present or absent, routine administra-tion of the diocese rested, not with him but with two subordinate officers who were tending to become permanent—the vicar-general (usually a member of the cathedral chapter) who acted as the bishop's deputy in all matters not requiring episcopal orders: and the official principal, who executed his office in the consistory court. Below them were local officers with limited jurisdiction—the archdeacons, whose duty was to hold yearly visitations of the archdeaconries and to preside in the archi-diaconal court; and the rural deans, charged with oversight of subdivisions of the archdeaconry. By the fourteenth century, however, very few archdeacons seem to have been personally active in the dioceses. The office was commonly regarded as a *beneficium*, the income from which might go to some wealthy

---

[1] A. Hamilton Thompson, 'The Registers of John Gynewell, bishop of Lincoln, for 1347–50', *Arch. Journal* lxviii (1911), 301–60.

[2] Trefnant's *Register* has been edited by W. W. Capes (Cant. and York Ser. xx).

[3] Langland is severely critical of these papal appointments:

'That bere bisshopes names · of Bedleem and Babiloigne
That hippe aboute in Engelonde · to halwe mennes auteres,
And crepe amonges curatoures · and confessen ageyne the lawe'

(*Piers Plowman*, B xv, 538, 557–8.)

pluralist, like William of Wykeham who was archdeacon of Lincoln for several years. Visitations could be bought off by the payment of procurations, collection of which was the business of such unattractive characters as Chaucer's *sumonour*.[1] Nowhere, perhaps, was the late medieval Church more open to criticism than in her toleration of these hangers-on of minor ecclesiastical officials, who took advantage of their position to prey on the pockets of the people and to exploit their credulity.[2]

Rectors of churches were among the most important contributors to both papal and royal taxes; but the term rector, or parson (*persona*), covered taxpayers of many different types. The rector of a small country parish might well be a manumitted serf, tilling the glebe with his own hands and distinguished from the peasantry around him only by his superior education, a man, perhaps, like Chaucer's parson who 'koude in litel thyng have suffisaunce'. Holders of single benefices worth six marks or less were exempt from taxation; but in a bad year even those whose assessment was rather higher often found difficulty in meeting their obligations. The protest of the clergy of Hereford diocese in 1346 that they were poor, their benefices were small, and their crops had failed,[3] reveals a common enough state of affairs. Yet some country rectors seem to have been prosperous. The rector of Whickham in Northumberland was joint lessee of a coal-mine at a rent of 500 marks;[4] some of the Yorkshire parish clergy devoted their leisure to cloth-making;[5] and it will be remembered that the parson of Trumpington, in the *Reeve's Tale*, gave his daughter a rich dowry when she married the wealthiest of the village craftsmen and was proposing to endow his granddaughter out of the profits of his cure.[6] Rich or poor,

[1] ' "Purs is the ercedekenes helle", seyde he.' *Canterbury Tales*, l. 658.
[2] For an excellent summary of late medieval diocesan administration, see A. Hamilton Thompson, *English Clergy*, pp. 40–71.
[3] Trilleck's *Register*, ed. J. H. Parry (Cant. and York Ser. viii), p. 275. In 1357 the bishop was ordered to distrain on the effects of certain clergy for arrears of the tenth (p. 354).   [4] R. L. Galloway, *Annals of Coal-Mining* (1898), p. 44.
[5] H. Heaton, *The Yorkshire Woollen and Worsted Industries* (1920), p. 21.
[6] *Canterbury Tales*, ll. 3983–6:

> 'For hooly chirches good moot been despended
> On hooly chirches blood, that is descended.
> Therfore he wolde his hooly blood honoure,
> Though that he hooly chirche shoulde devoure.'

See H. G. Richardson, 'The Parish Clergy of the Thirteenth and Fourteenth Centuries', *Trans. Royal Hist. Soc.* 3rd ser. vi (1912), 89–128.

such parsons were in a different class from the very numerous rectors who were seldom or never seen in the parishes. By definition, the rector was the individual or corporation who owned the greater tithe; and under the prevailing system of provisions and appropriations, the income from rectories often went to augment the revenues of monastic or collegiate churches, or of papal, royal, or baronial clerks who would not normally be resident.[1] Licences for non-residence could be obtained without difficulty for purposes of study at a university (the career of Wyclif affords the classic example) or for service in an important household, though there seems to have been some attempt to keep a check on casual absenteeism. When the rector of Stratford-upon-Avon was granted a year's leave of absence to visit his parents in Savoy, he had to swear that he would not 'turn aside to other places', that his only reason for absenting himself was his desire to see his relatives and to clear off his debts and that he would return within the year and henceforth reside in person.[2] But the tendency was to regard the parish church primarily as a piece of property; and towards the end of the century fraudulent and fictitious exchanges of benefices through the medium of brokers had become so common as to provoke from Archbishop Courtenay his well-known denunciation of 'choppechurches'. Courtenay's letter was addressed to the bishops, some of whom must undoubtedly have connived at these simoniacal transactions.[3]

A church appropriated by a corporate body, or having an absentee rector, was normally served by a vicar who would receive either a fixed stipend, or a proportion—often a third—of the revenue of the church, together with a house, garden, the small tithes, and the offerings. The majority of such vicars were secular priests, though from an early date the Premonstratensian canons had been privileged to hold cures of souls, and licences for this purpose were sometimes granted to other canons regular. After the Black Death the serious shortage of

[1] The number of alien rectors was relatively small, however. (Pantin, *English Church*, pp. 58–59.)

[2] Reynolds's *Register*, ed. R. A. Wilson (Dugdale Soc. ix, 1928), p. 89.

[3] Wilkins, *Concilia*, iii. 215–17. See A. Hamilton Thompson, pp. 107–9. There are many examples of a man exchanging one church for another and then, a day or two later or even on the same day, exchanging this for a third. Out of some 1,100 institutions in the diocese of Coventry and Lichfield between 1358 and 1385, 320 were by exchange. *Reg. Sede Vacante* and Stretton's *First Register*, p. viii.

secular clergy led to an increase in these licences and the rule
that a canon serving a parochial cure should not live alone, but
in the company of a *socius* of his own Order, was gradually
allowed to lapse.[1] Residence was obligatory for the vicar whose
office was usually permanent, though curates, or chaplains,
would be appointed on a temporary basis in an emergency, or
if the rector were for any reason incapable.[2] These vicars and
curates were drawn from the large body of unbeneficed clergy
which constituted the casual labour force of the Church; and,
if the number available for the parishes was tending to shrink,
this was the result, not only of a general shrinkage of population, but also of the counter-attraction offered by the chantries.
Service of a chantry (even if it included the maintenance of a
school) was much less onerous than service of a parish and the
financial rewards might be as good. It was in an effort to arrest
this diversion of clerical labour that Archbishop Zouche ordered
that chantries should not be filled until the parishes had been
supplied, and that the wage-laws fixed the stipend of a priest
with cure of souls at one mark more than that of a chantry
priest. But enforcement of these laws was never easy. Examples
are known of chantry priests who received more than many
vicars or rectors;[3] and both Langland and Chaucer seem to have
regarded chantry service as offering easy money to the idle or
irresponsible.[4] Yet the better type of chantry priest held a recognized and often permanent position; if not a freeholder, he was
at least a stipendiary, and he might be a member of a college or
confraternity, such as that endowed by the earl of Warwick at

[1] H. M. Colvin, *The White Canons in England* (1951), pp. 277–9. Mr. Colvin points
out that it was very exceptional for canons to act as vicars before the end of
Henry III's reign.

[2] e.g. under Droxford at Bath and Wells, a rector found illiterate at the time of
his institution was ordered to present a chaplain for examination by the bishop, to
act as his curate: another found insufficient 'in cantu et ad regimen animarum' was
ordered to nominate a chaplain for a year. Droxford's *Register*, ed. Bp. Hobhouse,
Som. Rec. Soc. (1896), pp. 200, 216.

[3] Kathleen Wood-Legh, *Church Life under Edward III* (1934), p. 122.

[4] 'Persones and parisch prestes · pleyned hem to the bischop,
That here parisshes were pore · sith the pestilence tyme,
To haue a lycence and a leue · at London to dwelle,
And syngen there for symonye · for siluer is swete.'

                                  (*Piers Plowman*, B, Prologue, 83–86.)

Chaucer admires the parson who did not set his benefice to hire

        'and ran to Londoun unto Seinte Poules
        to seken hym a chaunterie for soules.'

                              (*Canterbury Tales* (Prologue), ll. 509-10.)

Elmley, or by the duke of Gloucester at Pleshey. His status was very different from that of the vagrant priests and rootless clerks who made common cause with the rebels of 1381 or gathered round the lollard preachers in the alehouses or on the village greens.

The only monastic communities exempt from the Valuation of 1291 were the Military Orders (since the tax was nominally for a crusade) and poor nunneries and hospitals; but it seems certain that many monasteries were under-assessed. Dr. Rose Graham is of opinion that the tenth of 1291 was calculated only on the rental (probably the minimum rental) at which the monastic manors and granges might be let; and in an age when the religious farmed the greater part of their own lands at a profit, this bore little relation to their true income. At Durham, for instance, the temporalities were assessed at £620 and the spiritu- alities at under £700, whereas the bursar's rolls for each of the three years 1293, 1295, and 1297 show receipts amounting to over £3,000. Profits from the sale of wool were not included in the assessments, so that a great Cistercian house like Fountains, where the average annual profit from wool was probably not far short of £1,000, escaped with an assessment of its temporali- ties at £356. 6s. 8d.[1] Evasions of this order cannot have been kept secret and they may well have underlain much parliamen- tary clamour for increased taxation of the Church. For it was in the wide cornlands, rich pastures, and fine churches of the larger religious communities that the most spectacular manifes- tations of ecclesiastical wealth were to be found. Wyclif went so far as to assert that the monastic possessions alone would suffice to maintain all the poor of England;[2] and some at least of those who returned from the wars must have noted the contrast— often stressed by the popes in their appeals for funds—between the peace and apparent prosperity of the English communities and the ruin and desolation of their counterparts abroad. Yet there can be no doubt that, by the time of Wyclif, many Eng- lish houses were in serious financial difficulty. Of the Benedic- tine houses, only twenty-four stood in direct feudal relation with the Crown;[3] and most of these had achieved a separation

[1] *English Ecclesiastical Studies*, pp. 294–6. The average sale of wool at Fountains is stated by Dr. Graham to have been 76 sacks priced at from 21 to 9 marks.

[2] Knowles, *Religious Orders*, ii. 101–2.

[3] Helena M. Chew, *Ecclesiastical Tenants-in-Chief and Knight Service*, pp. 4–5.

between the lands of the abbot and those of the convent which had
the effect of freeing the latter from feudal services and incidents.
Even so, they were hard put to it to make ends meet; for a number of adverse factors were affecting the prosperity of the monasteries in a way which few contemporaries outside their walls
fully understood.

The Black Death was the most obvious, though not necessarily
the most fundamental, cause of financial and economic dislocation. We have no means of ascertaining what proportion of
monks perished; but in the monasteries, as elsewhere, the incidence of the plague seems to have been very uneven. Christ
Church, Canterbury escaped with the loss of only four monks;
but St. Albans lost its abbot, prior, subprior, and forty-six
monks within the space of a few days in 1349, Westminster lost
its abbot and twenty-six monks, and in a number of other houses
the losses were such as to produce a state of serious emergency.[1]
Yet recent work seems to point to the conclusion that the overall
effect of successive visitations of the plague was to accelerate
tendencies already at work and that the transition from the age
of high farming and direct exploitation of the demesne to the
age of rents and leases, which was almost complete by 1400
was already perceptible before 1348.[2] Abbots who became
rentiers were no longer in a position to exploit the possibilities
of their broad acres; the change marked the beginning
of the slow process which was ultimately to lose them the
initiative on their own estates.[3] Meanwhile, they had other
troubles to face. Urban riots broke out in a number of monastic
boroughs in 1327, notably at Bury St. Edmunds, St. Albans,
Dunstable, and elsewhere; and although, like the similar movements of 1381, they were unsuccessful, they were the cause of
serious financial loss to the houses concerned.[4] Hospitality
could be a formidable charge to the greater monasteries where
kings and magnates expected to be entertained. Richard II's
ten-day visit to Bury is said to have cost the abbey 800 marks;[5]
royal and noble guests were constantly arriving at St. Albans
and Westminster; in 1378 Gloucester abbey had to entertain

---

[1] Knowles, ii. 8–13.
[2] This question is discussed more fully below, pp. 315–42.
[3] Knowles, ii. 358.
[4] e.g. in 1330 the abbot and convent had licence to appropriate two churches 'in
consideration of the grievous losses sustained by them at the hands of the men of
Bury St. Edmunds'. *Cal. Pat. Rolls 1327–30*, p. 546.   [5] *Hist. Ang.* ii. 96–97.

the whole parliament. Patrons of lesser houses also expected free entertainment, not only for themselves and their retinues, but also for their horses and dogs. Kings, queens, and private patrons alike bombarded the religious with demands for pensions for clerks, corrodies for old or invalid servants, and for loans and gifts in money and kind. Corrodies, in particular, were coming to be regarded almost as a form of feudal service obligatory on the house.[1] If the abbot was among those bound by custom to attend parliament, this would inevitably entail considerable expense; so might a vacancy in the office of abbot unless careful precautions had been taken beforehand. To prevent the barony falling into the king's hands on the death of an abbot, it had become customary by this period for the more important houses to pay a fixed annual sum as a form of insurance. St. Albans paid fifty marks per annum to the Crown, and another twenty to the Curia to save the newly-elected abbot from the necessity of seeking papal confirmation in person.[2] The principle was doubtless sound, but the house lost heavily on the gamble when Abbot Thomas de la Mare remained in office for nearly half a century.

To offset such expenses the religious might seek to acquire new lands and rents or to appropriate churches; but none of these could be had without payment of substantial fines for breach of the Statute of Mortmain.[3] The prior and convent of St. Denys, for example, in order to acquire a messuage at Southampton worth two marks, had to pay a sum amounting to five times its annual value.[4] It became a common practice for certain houses to safeguard themselves against excessive fines by obtaining in advance licence to acquire lands and rents in mortmain up to a specified amount. Butley priory, in Suffolk, obtained such a licence for £10 in 1321 and another for £20 in 1365; but the concessions were less valuable than they looked, for it was the Crown which assessed the value of any property the monks might acquire.[5] Appropriation of churches (canonically

[1] Susan Wood, *English Monasteries and their Patrons in the XIII Century* (1955), pp. 106, 114–15. Corrodies were annual contributions of food or provisions levied on the house and often commuted for an annuity or pension.

[2] Knowles, i. 279.

[3] For illustration of the energy with which the law was enforced, see Helena M. Chew, 'Mortmain in Medieval London', *Eng. Hist. Rev.* lx (1945), 1–15.

[4] Wood-Legh, p. 67.

[5] J. N. L. Myres, 'Notes on the History of Butley Priory, Suffolk', *Essays . . . presented to H. E. Salter* (1934), pp. 195–7; see also T. A. M. Bishop, 'Monastic Demesnes

allowed only on grounds of poverty) brought the solid advantage of the greater tithe; but it brought also the obligation of paying the vicar, or curate, and of keeping the chancel in repair and furnishing it with books and other necessaries.[1] The greater houses managed to keep their heads above water; but many of the smaller were hard put to it to combine maintenance of their own economy with satisfaction of papal, royal, and patronal demands. Little wonder that his brethren should have remembered with affection and gratitude an abbot like William Clown who, during his rule at St. Mary-of-the-Meadows at Leicester, appropriated churches, acquired manors, rents, and possessions, secured exemption for himself and his successors from attendance at parliament, and arranged for limitation of the rights of the king's escheator during vacancies.[2]

Yet it would be a grave error to postulate an obviously impending collapse of monasticism in this century, or even a state of general decline. The Cistercian houses had lost many of their distinctive characteristics,[3] but the recrudescence of the Carthusians shows that the appeal of the more austere forms of the religious life was still powerful. Above all, the Benedictine houses had recovered the pre-eminence which had been theirs in an earlier age; though there was a falling-off from the numbers of that age, the greater houses recovered from the shock of the Black Death and retained a respectable complement of monks. A famous abbot like Thomas de la Mare of St. Albans, whom the Black Prince loved as a brother,[4] was a power in the land. A follower of the holy men of old in his personal integrity and in the depth and sincerity of his religious devotion, he was comparable to the greatest of contemporary bishops in his achievement as builder and administrator.[5] It was inevitable, in an age of emergent nationalism, that there should be attempts to force monastic 'possessioners' to make adequate contribution to the war; but the attacks of Wyclif and his followers on the religious life, as such, commanded no general support. Individual monks

and the Statute of Mortmain', *Eng. Hist. Rev.* xlix (1934), 303–6; and S. J. A. Evans, 'The Purchase and Mortification of Mepal by the Prior and Convent of Ely', ibid. li (1936), 113–20.   [1] Knowles, ii. 290–1.

[2] Knighton, ii. 125–7. See A. Hamilton Thompson, *The Abbey of St. Mary of the Meadows, Leicester* (1949), pp. 28–39.

[3] Edward III's foundation, St. Mary Graces, was beside the Tower of London.

[4] *Gesta Abbatum* (Rolls Ser.), ii. 375–8.

[5] His character and career have been vividly and sympathetically sketched by Professor Knowles, *Religious Orders*, ii. 39–48.

might be derided, as Chaucer derided them, and individual houses afford matter for scandalmongers; but the monastic profession was still respected; and, to a people distracted by the allurements of war and tormented by its hazards, the mellow buildings of the monks must often have looked like havens of piety, learning, and peace.

If the position held by the Mendicant Orders in the public esteem was much less secure, this was partly the consequence of their enforced dependence upon alms; for beggars are seldom popular. Within the carefully delimited areas attached to each house of friars, their procurators, or 'limitors', undertook the business of begging for the community, thus superimposing their claims on those of the purveyors, pardoners, summoners, and other birds of prey:

> When bernes ben full · and holly tyme passed
> Than comen cursed freres · and croucheth full lowe;
> A losel,* a lymitour · ouer all the lond lepeth,
> And loke, that he leue non house · that somwhat he ne lacche.†[1]

Moreover, the friars, in contrast with the monks, moved freely among the people, in the full glare of public criticism; and, whereas St. Benedict had vanished into the mists of antiquity, St. Francis had not long ceased to be a living memory. It was not the least of the friars' disabilities that men measured them against the saint who, more than any other, had captured the imagination of the later Middle Ages, and by that standard found them wanting. 'I haue seyne Charite also', wrote Langland,

> And in a freres frokke · he was yfounde ones,
> Ac it is ferre agoo · in seynt Fraunceys tyme.[2]

Neither he nor Chaucer had much use for friars; and even a sober critic like Gower, who admits the existence of some good friars, thinks that the mendicants are too numerous and questions whether their Orders are necessary to the Church.[3] That there were too many friars seems to have been generally agreed; and it was inevitable that Orders of uncloistered regulars should provoke some jealousy among the monks and that the secular clergy should resent the intrusion of the friars as preachers and

---

[1] *Pierce the Ploughman's Crede*, ed. Skeat (Early Eng. Text Soc.), pp. 22–23 (* a vagabond; † receive). [2] *Piers Plowman*, B xv, 225–6.
[3] *Vox Clamantis*, iv. 16–19.

confessors into the organization of the parishes and the loss of
fees that resulted from the licensing of their burial-grounds. The
friars' main spheres of usefulness—in royal and noble households, where they served as chaplains and secretaries, in their
own libraries and schools and, above all, in the universities[1]—
did not come within the purview of the ordinary man and were
not such as to appeal to him. Yet, however far the mendicants
fell short of an ideal standard (and it was probably by their
worst products that commonly they were judged), there can be
no denying that the religion and learning of the age would have
been immeasurably poorer for their loss.

A revised view of the fourteenth-century Church must suggest modification of some familiar strictures and commonly
held opinions. The system of papal provisions, though its potentialities were never fully realized, has been shown not to have
been evil in itself and to have been the means of promoting
many worthy men to office in the Church. Pluralism and non-
residence were not increased as a result of it and relatively few
aliens were provided to the parishes. Anti-papal clamour and
the anti-papal legislation of the period sprang from motives
that were far from disinterested and neither has much claim to
be regarded as the fruit of righteous indignation. In the business
of exploiting the resources of the native Church, the king, his
ministers, and the laity in general had nothing to learn from the
pope; although it is true that the anti-clericalism of the last
thirty years of the century derived much of its impetus from
speculation among the friars and in the Oxford schools. The dissolution of the alien priories resulted, not in the secularization
of their lands, but in the transference of most of them to other
religious foundations. Very few of the bishops of the period were
unlearned, incapable, or morally corrupt; and many of them
devoted their surplus wealth to the furtherance of education and
to the enlargement and beautifying of cathedral and other
churches. Many were in the king's service, but their absence
from their dioceses did not necessarily impede the smooth running of the diocesan machine. Royal and papal taxation bore
heavily on the clergy; but the poorest rectors were exempt from
the tenths and the richer churches were often in the hands of
well-to-do pluralists, or were appropriated by corporate bodies

[1] See below, pp. 503-4, 508-10.

who were bound to make themselves responsible for the cure of souls by appointing a vicar, or other deputy. The wealth of the monastic possessioners, though often more apparent than real, provoked envy and criticism, but there was as yet very little questioning of the value of the religious life well lived. The friars excited even sharper criticism, but they numbered among them some of the greatest scholars of the age and many men were still impelled to join their ranks. For, despite all appearances to the contrary, and despite the prevalence of a commercial attitude to benefices (freely castigated at the time by satirists and preachers), the age was genuinely religious; and the vitality of parochial and religious gilds, the multiplication of chantries, the crowded pilgrims' ways, suggest (though certainty in such matters must always elude the historian) that the hold of traditional catholicism on the minds and spirits of ordinary Englishmen was as strong at the end of the fourteenth, as in any earlier century. Few can have recognized, what we must recognize, that its defences had been breached.

The noisy protests at Carlisle notwithstanding, in 1307 hardly anyone in England questioned either the spiritual supremacy of the pope or the validity of his legislative acts for the Church as a whole, the entire apparatus of catholicism being taken for granted everywhere. By 1399 not only the papal *plenitudo potestatis*, but ecclesiastical authority and the whole medieval ecclesiastical system had been openly and strenuously challenged. John Wyclif is not to be brushed aside as a disappointed pluralist.[1] This he may well have been. But nothing that is said to his detriment (and much may fairly be said) should be allowed to obscure the fact that, before the end of the fourteenth century, and using its idiom, this unlikeable man of powerful intellect and narrow sympathies, had reached and stated almost all the conclusions subsequently held by the protestant reformers. Wyclif's influence in his own generation was limited; for he lacked the gift of winning men and the extravagance of his radical opinions served to unite all the forces of conservatism against him. None the less, the case for protestantism had been stated; and, from this point, once attained, though there might be temporary withdrawal, there was unlikely to be ultimate retreat.

[1] For Wyclif and the lollards see below, pp. 510-24.

XI

RURAL SOCIETY

WHEN the fourteenth century opened, colonization of medieval England had reached and, in many places, had passed its zenith; but the land was as yet only part subdued and Englishmen were still strangers to much of their own country. A very high proportion of the ground then under cultivation had been wrested from woodland of varying degrees of density and even in the period of 'high farming', the wild life of the woods pressed closely on the homesteads and the fields. Great royal forests, like the New Forest, Savernake, Arden, and Sherwood, the Lancastrian forests in Lancashire, Yorkshire, and Derbyshire, innumerable smaller private chases, the thick woodlands of Essex, Sussex, Buckinghamshire, Hertfordshire, and east Berkshire covered much of what has since become arable land. Coniferous trees flourish now in areas which in the Middle Ages were heathy wastes.[1] Parts of the Lincolnshire fenland and the Norfolk marshland had been enclosed and drained, but the time was not yet ripe for extensive projects of reclamation. On the moors of Devon and Cornwall little pastoral farms had been established, most of them by peasants, and in the north the plough was pushing its way up the lower slopes of some of the hills. But throughout the medieval period the struggle with nature was unremitting; and it needed only a sequence of such natural disasters as flood or drought, some relaxation of human effort, or a fall in the level of population, for the 'marginal' lands to revert to their primitive condition and for the frontiers of cultivation to contract.

The size of the population at the opening of the fourteenth century cannot be determined with anything approaching certainty. Between the time of Domesday Book, when it is thought to have been about one and a quarter, or one and a half, millions, and the poll-tax returns of 1377, we have nothing in the nature of a general estimate; and, for obvious reasons, the poll-tax returns (compiled only eight years after the last of three

[1] 'From rising ground, England must have seemed one great forest . . . an almost unbroken sea of tree-tops with a thin blue spiral of smoke rising here and there at long intervals.' W. G. Hoskins, *The Making of the English Landscape* (1955), p. 69. See also C. S. and C. S. Orwin, *The Open Fields*, 2nd edn. (1954), pp. 15-17.

serious visitations of the Black Death, which may have reduced
the pop lation by so much as a third), are of very little use as a
guide to the state of affairs seventy years earlier. Moreover,
there is no general agreement as to the conclusions to be drawn
from the returns themselves; for it is impossible to calculate the
number of beggars, who were exempt from the tax, or of persons
who succeeded in evading it.[1] Estimates of the population of
England before the Black Death vary from four million to two
and a half. In the face of so much uncertainty, all that can be
said with safety is that the population had been increasing,
though not, perhaps, steadily, since the time of Domesday Book
and is likely to have at least doubled itself by the end of the
thirteenth century, when it probably ceased to expand and may
possibly have begun to decline.[2] Even this minimum estimate
would represent a very substantial increase; and, if the two and
a half or three millions should be pushed up to three and a
half, or four, it seems probable that saturation point had nearly
been reached.[3] Not many more marginal lands could be
absorbed; for it was dangerous to cut down too many trees when
timber was used almost exclusively for fuel and extensively for
building and when the woods and waste lands were needed for
the pasturing of cattle and sheep. From a modern standpoint,
fourteenth-century England may look very sparsely populated.
Yet almost every rural community known to us today was already in existence; and many which were flourishing then have
since disappeared.[4] It was unlikely that there could be much
further increase of population within the limitations imposed by
a primitive agricultural technique.

The poll-tax returns are a better guide to the distribution of
the population than to its size; for, though there was some drift
towards London and one or two other large towns, it is unlikely
that the balance of the rural population altered greatly between

[1] Thus, J. C. Russell's figure of 2,232,373 (*British Medieval Population* (1948),
p. 146) is criticized by J. Stengers as being far too low, on the grounds that it does
not make nearly enough allowance for evasions. See *Revue Belge de Philologie et
d'Histoire*, xxviii (1950), 605.

[2] M. M. Postan, 'Some Economic Evidence of Declining Population in the later
Middle Ages', *Econ. Hist. Rev.* 2nd ser. ii. 3 (1950), 221–46.

[3] E. Perroy, 'Les Crises du XIVe siècle', *Annales, Economies, Sociétés, Civilisations*,
iv (1949), 168–9.

[4] M. Beresford (*The Lost Villages of England* (1954), pp. 337–93) gives the names
of numerous villages presumed to have been in existence in the thirteenth century
and to have since disappeared. W. G. Hoskins (*English Landscape*, p. 93) puts the
total of deserted villages at over 1,300.

1307 and 1377. The ancient distinction between highland and lowland zones was still of fundamental importance. North and west of a line drawn roughly from York to Exeter, there were no towns of any numerical significance and relatively few large villages; the country population was scattered and small. Within the lowland zone, south and east of this line, Surrey, Sussex, and Hampshire (the counties of the Weald and the New Forest) may have been little more populous than Shropshire or Hereford-shire; and it is unlikely that any town except London and, perhaps, York, had more than 10,000 inhabitants. Outside London, population, though nowhere dense, was densest in some of the chief wheat-producing areas, in East Anglia, south Lincolnshire, and the midlands between Sherwood and Oxford.[1] Communications were better than is sometimes supposed, for full use was made of the waterways; and the Gough Map (now preserved in the Bodleian Library and compiled almost certainly before 1350)[2] reveals an elaborate network of roads, by no means all of them Roman, with London and Coventry as the focal points, showing that the main outlines of the road system of the seventeenth century was already in existence in the fourteenth. Extensive use of carts for long-distance work as well as locally suggests that the road surfaces cannot have been impossibly bad. At all events, as Sir Frank Stenton has pointed out, the roads proved not inadequate for the needs of the age, administrative as well as economic.[3]

Recent work tends to stress disparity rather than analogy in medieval agrarian society and generalization becomes increasingly hazardous. The modern historian has to take account of variety and complexity in tenurial arrangements, peasant status, and types of husbandry. He must remember Kent, where there was little villeinage, East Anglia, where manor and village were rarely coincident, the widely scattered tenements of the old Mercian Danelaw, the primitive economy of the far north and

[1] H. C. Darby, *An Historical Geography of England before 1800* (1936), p. 232, fig. 30.

[2] See E. J. S. Parsons, *The Map of Great Britain*: a facsimile with an introduction (1958).

[3] F. M. Stenton, 'The Road System of Medieval England', *Econ. Hist. Rev.* vii (1936), 1–21. See also R. A. Pelham, 'The Gough Map', *Geographical Journal*, lxxxi (1933), 34; J. F. Willard, 'Inland Transportation in England during the Fourteenth Century', *Speculum*, i (1926), 328; 'The Use of Carts in the Fourteenth Century', *History*, xvii (1932), 248; Lady Stenton, 'Communications', *Medieval England* (1958), pp. 196–208.

the extreme west, the great pastoral farms of Shropshire and Lincolnshire, Yorkshire and the Cotswolds, as well as the nucleated settlements of the midlands, once thought of as 'typical' manors. He has to bear in mind the differences between the estates of undying ecclesiastical corporations (for which evidence is abundant) and those in lay hands (of which we know much less), and between large estates and small estates, as well as the more familiar if by no means simple distinctions between free and servile tenures. None the less, except, perhaps, in those remote and impoverished parts of the north and west where rural life may have remained much the same for generations, a pattern of change is perceptible, and most clearly perceptible in the champaign country of the south-west midlands, from Oxfordshire to Dorset. Its characteristics are a weakening of the customary obligations of dues and services and of customary restraints on the alienation and division of land and on peasant movement, its most obvious consequences an increase of leaseholds and the rise of a prosperous peasantry. In this chapter an attempt will be made to illustrate the process of change by a brief consideration of rural society at four distinct points in our period—in the early years of Edward II (*c.* 1307–14); on the eve of the Black Death (*c.* 1345–8); in the early years of Richard II (*c.* 1377–80); and in the last decade (*c.* 1389–99). The evidence is incomplete and often hard to interpret; but valuable recent studies of particular groups of estates and of general social tendencies have helped to illuminate many dark places and to enlarge our understanding of a complex and difficult subject.[1]

In the early years of Edward II we are in the last phase of the period of 'high farming', when many of the great landlords were producing for the market on a large scale by means of intensive exploitation of their demesnes. Surviving records afford abundant evidence of efficient estate management, particularly by ecclesiastical landlords, and of enterprising and up-to-date husbandry. By the opening of the fourteenth century formal baronial councils were assisting in the administration of many great estates. Such councils usually included four elements—local landowners and gentry, whose friendly co-operation and understanding of local conditions were invaluable; trained experts, often civilians, to counsel the lord on his rights at law and

[1] References to these studies will be found in the footnotes to the present chapter.

help him to avoid legal pitfalls; a judge or two, to safeguard his interests in the king's courts; and the indispensable permanent officials. But responsibility for determination of policy rested ultimately with the landlord himself; the elaborate hierarchy of officials which is to be found on most of the big estates in the fourteenth century, served to enhance his responsibility rather than to diminish it. The history of the Canterbury estates under Prior Eastry (1285–1331), or of those of the bishopric of Winchester under William Edington (1346–66), the Leicester chronicler's well-known panegyric on the achievement of Abbot Clown (1344–77) and, by way of contrast, the recurrent disturbances and relative poverty which lack of systematic policy produced at St. Albans, all serve to emphasize the key position held by the landlord, not least in a monastic community; while an important layman, like Thomas II, Lord Berkeley (1281–1321), exercised close supervision over all the affairs of his estates.

Financial and administrative arrangements differed widely from one group of estates to another and from one monastic house to another; but a tendency towards increased centralization is characteristic of the age of high farming.[1] Officials and senior monks working at the centre took responsibility, under the lord, for direction of estate management, auditing, and accounting; the auditors, often men of high standing, paid periodic visits to the manors where they kept an eye on the local officers whose integrity and capacity were fundamental to the smooth working of the whole seignorial economy. Some kind of administrative grouping was often found necessary for the control of widely dispersed possessions. From an early date, Christ Church, Canterbury, adopted what seems to have been a somewhat unusual system of monk-wardens, one in charge of each of the four 'custodies' (East Kent; the Weald and the Marshes; Surrey, Oxfordshire, and Berkshire; Essex, Norfolk, and Suffolk) into which the manors were divided. The wardens were required to pay twice-yearly visits to each manor in their respective custodies and Eastry entrusted them with the render of the annual liveries from the manors to the exchequer of the priory. Fruitful co-operation between the monk-wardens and the trea-

[1] Dr. R. A. L. Smith classified the Benedictine houses as (1) those through which all the funds, (2) only some of the funds, passed through a central receiving office, and (3) those in which there was no such office. The second of these systems is thought to have been the most common. See *Canterbury Cathedral Priory* (1943), pp. 26–27.

surers formed the basis of the central financial system at Christ Church, a system which, by the end of Eastry's term of office, was yielding a revenue of over £2,500. The bishop of Ely divided his estates into six bailiwicks, five of which were governed by sub-seneschals, the sixth (the Marshland bailiwick) by the constable of Wisbech castle.[1] On the estates of St. Swithun's, Winchester, however, the system of grouped manors was disappearing in the early fourteenth century, individual manors being entrusted to paid serjeants and visited once or twice a year by the steward of the priory who, here as elsewhere, was responsible for the manor court.[2] Paid serjeants were also in charge of the outlying manors of Durham, though the bursar retained control of a group lying within fairly easy reach of Durham itself.[3] It was probably only on small estates, like those of Owston in Leicestershire, where canons resided on the manors, that the religious made themselves directly responsible for the exploitation of the demesne.[4] The normal practice was to employ a salaried layman, a serjeant or bailiff, who, if he were permanently resident, maintained some degree of state among the villagers. At Forncett, for example, the bailiff received 52s. a year, a robe worth 20s., stabling and a daily peck of oats for his horse; his dwelling was kept in repair at the cost of his lord, the earl of Norfolk.[5] The position of the bailiff was in principle superior to that of the reeve who in origin was almost always a servile tenant, obligation to act as reeve being commonly included among the *servitia* owed to the lord and sometimes attached to a particular tenement on the manor. But the reeve's status was improving in the early fourteenth century. Capable and enterprising men were likely to be chosen for the office; and it was by no means uncommon for the same man to be retained for a period of twenty or more years, as was Walter le Notiere on the Westminster manor of Teddington from 1304–5 to 1326–7.[6] A reeve such as Stephen Puttock, who held office on the prior of Ely's manor of Sutton (Cambs.) in 1310, was clearly a man of

[1] E. Miller, *The Abbey and Bishopric of Ely* (1951), p. 263.

[2] J. S. Drew, 'Manorial Accounts of St. Swithun's Priory, Winchester', *Eng. Hist. Rev.* lxii (1947), 22.

[3] M. D. Knowles, *Religious Orders*, ii. 318.

[4] R. H. Hilton, *The Economic Development of some Leicestershire Estates in the 14th and 15th Centuries* (1947), p. 139.

[5] F. G. Davenport, *The Economic Development of a Norfolk Manor* (1906), p. 22.

[6] H. S. Bennett, 'The Reeve and the Manor in the Fourteenth Century', *Eng. Hist. Rev.* xli (1926), 360.

substance. Puttock was undoubtedly a villein; but he was also a sheep-farmer who had his own fold in which he kept other men's sheep and in ten years (1300–10) he had engaged in at least seven conveyancing transactions, three of these rendering him liable for rents totalling 36s. a year and two of them entailing the expenditure of nearly £15 in entry fines or purchase money. There was little to choose between such a man and the salaried bailiff or serjeant with whom he shared responsibility for exploitation of the manor in the interests of the lord. Other servile tenants had more limited official responsibilities. Such were the beadle, the cart-reeve, the reap-reeve, the hayward, and the woodward.

The underlying principle of medieval manorial accounting as Mr. J. S. Drew has demonstrated was that, whatever happened, the lord must suffer no loss. A production target was set for each manor and failure to reach it—whether the cause were fraud, negligence, error of judgement, or the normal hazards of agriculture—meant that the bailiff or other officer in charge of the manor had to make good the deficiency out of his own pocket. Thus, the grain crop was required to be x times the amount of seed sown, the wool fleeces to average a prescribed weight, each cow and ewe to yield milk for so much butter and cheese, each breeding female animal to produce the right number of young. No marketable commodity of any kind was so insignificant as to escape the keen eye of the auditors:

Neither doe theis Accompts scorne to discend soe lowe, as to declare, what money was yearly made by sale of the locks, belts and tags of the Sheep (as well as of the fleeces) of the hearbes of the garden, stubble from off the Corne lands, crops and setts of withy'es, of Osier rods, the Offall wood of old hedges, of butter, cheese and milke, dunge and Soile, of bran, nuts, wax, Hony, and the like.[1]

It is not unusual to find a defaulting bailiff having to pay in penalties two, three, or even four times the value of his annual stipend; and, as Professor Plucknett has shown, other perils beset him. Since the lord was under no obligation to give him a formal discharge, he might be called upon to account twice; or he might be rash enough to account before auditors whom the lord subsequently disavowed.[2] Some fortunate bailiffs were appointed with the express provision that they should not be

[1] *Lives of the Berkeleys*, i. 156.
[2] T. F. T. Plucknett, *The Medieval Bailiff* (1954), p. 26.

liable for the account; and many manorial accounts were, in
fact, submitted by reeves, though these too were sometimes
ready to pay a fine for exemption from what must have been a
heavy liability; for, in addition to the formal audit at Michaelmas, the accounting officer had to prepare the interim statement, known as the 'view of account', in the spring or early
summer.[1] Yet, since there were candidates for the office of both
bailiff and reeve, the system may have been less harsh in
practice than it looks to us. No doubt, the bailiff made good his
losses at the expense of the customary tenants, for the evidence
suggests that, however long he took to do so, he usually ended by
paying his debts; and where the bailiff was not accountable, it
was understood that the whole 'homage' was liable for the debts
of the reeve. Moreover, if the manor flourished beyond expectation, both bailiff and reeve stood a very good chance of lining
their own pockets. A shrewd and resourceful officer probably
did not need to dread the day of reckoning overmuch; for, like
Chaucer's reeve,

> Wel koude he kepe a gerner and a bynne;
> Ther was noon auditour koude on him wynne.
> Wel wiste he by the droghte and by the reyn
> The yeldynge of his seed and of his greyn.[2]

In the early years of the fourteenth century the lord and his
family, or the monastic community and its servants, were still
the first charge on the produce of the demesne. Farming for the
market was tending to restrict the older system of food liveries
but could not altogether supplant it.[3] 'Know, Sir', wrote Eastry
to the constable of Dover in 1323, 'that the half of our estates lie
so far away from us outside of the country in the direction of
Oxfordshire and Devonshire and elsewhere, that hence it be-
hoves us to sell our corn in those parts and to purchase other in
this district.'[4] Broadly speaking, this represented the policy of
most ecclesiastical landlords at this period. Food-farms were

[1] See N. Denholm-Young, *Seignorial Administration in England* (1937), pp. 131–51.
[2] *Canterbury Tales* (Prologue), ll. 593–96. Manorial accounts were drafted by
trained scribes who went the round of the manors for the purpose. See *Eng. Hist.
Rev.* xli. 364.
[3] It is, however, a symptom of increased technical efficiency and accounting zeal
that a system of double account is found on the Westminster and some other manors,
renders in kind to the hospice being valued and entered as sales on the account.
(I am indebted for this information to Miss B. Harvey.)
[4] R. A. L. Smith, p. 132.

commuted on the distant manors and demanded only from
those that were reasonably accessible; and the general tendency,
natural enough in an age of high prices, was to prefer cash
profits to food. Eastry divided the Christ Church properties
systematically into 'farm' manors, all of which lay in the two
Kentish custodies, and 'revenue' manors, most of which lay
farther afield, though there was also commercial farming in Kent
where the proximity of London and the continent offered opportunities too good to be missed. At the opening of the fourteenth century all the manors directly controlled by the bursar
of Durham were regularly sending up stocks of corn, dairy
produce, and vegetables to the priory; but before very long the
produce of some of these manors was being sold locally and
renders from them as well as from the outlying properties were
being made in cash. The canons of Leicester sold most of the
grain from their distant properties, much of it in small markets,
while they themselves bought foodstuffs, corn, wine, and wool in
the town of Leicester, at the east-coast fairs, and in the neighbouring villages. A proportion of the produce of the bishop of
Ely's manors always found its way to his table and we hear of
oats being sent by sea to York for John Hotham and his clerks,
when he was Edward II's chancellor. But the bishop's peripatetic existence often made it convenient for him, too, to buy
locally; and although much of his considerable income derived,
not from agriculture but from rents, he went in for commercial
transactions on a large scale, dealing with substantial merchants
who might buy up the produce of a whole group of manors and
sell it in distant parts of England. Thomas II, Lord Berkeley, on
the other hand, showed little interest in commercial farming and
looked to rents for the bulk of his income. He was conservative
in his dislike of bought food, spending Lent when possible on his
manor of Wike by Arlingham, 'for the better provision of fish',
and when he was in London for a parliament or other business,
employing two servants and four horses for the sole purpose of
fetching bread from his Essex manor of Wenden.

    Estate managers in the age of high farming had three principal
objects in view—to maintain and, if possible, to increase the
productivity of the soil; to extend the area of cultivable land;
and to enlarge and improve the flocks and herds. Up-to-date
methods of husbandry were employed in attempts to increase
the yield of corn per acre. Seed was sown more intensively. On

the Christ Church manor of Barksore, for example, the allowance of wheat per acre was raised from 3½ bushels to 4 in the
second decade of the century. Purchases of seed-corn appear in
many manorial accounts, notably those of Durham and Canterbury, and there were exchanges of seed between different manors
on the same estate.[1] The soil was manured and dressed with
lime and marl. Lord Berkeley had marl-pits to which some of
his lessees were allowed access; Henry Eastry spent £111. 5s. on
marl alone between 1285 and 1322. Wheat, as the highest-
priced crop, was the most desirable and attempts were made to
raise it wherever possible; but experience showed that some of
the marginal and newly reclaimed lands were better adapted to
the cultivation of oats. Oats accounted for 40 per cent. of the
corn sown on the Ely manor of Wisbech Barton between 1315
and 1322; in 1308, at the neighbouring manor of Coldham-in-
Elm, oats were sown on 630 out of 860 acres of land subsequently
described as *terra morosa et marisci*. At Canterbury, too, oats were
the cereal chiefly grown on the marsh manors. Eastry's summary
of his agricultural achievement shows that, in 1322, on the four
custodies, 2,677 acres were under wheat, 2,385 under oats, 1,510
under legumes, 1,434 under barley, and 367 under rye. Much
of this land was newly reclaimed, assarting on the scale then
practised representing a further attempt to stave off soil exhaustion. Eastry noted that he had already spent £360. 7s. on 'inning'
land and protecting it from the sea and there was similar activity
going on in nearly all the marshlands of England—in south
Lincolnshire and north Norfolk, in Holderness, in the Somerset
levels. William Clopton, abbot of Thorney from 1305–22, built
a house and offices in the midst of Thorney fen, enclosing them
with ditches; he also enclosed a large part of the fen 'to have as
arable or meadow with the lapse of time, if good fortune would
allow'.[2] But reclamation from fen and marsh was always precarious and this is probably the main reason why many landlords preferred to lease or sell such lands rather than take them
into the demesne. The Norfolk marshland was flooded twelve
times between 1250 and 1350; on the bishop of Ely's manors of
Terrington and West Walton, in 1316, 'magna pars terrarum

---

[1] E. M. Halcrow, 'The Decline of Demesne Farming on the Estates of Durham
Cathedral Priory', *Econ. Hist. Rev.*, 2nd ser. vii (1955), 346–7.

[2] H. C. Darby, *The Medieval Fenland* (1940), p. 50 (quoting from the Red Book
of Thorney, f. 460).

submersa est in mare'.[1] Fewer risks attended the kind of assarting practised by Lord Berkeley when, some time in the early
fourteenth century, he 'fell upon his Chace of Michaellwood
which had yealded to his ancestors or himselfe litle or noe profit
more than thornes and timber and improved out of the outskirts
or sides thereof . . . some hundreds of Acres'.[2] On the manor of
Westerham, given to Westminster abbey by Edward I in 1292,
over 500 acres had been cleared by 1300 and the removal of
further wood, undergrowth, and thorn added another 200 before the middle of the century.[3] At Lawling, in the heavily
wooded county of Essex, a survey made in 1310 showed that more
than 600 acres of woodland had been cleared in recent assarts;
and the Enstone charters show extensive assarting in Wychwood
forest in the early fourteenth century.[4] Moreover, on those
manors, chiefly in the midlands, where a three-field economy
was substituted for the older and simpler crop and fallow, the
area under crops could be increased by one-third.[5]

Expansion of stock-farming followed almost inevitably from
expansion of arable; for sheep manure was an essential adjunct
of grain cultivation, horses and oxen were necessary for ploughing and carting, cows and ewes for milk, butter, and cheese. But
shortage of animal feeding stuffs was a serious problem, reflected
in the much higher rents demanded for meadow than for arable
land, in the conflicts arising from the attempts of lords to increase
arable at the expense of the common pasture, and in the tendency
among peasants to overcharge with sheep or cattle such pasture
as they had.[6] The main incentive to sheep-raising, even in the
corn-growing areas, was, however, the high price that could be
had for wool. Christ Church, Canterbury, offers an impressive
example of mixed farming on the grand scale. In addition to his
8,373 acres of arable, the prior, in 1322, had 13,730 sheep (no
fewer than 10,000 of them agisted on the Kentish manors);
their manure was valued at £91. 6s. 8d., the milk of 6,000 ewes at
£96, the lambs were said to be worth £150 and their wool £50;
and the main wool crop of fifty sacks spelt an annual revenue of

[1] Miller, p. 96, n. 7.   [2] *Lives of the Berkeleys*, i. 158.
[3] T. A. M. Bishop, 'The Rotation of Crops at Westerham 1297–1350', *Econ.
Hist. Rev.* IX. i (1938), 39.
[4] R. H. Hilton, 'Winchcombe Abbey and the Manor of Sherborne', *Univ. of
Birmingham Hist. Journal*, ii (1949–50), 39.
[5] H. L. Gray, *English Field Systems* (1915), p. 403.
[6] *Vict. County Hist. Leicester*, ii (1954), 149, 164–5, 180.

£300. Enormous flocks of sheep are found elsewhere. Henry de Lacy, earl of Lincoln, had over 13,000, the fenland abbeys of Crowland and Peterborough between them had 16,300, the priory of St. Swithun's, 20,000.[1] By keeping sheep in a big way the landlord could at one and the same time enrich his fields, feed himself and his household, and add substantially to his income. Even if the sheep did not pay for all, it paid for much.

Exploitation of the demesne for commercial ends is the essence of high farming; but even at the opening of our period, few of the great landlords were indifferent to the profits to be had from rents at a time when these were high. Particularly was this true of the newly assarted lands to which, as had been seen, a certain element of risk often attached. Rents from such lands, since they entailed neither diminution of the demesne nor loss of customary services, could be reckoned as pure gain; and we find the astute monks of Christ Church adding to their rent-roll in the late thirteenth and early fourteenth centuries by the development of a system of competitive leaseholds for the tenants of the newly embanked lands on their marsh manors. About the same time, the bishop of Ely was drawing some 40 per cent. of his gross income of about £3,500 from variable and contractual rents; most of the newly reclaimed land in the fens and elsewhere went into tenancies, not into the demesne. But evidence from Durham suggests that higher rents could be had for demesne than for other land and by the beginning of the fourteenth century dispersal of the demesne had already begun on some estates, though for the most part only on a small scale and as part of a short-term policy. Those money rents which represented commutation of labour services fall into a different category; and the extent to which commutation had been adopted at the beginning of the period depended on local conditions and on the nature of the estate.

High farming was possible only for the great landlords and is seen at its most characteristic on the great ecclesiastical estates; but the tendencies which went to produce it were at work on a smaller scale and among smaller men. Not only landlords of the magnate class, but knights with an income from land of perhaps £10 or £20, free peasants with incomes of about 20s. and even villeins, were also striving to consolidate and enlarge their holdings, to increase their flocks and herds, to acquire rents, to

[1] Eileen Power, *The Wool Trade in English Medieval History* (1941), pp. 34–35.

farm for the market. The consequences are seen in a scramble for
both free and villein land and in the blurring in social practice
of the legal distinction between freeman and serf and between
free and servile tenements. Increasingly, the really significant
distinction is less between free and servile than between winner
and waster, between the man whose fortunes are on the up-
grade, whose descendants may swell the ranks of the yeomanry
and gentry of a later age, and the man whom economic pressure
or lack of enterprise are driving downwards and whose children
are likely to be the landless labourers of the future. 'Wherever
we look', writes Mr. R. H. Hilton, 'we find standing out from the
ordinary run of tenants with 15 or 20 acre holdings small groups
holding 100 acres or more.'[1] The roll of the Honour of Clare for
1308–9 shows an immense amount of traffic in land among the
earl's tenants;[2] and on the bishop of Ely's estates by the time of
Edward II, transactions in land are said to have become a flood.
Though the pace of change differed on different types of estates,
the process of dissolution of the old order is as unmistakable as
is the increasing stratification of the peasantry which was its
inevitable consequence.

Commutation of labour services, which was among the most
important of the factors making for stratification, might take
various forms—*ad hoc* sales of surplus works; commutation of
week-work, with the option to demand the services retained;
and full commutation of all services. The process was not uniform and the older view that commutation was a necessary
result of the growth of 'money economy' in other spheres, has had
to be abandoned in the light of more recent research. On the
contrary, labour services were often heaviest and lingered
longest on the great ecclesiastical estates where commercial
farming was practised intensively; though even on these estates
there had been many sales of works and further commutation
may not have been desired by such of the peasants as were interested in capital accumulation. On the whole, it was on the
smaller estates where the peasantry were least implicated in the
exploitation of the demesne, that commutation went farthest.
At the opening of the fourteenth century, on the Hampshire

---

[1] 'Peasant Movements in England before 1381', *Econ. Hist. Rev.*, 2nd ser. ii
(1949), 130.
[2] *Court Rolls of the Abbey of Ramsey and the Honour of Clare*, ed. W. O. Ault (1928),
p. xxix.

manor of Rockbourne, the property of a layman named John Byset, the only labour services still being exacted were nine days' reaping in the autumn performed by twenty-five *custumarii*; the remaining ten *custumarii* and thirteen cottars were paid for their work.[1] In eighteen villages of the Warwickshire hundred of Stoneleigh (a county without any really great estates), there were no labour services at all in eighteen villages and week-work is found only in two.[2] It is likely that from a very early date the smaller lords had been to a great extent dependent on permanent farm-servants (*famuli*) and on casual wage-labour; and the importance of the part played by this kind of labour, even on the large estates, is now recognized to have been much greater than used to be supposed. From his free tenants the lord could expect little beyond occasional *precariae*, or boon-services, mainly at harvest; and even where the full quota of labour services was exacted, the *famuli* were an essential element on any manor. No farm could be run on week-work alone. There had to be at least a nucleus of permanent full-time labourers to care for the animals and to act as blacksmiths, carpenters, and carters; and many lords employed a permanent staff of ploughmen during the ploughing season. Miss Page has concluded that on the Crowland estates the main burden of agricultural labour must have been carried by the *famuli*; several ploughmen, a carter, a shepherd, a maid, and, sometimes, a cowherd, a swine-herd, and a gardener were employed on many manors.[3] At Forncett we read of a *domus famulorum* (attached to the *curia servientis*) and of a maid being hired by the year to prepare pottage for the *famuli* who might number eight or nine plough-men, a carter, a cowherd, a swineherd, a dairymaid and, for three or four months, a harrower. Most of these were paid in kind, but the ploughman and the carter also had a yearly wage of 3s. On the Leicestershire manor of Lutterworth in 1322 the only labour used was that of the *famuli* supplemented by casual labourers who were paid by piece-work (*ad tascham*).[4]

[1] *Vict. County Hist. Hants.* v (1912), 419; *Cal. Inq. post Mortem*, iv. 287.

[2] R. H. Hilton, *Social Structure of Rural Warwickshire* (Dugdale Soc. Occasional Papers, no. 9, 1950), pp. 10–11.

[3] Frances M. Page, *The Estates of Crowland Abbey* (1934), p. 104.

[4] Two labourers were needed to look after a plough. One of these—the *tenor* or *conductor* who guided it—could be dispensed with when the ploughing season was over; but the *fugator* who drove the oxen was needed to look after them throughout the year. See M. M. Postan, *The Famulus* (Econ. Hist. Rev. Supplements no. 3), pp. 3, 15–16.

The full-time *famuli*, often holding in base serjeantry, were labourers working another man's land and were thus quite distinct from villeins who, whatever their commitments to their lord, lived by working their own holdings. On the Kentish estates of Christ Church, where there was no week-work, the *famuli* constituted the solid basis of a labour force which, by 1314, rested almost entirely on money-rent and wage-labour.

But, at the beginning of the fourteenth century, Kent, which knew little of villeinage, was still the exception. Elsewhere, the advance of commutation notwithstanding, villeinage was still a reality and from the lord's point of view an important one. A villein could still be defined as a man who owed week-work; for, though villein service, like feudal service, may have been beginning to look archaic, it was seldom altogether abandoned; and the incidents of villein, no less than of military tenure, were still, at least potentially, an important source of revenue. When, in 1307, Adam, son of Bartholomew the Pinder of Wisbech, wished to alienate his land and become a clerk, the jury empanelled in the halimote declared that if this were allowed the lord would lose not only the man but all his offspring (*sequela*). His losses would have included the entry fines payable on succession to the villein holding; the *heriot*, often by this date a money payment, payable on the death of a villein; the *leyrwite*, payable if his daughter were found pregnant before marriage and the *merchet* payable on her marriage and, sometimes, on the marriage of the villein himself, or of his son;[1] *chevage*, if for any reason the villein were to withdraw from the manor; and occasional tallages and the duties of suit of court and suit to the lord's mills.[2] And, although there was never anything like a slave-market in medieval England, it was undoubtedly possible for villeins to be sold or granted away.[3] Thus, there was every inducement to a lord to keep track of his villein tenants and to insist on maintenance of their servile status. If, like the bishop

---

[1] The object (seldom attained) of these controls seems to have been to prevent the mingling of free and villein land.

[2] If the lord had a fulling-mill, his tenants normally owed suit to this as well as to his corn-mill. See Eleanora Carus-Wilson, 'The Woollen Industry', *Camb. Econ. Hist.* ii (1952), 410.

[3] Miller, p. 141, n. 3. Sales of villeins with their land were common in Derbyshire in the first half of the fourteenth century. See *Vict. County Hist. Derby*, ii (1907), 165-6.

of Ely and many others, he failed to do so, this was because he was swimming against the tide. Land-hunger and the ambitions of the peasantry, both free and servile, were making it ever more difficult to maintain in practice the clear-cut distinctions of the lawyers. Within the villeinage itself, stratification was on the increase; the gulf dividing the prosperous villein from the free tenant who might be his neighbour was much narrower than that which divided him from the poorest of the serfs whose legal disabilities he shared.

For there was no sort of uniformity among the men whom the law classed together as villeins. Over some, who held no land or only free land, the lord's rights were so slight that they could readily be bought out for cash; some paid merely a token rent for such land as they held and were burdened with heavy services; others paid money rents in lieu of labour. Some held the virgate reckoned as the normal villein tenement; others held half a virgate, others (fardel-holders) a quarter, others (the cottars and the squatters) even less. In an age of commercial farming, such a society offered abundant opportunity for the rich to grow richer and the poor, in consequence, poorer. The scramblers for land among the peasantry did not respect the boundaries dividing free from villein tenements; and their example offered encouragement to the ambitious villein seeking to profit at the expense of his neighbours. Long before the age of high farming was over and the great landlords had accepted the position of *rentiers*, the ambitions of the peasantry were expressing themselves in action. At the top of the scale were the freeholders, many of them settlers on newly assarted land; these men were often eager to add to their possessions by the purchase of villein land and strangers were coming from outside with the same end in view. Next in standing and influence were those villeins whose lands were now held for rent or for the perform-ance of some administrative service; these men too were often buying land, sometimes free land. In theory, villein holdings were inalienable and impartible; and in the champaign country fragmentation of villein holdings may not have gone very far by the opening of the fourteenth century. But these restraints were becoming increasingly hard to maintain and in face of a keen buyers' market for land, many landlords had to accept the inevitable and content themselves with registration of such transactions in their courts and with the entry fines payable on certain transfers of villein land. As villeins bought free land and free men villein land, the unreality of the legal distinctions between them became ever more apparent. The 'complicated tangle' which resulted needed all the ingenuity of the lawyers to unravel it.[1]

Thus, the overall picture of the early years of Edward II is of a flourishing agriculture, most conspicuous on the great ecclesiastical estates where capable management and intelligent husbandry were exacting their maximum yield from many manors. On many of these estates customary labour was still being exacted and villein services were often heavy; but wage-labour, both permanent and casual, plays a large and increasingly important part; and money-rents and wage-labour predominate on the smaller estates. A population, which may have been static but was not yet markedly declining, was eagerly buying land; the example of the great landlords was being followed by knights, gentry, free peasants, and pushful villeins. Legal distinctions between freeman and villein, though still of the utmost importance to the lord because of the incidents, were becoming harder to maintain; and within the villein class itself there must have been some who were losers, though probably very few who were altogether landless. On many manors there had always been a reserve of younger sons and smallholders ready to take work wherever it could be found; and it may well have been with the aid of some of this casual labour that the prosperous peasant was able to take advantage of the market for his produce. The picture is, of course, not uniform for the whole country, nor were all the changes irreversible. For example, the years of flood and depression after 1314 forced the monks of Christ Church into a policy of demanding performance of those labour services which had yearly been commuted for a money rent; and on the abbot of Battle's manor of Hutton (Essex), both labour services and money-rents increased between 1283 and 1312.[2] Villeins were still subject to many restraints and disabilities and no general 'break-up of the manor' can be postulated for the reign of Edward II. But pointers in the direction of such a break-up are unmistakable long before the first onslaught of the Black Death.

[1] D. C. Douglas, *The Social Structure of Medieval East Anglia* (1927), p. 69.
[2] K. G. Feiling, 'An Essex Manor in the Fourteenth Century', *Eng. Hist. Rev.* xxvi (1911), 333.

By the fourth decade of the century, the age of high farming was well over. With falling prices for agrarian produce, rising rents and rising wages for agricultural workers, large-scale farming for the market lost many of its former attractions. Economic historians are as yet unable to offer any fully satisfactory explanation of these phenomena in the twenty or so years preceding the Black Death; but it seems possible that the population was beginning to push against the means of subsistence and may even have begun to decline; the drop in agricultural prices suggests a shrinking market. Despite the achievements of individual landowners, manuring and marling, so far from producing a general increase in the yield of grain per acre, may not have succeeded in staving off soil exhaustion; nor had there been any fundamental improvement in stock breeding; and some of the marginal assarts were going out of cultivation.[1] The great European famine of 1315–17 had caused temporary dislocation, the effects of which were felt for several years; and weather conditions and political upheavals had had disastrous effects on some of the most flourishing agricultural concerns, on the Christ Church manors, for instance, where the Kentish floods of 1324–6 killed well over 4,000 sheep, and at Durham, where disturbances on the northern border reduced the income of the priory from £4,526 in 1308 to £1,931 in 1340–1. The effects of this general decline are seen in an increasing tendency among landlords to shelve responsibility for their demesnes by leasing parts of them, reducing capital expenditure wherever possible, and pursuing still further the policy of commuting labour services for money rents. At Crowland, though demesne leases do not become regular until the second half of the century, twenty-one parcels were leased between 1326 and 1344. Certain of the English manors of the priory of Bec—Dunton (Essex) and Tooting (Surrey), for example— were leased almost continuously from 1322, some of them to manorial officials, some to other tenants, some to outsiders, among them a rich London goldsmith, Simon of Barking.[2] In 1347 the Black Prince leased 89 acres of the demesne of his manor of Berkhamsted for life to his esquire Robert de Kymbell

[1] R. H. Hilton and H. Fagan, *The English Rising of 1381* (1950), p. 22. A survey of 1340 shows that in some twenty-five Cambridgeshire parishes at least 5,000 acres had fallen out of cultivation. See *Vict. County Hist. Cambs.* ii. 71.

[2] Marjorie Morgan, *The English Lands of the Abbey of Bec* (1946), pp. 115–17.

and Christiana his wife, for an annual rent of £10. 15s. 10d.;[1]
and at Ely where, as has been seen, leasing of the demesne had
begun at an early date, the process was greatly accelerated. By
1348, 473 acres had been leased at Somersham; the income from
demesne leaseholds at Wisbech Barton rose from £2. 13s. 4d. in
1320 to £48. 10s. in 1345. The bishop of Ely was among the worst
sufferers from the agricultural recession of the period. The
agricultural income from his manor of Shelford, which had
averaged about £80 from 1319–23, dropped by more than half
between 1325 and 1333 and averaged only £10 between 1333
and 1346; at Shelford the price of wheat, which had ranged
from 10s. 6d. to 14s. a quarter in the early twenties, had dropped
to 5s. 6d. between 1325 and 1333 and in the late thirties and
early forties was little over 4s. Reduction of demesne costs
became imperative. The quantity of new seed bought was about
halved after 1325 and from 1333 there was no more large-scale
expenditure on the mill, fold, or house. Wage-bills were heavy;
but on the credit side the bishop could set the rising scale of
rents. One shilling an acre was paid for demesne at Great Shelford in 1336, anything from 8d. to 2s. 4d. at Somersham in 1348.

Declining prosperity tended to encourage further commutation of labour services, though a remarkable exception to the
general trend is to be found at Christ Church, where the rolls
show that, notwithstanding the decline of demesne farming
which began in the thirties, full labour services were exacted
from 1340 to 1390. But by contrast with those of other great
Benedictine houses, labour services on the Kentish estates of
Christ Church had always been light, the main burden being
borne by the *famuli*. Elsewhere, commutation went on apace
and the scramble for land continued; for the agricultural
recession did not discourage the smaller men whose interest lay
primarily in subsistence farming. At Durham certain of the
prior's free tenants were accused in the halimote of helping serfs
to escape so as to gain access to their holdings; at Ely the leasing
of complete villein tenements was becoming more common and
the income from sales of works was rising;[2] on the Winchcombe

---

[1] *Black Prince's Register*, i. 48–49. The conditions were that Robert and his wife
when they died should leave the house in good repair and the lands as well, or
better, tilled than when they received them, viz. ploughed for fallow and turned
over three times for the next winter season and eighteen of them manured by the
sheep.

[2] Miller, pp. 104–5. e.g. at Great Shelford, from an average of £2 in 1319–21 to

estates there seems to have been a permanent decision in favour
of money rents (except for a small number of boon works at
harvest), two decades before the Black Death. So long as rents
continued to rise and agricultural prices to fall, there were
powerful incentives to convert both the demesne and the villein
tenements into leaseholds; though at this period the majority of
leases were probably still short-term. No permanent change of
policy had yet been envisaged. Edward III's war taxation in the
decade 1335–45 rendered leasing still more attractive; for cash
was in greater demand than ever before and a swollen rent-roll
looked like the easiest way of ensuring an income. The lay landlord might hope also to offset what he lost on the sale of his corn
and wool by what he gained at the wars; but no such compensation was open to the ecclesiastical landlords, for many of whom
the Black Death came as the culmination of a period of shaken
prosperity which had lasted already for ten or twenty years.

A generation or so later, on the eve of the Rising of 1381, the
effects, other than the immediate effects, of this disaster were
beginning to be perceptible. A new and deadly form of plague
(pneumonic in one form and spread by direct contagion, pure
bubonic in another and spread by rats along lines of communication), reached this country in August 1348. It first
appeared in Dorset, spreading westward through Somerset and
Devon and north-eastward to London, where it was active by
November. From Norwich, where it appeared in January 1349,
it spread northwards. Few parts of the country escaped altogether but the violence of the first plague was spent by the end
of the year 1349. A second and less severe visitation, known as
the *mortalité des enfants*, followed in 1361–2, and a third in 1369.
Ten years later there was another outbreak in the north of
England. Though the plague of 1348–9 made the deepest impression upon contemporaries, partly because it was the first and
partly because it struck older and more prominent people, it is
likely that the most lasting damage resulted from the later
visitations. Men and women who were widowed in 1348–9
remarried as soon as possible, and may have raised as many
children as in the years before the plague; the failure of later
generations to maintain this standard was probably due to the

an average of £12 in 1340–2. The income from this source was about trebled at
Wisbech between 1320 and 1345 and at Somersham between 1327 and 1342.

high mortality among younger people. J. C. Russell is of opinion that a mortality of about 20 per cent. in 1348–50, followed by less serious losses in later plagues, may have reduced the population by half at the end of the century.[1] There must necessarily be a large element of guess-work in all such calculations; but it is fairly well established that about 44 per cent. of the beneficed clergy in the dioceses of York and Lincoln and nearly 50 per cent. in Exeter, Winchester, Norwich, and Ely perished; and it is possible that altogether about half the English clergy fell victim to the plague.[2] Studies of particular districts and manors tend to support the hypothesis of a very high rate of mortality. The total death-rate at Bristol has been estimated at between 35 per cent. and 40 per cent. of the whole population; some 70 tenants of the manor of Fingreth-in-Blackmore (which probably never had more than 65 at any given time) died during the first six months of 1349; a minimum of 740 people died in ten villages of the hundred of Farnham between Michaelmas 1348 and Michaelmas 1349, including 185 heads of households; the next year saw the death of a further 101 and the next, of another 58, making a total of 344 casualties in three years.[3] It is clear, however, that the mortality was very unevenly spread. There were 74 deaths on the St. Albans manor of Horwood and 71 at Abbot's Langley; but only between 5 and 8 at Shipton and Greenborough.[4] Not very many of the 'lost villages' can be shown to have disappeared as a direct result of the Black Death. There were some, like Tilgarsley (Oxfordshire), Middle Carlton Lincolnshire) and Ambion (Leicestershire)—over the site of which the battle of Bosworth was to be fought—which, having been completely depopulated, were never resettled: and others, like Coombe in Oxfordshire, which were resettled on new sites in the neighbourhood. But it is noteworthy that the plague of 1348–9 seems to have caused neither general panic, flight from

[1] *British Medieval Population*, pp. 229–35, 367.

[2] A. Hamilton Thompson, 'The Pestilences of the Fourteenth Century in the Diocese of York', *Archaeological Journal*, lxxi (1914), 97–154; G. C. Coulton, *Medieval Panorama* (1938), p. 496; Y. Renouard, 'Conséquences et intérêt démographiques de la peste noire de 1348', *Population*, iii (1948), 462.

[3] C. E. Boucher, 'The Black Death in Bristol', *Trans. Bristol and Glos. Arch. Soc.* lx (1938), 37; J. L. Fisher, 'The Black Death in Essex', *Essex Review*, lii (1943), 13–20; E. Robo, 'The Black Death in the Hundred of Farnham', *Eng. Hist. Rev.* xliv (1929), 560–2.

[4] A. Elizabeth Levett, 'Studies in the Manorial Organisation of St. Albans Abbey', *Studies in Manorial History* (1938), pp. 179–80.

the most badly affected areas, nor more than very temporary
dislocation of the wool trade.[1] For the most part, its effects were
seen in intensification of tendencies already at work, rather than
in cataclysmic dislocation of the existing order.

Nearly everywhere, falling prices and the labour shortage consequent on the plague stimulated desire to lease the demesne.
At Forncett, not only the arable but also some of the demesne
buildings—the sheep-house, the stables, and the chambers east
and west of the gate—had been leased by the end of Edward
III's reign; at Crowland, demesne leases became regular in the
second half of the century; in the seventies, the practice of leasing
out whole demesnes was adopted on a number of the east Kent
manors of Christ Church; a large number of demesne leasings
took place in the same period in the Honour of Clare and by
1380 most of the demesne had passed out of the lord's hands.[2]
On most estates leasing of customary holdings followed a
similar pattern, these customary leases, usually for life, being
often granted to joint holders. The men who took up leases of
servile tenements were either the thriving peasantry of the
manors or fortune-hunters from outside, some of whom were
prepared to go to considerable lengths in order to win a stake in
the land. At Sherborne, where the majority of holdings had
formerly been single virgates, one or two much larger holdings
had been created by 1355 and many new names appear on the
rolls. Henry atte Halle had five virgates and two other tenants
had three apiece. Even more remarkable are the cases of free
men being willing to contract away their freedom in order to
win possession of villein land. The *adventitii* who, at St. Albans,
took up such land, often by marriage with a villein heiress, were
forced to make sealed contracts with the abbot to pay merchet
and heriot and to be obedient to him both in body and goods;
and at Baslow, in Derbyshire, we hear of three men who took
'native land' by the accustomed services. On the whole, there
seems to have been no great difficulty in finding tenants. At
Fingreth heirs were eventually found for most of the vacant
holdings, though nine of them remained in the lord's hands;
thirty-six of the fifty-two vacant holdings in the hundred of

[1] Wool exports seem to have been normal again by Sept. 1351. See H. L. Gray,
'The Production and Export of English Woollens in the Fourteenth Century', *Eng.
Hist. Rev.* xxxix (1924), 17–18.
[2] G. A. Holmes, *The Estates of the Higher Nobility in the Fourteenth Century* (1957),
p. 92.

Farnham were filled; at Crowland many plots were taken up at once on the old terms and a roll of 1361 reveals no startling changes since 1339; most of the Durham holdings which fell vacant as a result of the plague passed immediately to new holders. In normal times many of those who so readily took up vacant holdings would have had to wait much longer for them, or might never have secured them at all; so that, although the problem of finding tenants was solved without undue difficulty, the labour problem was further aggravated, in consequence.

. Many landlords reaped substantial profit as a direct result of the plague. In the hundred of Farnham entry fines, which in a normal year might average between £8 and £20, reached the astonishing figure of £101. 14s. 4d. in the year 1348–9; and the cattle collected as heriots—26 horses and a foal, 57 oxen, 1 bull, 54 cows, 26 bullocks, 9 wethers, and 26 sheep—proved something of an embarrassment to the reeve. Since to sell in an over-crowded market must entail acceptance of very low prices —Knighton tells us that horses which in normal times would have fetched 40s. were being sold in 1349 for 6s. 8d. and that the price of other stock had fallen proportionately[1]—the reeve reserved some meadows 'for the multitude of the lord's cattle which come from heriots', enlarged the stables, and at the end of the summer ordered an unusually large slaughter of oxen and cows to provide salted meat for the bishop of Winchester's table. The chance of such profits was not to be thrown away lightly; and, in any event, alienation of the demesne had not proceeded far enough on many estates for villein labour to have become wholly redundant. Moreover, the sharp rise in wages which followed the plague (the 33 per cent. rise of 1340–8 was succeeded by a 60 per cent. rise in the next decade),[2] coupled with the inevitable shortage of labour on many manors, encouraged lords to try to hold on to much of what they had in the way of customary services. And it has always to be remembered that by the middle of the century, lords were by no means the only employers of labour. The prosperous peasant who was building up a compact holding also might need help in the farming of his

---

[1] Knighton, ii. 62.
[2] Lord Beveridge, 'Wages in the Winchester Manors', *Econ. Hist. Rev.* vii (1936), 31; see also 'Westminster Wages in the Manorial Era', ibid., 2nd ser. viii (1955). A comparison of the figures for these two estates shows a higher rate of wages prevailing on the Westminster manors and that the effect of the plague in raising wages lasted longer in London than outside.

land. Thus, serf-owning lords had strong incentives to try to constrain bondmen to continue to work on the old terms; and employers of all ranks wished to keep down the wages of hired labourers. The labour legislation which followed the first visitation of the Black Death was largely the result of panic at the prospect of soaring wage-bills.

The ordinance of 1349 and the more specific statute of 1351 sought to ensure a supply of cheap labour by pegging wages at pre-plague rates and insisting that all landless men of sixty years and under must accept employment at these rates, their own lords having the first claim on their services. Reapers, mowers, and other workmen and servants were forbidden to leave their masters until their contracts had been fulfilled and other masters were forbidden to receive them; no master was to offer wages above the customary rates. An attempt was made to regulate prices; but, since specific regulation applied mainly to goods sold directly by the makers, it ran contrary to the interests of craftsmen and artisans; all that was laid down about the price of foodstuffs was that they should be 'reasonable'.[1] The demand for labour legislation came from the commons in parliament; and a policy of wage restraint would indeed have been advantageous to the small employer in a period of labour shortage, for it would have enabled him to compete on equal terms with his richer neighbours for such labour as there was. Vigorous and largely successful attempts at enforcement were made for some years after 1351; since the penalties levied were to be applied to lightening the burden of subsidies, there was strong inducement to execute the law in all its rigour. But, in the long run, the intention of the statute was defeated, partly by the great landlords who discovered that it hampered them in the competition for labour, and partly by the refusal of the labourers themselves to accept its limitations. Nor was this the end of the landlords' difficulties. They had to deal, not only with recalcitrant wage-labourers, but also with recalcitrant bondmen who, in the general dislocation resulting from the plague, redoubled their efforts to throw off the burdens and restraints of serfdom which hampered them on the road to prosperity and full enjoyment of the opportunities afforded by the growth of leaseholds. Thus, peasants who were united in little else found themselves united

[1] *Statutes* 23 and 25 Edw. III (st. 2). See Bertha H. Putnam, *The Enforcement of the Statutes of Labourers 1349–59* (1908), p. 219.

in hostility to a government whose policy offended both those desirous of taking advantage of soaring wages and those who were becoming commodity producers and found themselves impeded by labour rents and by the incidents of servile tenure; while abductions of serfs from one another's manors by the lords embittered peasants of all grades.[1] To make matters worse for the landlords, the rents whereby they had hoped to offset the decline in agricultural prices began to fall after the middle of the century. They proved hard to collect; and as repairs were neglected and farm buildings collapsed, land decreased in value. After 1350 presentments in the halimote at Durham for failure to carry out repairs became common; at Forncett, by 1376, some of the manor court buildings had fallen into decay. Moreover, the policy of short leases removed the incentive to farm for posterity and tempted farmers to look for quick returns even at the cost of exhausting the soil. And there were disastrous murrains of cattle in 1348, 1363, and in 1369, when the harvest was the worst for half a century and both the wool and the cloth trades began to slump.

Thus, it is not surprising that the generation between the first visitation of the Black Death and the Rising of 1381 should have seen social disturbance in many parts of the country. Disaffection was not universal and resistance to the policy of the government and of the landlords did not develop simultaneously in all districts. On the Winchester manors, for example, there is no trace of any attempt to transgress the labour laws; leases had made some, but not much progress by 1376; and the practice of leasing the whole demesne or manor was still practically unknown.[2] On the Leicester estates, too, there seems to have been remarkable stability after the middle of the century, possibly because a prosperous peasantry was now able to expand at the expense of the less fortunate whose holdings had lapsed to the lord and been relet at cheap rates. A roll of 1361 from Crowland reveals no startling changes; signs of unrest do not appear before the sixties. But thereafter, cases of waste and destruction are numerous, the rolls recording many cases of trespass by offenders with strange names. There was a steady exodus of bondmen from the manors, though, perhaps because of the growth of the

[1] E. A. Kosminsky, *Studies in the Agrarian History of England in the Thirteenth Century* (1956), p. 317.

[2] A. Elizabeth Levett, *The Black Death on the Estates of the See of Winchester* (1916), pp. 101, 130.

composite holding, there was still no difficulty in finding tenants. At St. Albans discontent was enhanced by proximity to London, by excessive subdivision of holdings and the lack of adequate pasture, as well as by migrants from outside; cases of default of service became common at Barnet from 1354 and there was a steady drain of villeins to London. At Forncett, some new payments appear—to be 'unburdened' of a holding, to be excused all works and customs, to live outside the demesne; the whole value of this manor had probably decreased by about £30 in 1375–6. The Black Prince's register reveals many cases of fugitive villeins. Sir Richard Damory asks the prince to return certain bondmen of his said to be dwelling on the prince's manors of Beckley and Horton; the prince's serjeant-at-arms is ordered to seek out all bond tenants of the manor of Ippesden who have withdrawn and to bring them back; one of his clerks is ordered to seize certain bondmen of the manor of Little Weldon who have withdrawn.[1] At Tottenham in Middlesex in 1351 the justices of labourers were driven from their sessions; and in Northamptonshire and the Holland division of Lincolnshire, attempts were made to kill them.

In the years immediately preceding the Rising of 1381 the tempo of resistance seems to quicken. At Coleshill (Berks.) in 1377 a tenant named Henry Jordan was amerced because he refused to work for his lord and reaped his own corn at the time of the great demesne reaping; Thomas Jordan went and disturbed those who were performing their proper boon-work; Robert Symmings, a bondman, had left the demesne and the whole homage was in mercy for not producing him. The hay on six acres of meadow had been spoilt because the villein who should have carted it had defaulted. Tenants who used to work in the autumn did nothing this year because of a *magnus rumor* among the serfs.[2] At Harmsworth (Middlesex), a manor of the abbot of St. Katherine's, Rouen, the years 1377–80 saw serious disturbances. Men were fined for missing the haymaking and reaping and for reaping their own corn on the day of the great *precaria*. Robert Baker upbraided the jury in full court for returning a false verdict and Walter Breuer disturbed the court with his scornful words and would not be prevailed on by the steward to behave himself reasonably. Next year (1378) one of the tenants opened the lords' sluices so that the hay was flooded.

[1] *Register*, iv. 42, 64, 96.   [2] *Vict. County Hist. Berkshire*, ii (1907), 189.

In 1379 Thomas Reynolds and Nicholas Herberd were amerced for abusing the lord's servants; by 1380 Reynolds had forfeited his land and had to pay the large sum of £5 to recover it.[1] That these were symptoms of more than purely local discontent is evident from the petitions presented to Richard II's first parliament. The commons declare that they are impoverished by the outrageous wages demanded by labourers and ask for inquiry by the justices twice a year, strict enforcement of the wage laws and prohibition of holidays with pay. Villeins and tenants in villeinage in different parts of England, another petition declares, by counsel and abetment of certain persons have purchased exemplifications of Domesday Book in the king's courts and, misunderstanding them, are withdrawing their services and have confederated and allied against the lords and their officers, promising to help one another when they are distrained, collecting great sums of money, and threatening the officials with death. Failure to reap the corn is causing grievous loss. The petitioners remind the government of the aftermath of the French *Jacquerie* and hint darkly at the possibility of civil war or mass treason if a foreign enemy should land.[2]

Thus, there was plenty of inflammable material ready to the hands of Wat Tyler and John Ball, although, as will be suggested in a later chapter, it is by no means certain that rural and urban discontents need have fused into general revolt had it not been for the poll-tax and the ineptitude of the government during Richard II's minority.[3] Once the general revolt had begun, however, all the accumulated social grievances of the peasantry found ready expression; and Kosminsky sees in the demands of the Essex rebels for abolition of serfdom, for labour services on a basis of free contract, low rent for land and freedom for peasant trade, the programme of the upper and middle ranks of the peasantry; and in the demands of the Kentishmen for restoration of usurped common rights, the division of church lands among the peasantry and repeal of oppressive legislation, that of the class of landless, or nearly landless, labourers whom he believes the lords and the prosperous gavelkinders to have been ruthlessly exploiting.

The Revolt caused widespread panic which long outlasted its

---

[1] *Vict. County Hist. Middlesex*, ii (1911), 82.
[2] *Rot. Parl.* iii. 17, 21.   [3] See below, pp. 422-3.

suppression; but it had no perceptible effect on the evolution of rural society. Flights of bondmen, refusal of services, 'confederations and conspiracies' continued as before. Exodus from the manor of Crowland went on, twenty-one bondmen fleeing between 1365 and 1380, sixteen between 1380 and 1400. On the bishop of Hereford's manor of Bromyard in the last decade of the century, several customary tenements escheated because the heirs failed to appear in court to claim them; and on his manor of Eaton customary tenants were forfeiting their holdings rather than accept office as reeve.[1] Numerous payments of chevage elsewhere illustrate the persistence of the urge towards freedom. At Forncett, John Rougheye agreed to pay 18d. a year for licence to dwell outside the demesne. At Ogbourne St. Andrew in 1389 the tenants formed a confederacy in support of their claim to the privileged status of sokemen of the ancient demesne which they and their ancestors had been pursuing sporadically since the beginning of the century. An inquiry held before the justices of the king's bench in Essex this same year revealed widespread discontent among the labourers, manifesting itself in a desire to shake off villein tenure, a preference for casual work by the day, a demand for higher wages, and the refusal of some men to serve at all. The Cambridge statute of 1388 represented a reactionary effort to restrict the mobility of labour by forbidding labourers to move about unless furnished by the justices with sealed letters patent, to conscript the children of labourers for service on the land, and to maintain statutory regulation of wages.[2] But the inquiries which accompanied attempts to enforce the statute soon made it clear that strict enforcement was impossible; the commissions issued to the justices of the peace in 1390 were tantamount to an admission that there could be no return to the wage conditions of the past. The demands of the labourers, as Miss Kenyon has shown, were often exorbitant, but their desperate need of labour forced many employers to concede them.[3] So strong was the workman's preference for casual labour that vast sums had to be paid for the favour of a yearly contract. Two ploughmen at Sheering, for instance, demanded a rise of from 28s. to 40s. per annum with

[1] A. T. Bannister, 'Manorial Customs on the Hereford Bishopric Estates', *Eng. Hist. Rev.* xliii (1928), 223–4.

[2] *Statutes* 12 Ric. II, cc. 3, 4, 5.

[3] Kathleen M. Kenyon, 'Labour Conditions in Essex in the Reign of Richard II', *Econ. Hist. Rev.* iv (1934), 429–51.

free clothing and corn (the maximum wage allowed by the statute being 10s. without either), and when their demand was refused, they threw up the job. In the hundred of Barstable ten labourers 'refused to serve except by day, taking each of them from divers men at Orsett and elsewhere in the hundred, two-pence a day and a free dinner'. At Widdington a man who had been 'a good ploughman for the greater part of his life' had by 1389 become a common labourer for the sake of greater profit and took twopence a day and his food in winter and fourpence at harvest, refusing to work on any other terms. Competition for labour from trades and industries further stimulated the demand for higher wages and explains the statutory provision that labourers' children were not to be set to learn a trade if they were wanted on the land. A ploughman named John Pretylwell was enticed away by a maltmaker who offered him 26s. 8d. a year, with food and clothing; two labourers in East Tilbury became shipmen *pro maiori lucro* and others on the Essex coast abandoned the land to become *draggatores de oystres*. Free meals —a natural demand when the price of food was high and a convenient device for attracting labour without breach of the statute—imposed a heavy burden on employers, for the workers' appetites were large. The Waltham custumal shows that for his midday meal a boon-worker at harvest could expect to share with a friend bread, ale, pottage, and either a dish of beef and a dish of pork or mutton, or a dish of fish and five herrings; in the evening they could expect bread, ale, and either four herrings, or milk and cheese.

It was probably the prosperous leaseholders of the demesne who were prepared to pay these very high wages; for by the last decade of the century landlords nearly everywhere had ceased to farm their own land. Though the production of barley and oats had risen slightly, the yield of wheat per acre in some of the principal corn-growing areas in 1396-7 was only 0·76 quarters (the London quarter of 8 bushels being taken as the standard measure), as against 1·38 in 1299-1300—a significant index of the decline of high farming.[1] The most revolutionary change was probably that made at Christ Church, Canterbury, in the priorate of Thomas Chillenden (1391–1411), who broke com-pletely with the conservative policy of his predecessors and with-in half a dozen years established leaseholds everywhere on the

[1] N. S. B. Gras, *The Evolution of the English Corn Market* (1915), p. 216.

Christ Church estates, the needs of the monks being provided
for by a system of food-rents from the corn-growing manors of
Kent. Chillenden was an astute man of business and in the
role of prior-treasurer he assumed undivided control over the
finances of the monastery. Leases were commonly granted to
manorial serjeants whose experience fitted them to manage the
land to the best advantage; and his vastly swollen rent-roll
enabled Chillenden to reconstruct and complete the nave of the
cathedral and to undertake building projects on many of his
manors. Similar developments were probably taking place on
the majority of ecclesiastical estates. At Owston income from
rents rose from 68 per cent. in 1386 to 76 per cent. at the beginning of the fifteenth century; by 1400 the canons were farming
the demesne in three villages only. On the Battle abbey manor
of Hutton the acreage of crops on demesne shrank from 347
acres in 1341–2 to 254 in 1362–3 and to 166 in 1389–90. At
Soham, in Cambridgeshire, leasing of the demesne was complete by 1390, at Chatteris by 1397. By the end of the century
demesne tillage had ceased on almost all the widely scattered
Lancastrian estates;[1] and even a conservative landlord like
Thomas IV, Lord Berkeley, who, till about 1385, had walked in
the ways of his ancestors, when he saw the times beginning to
alter, altered with them. '. . . then instead of manureing his
demesnes . . . in each manor with his own servants, oxen and the
like under the oversight of the Reeves of the manors . . . This
lord began to sell his meadow grounds.'[2]

Serfdom did not end with the fourteenth century. But the
bonds which held the villein to the soil were loosening almost
everywhere and the demarcation of the agricultural population
into the three main strata—freeholders, tenant-farmers, and
landless, or nearly landless, labourers—which was to form the
characteristic pattern of English rural society until the end of
the eighteenth century, is already clearly foreshadowed. By and
large, the great landlords had delegated their responsibilities as
farmers to their tenants; henceforward, their main interest in
their estates would be financial and administrative. For the lay
landlord the change had obvious advantages; and where a great
ecclesiastical landlord put his rents into building or education
he was justifying his existence as a *rentier*. But for the monks as
a whole, the developments which had established themselves at

[1] Holmes, *Estates of the Higher Nobility*, p. 116.   [2] *Lives of the Berkeleys*, ii. 6–7.

the end of the period were pregnant with sinister if unforseeable possibilities. The monastic landlord who had accepted the position of a figure-head had gone far towards making himself look redundant.

Questions as to the standard of living and mental climate of the peasantry, their relations with the lord and his officials, the degree of effective, as distinct from legal, independence which they enjoyed, are more easily asked than answered. There is general agreement that the Revolt of 1381 owed much of its impetus to men who were rising in the world and striving to be free from archaic restrictions. We know all too little, however, of the condition of the mass of the peasantry. There may have been much real poverty in the countryside and some fardel-holders and cottars must often have been living near to the margin of subsistence. Chaucer's 'povre wydwe, somdeel stape in age'[1] had three large sows, three cows, and a sheep for the support of herself and her two daughters; their diet consisted mainly of milk, brown bread, broiled bacon, and sometimes an egg or two. And Langland has left us a poignant picture of the lot of the poor man's wife:

> Al-so hem-selue · suffren muche hunger,
> And wo in winter-tyme · with wakynge a nyghtes
> To ryse to the ruel · to rocke the cradel,
> Bothe to karde and to kembe · to clouten and to wasche,
> To rubbe and to rely · russhes to pilie,
> That reuthe is to rede · othere in ryme shewe
> The wo of these women · that wonyeth in cotes.[2]

Even those bondmen who were somewhat better off were often subject to heavy labour services, to customary rents and pay-ments in kind, as well as to such miscellaneous external demands as plough-alms, light-scot, and procurations to the Church, and royal taxes, tallages, and purveyances. Yet the impression con-veyed by some modern writers of a population of agricultural slaves ruthlessly exploited alike by Church and State, by land-lords and by their more prosperous brethren, accords better with Marxist preconceptions than with the facts of English history. In the fourteenth, as in every subsequent century before our own, many country people were all too well acquainted with poverty; but few of them could be equated with slaves, either

---

[1] *Canterbury Tales* (Nun's Priest's Tale), i. 2821. [2] *Piers Plowman*, C x. 77-83.

ancient or modern, and unrelieved misery can hardly have been
general. The degree of spiritual consolation which may have
been afforded by the ministrations of the clergy and the sacraments of the Church is not measurable, of course; but we do less
than justice to our medieval forefathers if we leave out of all
account the Christian framework in which their lives were set.
It was almost inevitable that a landowner should seek to increase
his profits and to improve his estate for his heirs; but his religion
enjoined charity, forbade him to despise the poor, and reminded
him that the man who pulled down his barns to build greater
might yet be a fool in the eyes of God.[1] It may have been merely
lip-service to an ideal of Christian brotherhood which constrained the duke of Lancaster to refer to some of his reeves as
his *bien amez tenantz en bondage*; but such service was better than
none.[2] Nor is there any reason to suppose that the average landlord was ruthlessly inhumane. On the contrary, the records
yield numerous examples of acts of grace and clemency. The
abbot of Ramsey extended his pardon to one of his bondmen who
had been ordained without his licence and allowed him to go to
school and take further orders, asking in return only ten psalms
for the soul of his predecessor. Another of his tenants who
had committed the serious offence of marrying without leave,
withdrawing from the manor, and 'making herself free', was
pardoned, *quia pauper*. A bishop of Chichester (William Rede),
remembered the tenants of his manors, both free and bond, in
his will.[3] Miss Page's study of the administration of the Crowland
estates led her to the conclusion that it was 'on the whole, a
tribute to the ecclesiastical landlord'. Two appeals from his
serfs to the abbot of Crowland were met by an order for investigation and restitution; he reduced tallage during the 'hard
years' at Dry Drayton; entry fines were often remitted; and there
was general concern for the poor and sick, widows and the
infirm being granted a part of the holding, a 'shelter', and an

[1] Cf. *Piers Plowman*, B vi. 46–51:

> 'And mysbede nou3te thi bonde-men · the better may thow spede;
> Thowgh he be thyn vnderlynge here · wel may happe in heuene,
> That he worth worthier sette · and with more blisse,
> Than thow, bot thou do bette · and lyue as thow shulde;
> For in charnel atte chirche · cherles ben yuel to knowe,
> Or a kni3te fram a knaue there · knowe this in thin herte.'

[2] *John of Gaunt's Register 1379–83*, i. 48 (no. 125).
[3] F. M. Powicke, *The Medieval Books of Merton College* (1931), p. 89.

allowance of grain. It has been suggested that the abbot of Crowland may have been exceptionally kindly;[1] but the registers of both the Black Prince and John of Gaunt present not dissimilar pictures. The Black Prince ordered redress of his tenants' grievances and licensed schooling and ordination;[2] and we find Gaunt personally commanding repair of his tenants' dwellings, excusing them rents and tallages, licensing a bondwoman to go on pilgrimage and a bondman to be ordained.[3] For the princes of the blood were not inaccessible to the poor; and the tenants of the abbot of Vale Royal on his manor of Darnhall did not hesitate to seek out the king himself and even Queen Philippa, in the course of their protracted struggle for freedom. It was not the great landlords and their councillors, so much as their stewards, serjeants, and bailiffs who were the real oppressors of the poor; but Magna Carta (c. 38) had limited the bailiff's power to put a man on trial;[4] and complaints against offending officials were often sympathetically received. Under Thomas II, Lord Berkeley, the jury of Berkeley hundred made many preferments against the bailiff for wrongful imprisonment and for setting both men and women in the stocks without cause. On the abbot of Bec's manor of Weedon a rebel bondman who undoubtedly had been guilty of subversive activities, was none the less awarded the very large sum of 40s. as damages against the abbot's bailiff who had thrashed him.

For it is not to be supposed that nothing but seignorial charity stood between the peasant and oppression or injustice. The extent to which the custom of the manor or of the township afforded real protection must have differed from place to place and from time to time; and the will of the lord was obviously powerful. But the capacity of the homage to act as an organic body is no less obvious; and distinctions of legal status seem to have been ignored among customary suitors to the manor court. On the Winchester estates the homage is found leasing pasture collectively, bargaining with the lord and with groups of villeins, and choosing the manorial servants. In 1368 the tenants of Long Sutton, a manor of the Duchy of Lancaster, agreed to pay an annual rent of £162. 4s. 5d. for 1,000 acres of meadow and pasture 'to have and to hold by the aforesaid homage and their

---

[1] *Vict. County Hist. Hunts.* ii. 76.   [2] *Register*, iv, nos. 8, 268, 377, 442.
[3] *Register 1379–83*, p. xxi; nos. 381, 721, 1242.
[4] See Plucknett, *Medieval Bailiff*, pp. 8–13.

heirs in bondage, to be divided among themselves according to the condition and ability of each of them . . .'.[1] Men of one vill may collectively sue the men of another and win damages; and evidence drawn from the Northamptonshire townships of Harlestone and Easton Neston suggests that in villages where there was more than one lord the sense of community may even have been stimulated.[2] The remarkable petition presented by the tenants of the Christ Church manor of Bocking during the priorate of Henry Eastry reveals the petitioners, not only as keenly aware of their rights and of their common interest, but also as well able to defend them. For, in their formal protest to the prior against an over-zealous steward, these peasants plead the Great Charter and common and statute law as well as the custom of the manor. To each point in the petition Eastry returned a conciliatory answer, 'that the deed had not been done by him nor by his wish and that in future he would not suffer such evil to be done to any tenant of the vill but that they should be maintained in their customs in all matters'. He could be harsh enough in defence of his legitimate rights, as his tenants of Monks Risborough who had spent seven years in his prison at Canterbury had good cause to know. But his handling of the Bocking case suggests that his concept of rights was neither one-sided nor limited to those who were in a position to enforce them in the public courts.[3] Flagrant violation of manorial custom might, indeed, often be dangerous; for discontented or rebellious bondmen seem to have found no difficulty in enlisting help from outside. The Norman-French wording of the Bocking petition betrays the hand of a lawyer; and other manorial records afford instances of influential persons showing readiness to co-operate in peasant movements. A knight named 'Hildebrand of London' supported the tenants of Ogbourne in their attempt to withhold rents and services from the abbot of Bec; and the rebels of Weedon and Hooe had the support of powerful persons (*genz de pors*). When Sir Roger Bacon, in 1381, threw in his lot with Geoffrey Litster, he was by no means the first of his kind to champion peasant claims.

[1] Power, *Medieval English Wool Trade*, pp. 38–39.
[2] Helen Cam, 'The Community of the Vill', *Medieval Studies presented to Rose Graham* (1950), p. 11; Joan Wake, 'Communitas Villae', *Eng. Hist. Rev.* xxxvii (1922), 406–13.
[3] J. F. Nichols, 'An Early Fourteenth-Century Petition from the Tenants of Bocking to their Manorial Lord,' *Econ. Hist. Rev.* ii (1930), 300–7.

Like all medieval societies, the society of fourteenth-century England was frankly and unashamedly hierarchical. For illustration of the hold which notions of 'degree' had over the mind of the governing classes we need look no farther than the sumptuary legislation of 1363 with its strict classification of the lay population, as *garçons* (agricultural workers, and others with goods worth less than 40s.), *gens de mestere* (merchants, burgesses, and artificers), *yomen, valletz, esquiers, chivalers*, and *seigneurs*, and its regulation of the food and clothing appropriate to each estate.[1] Yet such measures were bound to prove abortive and only a society in ferment could have called them forth. Stratification of classes was never rigid. The Church and, to a lesser extent, the towns and the army offered careers open to talent. William of Wykeham, bishop of Winchester and chancellor of England, is thought to have been the son of a serf. Simon de Paris, alderman of London, came of Norfolk bondmen; and the statute of 1406 making it illegal to take apprentices from poor families—a measure which aroused strong opposition in London—tells its own tale.[2] The great Sir John Hawkwood, whose exploits as a free-lance captain in Italy won him the hand of a daughter of the Visconti and a splendid equestrian monument in the Duomo at Florence, was the son of an Essex tanner and had himself been a tailor's apprentice. Expanding industries in the countryside opened up new opportunities for full-time or casual employment to the peasant and his family.[3] Such restrictions on mobility as existed—restrictions imposed by law and custom, by natural conditions and by social convention—were not absolute. Society as a whole was mobile, active, and fundamentally healthy. No power known to medieval man could have prevented the able and enterprising peasant from going up, the slack, the feeble, and the unlucky from going down. Some of the greatest Englishmen of the future were to spring from medieval peasant stock; and, although for the majority of the rural population, late fourteenth-century England may have looked far from merry, there are some grounds for believing it to have been prosperous and none for supposing that, even for the poorest, it was a land denuded of charity or hope.

---

[1] *Statutes* 37 Edw. III, cc. 8–15.

[2] See Sylvia L. Thrupp, *The Merchant Class of Medieval London* (1948), pp. 211, 215.

[3] For fuller discussion of the rural industries see below, pp. 363–73.

The hypothesis that late fourteenth-century England, particularly in the last decade, was in a state of overall prosperity derives from a variety of sources, some of which will be considered in the ensuing chapter. Though this prosperity obviously owed much to wool production and to the expansion of certain industries and trades, there can be no doubt that it would have been wholly inconsistent with a serious decline in arable farming; for it was in arable farming that the great majority of the people were still employed and from the arable farms that the nation derived its food supply. Yet many aspects of the picture of agricultural development sketched in this chapter have been sufficiently depressing; and we seem thus to be faced with a paradox demanding explanation. Since there is no denying that lands were going out of cultivation and villages being abandoned, or that a static or already declining population had been drastically reduced by a series of devastating plagues, the inference is irresistible that total productivity must have been less at the end than at the beginning of the century. Further, it would appear to be fairly well established that, even if the landlord class was finding it possible to augment its resources from the profits of war, investment in industry, or astute manipulations of the land law, the income which many landlords were drawing from their arable property was tending to contract, often seriously. Yet, if the area of land under cultivation was smaller, so was the number of mouths to be fed. Even on a conservative estimate of the mortality resulting from the Black Death, it cannot be supposed that shrinkage of cultivated land kept pace with the shrinkage of the population; and the answer may be that there was more wealth to go round. We cannot imagine, of course, that this surplus wealth (if such there was) was evenly distributed. Evidence already examined has suggested that the greater wealth of some sections of the peasantry may have been matched by the increased poverty of others; and there may well have been regions where the violence of the plagues caused total and lasting dislocation. But even when allowance has been made for the possible (though by no means proven) existence of a 'submerged proletariat' of agricultural workers and for the melancholy condition of districts like the Norfolk marshland, it remains impossible to ignore or explain away the overwhelming impression which meets us in so many quarters that England was in a state of high prosperity in the latter years of Richard II.

This prosperity may have been a temporary and short-lived phenomenon, not incompatible, perhaps, with a depressive trend over a much longer period. But even a decade of prosperity seems to point inescapably to the conclusion that agriculture, which was the basis of the national economy, must have been maintaining, at the very least, a sufficient level of productivity.

XII

TRADE, INDUSTRY, AND TOWNS

A NATION of farmers, fishers, and graziers will normally be, in the main, an exporter of raw materials and an importer of manufactured goods; and this, the traditional picture of English economy, did not begin to alter substantially until the second half of the fourteenth century. When our period opened, raw materials still constituted the bulk of the country's exports. Agriculture was the basic industry and the volume of the export trade in corn was often considerable; for example, in 1334, seven merchants were licensed to export 52,000 quarters of wheat to Gascony; and prohibition of corn exports for some years after 1391 provoked a complaint in parliament that the resulting fall in prices was making it impossible for tenants to pay their rents.[1] This does not mean, however, that there was a regular surplus of food. Sea transport was cheaper than transport by land and corn might be shipped from the east coast ports to the Low Countries while there was shortage in the western shires.[2] Gascony was a principal market for dried and salt fish; and salt was exported in bulk to Norway, the Netherlands, and Normandy until about the middle of the century, when competition from the rising woollen industry killed the salt trade by luring the salters from their 'sodden, evil-smelling and often wind-swept marshes' to more congenial occupations.[3] The coal trade, however, was beginning to expand, Newcastle exporting some 7,000 tons so early as 1377; most of the vessels which came to the Tyne in the nineties took away coal as part of their return cargoes.[4] Tin, metal wares, leather goods, and dairy produce also found markets overseas. But the 'sovereine marchandise' of the realm, the pre-eminent export, was wool. Damp and salt-laden island air and a wide variety of pasture—marshes, rich grasslands, hills, moors, and chalk downs—favoured the raising of many different breeds of sheep. Little is known of

---

[1] *Cal. Pat. Rolls 1330–34*, p. 359; Nelly J. M. Kerling, *Commercial Relations of Holland and Zeeland with England* (1954), pp. 104–5; *Rot. Parl.* iii. 320.

[2] See Sir George Clark, *The Wealth of England* (1946), p. 2.

[3] A. R. Bridbury, *England and the Salt Trade in the later Middle Ages* (1955), pp. 26–38.

[4] L. F. Salzman, *English Trade in the Middle Ages* (1931), pp. 281–2.

these breeds today. But it is well established that English wools ranged from the coarse products of Devon and Cornwall (which were seldom exported), through the middle-grade wools of the south downs, the midlands, East Anglia, and Yorkshire, up to the two most highly priced grades—the fine short wool of the Ryeland sheep, named from the tract of country between the Severn and the marches of Wales, and the fine long wool of the Cotswolds, Lincolnshire, and Leicestershire.[1] Great landlords, both lay and ecclesiastical, merchants and financiers, enterprising townsmen and resourceful country folk all drew much of their wealth from wool. It was wool that gave England her reputation in the markets of Europe, drawing buyers from Italy, Spain, Flanders, and many other lands; and, in a sense, it was on the pastures of England that Crécy and Poitiers were won. For the importance of the wool trade was as much political as economic; and the reign of Edward I had already seen a foreshadowing of those conflicts over the organization of the trade which were to determine its fortunes in the time of his grandson.

Surviving figures in the enrolled customs and subsidy accounts seem to place it beyond doubt that expansion of the wool trade was best favoured by a system which imposed the minimum of restraint on commercial enterprise; and, on a long view, expansion of the trade must obviously be in the interests of the whole nation. Yet it was difficult, both for rulers bred in the mental climate of the age of chivalry and for merchants intent on the amassing of private fortunes, to take long views on such matters; and, even had they been alive to the probable economic consequences of some of their actions, it is doubtful if many of them would have judged ultimate contraction of the trade too high a price to pay for victory over the French and Scots or over commercial rivals at home and abroad. The history of the wool staple in this period points inescapably to the conclusion that considerations of immediate profit, political or economic, were the determining factors governing the direction of the trade.

The first compulsory wool staple was established at St. Omer in the spring of 1314. An ordinance of 20 May 1313 had arranged for replacement of the 'preferential staple' at Antwerp by a compulsory staple, through which must pass all wools destined for Flanders, Brabant, or Artois, the location to be in the hands

[1] Eileen Power, *The Wool Trade in Medieval English History* (1941), pp. 21–23.

of the merchants of the staple and their mayor.[1] It was some of these merchants who had persuaded Edward II that a compulsory staple would benefit both themselves and him; and there is reason to believe that for this concession the merchants paid a substantial sum to the king who could also look forward to drawing fines from recalcitrant shippers and to raising loans from others, in exchange for exemption from the staple regulations.[2] These were valuable assets to Edward II, whom the Ordainers had lately deprived of the services of the Frescobaldi; and it was consistent with Thomas of Lancaster's general policy of thwarting the king at every point that, at Sherburn-in-Elmet in 1321, he should have taken up the cause of the aliens who were the principal sufferers from the new system.[3] Yet the native merchants themselves were divided on the whole issue; and a merchant assembly at Westminster in April 1319 produced a recommendation that there should be two English staples, one to the north and one to the south of the Trent, in place of the foreign staple.[4] Such a plan may have commended itself to the London capitalists and a few others; but it did not meet the wishes of the growers and smaller merchants who were opposed to a scheme that was likely to mean concentration of the trade in London and a single northern port.[5] In the event, the war of St. Sardos killed the St. Omer staple; and, although another was established at Bruges, the position was sufficiently precarious for the younger Despenser, backed by a body of merchant opinion, to launch an assault on the whole system of foreign staples. An ordinance issued at Kenilworth on 1 May 1326 established home staples at nine English, two Welsh, and three Irish towns.[6] In the critical summer of this year such a change may have looked like a short cut to the popular support

[1] A Shrewsbury burgess named Richard Stury was described, in Feb. 1313 as 'mayor of the merchants of our realm' (*Foedera*, ii. 202); after the ordinance of that year, he was called in addition 'mayor of the wool staple'. *Cal. Pat. Rolls 1313–17*, p. 15.

[2] R. L. Baker, 'The Establishment of the English Wool Staple in 1313' (*Speculum*, xxxi (1956), 444–53). The author demonstrates the inherent improbability of Tout's view that the staple of 1313 was the work of the Ordainers.

[3] Bridlington, p. 63.

[4] A. E. Bland, 'The Establishment of Home Staples, 1319', *Eng. Hist. Rev.* xxix (1914), 94–97.

[5] Tout, *Place of Edward II*, pp. 230–1.

[6] *Cal. Pat. Rolls 1324–27*, p. 269. The staple towns were York, Newcastle, Lincoln, Norwich, London, Winchester, Exeter, Bristol, Shrewsbury; Carmarthen, Cardiff; Dublin, Drogheda, and Cork.

which the Despensers so greatly needed; but it was unwelcome
to the bulk of the English merchants outside the staple towns and,
after the fall of Edward II, Isabella and Mortimer first suspended
the home staples, in return for loans and then, in the Northampton parliament of 1328, abolished them altogether.[1]

The beneficial effects of this restored freedom of trade are reflected in the export figures which remained high between 1328
and the outbreak of the Hundred Years War, rising from some
31,000 sacks to some 37,000 in 1339.[2] But dispersal of the wool
trade among a multiplicity of markets was ill adapted to meet
the needs of Edward III on the eve of his conflict with Philip VI.
For Edward, the value of the trade lay in its potentialities as
a diplomatic lever, as security for the loans without which he
could not finance his wars, and as a source of ready cash for the
payment of his allies and his troops; hence his desire to canalize
it so far as possible and to entrust its management to a few
monopolists with whom he could do business. It is significant of
the growing strength of Edward's position in the country that,
on the opening of hostilities in 1337, he should have been able to
restore the compulsory foreign staple as an instrument of his war
finance, first at Antwerp and then, from 1340, at Bruges; and it
was on the basis of this monopoly that he negotiated his loans
from the Bardi, the Peruzzi, and the native financiers.[3] The
victor of Crécy and Calais could afford to ignore persistent
grumbling from his subjects, both inside and outside parliament.
Yet, as the fifth decade of the century drew to its close, the compulsory foreign staple began to lose many of its attractions, even
for the king. The action of the Flemings in refusing access to the
Bruges staple, not only to foreigners but also to buyers from the
smaller industrial centres within Flanders itself, kept the price of
wool low and, in 1348, after making formal protest to Bruges,
Edward arranged for a temporary transfer to Middelburg in
Zeeland. Moreover, the narrow seas were infested with pirates;
the interruption of trade and of the war by the Black Death

---

[1] *Statutes* 2 Edw. III, c. 9.

[2] Figures kindly supplied by Professor Carus-Wilson. See also H. L. Gray, 'The
Production and Export of English Woollens in the Fourteenth Century', *Eng. Hist.
Rev.* xxxix (1924), 15. Complaints made in 1332 that certain merchants had set up
a staple at Bruges led to the re-establishment of home staples, but they were again
abolished in the York parliament of Feb. 1334. *Cal. Pat. Rolls 1330–34*, p. 283; *Rot.
Parl.* ii. 377.

[3] Above, pp. 155–7.

reduced both the possibility of raising extraordinary supplies on a large scale and the immediate need to do so; and, above all, the ruin and discredit of such financial houses as Chiriton and Swanland had produced violent reaction against monopolists. Thus, there were powerful inducements to a drastic change of policy; and in 1353, in return for a wool subsidy for three years, Edward conceded the Ordinance of the Staple, which was confirmed by statute in the following year.[1]

These acts went far beyond their predecessors; for they not only swept away the foreign staple in favour of fifteen staple towns at home (ten in England, one in Wales, and four in Ireland),[2] but they also left the aliens free to buy where they liked, while prohibiting English merchants from engaging in the export trade. Clearly, the intention was to prevent a recurrence of monopolies; but, though profitable to the wool-growers, whose influence in parliament was powerful, the act was bound to defeat its own ultimate objective—expansion of the trade in such a way as to secure wider distribution of its profits among Englishmen—since it left the alien middleman in possession of the field. When the price of wool began to fall it became evident that the system could not last; and the conclusion of peace with France in 1360 encouraged the commons to look for an end of the *maltolte* and for a return to freedom of trade.[3] Yet there were limits to the concessions which Edward III was prepared to make. All the revenues of the realm, the chancellor explained to the magnates in 1365, would not cover half the king's expenses;[4] and though this speech need not be taken at its face value, no king in Edward's position could afford to dispense altogether with loans. A new solution to the problem now suggested itself. It was at Calais that the peace treaty had been ratified and it was of the utmost importance to bind this cherished conquest firmly to the English Crown. Already in 1348 Calais had been made the staple for tin, lead, and cloth; plans for a wool staple there had been mooted in 1361; and as a working compromise they had obvious advantages. A staple at Calais would maintain the wool market at the gates of Flanders, but

---

[1] *Statutes* 27 Edw. III, st. 2; 28 Edw. III, cc. 13, 14, 15.

[2] Newcastle, York, Lincoln, Norwich, Westminster, Canterbury, Chichester, Winchester, Exeter, Bristol; Carmarthen; Dublin, Waterford, Cork, and Drogheda.

[3] G. Unwin, *Finance and Trade under Edward III*, pp. 243-4.

[4] *Rot. Parl.* ii. 285.

on English soil; the aliens might be left in control of the direct trade to Italy and Catalonia, and the rest of the trade so canalized as to leave the king free to raise his loans; bullion need not be exported for the payment of the garrison. Thus, in 1363, the Calais staple was established, under the control of twenty-six merchants *de valentioribus regni*,[1] the Company of Merchants of the Staple; though the home staples retained their mayors and courts for the recovery of debts and other pleas under the law merchant and continued to be the only English towns from which wool might be shipped. The Company of Merchants of the Staple was much more broadly based than the older private companies; but even this quasi-monopoly was regarded with some suspicion and we find the commons demanding that the ordinance regulating the wool custom be observed without any trickery on the part of 'the new company of merchants now dwelling at Calais'.[2] On the renewal of the war in 1369 the Calais staple was temporarily suspended; and, though it was restored in August 1370, there seems to have been difficulty in enforcing its use. Complaints in 1373 and the outbursts in the Good Parliament of 1376 reveal a complete volte-face in the attitude of the commons who now constituted themselves the spokesmen of the London capitalists, demanding a re-establishment of the Calais staple and directing their attacks against those who, like Latimer and Lyons, were known to be evading it and to be encouraging the king to license others to do the same.[3] An ordinance of 23 July 1376 met these complaints by affirming that the staple should be at Calais; and a subsequent statute (1378) eliminated the need for individual licences and, by allowing free export through Gibraltar, recognized Italian control of the Mediterranean trade.

Thus, Richard II's government inherited the Calais staple at a time when the position both on land and at sea was deteriorating rapidly. The commons' petition in his first parliament that, if Calais should be attacked, another location should be found, betrays an understandable nervousness, unlikely to be dispelled by the series of devastating raids which fell upon the south coast in the early years of the reign.[4] Once again, the staple had to be

---

[1] Knighton, ii. 117.
[2] 'saunz fraude, engine, subtilte ou ordinance au contreire', *Rot. Parl.* ii. 276.
[3] Ibid., pp. 318, 323, 326; *Anonimalle Chronicle*, pp. 81–82, 85–86.
[4] *Rot. Parl.* iii. 23, 24.

put to diplomatic uses; its removal to Bruges was made a condition of the projected Anglo-Flemish alliance of 1382 and only the disaster of Roosebeke and the seizure of Bruges by the French in the autumn prevented the execution of this plan and led to the substitution of Middelburg, where the staple was established in 1384. Thereafter, the interaction of economic and political considerations in determining the location of the staple becomes increasingly obscure. No doubt, the single continental staple, whether at Middelburg or Calais, continued to appeal to the group of prosperous Londoners who might hope to profit by it;[1] but, for the bulk of the merchants, and even for the king, its advantages were no longer so obvious, particularly after the failure of the Norwich crusade had left Flanders at the mercy of the French. Pole was probably sincere in his plea to the parliament of 1385 that it would be to the manifest benefit of the king and the kingdom if home staples were to be restored.[2] Parliament accepted his plea and ordered the removal of the staple to England; but no action seems to have been taken, perhaps on account of pressure from Nicholas Brembre, the king's agent in the city, whose associates there preferred maintenance of the Middelburg staple. After Pole's fall the Cambridge parliament of 1388 ordered removal to Calais, a decision which may have been influenced by the fact that the captain of Calais, Sir William Beauchamp, was hand-in-glove with the lords appellant.[3] But with the beginning of Richard II's independent rule, the voices of those who favoured home staples were raised again; and by the end of 1390—perhaps as a bid for popular favour—Richard had ordered a return to the arrangements of 1353. Parliament allowed him his *quid pro quo* in the shape of the wool subsidy for three years at the high rate of 43s. 4d. per sack, but only on the explicit condition that if the staple were again taken out of the realm before the three years were up, collection of the subsidy should cease.[4] Richard may have intended to keep his promises; but he was now confronted with a situation not unlike that which had faced his grandfather thirty years before. He was seeking for peace with France; and, unless Calais were

---

[1] London merchants took over 3,000 sacks to Calais in the autumn of 1386 and then diverted them to Middelburg to swell the total of 18,545 sacks which entered the port from England this year. F. Miller, 'The Middleburg Staple 1383-88', *Camb. Hist. Journal*, ii (1926), 63–66.

[2] *Rot. Parl.* iii. 203.

[3] *Statutes* 12 Ric. II, c. 16.   [4] *Rot. Parl.* iii. 268; 278–9.

to be abandoned to the French (a possibility which no one in England was prepared to contemplate), the interests of her merchants must be made to depend on maintenance of the English connexion and the garrison must be assured of its wages. Thus, by the summer of 1392 the staple was back at Calais, where it remained; and the privileges of the town were obstinately upheld by the king who, in 1394, refused requests from the tinners of Cornwall and from five of the southern shires to be allowed to send their goods elsewhere.[1]

Even when full allowance is made for the effects of war taxation and of the fluctuations of the foreign market, the export figures published by Professor Carus-Wilson suggest that the prosperity of the wool trade was closely bound up with the vicissitudes of the staple system.[2] It can hardly be mere coincidence that with the establishment of home staples in 1353, exports should have soared to well over 40,000 sacks, reaching an annual average of 32,000 for the decade 1350–60; nor that, after the transference of the staple to Calais, they should have begun to drop, the average for the decade 1390–9 being only 19,000 sacks. The effects of this decline on the customs revenue were serious enough; that they were not utterly disastrous for the nation as a whole was only because, while exports of raw wool had been declining, exports of woollen cloth had been rising, until by the close of the century, the two were about level and the first chapter had already been written in the long history of England's pre-eminence as an exporter of manufactured woollen goods.

For a rising cloth industry there could be no more effective protection than a heavy export tax on wool; and, given internal conditions favourable to the woollen industry, expansion of the cloth trade may be regarded as an almost inevitable consequence of the contraction of exports of the raw material. At the beginning of the fourteenth century, when the woollen industry is known to have been in difficulties, it is likely that England's export trade in cloth was small, though it does not appear to have been negligible. The English had long been known as makers of speciality cloths and there is evidence, for example, that alien exporters took cloths to the value of £2,000 out of the

---

[1] *Rot. Parl.* iii., pp. 319, 322.
[2] Eleanora Carus-Wilson, *Medieval Merchant Venturers* (1954), p. xvi and table facing p. xviii.

port of Boston between February and Michaelmas 1303.[1] But
the scale of imports averaging, perhaps, some 10,000 cloths a
year, and a measure like that incorporated in the Kenilworth
ordinance of 1326, forbidding commoners with incomes from
land of less than £40 to wear imported cloth, suggest that exports
cannot have been high. Precise figures are not available, for the
petty custom of 1303 was payable by aliens only and even this
was twice suspended in Edward II's reign.[2] But from 1347 when,
despite parliamentary protests, a small graduated duty was imposed on all exported cloth (a significant index of the growing
importance of the trade) it becomes possible to reach some tentative conclusions.[3] For the years immediately preceding the
Black Death an export figure of some 6,000 cloths has been suggested; but for five years after 1348 the total never exceeds
2,000 cloths and once it sinks below 700. Signs of recovery appear
in 1353, and by 1366–8 the average number of cloths exported
was in the region of 16,000. In the town of Bristol alone cloth
exports, which in 1353–5 had averaged 1300, by 1367 were
near 8,000. The year 1369 opened a period of depression lasting
until about the end of Edward III's reign; thereafter, except for
a brief period of recession between 1385 and 1388, the upward
trend is unmistakable, reaching an average of about 43,000
cloths in the years 1392–5.[4] Apart from a few speciality cloths,
imports in the meantime had withered away.

For explanation of these phenomena we have to look, not
only to the wool tax and to conditions favouring cloth production at home, but also to the state of the foreign market. Recession after the Black Death, when corn prices were rising and
panic and confusion widespread, is hardly surprising; there can
have been few incentives anywhere to lavish expenditure on
cloth. Moreover, famine in 1347 and the ravaging of the Gascon
vineyards by the French and English armies between 1345 and
1348 had gone far to destroy the purchasing power of England's
principal customer and hence of the prosperity of Bristol which
enjoyed a virtual monopoly of the Gascon trade. Recovery between

[1] H. C. Darby, *Historical Geography of England*, p. 316.
[2] In 1309–10 and 1311–12.
[3] It was argued that since the king already had the wool custom it was reasonable
that he should also have a custom on cloth. *Rot. Parl.* ii. 168.
[4] Professor Carus-Wilson's figures (which do not include worsteds); *Merchant
Venturers*, pp. 165–73; *Camb. Econ. Hist.* ii. 416. Most of the cloths exported to the
Netherlands were unfinished cloths, the dyeing and finishing being done abroad.

1353 and 1369 is explicable, partly by the retreat of the war from
Gascony and the reconstruction of her devastated countryside
which reached its climax in 1368;[1] partly by a succession of
better harvests; and partly by the conclusion of general peace
in 1360, with consequent relaxation of pressure on merchant
shipping and lessening of dangers at sea. Renewed depression
after 1369 coincides with the resumption of fighting, the advance
of the French armies into Gascony, reappearance of the plague,
floods, famine, and soaring prices, all of which combined to produce economic, no less than political, tension in the last years of
Edward III. The causes of the upward trend which followed
these bad years, the trend which, apart from some brief set-
backs, was to prove lasting, are rather more complex. It is true
that there were no more great continental campaigns, that there
was slow recovery in Gascony, that the worst virulence of the
plague appeared to be spent. But expansion of the Bristol trade
was not proportionate to that of the country as a whole, a fact
which suggests that recovery in Gascony was not the sole, nor
even, perhaps, the main stimulus at work. Other markets were
now opening up; it was in the last quarter of the century that
English producers began to penetrate the Low Countries (where
shortage of raw material and political confusion had almost
eliminated their rivals in the old cloth towns), the Mediterranean, whither the Italian merchants, trading mainly with
London and Southampton, were beginning to carry English
cloths in increasing numbers,[2] and, above all, those regions in
the Baltic and North Sea area, formerly the jealously-guarded
preserve of the Hanseatics. All these markets were ultimately to
prove of more importance than Gascony; and it is in their
struggle with the Hanse that the English first appear as serious
competitors for dominance of the cloth market of western
Europe.

Woollen cloth was likely to find a ready market in the inclem-
ent regions of the north; but attempts to expand in this direction must bring the English into conflict with the interests of the
Hanse. These merchants of the north German towns had long

---

[1] R. Boutruche, *La Crise d'une société* (1947), p. 207.
[2] Exports of cloth from Southampton were 266 per cent. higher in 1392 than
forty years before (Darby, p. 276). A recent study shows a Tuscan merchant in the
last decade of the century importing scarlet *berrette* dyed in England, Essex cloth,
Guildford cloth, and unbleached cloth from the Cotswolds ('Chondisgualdo') and
Winchester. See Iris Origo, *The Merchant of Prato* (1957), pp. 72–73.

held a highly privileged position in England itself; in London they had their own gildhall, the *gildhalla Theutonicorum*, in Thames Street, adjoining the Steelyard allotted to the merchants of Cologne.[1] Though, like other aliens, they paid the triple duty on wool—the ancient and new customs and the subsidy—they were altogether exempt from the duty on cloth; thus, the Crown had nothing to gain from attempting to protect their interest in the trade. Their own object was to maintain a rigid monopoly in the north European markets from Novgorod to Bruges and it was in order to defend this monopoly, if necessary by force, that the Hanseatic League had been organized for war against Denmark in 1367. But the English merchants (many of whom were also clothmakers) were not prepared to leave the handling of their wares entirely to the Germans. They wished to strike out for themselves and to establish their own contacts in the northern ports, above all, in Prussia, the chief distributor of English cloth in Poland and west Russia. The establishment of direct connexion by sea between the towns of the Zuyder Zee and the Baltic facilitated such enterprises, while at the same time threatening the dominant position of Lübeck on the Jutland peninsula. Signs of tension were beginning to appear in the middle seventies;[2] and the government of Richard II's minority tried to make maintenance of the Hanse privileges in London contingent on similar privileges being allowed to English merchants in the Hanse towns; they refused to confirm the Hanseatic charters until 1380, when the English right to trade in their territories was formally admitted. But the 'principle of reciprocity' proved difficult to apply, for the English authorities persisted in attempts to whittle down the privileges of the Hansards. Prussia continued hostile; and piratical activities, culminating in the capture of a Hanseatic fleet off the mouth of the Zwin in 1385 and in a series of mutual reprisals, led to a temporary stoppage of all Anglo-Hanseatic trade and to the removal from Danzig of the colony of English merchants already established there. Such a stoppage was to the advantage of neither party and an agreement in 1388, which reaffirmed the principle of reciprocity, allowed the English in Danzig to resume business.

[1] Alice Beardwood, *Alien Merchants in England 1350–1377* (1931), pp. 14–16. (The Steelyard was on the site of the present Cannon Street Station.)
[2] e.g. the Hanse merchants' complaint in June 1378 of molestation at the hands of London citizens. *Cal. Letter-Book H*, p. 101.

But the Prussians continued to regard them as suspect and to restrict their rights of residence; and when the century ended, the conflict, so far from being settled, was about to enter on a phase of still more acute bitterness.[1]

The opening of the northern trade enhanced the prosperity of many English towns, most of all, of London. Bristol controlled the Gascon trade; Southampton shared the Italian trade with the capital; but the northern trade had its natural exodus from the east-coast ports. Analysis of individual sailings in the early nineties has shown that the greater part of the northern trade was being carried by London and Hull, the trade of the latter having increased nearly fourfold since the sixties; while the drapers of Coventry were sending large consignments of cloth through the east-coast ports to the Baltic.[2] The merchants of these and other cloth-producing and cloth-exporting centres had no cause to lament the partial disruption of the wool trade, which had gone far to eliminate their rivals in the Low Countries and in Italy, had brought undreamed-of prosperity to their localities, and had altered and enlarged their whole concept of trade by forcing them into active competition with the aliens on their own ground and into the kind of hazardous adventure which had been familiar to their Anglo-Saxon and Scandinavian forefathers and was to win still greater renown for their descendants in the sixteenth century.

Imports from the continent to England included certain raw materials. From Prussia were obtained naval stores, like timber, pitch and tar, and, in times of dearth, corn. The virtual cessation of salt production meant that this essential commodity had to be brought from abroad, chiefly from the Bay of Bourgneuf on the southern borders of Brittany.[3] Luxuries, like oil, rice, almonds, and, towards the end of our period, oranges, came from the Mediterranean; and such foreign manufactured goods as tiles and stained glass, straw hats, and soap found a market here. But no other import could compare in value with wine; and it was from the vineyards of Gascony that the extensive needs of the Church, the royal household, the nobility and gentry, and the well-to-do burgesses were regularly supplied. Rhenish wines

[1] Eileen Power and M. M. Postan, *Studies in English Trade in the Fifteenth Century* 1933), 91–101, 105–9.
[2] Carus-Wilson, *Merchant Venturers*, pp. 258–9.   [3] Bridbury, pp. 55, 114.

were consumed elsewhere in northern Europe and, to a limited extent, in England; but the Gascons possessed what amounted to a monopoly of the English market and of the 90,000 to 100,000 tuns of wine exported annually from Bordeaux at the beginning of the century, between one-fifth and one-quarter came to this country.[1] The price was low, averaging about £3 a tun wholesale and 5d. to 3d. a gallon retail;[2] and the king was a principal gainer from this enormous trade. His butler bought some 2,000 tuns a year for the royal household; alien importers paid 2s. a tun on every tun of wine brought into the country; and native importers were subject to a prise of two tuns from every ship carrying twenty tuns or more, and of one tun from those carrying between ten and twenty, the butler paying 20s. as freight for every tun thus prised.[3] In so far as economic considerations were in his mind at all, Edward III must have hoped that victory over the Valois would serve to consolidate the commercial links already existing between England and Bordeaux and even, perhaps, to expand the English market for Gascon wines. If such were his hopes, they were doomed to disappointment; for a combination of war and natural disasters disrupted the trade for long periods, and recovery, when it came, was only partial.

Several causes contributed to this disruption. First, and most obvious, was the damage caused to the vineyards by the campaigns within Gascony itself. In the early years of the war, between 1337 and 1340, fighting in the Dordogne valley and the region of Entre-deux-Mers (the productive hinterland known as the Haut Pays), led to the virtual disappearance of the wines of this district. In 1345, though the Bordelais itself was not invaded and Henry of Derby succeeded in regaining control of many of the wine-producing areas in Poitou, the Agenais, Saintonge, and Périgord, any advantage which might have accrued to the trade was more than outweighed by the devastation resulting from the war. Finally, on the renewal of the war in 1369, when

[1] Margaret James, 'The Fluctuations of the Anglo-Gascon Wine Trade during the Fourteenth Century', *Econ. Hist. Rev.*, 2nd ser. iv (1951), 175–6.

[2] F. Sergeant, 'The Wine Trade with Gascony' (*Finance and Trade under Edward III*), p. 280. In 1309 there were said to be 354 taverners in London alone. *Ann. Paul.*, p. 267. The maximum price of Gascon wine was fixed at 4d. a gallon in 1342. *Cal. Letter-Book F*, p. 83.

[3] At the opening of the period the Cinque Ports were exempt from this prisage. London gained exemption in 1327, York in 1376.

the French armies swept into Gascony and up to the very gates
of Bordeaux itself and the peasants fled panic-stricken leaving
their grapes to perish on the vines, the trade came almost to a
standstill.[1] The ravages of nature had also to be reckoned with.
Following hard on the heels of Derby's campaign, the Black
Death produced a severe economic crisis; reappearing in 1373,
it was succeeded by a harvest so bad as to cause famine throughout the Bordelais. Such disasters were intermittent and the re-
cuperative power of medieval trade showed itself in short
periods of renewed prosperity from 1340–5 and from 1360–9;
but further damage resulted from the threat to lines of communication arising from almost incessant war and piracy at sea.

The insecurity of the seas made it necessary to insist that the
wine ships should travel in convoy, a system which proved
damaging to the trade in two respects; it restricted free enterprise and it forced up costs. The wine fleets usually set sail with
their convoys in late summer and early autumn; this meant concentration of an abnormal demand within a short season of the
year. In order to prevent merchants and vintners from reim-
bursing themselves at the expense of their customers, stricter
regulation of retail prices became necessary; but the dealers
resisted the restrictions and prices rose in spite of them, reaching 8d. a gallon in 1363 and causing renewed agitation on the
part of consumers.[2] To meet the cost of protecting merchant
shipping in time of war special subsidies were levied, rising
from the modest rate of 6d. per tun in 1340 to 2s. in 1371. These
subsidies, together with those levied on merchandise, should
have been applied strictly to the protection of merchant shipping; and when, in October 1350, a heavy tax of 3s. 4d. on the
tun was imposed and collected and no protection given, the
commons demanded that the subsidy should cease; to their
grant of tunnage and poundage in 1373 they attached the condition that it should be levied only for the duration of the war.
But the government gave them little satisfaction and from this
time onwards, though with some variations in the rate, tunnage
and poundage became a regular parliamentary grant.[3] Meanwhile, purveyance afforded the pretext for many forms of extor-

---

[1] *Merchant Venturers*, p. 253.
[2] In 1383 a Londoner was imprisoned for selling wine at 8d. a gallon. *Cal. Plea
and Mem. Rolls 1381–1417*, p. 41. In Feb. 1387 a royal pardon had to be sought for
vintners and taverners who had sold Gascon wine illegally. *Cal. Letter-Book H*, p. 308.
[3] *Rot. Parl.* ii. 229, 317.

tion, the evil practices of the king's butler and his deputies being condemned by statute in 1351 and 1369.[1] Ceaseless piracy on the high seas and the desperate attempts of the government to prevent the flight of bullion to Gascony added to the instability of a trade which the war policy of Edward III had brought near to ruin.[2]

Signs of recovery became evident, however, in the last twenty years of the century, when the worst depredations on sea and land had ceased, when the convoy system was being more effectively organized, and the Gascon vineyards were being re-planted. Richard II's government spared no effort to capture a monopoly of the Gascon exports and when, in 1387, certain Gascons proposed to send their wines to the Middelburg staple, the king threatened them with confiscation and death. These efforts met with some success and by the end of the reign England had a virtual monopoly of the Gascon trade. But Bordeaux sold to England only at the expense of her other customers; the total volume of her exports had been permanently reduced. Nor was it found possible to lower freight charges to their pre-war level, or to bring back the days of cheap and abundant wine. The near-doubling of the price of wine (from £3 to £5 or £6) was among the most melancholy economic consequences of the first phase of the Hundred Years War.

By far the most impressive industrial development of the fourteenth century was the expansion of cloth-making. When the century opened, this ancient craft was suffering a depression of unprecedented severity; when it closed, England was well on the way to establishing herself as one of the foremost cloth-making nations of Europe. For explanation of this remarkable phenomenon we have to look to several causes—to technological developments, to government action, to changing conditions abroad. Deliberate fostering of the industry had been attempted both by Henry III and by Edward I and was to be attempted again by Edward II and Edward III. But it was not until the second half of the fourteenth century that economic and political factors coalesced so as to produce a state of affairs in which the native woollen industry could flourish as never before.

Symptoms of decline are evident in the early years of the

[1] *Statutes* 25 Edw. III, st. 5, c. 21; 43 Edw. III, c. 3.
[2] Margaret James, 'Les Activités commerciales des négociants en vins Gascons en Angleterre durant la fin du moyen âge', *Annales du Midi*, lxv (1953), 43.

century. The weavers, who had been pioneers in the formation
of craft-gilds, had lost their former pre-eminence and at Lincoln,
home of the famous 'scarlets' where, in the time of Henry II
there had been more than 200 of them, by 1321 there were no
weavers and only a few working spinners. The woollen industry
of Northampton which, under Henry III, had employed 300
men, was nearly extinct by 1334; in 1323 all the weavers of
Oxford were dead and none had taken their place; the weavers'
gild of Winchester asserted that large numbers of cloth-makers
had left the town; the men of Leicester complained that the extortions of Thomas of Lancaster had so injured the trade that
there was only one fuller in the town, 'and he a poor man'. In
London where, by 1321, the number of working looms had been
reduced from 380 to 80, the adoption of restrictive practices
tells the same tale.[1] Complaints which are often associated with
claims for relief from taxation may justly be viewed with some
suspicion; but the weavers' plaints are too closely related in
point of time to be dismissed as baseless, even had Edward III
himself not referred to the decay of their art.[2] Partial explanation of the decay of the cloth towns may be found in the high
demand for English raw wool abroad, in the markets of Germany, Flanders, and Italy; exports on the scale attained at the
turn of the century can have left little room for the maintenance,
let alone the expansion, of a native industry. There must always
have been cloth-making in the villages, but import figures suggest that at this period those who could afford to do so preferred
to buy cloth manufactured abroad. A second and even more
powerful cause of decline may be found in the restrictive and
monopolistic practices of the weavers' gilds, in the limitations
imposed on numbers of apprentices, on holidays and hours of
work, on wages and prices, and on individual enterprise; and,
even more, in the hostility shown to the weavers by the municipal authorities of London, Winchester, Oxford, and other towns
where intense jealousy of a rising craft had manifested itself in
attempts to bar them from the freedom of the city or borough
and to impose other crippling disabilities.[3] When the municipal
authorities abandoned their hostility it was already too late; for

[1] J. W. F. Hill, *Medieval Lincoln* (1948), p. 326; *Rot. Parl.* ii. 85–86; *Camb. Econ.
Hist.* ii. 409; L. F. Salzman, *English Industries of the Middle Ages* (1923), p. 202.
[2] *Cal. Pat. Rolls 1330–34*, p. 363.
[3] A. L. Poole, *Domesday Book to Magna Carta* (vol. iii of the present History),
p. 87.

a technological revolution had taken place which had largely transformed the industry from an urban to a rural craft.

The essence of this revolution was mechanization of the processes of fulling—a change which has been described as being as decisive in its way as the mechanization of spinning and weaving in the eighteenth century.[1] When fulling-mills were introduced into western Europe water-power was substituted for the human foot. We hear of two on the Templars' English estates in the reign of King John and by the fourteenth century they were springing up all over the country. The old-established fullers in the clothing towns who saw a threat to their livelihood in this simple form of automation, tried to insist on the maintenance of foot-fulling; but like other opponents of the relentless march of technology, they found themselves fighting in a hopeless cause. Fulling-mills soon revealed themselves as a profitable form of investment able to attract capital from both town and country. In 1298 and again in 1310 complaints were heard in London of fullers who were sending out work to the mills at Stratford and elsewhere, when such cloth ought to have been fulled 'by the feet of the men of the craft or their servants in their houses within the city'.[2] Henry the Walker, in 1323, confessed to the merchant gild of Leicester that he had taken cloth out of the town to the fulling-mill, in defiance of a gild regulation of 1260 that no gildsman should keep a fulling-mill outside the town; and, so late as 1346, it was forbidden to take 'raucloth' out of the town of Bristol to be fulled.[3] But it was inevitable that the new process should ultimately affect the location of the industry; for fulling-mills demanded swift currents and it was on the lower and slower courses of the rivers that most urban communities had grown up. Moreover, the remote upland valleys where the necessary water-power was available, often lay in wool-growing districts, like the West Riding, the Welsh marches, the Cotswolds, and the Berkshire and Wiltshire downs. The twofold

[1] Eleanora Carus-Wilson in *Camb. Econ. Hist.* ii. 409–13; see also *Merchant Venturers*, pp. 183–210. The object of fulling was to shrink the cloth, thicken and felt it till the fibres were so interlaced that the woven pattern was indistinguishable. The primitive method was to cover the cloth with soap or fuller's earth and then trample it under foot.

[2] Frances Consitt, *The London Weavers' Company*, i (1933), 7; *Cal. Letter-Book D*, pp. 239–40. By the end of Edward III's reign there were also fulling-mills at Wandsworth, Oldfield, and Enfield. *Cal. Letter-Book H*, p. 37.

[3] *Records of the Borough of Leicester*, ed. Mary Bateson (1899), i. 347; *The Little Red Book of Bristol*, ed. F. B. Bickley (1900), ii. 7.

attraction of fine wool and swift, clear water served to draw to these regions groups of weavers and other textile workers, many of whom were doubtless already eager to be free of the cramping restrictions of borough and gild. Where the lord of the manor had taken the initiative in erecting a fulling-mill he might attempt to establish a monopoly; but it was normal for such mills to be leased and enterprising weavers and fullers could draw on abundant supplies of labour from among the younger sons of the tenantry, the women and the smallholders, at wages which were generally lower than those demanded in the towns. The history of the fulling-mills in the Grasmere neighbourhood affords an interesting example of the expansion of the industry in a remote district. By the end of the fourteenth century there were three fulling-mills at Troutbeck, each of them leased from the lord by two or three men in partnership. An additional mill had been planted in Loughrigg and the rent of the original Grasmere mill had doubled since 1324.[1]

Thus, in the fulling-mill we may perceive an important part of the explanation of the rise of a rural, at the expense of an urban, woollen industry. But the fulling-mill could not save the industry from disaster so long as the bulk of fine English wool was going overseas and imports of foreign cloth were unrestricted. It needed government action, prompted by the exigencies of a great war, not only to arrest decline but also to make possible the unprecedented expansion of the second half of the century. Deliberate government action was probably much less important than the incidental results of Edward III's war taxation; but, whatever their underlying motives, the kings of the period did express awareness of the need to foster the native industry; and Edward III, in particular, took positive action designed to further this end. Despenser's Kenilworth ordinance had already foreshadowed a policy of sumptuary legislation coupled with relaxation of the assize of cloth and offers of franchises to cloth-workers.[2] This policy was reaffirmed in 1327; and four years later, Edward III granted letters of protection to John Kempe of Flanders, 'weaver of woollen cloth' and to the 'men, servants and apprentices' whom he had brought with him

[1] M. L. Armitt, 'Fullers and Freeholders of the Parish of Grasmere', *Trans. Cumberland and Westmoreland Ant. and Arch. Soc.*, N.S. viii (1908), 139. For a list of rural fulling-mills see Carus-Wilson, *Merchant Venturers*, pp. 195-7.

[2] Tout, *Place of Edward II*, p. 235.

to exercise his craft in England.[1] Similar letters of protection were granted to other foreign cloth-workers within the next few years; and when war with France became imminent, two further measures were taken. In August 1336 an embargo was laid on the export of English wool to Flanders; and in the following spring, letters of protection and sufficient franchises were offered in general terms to all foreign cloth-workers wishing to settle in England.[2]

Periodic embargoes on wool exports were purely diplomatic manœuvres, wholly inconsistent with Edward III's long-term policy of financing his wars by means of the taxation of wool and of loans raised on the security of the wool customs; and sumptuary legislation could never be satisfactorily enforced. We may be sure that the woollen industry owed little to either of these expedients; but its debt to the alien immigrants is more difficult to assess. Some scholars have seen in their advent the solution to the problem of a decaying industry.[3] Edward III, it has been suggested, succeeded where his predecessors had failed, because the Flemings were becoming increasingly discontented with conditions at home; and we are reminded of Thomas Fuller's tale of how the king sent secret emissaries into Flanders to underline the contrast between Flemish wretchedness and English prosperity, and to tell the Flemings 'how happy they should be if they would but come over into England, bringing their Mystery with them, which would provide their welcome in all places. Here they should feed on fatt Beef and Mutton till nothing but their fulnesse should stint their stomacks: yea they should feed on the labours of their own hands'; and of how, as a result of the arrival of a large number of Flemings, 'English Wool improved to the highest profit'.[4] That the king's offers attracted many foreign weavers to this country is not in doubt; but their numbers cannot be assessed, even approximately, nor is it possible to discover the nature or extent of their influence on the native industry. Flemish textile workers were arriving in York about the middle of the century and by 1360 they had established a colony in the city; but expansion of the industry was already well under weigh before they arrived and York textile workers as a whole included more from the single city of Lincoln than

[1] *Foedera*, ii. 823, 849.   [2] *Statutes* 11 Edw. III, c. 5.
[3] Salzman, *English Industries*, p. 203; Lipson, *Economic History* 10th edn. (1949), i. 453; Consitt, pp. 33–38.   [4] *Church-History of Britain* (1655), pp. 111–12.

from all the Low Countries; nor do the Flemings appear to have
contributed anything to the development of the cloth industry
in the West Riding during this period.[1] The local records do not
suggest any great influx of aliens to Norwich; and there is no
reason to suppose that the famous Thomas Blanket of Bristol
(who flourished about 1336) was a recent immigrant.[2] There is
little evidence of alien immigration elsewhere in the south-
western counties—Hampshire, Wiltshire, Dorset, Somerset, and
Gloucester—where the rural industry was largely concentrated;
and though a few Flemings probably found their way into different parts of the country[3] and London was always honeycombed
with aliens, it seems likely that the influx of foreigners generally
should be regarded as a symptom rather than a primary cause of
expansion and that their influence showed itself chiefly in the
introduction of new varieties of cloth and possibly of some technical improvements. Edward III consistently protected the
aliens against the hostility of the native clothworkers;[4] but it is
not altogether easy to credit him with perception of the long-
term possibilities of the situation, with awareness of the wealth
of raw material available at home, of the germs of decadence in
the Low Countries, of the great expansion of European commerce.[5] If he kept the export duty at a low figure this may well
have been, either because to raise it would have been inexpe-
dient politically, or because he was not fully alive to the prevailing trend. Edward may well have appreciated the benefits
of cheap and abundant home-made cloth; but the spectacular
expansion of the trade did not come till the next reign and it is
perhaps significant that he appears to have made no use of the
woollen industry as a propaganda weapon with which to fend
off parliamentary criticism of his staple policy.

[1] H. Heaton, *The Yorkshire Woollen and Worsted Industries* (1920), pp. 15–17. The
writer points out that there were Flemings in York in the thirteenth century and that
the mayor of the city who was killed at Myton in 1319 was named Nicholas le
Fleming. The le Flemings had long been people of standing in York. See *Vict.
County Hist. Yorkshire*, iii. 437.

[2] *Records of the City of Norwich*, ed. Hudson and Tingey (1906), ii. lxvi; Gray,
Eng. Hist. Rev. xxxix. 22–23.

[3] The surname Fleming appears in Lincolnshire, Devon, and Somerset in the
reign of Edward I. See Bardsley, *Dictionary of English and Welsh Surnames* (1901).

[4] e.g. his order of 12 Oct. 1344 to the mayor and sheriffs of London to make
proclamation against attacks on foreign cloth-workers. *Cal. Letter-Book F*, p. 111.

[5] As argued by H. E. de Sagher, 'L'immigration des tisserands flamands et
brabançons en Angleterre sous Edouard III', *Mélanges . . . Henri Pirenne* (1926),
pp. 109–26.

The unknown quantity in the history of late medieval cloth production is the domestic market. Until 1353 no duty was payable on cloths exposed for sale at home; after that date the royal aulnagers were made responsible for the collection of a small tax on such cloths, and some of their accounts have been preserved. But there is evidence that this tax was deliberately evaded;[1] and the fragmentary and unsatisfactory nature of the evidence has been authoritatively demonstrated.[2] In any event, the aulnagers can seldom have touched the large quantities of coarse woollen cloth which must have been produced in the villages from the wool of peasant flocks, spun in the cottages and made up into cloth by local weavers;[3] or that which was woven surreptitiously in the towns, in defiance of gild regulations.[4] No direct evidence remains of trends in this kind of production; and the basic question of whether the century saw an expansion of total cloth productivity therefore remains unanswerable. The Black Death must have given rise to labour problems in many villages; and it would seem that the decline in population must almost certainly have reduced the total demand for home-made cloth by the end of the century; for, though there may have been more rich individuals among the peasantry, we do not know that they spent much of their surplus wealth on clothes and bedding and it may be more likely that most of them spent it on food.[5] Reiterated complaints about shortage of pasture, the high rents payable for it, and the demands for restoration of commons in 1381, may point to expanding peasant flocks; but they tell us nothing of the amount of cloth being produced by peasants for peasants. We may, indeed, postulate a rising demand for cloth in some quarters. Under Edward III the needs of the armed forces were considerable; 2,000 pieces of cloth were bought for the navy alone in 1337. Soaring wages and the

[1] See the Bristol ordinance of Richard II's time, asserting that 'cloths, half cloths and remnants have been sold in inns, chambers and in other places in secret', so that the king had been defrauded of the petty custom. *Little Red Book of Bristol*, ii. 71.

[2] *Merchant Venturers*, pp. 279–91.

[3] On the manor of Minchinhampton it was customary to allow a weaver living by his craft and needing a dwelling, to make an enclosure, 'the weaver's assart', from the waste. See C. E. Watson, 'The Minchinhampton Custumal', *Trans. Bristol and Glos. Arch. Soc.* liv (1932), 266.

[4] e.g. those in force at Bristol forbidding night work and the placing of weavers' instruments in solars or cellars out of sight of the people. *Little Red Book*, i. 3–4.

[5] M. Mollat and others, 'L'Économie européenne aux deux derniers siècles du moyen âge', *Relazioni*, vi (1955), 860–2.

growing wealth and population of London and one or two other towns must have created flourishing markets there, particularly, perhaps, for the middle-grade cloths which were the characteristic products of English looms of the period.[1] And, though the standards of dress and furnishings among the royal family and the aristocracy may have been no higher at the end of the century than in the great days of demesne farming (when they were very high indeed), a far larger proportion of the cloth used by the very rich was of English manufacture, purchases for the royal family, for instance, being made mainly from native cloth-dealers in London and some provincial centres.[2] But whatever stratum of society we contemplate, we are still faced with the question whether the shrinkage in population must not have served to counteract any rise in individual demands. It is hard to resist the conclusion that, unless the demographers are wrong, there must have been overall contraction; for, if we are to postulate anything like a halving of the population by 1400, no increase of prosperity among the survivors can readily warrant assumption of an increase of total productivity.

By this date the woollen industry, as has been seen, had lost much of its urban character; and two other important classes of industrial workers—the miners and the masons—were likewise wholly, or partly, independent of the towns. The miner was in a peculiar position; for he enjoyed rights and privileges deriving in the main, not from specific grants, but from ancient usage; and all over Europe the world which he inhabited was a free world, by contrast with that of the peasant. Wage-labour was common enough and most of the miners were probably poor men; but, except in some of the king's mines where a system of impressment was in use, their tenure was generally free and in many parts of the country rights of mining were open to all comers. An inquiry held at Ashbourne in 1288, for example, confirmed the miners in their free tenure of the small mining properties which they worked;[3] and in the south-west, miners were exempt from tallages, tolls, and subsidies throughout the land. The men of the stannaries had their own law and their own courts and were free from pleas of villeinage and from royal

[1] In York alone, 3,256 cloths (excluding kerseys) were manufactured between Sept. 1394 and Sept. 1395. *York Memorandum Book*, ed. Maud Sellers (Surtees Soc. cxx, 1911), i. xxix.

[2] *Merchant Venturers*, pp. 242, n. 3, 243.   [3] *Vict. County Hist. Derby*, ii. 331.

pleas, except those touching life and limb; in all other pleas
they were answerable only to the warden. In Cornwall alone
about a thousand persons were entitled to rank as tinners, that
is, as free men with extensive rights of prospecting in the county.
The work was hard and unpleasant and where wage-labour
was in use the wages were no higher than on the farms;
but the hazards of a 'doubtfull and daungerous occupasyon'
were in some degree compensated for by a system of partial
self-government and by the growth of self-conscious political
units.[1]

Tin, lead, iron, and coal were the principal commodities
mined in late medieval England; and men of all classes, from
the king downwards, found investment in mining a useful means
of augmenting their incomes. When supreme control of all the
stannaries was vested in the duke of Cornwall in 1338, the annual
output of tin had reached 700 tons and the revenue taken by the
Crown from coinage dues was well over £2,000.[2] And, though
the Black Death dealt the industry a blow from which it was
slow to recover, one Abraham the tinner, in 1357, was said to be
employing 300 persons on his works. Lead-mining, at the opening of our period, was valued chiefly for the silver it produced
and was being carried on actively in the neighbourhood of Bere
Alston in Devon and to a lesser extent in the Mendips; but the
boom period in Devon was over by about 1340 and when, half
a century later, lead was needed for the repair of the great hall
at Westminster, the king's contractors obtained it from Derby,
Nottingham, and Yorkshire.[3] The lead-mines within the royal
forest of Mendip were leased to the bishop of Bath and Wells
who drew from them a revenue of one-tenth of the lead produced;[4] in Derbyshire many local men who combined the functions of merchant and smelter grew rich on the proceeds. The
Forest of Dean was still the most important iron-producing district, though London was beginning to draw its supplies from
the Weald. Almost all the English coalfields were already being
worked to some extent, though coal-mining was limited to

[1] Camb. Econ. Hist. ii. 441–56; L. F. Salzman, 'Mines and Stannaries', English
Government at Work, iii. 67–104; H. Finberg, Tavistock Abbey (1955), pp. 167–91.
[2] Tin had to be taken to an appointed centre and there weighed and stamped
before it might be sold; even in the least prosperous period, revenue from these
dues never fell below £1,000.
[3] Vict. County Hist. Derby, ii. 346.
[4] J. W. Gough, The Mines of Mendip (1930), pp. 50–52.

surface workings until about the middle of the century, when vertical pits with tunnels for draining first began to be sunk.[1] Thomas Hatfield, bishop of Durham, leased the mines on his manors of Whickham and Gateshead for the enormous rent of 500 marks, the lessees being allowed to draw some twenty tons a day; and the monks of Durham and Finchale also looked to the coal trade for augmentation of their revenues. Edward III licensed the men of Newcastle to dig for coal outside the walls and, in offering the miners of Gateshead his special protection in 1368, expressed the view that it would be valuable to have coal taken to all parts of the kingdom. Coal was still used mainly for smith's work; but with the increase of chimneys towards the end of the century it was beginning to be burned in private houses also.

The building industry was essentially capitalist in character, the principal employers being the monarchy and the Church; the masons who moved about the country undertaking work for contractors tended to form temporary associations based on the lodges, or workshops, where they were employed.[2] There were masons' gilds in London and some other large towns; and subordinate workers in the industry, such as carpenters, glaziers, and painters, could doubtless find enough work in the towns to justify similar associations. But, on the whole, the freemasons (so-called because they worked freestone, as distinct from rubble), the rubble-workers and the layers, setters, or wallers, who placed the stones in position, organized themselves in lodges, rather than in gilds.[3] The arrangement made at St. George's, Windsor, when the work on the chapel was finished in 1367, shows that some at least of these lodges must have occupied more than temporary hutments. Workers in the building industry were subject to impressment by the Crown and by other employers licensed by the Crown; but the attempts to conscript labour for Edward III's large-scale reconstructions at Windsor castle show that the system did not always work smoothly. In 1360 thirteen sheriffs were ordered to send 568 masons, but it appears that the total number employed in any one year never exceeded 300; and the tendency of conscripts to withdraw in pursuit of higher wages led the sheriff of York to

[1] R. L. Galloway, *Annals of Coal-Mining* (1898), p. 56.
[2] L. F. Salzman, *Building in England down to 1540* (1952), pp. 30–31.
[3] D. Knoop and G. P. Jones, *The Medieval Mason* (1933), pp. 3–4.

supply the men he impressed with red caps and liveries, 'lest they should escape from the custody of the conductor'.[1]

Though the miners and rural craftsmen generally were little affected by it, the tendency among other workers and townsmen to organize themselves in fraternities, misteries, and gilds is among the most remarkable social phenomena of the later fourteenth century. A government inquiry instituted in 1389 with a view to discovering their origin, usages, and the extent of their property, reveals substantial numbers of them in almost every English town.[2] The few who could plead the authorization of a royal charter were the exceptions and, chartered or not, all had originated as voluntary associations, serving many different purposes but not readily classifiable in distinct categories. Some, of the type commonly known as parish gilds, were purely religious in character; and the gilds formed by those engaged in a particular trade, craft, or occupation almost always undertook at the very least, the maintenance of an altar light and often professed other religious and charitable intentions. Some gilds were predominantly associations of traders, others of craftsmen; but no hard-and-fast line divided the one from the other in an age when the master-craftsman commonly bought his own raw materials and sold his products in his shop or stall. Again, merchant or mercantile associations might be limited to those, like the drapers and the vintners, who dealt only in a single line of goods, or they might include those, like the mercers and the grocers, who handled a variety of articles. In a great city like London each merchant group was rich and powerful enough to form a single company; but in the smaller towns there might well be no more than one or two federated societies (comprising, perhaps, grocers, haberdashers, ironmongers, apothecaries, and goldsmiths), successors to the gild merchant of an earlier age.[3] These late medieval associations of merchants tended to be wealthier and more influential than those formed by groups of craftsmen engaged in a single industry, but they were not regarded as essentially different in kind and thus they are more closely analogous to the craft gilds than to the merchant gilds

[1] Knoop and Jones, 'The Impressment of Masons for Windsor Castle, 1360–63', *Economic History*, iii (1937), 350–61.
[2] Some of the returns (made in English) to the inquiry were printed by T. and Lucy T. Smith for the Early Eng. Text Soc. *English Gilds* (1870).
[3] C. Gross, *The Gild Merchant* (1890), i. 127–9.

of the eleventh and twelfth centuries.[1] However composed, the gilds served many useful purposes. The system of apprenticeship ensured maintenance of high standards in trade and industry and offered young men (and a few young women) both a technical training and an education in citizenship; the gilds regulated holidays and hours of work and, to some extent, wages and prices; their plays and pageants afforded opportunities for recreative, even for creative, activities; they assumed responsibility for their members in poverty, sickness, and old age, helped to bury them, remembered them in their prayers, and cared for their widows and orphans. Many gilds were benefactors of the Church, education, and the arts; the gilds of St. Mary and Corpus Christi in Cambridge undertook the foundation of a college. The defects of the system (though these were less apparent in the fourteenth than in later centuries) were to be seen in a tendency to damp down individual enterprise, to discourage new entrants, to maintain a rigid monopoly, not only against aliens but also against 'foreigners' from the countryside or from neighbouring towns (which often had the effect of forcing up prices), and, through their mutual rivalries, to foment urban unrest. The government inquiry of 1389 was itself the outcome of a nervous dread of secret fraternities. Yet whatever their failings, the multiplication of gilds in the second half of the century shows that here, as elsewhere in western Europe, they must have offered some kind of satisfaction to the economic, social, and spiritual aspirations of the medieval bourgeoisie.

The history of the London gilds affords by far the most impressive English example of the development of this form of association in the fourteenth century. London, as is well known, never had a merchant gild of the type that was common in many English towns from the eleventh to the thirteenth century; and a few of the London craft gilds were already old when Edward II became king; the weavers and the bakers, for example, had a long history. But the great days of the weavers lay in the past and the high claims made for their court were successfully challenged by the king's judges during the *Iter* of 1321.[2] The most influential gilds of the new century tended to be associations

---

[1] At Beverley, for example, the merchant, or mercers' gild was the largest and took precedence of the other gilds; but its members were reckoned among the craftsmen and enjoyed no special superiority. A. F. Leach, *Beverley Town Documents* (Selden Soc., 1900), p. xliii.

[2] G. Unwin, *The Gilds and Companies of London* (1908), p. 44.

of traders who were primarily wholesalers, reaching far outside the city alike for their commodities and their customers.
Prominent among them were the fishmongers, purveyors of a
basic necessity of medieval diet, who had their estates on the
Thames and the Lea and rode out in companies to bargain for
fish in the ports of East Anglia. Individual fishmongers might
range from great merchant princes, like Hamo de Chigwell,
mayor of the city in the latter years of Edward II and successful
defender of the privileges of the fishmongers' court, or William
Walworth, whose services in the rising of 1381 won him a
knighthood, down to poor street-hawkers; but the fishmonger
came to be regarded as the type of wealthy capitalist, a principal object of popular hostility and of the jealousy of the smaller
crafts. Rivalling the fishmongers in wealth and influence were
the grocers, whose company came into existence in 1345 as an
amalgamation of pepperers, canvas dealers, and spicers and
against whom the statute of 1363, enacting that each merchant
confine himself to one type of wares and each craftsman to one
mistery, was principally directed.[1] The other influential gilds
of the fourteenth century, forerunners of the greater London
Companies, were the mercers, the drapers, the vintners, the
goldsmiths, the skinners, and the tailors.[2] Within these greater
gilds there tended to be increasing differentiation between the
merchants, or liverymen, who directed policy, and the ordinary
traders or craftsmen, the journeymen, who came to be known
as the yeomanry;[3] while altogether outside the ranks of the
great companies were the shifting groups of many lesser crafts
and trades. Some of these, like the fusters who made saddle-
trees, were piece-workers; some, like the hatters and the pouch-
makers, were primarily dealers, selling articles made locally;
some, like the woolmongers and the leather-sellers, were small
traders in raw materials; some, like the bakers, the brewers, and
many metal-workers, were engaged in simple forms of manufacture, selling their products direct to the consumer. The tendency of the greater gilds to break up into separate fraternities
and of the lesser ones to coalesce, makes calculation of numbers
difficult; the picture is of 'an ever-moving kaleidoscope'; but

---

[1] *Statutes* 37 Edw. III, cc. 5, 6. The act was repealed, so far as it affected the
grocers, in the following year. *Cal. Letter-Book G*, p. 179.

[2] The ironmongers, salters, and haberdashers were added in the next century.

[3] Sylvia L. Thrupp, *The Merchant Class of Medieval London* (1948), pp. 29 ff.

there were at least fifty gilds of all kinds in London at the end of Edward III's reign and more than double that number half a century later.[1]

Such ebullient social energy was not easily containable within the rigidly oligarchic framework of the city constitution. The most important underlying causes of tension were, on the one hand, the desire of the *potentiores* to retain control, not only of the trade, but also of the industry of the city; and, on the other, the reluctance of the less well-to-do to leave all direction of policy and the choice of the mayor, aldermen, and other officers to a closed oligarchy. At the opening of the fourteenth century the ancient rights of the citizen community—the right of the men of the wards to elect the aldermen, the right of the community to elect the mayor and sheriffs and to approve legislation—had long been tending to atrophy for want of precise definition; and effective power now rested with the body of twenty-four aldermen who assisted the mayor. Rapid multiplication, particularly of the smaller gilds, opened up a new means of bringing pressure to bear on the governing clique; and the civic storms which blew up at intervals during the century, owed much to the activities, both overt and underground, of the less wealthy gildsmen, craftsmen, and traders alike. The key position held by the capital politically meant that periods of national unrest tended to promote subversive movements within the city; the reign of Edward II and the troubled years of the seventies and eighties well illustrate this interaction. Thus, it was in the presence of the Ordainers that, in August 1312, a meeting of twelve citizens from each ward proceeded to lay down the principle of the responsibility of the aldermen to the men of their wards. Seven months later 'the good men of the commonalty' declared that the mayor and aldermen were responsible collectively to the citizens at large; and later in the same year (1313) there was an attempt to set up a council of citizens representative, not of the wards, but of the crafts. This method failed to establish itself; but in 1319 the citizens secured a royal charter which might have been a landmark in the constitutional development of London. Under it, the aldermen were to share custody of the common seal with three elected commoners; aldermen were to be elected annually and to be ineligible for

---

[1] *Cal. Plea and Mem. Rolls of the City of London, 1364–81*, ed. A. H. Thomas, p. xxxvi.

immediate re-election; and membership of a mistery became, for most 'foreigners', the only road to the freedom of the city.[1] But it proved impossible in practice to break the monopoly of power enjoyed by the aldermen who had the support of the greater gilds; and indeed, few employers, great or small, were favourable to the organization of wage-earners and other dependent workers or wished to see them active in city politics.[2]

Under Edward III less was heard of civic radicalism. Not until 1376, a year when political passions were running high, when a number of prominent Londoners had been convicted of fraud in parliament and the city was seething with unrest, do we meet with another serious attempt at constitutional reform. Under the influence of John of Northampton, a draper of moderate fortune, who had built up a party from among the lesser crafts and minor members of the greater, it was agreed to combine administrative and electoral powers in a single common council, elected from the misteries and bound to meet with the mayor and aldermen at least eight times a year, an attempt also being made to restore the rule forbidding re-election of aldermen. Such temporary success as these reform measures achieved owed much to Northampton's anti-monopolistic policy, above all to his attack on the fishmongers and drive for cheap fish, which commanded sympathy in many quarters, both inside and outside the city; and thoroughgoing reaction was staved off until the end of his second year of office as mayor, in 1383. But even before this date much of the scheme of constitutional reform had been whittled away. Re-election of aldermen after a year was conceded; membership of their court was not modified significantly; elections to the common council tended to revert to the wards. When the grocer, Nicholas Brembre, succeeded Northampton in the mayoralty, the minimum of eight meetings for the common council was reduced to four and the right of election was formally restored to the wards; by 1394 the aldermen had been made irremovable during good behaviour. Their controlling influence had been completely restored and the stage set for the oligarchic régime of the ensuing century. Northampton's extravagance and violence and the widespread fear of

[1] Text in *Liber Custumarum*, pp. 268–73; summary in *British Borough Charters, 1307–1660*, ed. M. Weinbaum (1943), pp. 74–75.

[2] e.g. in 1396 a religious fraternity of serving-men and saddlers was suppressed by the City at the request of the master-saddlers who feared it might be used to force up wages. Smith, *English Gilds*, pp. cxlvi–cxlvii.

social revolution engendered by the Rising of 1381 alarmed those with vested interests in the larger companies; and Northampton's association with John of Gaunt aroused much popular hostility. After his fall, all that remained of the radical movements of the century was the common council which London alone among English towns had succeeded in establishing as a permanent feature of the constitution—a body ineffective in practice for years to come but, at least in principle, a second council capable of acting as a check on the mayor and aldermen.[1]

Restoration of the aldermen's controlling influence meant that direction of the activities of the gilds rested ultimately with the civic authorities. Gild ordinances were commonly registered and rules relating to apprenticeship enforced by the city—a necessary precaution, since it was from the ranks of the apprentices that future freemen were likely to be drawn. It was the civic authority which, in 1308, recognized the election of a barbers' supervisor, which, in 1350, granted a petition from the shearmen for the election of officers to rule their mistery, which, in 1387, forbade the enrolment of foreigners as apprentices. In 1367 the city fixed wages for the entire building trade; and poor goods exposed for sale were commonly brought before the mayor and aldermen. From time to time the members of the crafts had also to reckon with the Crown, sometimes as a source of privileges which it was beyond the power of the city to bestow, sometimes as a hampering or restraining force. It was by virtue of a royal charter that London girdlers and saddlers gained the right to enforce their regulations throughout the realm; but it was also the Crown which attempted to hold every craftsman to one mistery and, so late as 1390, was ordering the London tanners and shoemakers to keep apart from one another.[2] Time was to show the impossibility of preventing such amalgamations: but the attempt to prevent them affords interesting evidence of the nervousness engendered in the mind of authority by the widespread contemporary movement towards voluntary association.

The capital was precocious in more senses than one. Its population of perhaps 35,000 outnumbered those of York, Plymouth,

[1] J. Tait, *The Medieval English Borough* (1936), pp. 303–16; Ruth Bird, *The Turbulent London of Richard II* (1948), pp. 30–43; Thrupp, pp. 53–80.
[2] Stella Kramer, *The English Craft Gilds* (1927), pp. 7, 52; E. F. Meyer, 'English Craft Gilds and Borough Governments of the later Middle Ages', *University of Colorado Studies*, xvi (1927–9), 323–62.

Bristol, and Coventry combined.[1] It had commercial dealings
with almost all parts of the continent and was capturing an
increasing share of the cloth trade, about one-third of all the
woollens exported being shipped from its wharves by the end of
the century, when most of its greater companies had secured
royal charters. During the law terms, and when parliament and
convocation were in session, its population was swollen by students, retainers, and clerks, as well as by magnates, ecclesiastical
and lay, an increasing number of whom maintained permanent
town houses. Though most of the government offices were established in Westminster, the mint and the armouries were at the
Tower and by 1361 the great wardrobe had found a permanent
home near Baynard castle, on the east bank of the Fleet river.
Already predominant economically, London under Edward III
and Richard II was rapidly becoming a true capital, the social
and literary, as well as the political and administrative centre of
England and the only English town in any way comparable
with the great cities of northern Europe.[2] Yet, for all its pride,
its magnificence, and its bustling activity, Chaucer's London
still retained some of the characteristics of a country town; when
the bells of Bow church (St. Mary de Arcubus or St. Mary Arches)
sounded the curfew, it was to warn not only city workers, but
also those benighted in the fields.[3] It seems likely that only the
proximity of the open countryside can have saved from deadly
disease the closely packed inhabitants of a town where open
channels for liquid refuse ran through the streets into the streams
and ditches, where pigs rooted in the garbage, where butchers
slaughtered their beasts in Fleet Street, and lepers were often
found wandering at large. Yet conditions may have been no
worse, perhaps even better, than in some later centuries;[4] and it
is certain that neither dread of disease, nor the turbulence of the

[1] The population of these towns in 1377 is given by J. C. Russell as: London,
34,971; York, 10,872; Bristol, 9,518; Plymouth, 7,256; Coventry, 7,226. No other
English town had a population of more than 6,000 (*British Medieval Population*,
pp. 142–3).
[2] T. F. Tout, *The Beginnings of a Modern Capital* (Raleigh Lecture, 1923).
[3] 'And forasmuch as in the city of London . . . a certain bell called Bowbell . . . is
solemnly rung whereby all working in the city and benighted in the fields may be
able to betake themselves more quickly to the said city for shelter. . . .' *Records of
the Borough of Northampton*, ed. C. A. Markham and J. C. Cox (1898), i. 252–3.
[4] See E. L. Sabine, 'Butchering in Medieval London', *Speculum*, viii (1933),
335–53; 'Latrines and Cesspools of Medieval London', ibid. ix (1934), 303–21;
'City Cleaning in Medieval London', ibid. xii (1937), 19–43.

streets, nor the citizens' notorious suspicion of strangers, served
to arrest the steady flow of countrymen and smaller townsmen
to London which, alone among English cities, recruited her
population from every corner of the land.[1]

The records of some other towns tell a very different tale. It
has been postulated that after 1377 the urban populations, outside London, ceased to expand;[2] and though this hypothesis
may need some modification, it is undoubtedly true that not all
the towns were prospering. Lincoln affords a melancholy example of a once noble city which in this century had fallen on
evil days. She suffered badly from the Black Death, and the
transfer of the wool staple to Boston in 1369 was significant of
her general decline. Four years earlier Edward III had harshly
informed the citizens that because of the uncleanliness of their
streets, 'the evil name of them and their city grows worse and
worse';[3] they were said to be refusing to contribute to the cost
of the new gildhall; and a former mayor and parliamentary
representative was named among the insurgents of 1381.[4] At the
end of the century both Lincoln and Yarmouth complained that
their inhabitants were departing because of inability to pay the
farm.[5] In his charter to Newcastle (1357) Edward III referred
to the great impoverishment of the town;[6] after a period of
prosperity in the seventies, Leicester began to decay;[7] and the
ancient confederacy of the Cinque Ports (Winchelsea alone
excepted) was already in decline before the outbreak of the
Hundred Years War.[8] Yet the evidence as a whole hardly warrants us in projecting back into the fourteenth century the
general decay of the old corporate boroughs which had become
evident by the end of the Middle Ages. Apart from normal fluctuations of trade and industry and the incidence of plague, many
towns seem to have maintained themselves in a state of reasonable, and a few of mounting, prosperity. York, the second city
in the kingdom and a shire incorporate by 1398, with an un-
rivalled position on a great tidal river, was well established as

[1] London surnames afford evidence of migration from all parts of the country,
even from the northern border shires. See Thrupp (App. C).
[2] J. C. Russell, pp. 282–3.   [3] *Cal. Pat. Rolls 1364–67*, p. 89.
[4] Hill, *Medieval Lincoln*, pp. 254–8.   [5] *Rot. Parl.* iii. 438.
[6] *Cal. Charter Rolls*, v. 154.
[7] *Records of the Borough of Leicester*, ii. 29.
[8] J. A. Williamson, 'Historical Geography of the Cinque Ports', *History*, xi
(1926), 97.

a large commercial and industrial centre. Bristol, a shire by 1373, had long since outgrown its original city walls and even before the end of Edward III's reign, it was said that all the corn produced within ten leagues did not suffice to feed the population of a great port which was fast becoming also a notable centre of industry.[1] The fortunes of Southampton fluctuated and its very small population of 1,728 indicates a period of slump; but the rise of the Cotswold flocks had shifted the main source of the Italian wool supply from east to west and, although the Venetian carracks continued to prefer London, Southampton gradually established itself as the principal centre of Genoese shipping and as a market for the sale of the alum, wood, and fruits absorbed by the towns and villages of the south-western shires.[2] Plymouth, in 1377 the fourth largest borough in England, derived its prosperity from fish, tin, and a general carrying trade. Extensive building projects at Norwich under Richard II, the existence of no fewer than thirty-eight gilds at Beverley in 1390, the recovery of Ipswich in the nineties after a long period of depression, all point to a fair degree of prosperity. Moreover, the century saw the rise of more than one town of hitherto limited importance. Edward I's borough of Kingston-on-Hull by the nineties was sharing with London the greater part of the northern cloth trade. Coventry, already by 1377 the fifth town in the country and nearly as large as Plymouth, had gained independence of its joint overlords, the prior and Queen Isabella, obtained a charter in 1345 and a mayor three years later, and was thus free to take full advantage of its nodal position in the road system and of its own rapidly expanding drapery business to enlist the royal family among its customers and to raise the most splendid perpendicular tower in England.[3] The episcopal cities of Salisbury and Winchester by the middle of the fourteenth century were among the foremost industrial communities in the country.

Social and constitutional developments in the smaller towns followed broadly the same lines as in London—multiplication of gilds and sporadic efforts, almost always unsuccessful, to throw off or loosen the political stranglehold of executive officers

---

[1] *Cal. Close Rolls 1374–77*, p. 324; Carus-Wilson, *Merchant Venturers*, p. 4. (The figure of 10 leagues is not to be taken literally.)

[2] Alwyn Ruddock, *Italian Merchants and Shipping in Southampton 1270–1600* (1951), pp. 41–55.

[3] M. Dormer-Harris, *The Story of Coventry* (1911), pp. 57–72.

or of a narrow oligarchy of the wealthier citizens, matched by the determination of the dominant class to retain control of urban policy, above all in the economic sphere. Relations between the borough authorities and the gilds at this period are often hard to trace, for the records of many craft gilds are scanty and incomplete; but it is sufficiently clear that in most towns a governing body composed of the wealthier citizens, the *potentiores*, exercised fairly close supervision over the activities of the crafts. A successful compromise seems to have been achieved at York, where the mayor and council of twenty-four (sometimes enlarged to forty-eight) made themselves responsible for general supervision of trade, while leaving the powerful craft gilds a fairly free hand in the management of their own affairs.[1] At Norwich, on the other hand, the civic authorities found control of the gilds difficult, perhaps because ancient popular traditions had lingered longer here than elsewhere. Norwich had no mayor until the fifteenth century and executive authority rested with the four bailiffs assisted by an elected council of twenty-four, on whom, because 'many of the commune of the town have of late been very contrarious', power to make ordinances for the good government of the city was conferred by royal charter in 1380.[2] But the 'Forty-Eight' (reduced to forty by the charter of 1373) who governed Bristol after 1344, kept a tight rein on local trades and industries. Gild regulations had to have the sanction of the borough authorities and the ordinances of the weavers, cobblers, and others were enrolled among the municipal records; when, in 1395, the barbers needed new ordinances, these were drafted by the municipal authorities, with the consent of twelve barbers, 'assembled for the government of their craft'. The borough also registered indentures of apprenticeship and often regulated prices and hours of sale.[3] Leicester, as a seignorial borough in the hands of the earls and dukes of Lancaster, was less 'free' than the majority of royal boroughs; it was not until 1375 that John of Gaunt leased the town to the mayor, bailiffs, and burgesses. But even here it is possible to trace back into the previous century a closed council of a mayor and twenty-four jurats whose authority over the gilds was rarely contested.[4] At

---

[1] *York Memorandum Book*, pp. xxxiv–xxxv.
[2] *Cal. Charter Rolls*, v. 264; *Records of the City of Norwich*, I. xxv–lv.
[3] *Little Red Book*, pp. 3–4, 42, 69–71.
[4] *Records of the Borough of Leicester*, I. ix–x, II. xlv.

Beverley, by the second half of the century, effective power had become concentrated in the hands of twelve keepers, or governors, an oligarchy of the ruling class, which managed the common property of the town and levied its taxes.[1] Generalization is hazardous, for no two towns have the same history; but certain factors undoubtedly helped to ensure the easy triumph of oligarchy in the late medieval borough. Concentration of economic power in the hands of a few rich gilds, or rich individuals, is the most obvious; this power was reinforced by the acute class-consciousness of medieval burgesses, by their deep-seated respect for wealth, their apparent readiness to accept designation as *mediocres* or *inferiores*, over whom the *potentiores* enjoyed a natural ascendancy. Moreover, fragmentation of a small population into numerous gilds did not make for concerted political action; the fact that many towns, from London downwards, found it necessary to impose fines for non-attendance at such civic assemblies as did exist, points unmistakably to political apathy. Apathy may, indeed, lie at the root of the matter. For it is unlikely, on the face of it, that many of the inhabitants of boroughs, which by modern standards were merely large villages, were greatly concerned with constitutional issues, as such; and it may be of some significance that it is at Lincoln, a town in a state of commercial decline, that we find the clearest evidence, at the end of the century, of bitter quarrels over the election of the mayor and bailiffs and the respective rights of the commonalty and the *potentiores*.[2] More fortunate citizens and burgesses elsewhere were content to let such questions well alone.

[1] *Beverley Town Documents*, p. xxxviii.
[2] Hill, *Medieval Lincoln*, pp. 259–61. (This particular quarrel was composed by Gaunt with the aid of Bussy.)

XIII

THE GOOD PARLIAMENT AND THE
PEASANTS' REVOLT (1371-81)

THE year 1371 saw the opening of a disastrous decade. As Edward III grew older public confidence in his government began to be shaken; and the succession of a minor was unlikely to restore the stability of the throne. General uneasiness, rising at times to panic, manifested itself in three open assaults on the executive. The first and least sensational of these attacks resulted in the substitution, in 1371, of lay for clerical ministers of state; the second, in 1376, in the impeachment of the king's chamberlain and a number of lesser officials; the third, in 1381, in the murder of the chancellor and treasurer and the indiscriminate massacre of certain officers of the law and minor civil servants. Much relating to these outbreaks still remains obscure; but it is not difficult to perceive the main causes of the general decline in public morale characteristic of the period. Edward himself was past the days of his greatness and though, so late as 1372, he again attempted to take the field in person,[1] his health and intellect were failing and he was becoming increasingly subject to the influence of a group of unpopular courtiers and of his mistress, Alice Perrers. The Black Prince could still command respect, but he was already a sick man and was little fitted, either by training or temperament, to assume the role of saviour of society. Lionel of Clarence had died in 1368 leaving as his sole heir his daughter Philippa, wife of Edmund, earl of March, a young man who did not attain his majority till 1373. John of Gaunt, increasingly preoccupied with the claim to the Castilian throne acquired through his second marriage, was much abroad and lacked the prestige of the successful soldier which Edward III's subjects prized above all else. Edmund of Langley, earl of Cambridge, was a man of no ability and Edward's youngest son, Thomas of Woodstock, was not yet of

---

[1] After the loss of Pembroke's fleet off La Rochelle Edward embarked in the *Grace Dieu* at Sandwich and appointed the five-year-old Richard of Bordeaux as keeper of the realm. He remained aboard for over six weeks but in Oct. contrary winds forced him to abandon the projected expedition. *Hist. Ang.* i. 315.

POLITICAL UNREST 385

an age to make his influence felt. Little was now to be expected
from the royal family and discontent with the administration
ran riot. In the parliament of 1371 the lords drove the bishop of
Winchester from the chancery and the bishop of Exeter from
the exchequer; but with a judge (Sir Robert Thorp) as chancellor and a lay baron (Sir Richard Scrope) as treasurer, the burden of taxation was not alleviated and the war prospered no
better.[1] The motive for summoning the great council to Winchester in that year was said to be fear of revolt among the merchants of London, Norwich, and other towns.[2] The parliament
of 1372 gave expression to its mistrust of the expert by enacting
an ordinance excluding lawyers and sheriffs. In the parliament
of 1373 the commons were restless and critical, refusing to grant
a subsidy until they had been allowed consultation with a committee of the peers. The king's chamberlain, Lord Latimer, was
thought to be defrauding him, customs officers were said to be
exceeding their powers, disreputable merchants to be enriching
themselves at the expense of the nation. Even nature seemed
hostile; for the failure of the harvest of 1369 had forced wheat
prices up to famine level in the following year; the *calores nimii*
of 1374–5 were accompanied by renewed visitations of the
plague; and foreign trade was slumping. An overtaxed and
leaderless people, at once war-weary and bellicose, was becoming ripe for revolt.

Failure abroad and divergence of aims among the generals
served to intensify domestic discord. Gaunt's abortive enterprise of 1373, following hard upon the loss of Pembroke's fleet,
converted him to the desirability of coming to terms with
Charles V; a suspension of hostilities would afford him opportunity to restore and reorganize his depleted resources and to
review the position in Spain. But the negotiations were much
protracted; and by midsummer 1375, when a truce was at last
achieved, Latimer, with Gaunt's approval, had already arranged to abandon the Breton fortress of Bécherel, committed
to him some years earlier; while the Norman stronghold of St.
Sauveur, of which Latimer was also captain, was handed over as

[1] Thorp, who was chief justice of the common bench, was succeeded on his
death in 1372 by the chief justice of the king's bench, Sir John Knyvet. Scrope, a
retainer of Gaunt, had married a daughter of William de la Pole. Tout describes
him (*Chapters*, iii. 277) as 'a strong representative of the educated new nobility that
was now wresting from the clergy their monopoly of high office'.
[2] *Hist. Ang.* i. 313.

a consequence of the truce. This truce of Bruges was widely unpopular in England, partly because of the display and extravagance which accompanied the negotiations, partly because
Charles V's reason for agreeing to it was rightly believed to be
reculer pour mieux sauter.[1] Moreover, its conclusion coincided with
a promising assault on Quimperlé, led by March, Despenser,
and Cambridge, acting in conjunction with John de Montfort.
None of these magnates seems to have been cognizant of what
was afoot at Bruges; and news of the truce which cheated them
of the prestige and profits of victory was received with anger and
dismay. Already an unsuccessful soldier, Gaunt was now regarded in many quarters as a short-sighted and irresponsible
diplomat.[2]

Meanwhile, their conduct of Irish affairs was doing nothing
to restore the credit of the government. It was in the hope that
he might recover the revenues due to the English Crown that
William of Windsor, husband of the king's mistress, was sent
to Ireland as lieutenant in 1369.[3] As an English official of no
social consequence, Windsor was unlikely to commend himself
to the Anglo-Irish magnates or to the 'obedient shires' of the
Pale; and odium was inseparable from the policy of heavy and
sustained taxation to which he was committed. Complaints
against his administration soon began to filter through to England and by 1371 these had become so vociferous that Edward III was forced to recall him and to substitute a former
chancellor of Ireland, Sir Robert Ashton, who was instructed
to stay the levy of unlawful taxes, to bring the whole matter of
taxation before the Irish parliament and to look into Windsor's
accounts. Within a few months of Ashton's arrival, and doubtless at his instigation, an embassy from the Irish parliament
appeared before the king and his council to explain the perils
with which Ireland was threatened and to demand the help of
the earl of March. March, to whom had descended Clarence's

---

[1] So Walsingham (*Hist. Ang.* i. 318), who refers also to 'horribiles expensas et in-
credibiles'. The York chronicler writes of 'graunt despens et graunt riot . . . pur
reveler et dauncer' and puts the expenditure at over £20,000 (*Anon. Chron.*, p. 79).

[2] *Hist. Ang.* i. 319. See C. C. Bayley, 'The Campaign of 1375 and the Good Parliament', *Eng. Hist. Rev.* lv (1940), 370–83. I am indebted to Dr. G. A. Holmes for
allowing me to make use of his unpublished dissertation, 'The Nobility under
Edward III' (Cambridge, 1952), which includes an excellent analysis of the events
leading to the Good Parliament.

[3] For a detailed study of Windsor's administration in Ireland see Maude Clarke,
Fourteenth Century Studies, pp. 146–241.

earldoms of Ulster and Connaught as well as the extensive possessions of his own family in the south-east, was by far the richest
and most powerful of the absentee Anglo-Irish lords and Edward III promised to send him to Ireland as soon as convenient.
Preparations for his departure were taken in hand and then,
somewhat inexplicably, abandoned.[1] March himself may well
have preferred the prospect of campaigning in Brittany to that
of attempting to reconcile the Irish; and, by 1374, Edward felt
strong enough to reappoint Windsor and to order the collection
of the disputed subsidies. Windsor was maintained in Dublin by
the favour of the court for another two years; but charges of
corruption multiplied and it seems reasonable to suppose that
the 'great roll of indictments' presented to the English council
about the time when the Good Parliament assembled, had the
support both of March and of Ashton, who had lately been
appointed treasurer of England.

The danger of summoning parliament in an atmosphere of
mounting discontent was obvious to the king's advisers and none
was held in 1374 or 1375. But the inevitable accumulation of
judicial business and the financial necessities of the government
made further postponement impossible and at the end of April
1376 the estates assembled at Westminster.[2] The king was
present at the opening ceremony in the Painted Chamber of the
palace but he soon withdrew, leaving to John of Gaunt the
thankless task of presiding over the assembly. In his opening
address the chancellor, Sir John Knyvet, declared the cause of
summons to be the peace and defence of the realm and the
maintenance of the war and urged lords and commons alike to
consult diligently on these matters. The next day (30 April) the
lords assembled in the White Chamber of the palace, the commons in the chapter-house of the abbey. All the available evidence points to a state of unusual excitement and tension among

[1] For references to March's intended departure for Ireland see *Cal. Pat. Rolls
1370–74*, pp. 337, 353, 373 (Sept. to Dec. 1373). On 16 Apr. 1374 March 'staying
in England' (p. 426) nominated attorneys in Ireland.

[2] The chronology and history of the Good Parliament may be reconstructed
from the official roll (*Rot. Parl.* ii. 321–60), supplemented by *Chronicon Angliae*,
pp. 68–101, and by the invaluable evidence of the *Anonimalle Chronicle* of St. Mary's,
York. Secondary accounts will be found in Stubbs, *Constitutional Hist.* ii. 448–55;
Tout, *Chapters*, iii. 292–307; Steel, *Richard II*, pp. 23–31; Plucknett, 'The Impeach-
ments of 1376', *Trans. Royal Hist. Soc.*, 5th ser. i (1951), 153–64; and Wilkinson,
Constitutional Hist. ii. 204–26. For the chronology, see especially, J. G. Edwards,
The Commons in Medieval English Parliaments (1958), pp. 36–38.

the latter. One hundred and forty-six *communes petitions* were drafted, ranging over almost the whole field of administration and public law. There were complaints against the Church, against foreign merchants and brokers, and the jurisdiction of the household courts. A number of the petitions dealt with abuses in local government and with the disorderliness and malice of labourers and beggars; others demanded the reform of naval administration and the better protection of the Welsh and Scottish borders; there were requests for annual parliaments and for the election of knights and sheriffs 'by the common choice of the better people of the shire'. For the background to these complaints we have to turn to the debates in the chapter-house (happily preserved in the *Anonimalle Chronicle*) which reveal a high pitch of indignation at past grievances and the prospect of future taxation, tempered, however, by common sense, awareness of mutual responsibility, and the need for orderly procedure. The commons began by imposing an oath of mutual loyalty and secrecy. Speakers then moved in turn to the lectern in the middle of the room where they might be heard by all. The first of these, *une chivaler del south pais*, went straight to the point by declaring that the commons were too poor and feeble to endure further taxation; what had been granted in the past had been wasted or embezzled. Instead of discussing a grant of supplies, let them consider how the king might govern the realm and maintain the war out of his own resources. The next speaker suggested that restoration of the staple to Calais would relieve the Crown of an annual charge of £8,000 for its defence. A third speaker emphasized the futility of isolated action by the commons who could not expect to carry out any policy with honour and profit unless they could enlist the support of the lords; he proposed that the king and his *sage conseil de le parlement* should be asked to assign certain prelates and secular peers, named by the commons, to advise them and to hear their debates. This proposal was adopted unanimously and after some further discussion, Sir Peter de la Mare, one of the knights representing the county of Hereford, summed up the debate by recapitulating what had been said. Because he spoke well and wisely and told his audience much that they had not known hitherto, he was chosen 'davoir la sovereinte de pronuncier lour voluntes en le graunt parlement', in other words, to act as their Speaker. The debate had continued for over a week when a message came from the king asking for a speedy answer to his request for supplies.

No doubt rumours of the debate had already begun to percolate through the walls of the chapter-house. Gaunt is said to have been *a tresgraunt male ease* when he asked for the name of the commons' spokesman. Many of their number had been excluded from the parliament chamber and, though this seems to have been customary, de la Mare chose to interpret it as an attempt to destroy their solidarity and refused to speak until all his companions had been admitted. He then made his statement, praying for the aid of twelve peers to assist the commons in amending the grievous defects of the realm. When the lords agreed that the request was reasonable, he named four bishops, four earls, and four barons, presumably those who were expected to be most sympathetic to the commons' complaints. They included the earls of March, Warwick, Stafford, and Suffolk, Henry Percy, and two aristocratic young bishops, Courtenay of London and Despenser of Norwich.[1] The matter was referred to the king who at once gave his consent. The appointment of a similar committee in 1373 had produced a grant of supplies and doubtless there were other reasons why it was thought politic to comply. After consultation with the lords' committee, the commons again appeared in the White Chamber on 12 May, when the Speaker reported that in their opinion the burden of taxation was the fault of certain councillors and royal servants who were enriching themselves at the king's expense. In reply to Gaunt's apparently surprised question, 'Who have made such profits and in what way?' de la Mare named Lord Latimer, the king's chamberlain, and a London merchant named Richard Lyons. Witnesses were called, among them two former treasurers, Brantingham and Scrope, and at the end of the day, the Speaker amid loud cheers demanded the arrest of Lyons. What

[1] *Rot. Parl.* ii. 322, gives the names as the bishops of London, Norwich, Carlisle, and St. Davids, the earls of March, Warwick, Stafford, and Suffolk, Lord Percy, Sir Guy Brian, Sir Henry Scrope, and Sir Richard Stafford. *Anon. Chron.* (p. 84) substitutes the bishop of Bath for St. Davids and Roger Beauchamp for Scrope. In Tout's view (*Chapters*, iii. 295, n. 2) 'the official record is much to be preferred'; but the association of both Houghton of St. David's and Scrope with Lancaster raises doubts. Harwell of Bath and Wells had been the Black Prince's chancellor in Aquitaine and Richard Stafford was a member of his household. The four earls were all great landowners. Despenser of Norwich had a personal grievance, for he complained that certain *Privez entour le Roi* had ousted him from an advowson in his diocese. *Rot. Parl.* ii. 330.

happened in the interval is obscure, but on 24 May the lords
sent for the commons who now demanded the removal of Alice
Perrers from court and the reform of the council.[1] These demands were transmitted by the lords to the king who agreed to
them; and new councillors were named and sworn in parliament, the first known example of such a procedure. Among the
nine so named were several members of the liaison committee,
a piece of planning which, as Professor Plucknett suggests, may
well have been inspired in the March circle.[2] De la Mare then
addressed himself once more to the cases of Latimer and Lyons,
asking for a judgement on the latter and for the arrest of the
former until he had made satisfaction. He seemed to assume that
their guilt was proved and that no further proceedings were
necessary.

Lord Latimer, however, was not to be disposed of so easily.
He was no mere court jackal but a man of noble rank and
mature years, with military, administrative, and diplomatic experience behind him.[3] It was he who challenged de la Mare's
assumptions, demanding, first, a formal trial and later raising
the crucial question of the authority by which he was arraigned.
Despite a heated protest by the bishop of Winchester, a trial of
some sort seems to have been allowed; and to Latimer's demand
for an accuser, de la Mare made the historic answer that he and
his fellows would maintain all their charges in common—thereby stumbling, almost by accident, on what was to develop into
the famous procedure of impeachment.[4] The trial was long and

---

[1] Alice Perrers was said to have drawn £2,000 or £3,000 a year from the royal
coffers, to have used maintenance and bribery in the law-courts, and to be bringing
dishonour and ill-fortune on the king and the realm.

[2] Courtenay, March, Stafford, Percy, Guy Brian, and probably Roger Beauchamp were members of both bodies. The other new councillors were the archbishop
of Canterbury (Sudbury), the bishop of Winchester (Wykeham), and the earl of
Arundel. *Anon. Chron.*, p. 91.

[3] Born *c.* 1330, Latimer fought with the Black Prince at Crécy and was given the
Garter in 1362. He was lieutenant and captain-general in Brittany, fought at Auray,
shared in the *chevauchée* of 1373, and was an ambassador to Portugal. In 1372 he
became constable of Dover and warden of the Cinque Ports. *Complete Peerage*, vii.
470-5.

[4] For differing views of the origin and nature of impeachment see Maude Clarke,
Fourteenth Century Studies, pp. 242-71, and B. Wilkinson, *Studies in Const. History*,
pp. 82-107. Professor Plucknett's arguments in *Trans. Royal Hist. Soc.*, 4th ser. xxiv.
47-71 and 5th ser. i. 153-64, appear more convincing than either of these and may
be regarded as the last word on the subject. He seems, however, to exaggerate the
importance of Wykeham's one recorded intervention by referring to a 'debate'
between him and Latimer. Gaunt himself later suggested that Wykeham had spoken

difficult but the commons held tenaciously to their course and every time that Latimer made a point in his own defence they produced an answer. After he had been dealt with, a number of lesser persons were handled in the same way. These included John Neville, lord of Raby, William Elys of Yarmouth, a customs official, John Pecche, a London wine-merchant, and Adam Bury, a former mayor of Calais, all of whom were charged with various corrupt practices. The commons throughout maintained their role of prosecutors, though always careful to insist that they were acting on behalf of the king. As subsequent events were to show, the procedure of impeachment was not yet established on its historic lines, but the trials in the Good Parliament mark an important stage in its development. The guilt of the accused may have been assumed rather too easily by later historians. Latimer put up a very plausible defence. It has been shown recently that he was not personally responsible for the loss of Bécherel and St. Sauveur;[1] on the charge of encouraging the issue of licences for evasion of the staple, he said truly that such licences had been common long before his day; and, though he probably made what he could out of his office and position, he may well have been sincere in his assertion that only when the situation was desperate did he agree to the king's borrowing at exorbitant rates of interest. Lyons denied that he had made a 50 per cent. profit on a loan to the king and, if he was referring to the transactions of August 1374, it seems that his statement was correct.[2] John Pecche, too, was probably speaking the truth when he denied that he had made illegal use of his monopoly of sweet wines and alleged that the mayor and aldermen had approved his actions.[3] William Elys incurred censure as an associate of Lyons but the charges against him were bound up with a feud of long standing between two groups of East Anglian merchants. The case against Neville was evidently flimsy and that against Adam Bury, the only one of the accused who failed

on impulse, 'en le chalure de iour et saunz bone foye' (*Anon. Chron.*, p. 99), and the evidence hardly seems to warrant the suggestion that Wykeham was concerned to champion one method of judgement against another.

[1] By C. C. Bayley in *Eng. Hist. Rev.* lv. 370–83.

[2] Dr. Holmes has pointed out that a syndicate represented by Lyons and John Pyel lent the king 20,000 marks in cash and paid off royal debts to the value of 10,000, so that a false charge of a 50 per cent. rate of interest could easily have arisen. The king acknowledged debts to Lyons in July and Dec. 1373, Feb. 1374, Sept. 1374, and 5 Apr. 1376 (*Cal. Pat. Rolls 1370–74*, pp. 323, 383, 411; *1374–77*, pp. 5. 254).

[3] Bird, *Turbulent London*, p. 21.

to appear, was in the nature of a side-issue. Though it is easily understandable that the commons should have believed that they were being cheated and their money wasted, it is possible that the men they attacked, while not indifferent to their own interests, were also doing their best to help the king in his chronic financial predicament. The commons remained obstinately blind to the reality of this predicament. Edward's request for a tenth and fifteenth was disregarded; only the wool subsidy was granted and that for a bare three months and on condition that it should then cease for three years. When, on 10 July, the longest parliament that had yet sat in England was at last dissolved, the prospects for the government had seldom looked blacker.

We can peer but dimly into the political and personal ramifications which may have underlain the actions of the commons in the Good Parliament. Much seems to have been hoped from the Black Prince, whose death during the session was regarded by Walsingham as a blow to the popular cause; but however deeply the prince may have deplored the scandals at court, his affection for his 'very dear and well beloved brother of Lancaster' remained undiminished till the day of his death, when he named him as his principal executor. The earl of March is often regarded as the leader of the movement against the courtiers. He was displeased at the turn events had taken in France and Ireland, he may well have been jealous of Gaunt, and de la Mare was his steward. Yet, apart from his nomination as a councillor and as a member of the liaison committee, there is little positive evidence of the part played by March in the Good Parliament. He was still a young man and though he may have been willing to support the demands of his sturdy and resourceful steward, it is unlikely that he inspired them. He himself owed everything to Edward III and he can hardly have dissociated himself altogether from the tradition of royal solidarity so carefully built up by the old king. Henry Percy, whom Walsingham speaks of as active in support of the commons, may have had a grievance against Gaunt, who for some years past had been trying to break the monopoly of power enjoyed on the Border by the northern lords. But Percy can hardly have approved the attack on Neville of Raby, who was his son-in-law, and the readiness with which he allowed himself to be won over by Gaunt soon after the dissolution does not suggest that his support of the opposition sprang from strong conviction. There appears, indeed, to be no sound reason for rejecting what was clearly the opinion of the chroniclers, that the initiative came from the commons themselves. The parliamentary knights included many old soldiers, men of wealth and standing in the shires, with long experience of local government; among the burgesses were powerful groups of merchants to whom evasions of the staple and abuse of monopolies were matters of vital and direct concern.[1] March and his friends may have seized the opportunity to ventilate their grievances about the Breton war; but the bulk of the charges had a fiscal or commercial basis and are most likely to have sprung from the commons who, as the principal victims of government taxation and the principal sufferers from the trade depression, needed little persuasion to concentrate their wrath on the men whom they believed to be wasting or embezzling their money and goods.

Gaunt's unpopularity is not hard to understand. The failing health of his father and eldest brother had made it inevitable that he should be regarded as the foremost active representative of the government, the man responsible for its mistakes. His own military failures, particularly the disaster of 1373, had done him no good. And although his vast wealth and territorial influence set him above any suspicion of petty pilfering, it was doubtless calculated to arouse envy and mistrust. Walsingham dwells with some relish on the kind of unsavoury rumour that was in circulation in 1376. Gaunt lived in open sin with his daughters' governess, Katharine Swynford; he had poisoned his first wife's sister for the sake of her inheritance and was seeking to destroy his nephew Richard by the same means; he was plotting with Charles V to secure a papal bull declaring Richard illegitimate. Even the duke's royal birth was denied in a story, said to be sponsored by William of Wykeham, that he was a Flemish changeling smuggled into the abbey at Ghent in place of a daughter born to Queen Philippa. Parliament's explicit recognition of Richard of Bordeaux as heir-apparent after his father's death, seems to have been carried out in such a way as to encourage the suspicion that Lancaster had designs on the throne.[2]

---

[1] It is significant that, despite the subsequent royal pardon, none of the London merchants impeached in the Good Parliament was ever readmitted to the city's franchises.

[2] *Rot. Parl.* ii. 330.

These ugly slanders are almost enough in themselves to explain Gaunt's determination to throw in his lot with the king and the court party; but it is probable that he was also influenced by other considerations which, if not disinterested, were in no way discreditable. His ambitions in Castile had damped his enthusiasm for the French war; he resented the political activities of prelates like Wykeham and Courtenay; as a great prince of the blood he was outraged by the presumption of the commons in impeaching the king's servants and attempting to control the personnel of his council. He may well have felt distress and perturbation at the spectacle of his father, harrassed on his deathbed by the clamour of his subjects and deprived of the society of the mistress on whom he had become so fatally dependent. Above all, he can hardly have failed to recognize that the loyalty of the nation and its confidence in the monarchy had been severely shaken. Danger to the whole royal family threatened from the action of this parliament; the tragedy of Edward II was still recent history. If the effective power of the Crown was to be maintained, it was necessary, first, to punish the rebels and then to rehabilitate the royal dignity.

Thus, in the last months of the reign, Lancaster concentrated his energies on the task of restoring the prestige of the Crown and teaching its critics a lesson. In order to reverse the proceedings of the Good Parliament, a meeting of the great council was held at Westminster in the autumn of 1376. William of Wykeham was examined by this body on a number of charges relating to his conduct as chancellor ten years earlier and, after two days' trial, he was sentenced to forfeiture of his temporalities and forbidden to come within twenty miles of the court. His household was dispersed and the scholars of his new college in Oxford sent 'weeping and with simple cheer' to their friends. To victimize so highly respected a figure was unwise; but Wykeham's great wealth made him an obvious target and his conduct in the Good Parliament had suggested that he might still be dangerous. Peter de la Mare was arrested and thrown into Nottingham gaol; March was induced to resign the office of marshal; Suffolk was removed from the admiralty; Alice Perrers returned to court and Latimer resumed his seat on the council. The nine new councillors had already been dismissed, for the late parliament was declared to be no true parliament and all its acts were annulled. At the same time the solidarity of the royal family and Gaunt's loyalty to it were affirmed when Richard of Bordeaux was created prince of Wales, duke of Cornwall, and earl of Chester; a few days later Richard received Wykeham's temporalities. His mother, the princess Joan, was confirmed in her dower and other rights; and the youngest of Gaunt's brothers, Thomas of Woodstock, was made constable of England. Henry Percy was won over by the grant of the marshal's staff, Gaunt thereby securing the support of the greatest of the northern marchers. When Edward III's last parliament met in January 1377 the Speaker was Sir Thomas Hungerford, one of Gaunt's officials; and, though there is no evidence that this parliament was packed, it showed itself, on the whole, remarkably complaisant.[1] Lancaster's objective seems to have been twofold—restoration of the prestige of the Crown and humiliation of his episcopal opponents by the adoption of a hostile attitude to clerical privilege and papal claims. Prince Richard was introduced as the formal president of parliament; and the chamberlain (Sir Robert Ashton) gave formal notice that the king was determined to resist papal usurpations contrary to his rights and those of his realm. In his opening oration, the bishop of St. David's, as chancellor, referred to the king's recovery of health in his jubilee year and to the many other blessings bestowed by God upon the dynasty. 'No Christian king nor any lord in the land had so noble and gracious a wife or such sons . . . as the king has had. For all Christians had the king in dread, and by him and his sons the realm of England has been reformed, honoured and enriched as never before.' The king's grandson and heir was sent to preside over the assembly, as God had sent his beloved Son to the world. In return for all these mercies, subjects were bound to rejoice and feast; they must drive out rancour and be in charity with all men; they must offer good-will to their prince and obey his commands; and they must produce supplies for the French war which would soon be renewed on the expiry of the truce. The whole speech was an elaborate piece

[1] Not entirely so, however; for the commons demanded the release of de la Mare and the petitions include a protest against the annulment of statutes except by parliament. On the question of packing, see J. C. Wedgwood, 'John of Gaunt and the Packing of Parliament', *Eng. Hist. Rev.* xlv (1930), 623–5; H. G. Richardson, 'The Parliamentary Representation of Lancashire', *Bull. J. Rylands Library*, xxii (1938), 175–222. Mr. K. B. McFarlane has pointed out that Hungerford had been simultaneously in the service of Gaunt and the Black Prince (*Bull. Inst. Hist. Res.* xx (1947), 176).

of propaganda for the monarchy; it represented Lancaster's *riposte* to the claims of the Good Parliament.[1]

The estates responded well. A poll-tax of 4*d.* from all lay persons over the age of fourteen was granted and a petition entered for the pardon of all persons 'impeached without due process of law'. A colossal bribe, in the shape of a general pardon for all civil and criminal offences, in honour of the king's jubilee, doubtless helped to quell any latent opposition. Wykeham alone was excepted from the offer; but his fellow-bishops were not prepared to suffer this insult without protest. Led by Courtenay, the convocation of Canterbury refused to make any grant to the king until the bishop of Winchester had been allowed to join his brethren; when he was admitted, they offered a poll-tax of 1*s.* from every beneficed clerk and of 4*d.* from all other ecclesiastics. The question of supply was thus settled satisfactorily; but the tension persisting between Gaunt and the bishops prevented a peaceful denouement. Rightly offended by Lancaster's refusal to pardon Wykeham, Courtenay offered what was, in effect, a challenge by summoning the duke's protégé, John Wyclif, to appear before convocation on 19 February, on charges which probably related to his denial of the Church's rights to endowments and of her power of excommunication. Recognizing that the attack was directed at himself, Gaunt engaged four doctors of divinity to aid Wyclif in his defence, while he and Percy attended the trial in the Lady Chapel of St. Paul's, with an armed following. It soon became clear that the duke did not intend the proceedings to go forward. A bitter wrangle arose between him and Courtenay, the insults offered to the bishop caused a general uproar, and the trial broke up in confusion. The temper of the Londoners had already been roused by the scandals of 1376 which had been followed by changes in the constitution of the city;[2] but internal factions were forgotten when it became known that the new marshal was proposing to extend his jurisdiction within its territory. An angry mob besieged Gaunt's palace of the Savoy, hung his arms reversed (the sign of treason) in Cheapside, and hunted from the streets men wearing his livery. Gaunt and Percy were forced to flee for their lives to Kennington, where the princess Joan received them and offered her services as mediator. In the meantime, though Courtenay was able to quell their rioting, the

---

[1] *Rot. Parl.* ii. 361–3.   [2] Above, p. 377.

Londoners had hung the duke's arms reversed on the doors of St. Paul's and Westminster Hall and were circulating libels to the effect that he was the son of a Flemish butcher. Gaunt demanded that an example should be made of the citizens who had insulted the king, his sons, and all his lineage; and it was only after long negotiations that peace was made, the mayor deposed, and the Londoners compelled to set up on a marble pillar in Cheapside a gilded shield bearing the arms of Lancaster. Two days later (22 February) a deputation from parliament visited the king at Sheen to receive his pardon and, on the day following, parliament was dissolved. Gaunt had carried through his policy of reprisal and restoration, but at the cost of drawing upon himself the hatred of the capital and of some of the more powerful of the prelates. He was evidently alive to the need to strengthen his position where he could; for it was on 28 February that he obtained a grant for life of palatine powers in the duchy of Lancaster. The closing months of the reign passed, however, without any further outbreak of disorder and the duke continued to act as regent, in fact, if not in name. William of Wykeham enlisted the help of Alice Perrers to secure a pardon from the king and restoration of his temporalities; and when Edward III died on 21 June 1377, the political outlook was much less menacing than it had been twelve months before. Thanks to Gaunt's vigorous action, the royal power and dignity had been largely restored. His very unpopularity had served to stimulate loyalty to the young and innocent heir to the throne; and it may be conjectured that it was in these last months that he began to instruct his nephew in the powers and duties of a king.

When Edward III was known to be dying the Londoners sent a deputation to Kennington, where Prince Richard was living with his mother. Their leader, John Philipot, lamented the imminent death of the old king and requested the favour of his successor towards the city, begging for his presence in London and asking him to compose the unhappy quarrel between the Londoners and the duke of Lancaster. The citizens had good reason to be apprehensive; for they may well have believed that Gaunt was only awaiting the death of Edward III in order to avenge the insults which had been offered to him. But reprisals formed no part of Gaunt's programme in the new reign. A message from the new king which, since Latimer was one of the

royal emissaries, must almost certainly have been inspired by
Gaunt, promised them a royal visit and informed them of the
duke's readiness to submit himself to the king's will in the matter
of his quarrel with the citizens; a pacification could be arranged
if the Londoners would show a pacific spirit. Suspecting a trap,
they hesitated; but when the royal messengers swore that
nothing should be done to the prejudice of the city, they yielded
and sent a second deputation to the royal manor of Sheen,
where Richard II, his mother, and his uncles were gathered to
mourn the dead. Lancaster saw and seized the opportunity
offered him by this solemn occasion. Falling at the young king's
feet he begged him to take the matter into his own hands and to
pardon the citizens, as he himself was ready to pardon them. It
was a clever move; for the Londoners' suspicions appeared to be
allayed and a potentially dangerous enemy reconciled, while the
credit for the reconciliation fell to the new king. To enhance his
nephew's prestige still further, Gaunt allowed himself to be
publicly reconciled to the bishop of Winchester; and Walsingham's comment is just what he would have desired: 'O happy
auspice that a boy so young should of his own accord (*nullo
impellente*) show himself so solicitous for peace; that with no one
to teach him he should know how to be a peacemaker!'[1] Finally,
the pardon and release of Peter de la Mare gave widespread
satisfaction[2] and removed the last obstacle to the establishment
of the atmosphere of peace and good-will essential to the success
of the coronation. Gaunt saw to it that Richard did not forget
his promise to visit London where, a few days before the coronation, he was joyfully received.

The pageantry of the coronation on 16 July was Lancaster's
work. As steward of England he presided over the court of
claims; and after the ceremony, with his own hands he delivered
into the chancery an elaborate record of the whole proceedings.[3]
As steward and marshal he and Percy rode at the head of the
splendid procession through the city to Westminster and even
the unfriendly Walsingham admits that the citizens received
them warmly. In the procession from the palace to Westminster
Abbey, Lancaster, carrying the short sword *Curtana*, walked
beside the young king, 'fair as another Absalom'. Before them

---

[1] *Chron. Ang.*, p. 150.
[2] Walsingham compares his welcome with that accorded to Becket on his return
from exile.   [3] *Cal. Close Rolls 1377–81*, pp. 1–5.

walked the two other royal uncles and the earl of March; behind, came the archbishop of Canterbury, with Gaunt's late
episcopal opponents, the bishops of London and Winchester.
There were significant variations in the coronation order itself.
The oath of 1308 was modified by the addition of the words *iuste
et racionabiliter* after the king's promise to maintain the laws promulgated with the assent of his magnates on the demand of the
people. The archbishop's question to the people, whether they
would give their will and consent to the new king was placed
after, instead of before the oath, thus underlining its significance
as an act of recognition and allegiance to a king *de jure* and
blurring ancient notions of election. Further, the rubric directing the peers to touch the crown after the coronation interprets
this symbolic act as binding the lords to help in easing the burdens of the royal office.[1] The creation of four new earls[2] marked
the climax of a notable occasion, intended to demonstrate the
sanctity and magnificence of the hereditary monarchy, the devotion of his relatives to the boy king, the ending of old quarrels,
and the fair prospects which lay before the nation under the
leadership of the royal family. Once again, it is Walsingham who
supplies the appropriate comment: 'It was a day of joy and
gladness . . . the long-awaited day of the renewal of peace and
of the laws of the land, long exiled by the weakness of an aged
king and the greed of his courtiers and servants.'[3]

But as the acclamations died away the duke of Lancaster
found that he had some hard realities to face. The accession of
Richard II coincided with a renewal of the war and of French
and Spanish raids on the English coasts; so that the problem
of government during the minority was inseparable from the
problem of war finance. A regency in the hands of the royal
uncles, supported by a group of friendly magnates, would
doubtless have appealed to Gaunt as the best means of preserv-
ing the restored prestige of the monarchy; but the objection to
such a plan was his own unpopularity with the three groups
whose co-operation was indispensable for the financing of the

[1] The revised *ordo* of 1377 emanated from Westminster and is associated with
the name of abbot Litlyngton; see P. E. Schramm, *A History of the English Coronation*, tr. by L. G. Wickham Legg (1937), p. 80.
[2] Thomas of Woodstock became earl of Buckingham, Henry Percy, earl of
Northumberland, Thomas Mowbray, earl of Nottingham, and Guichard d'Angle
(one of the king's tutors), earl of Huntingdon.   [3] *Chron. Ang.*, p. 155.

war—the higher clergy, the London capitalists, and the commons in parliament. The duke had already embroiled himself with the bishop of London and his name was suspect in ecclesiastical circles on account of his treatment of Wykeham, his anti-clerical proclivities, and his patronage of John Wyclif. It seemed improbable that the prelates would be ready to persuade their clergy to show generosity to a government under Lancastrian control. The second group, the London capitalists, could be relied on for financial help because of their vital interest in the defence of the coasts and the safety of the seas; but it was likely, while the king was under age, that they would demand some control of policy as the price of their support. No formal reconciliation could wholly obliterate the memory of their recent quarrel with Lancaster and they must have been well aware how the first nobleman in England was likely to view the interference of tradesmen in affairs of state. With the third group, the commons in parliament, the London capitalists had weighty influence; and, though the commons had less coherence and were in some degree more malleable than either prelates or merchants, the events of 1376 had shown that they could be dangerous. In the uneasy relations subsisting between these three groups and the duke of Lancaster is to be found the clue to the complex politics of Richard II's early years.

Gaunt began by walking warily. For reasons which remain obscure, he was not included in the king's first council nominated by the magnates three days after the coronation.[1] It is difficult to believe that he had no say in its composition and his aim may have been to avoid provoking criticism by claiming a place on what could be only a short-lived caretaker government. His interests were represented by his chancellor, the bishop of Salisbury, and by Lord Latimer and he may well have hoped that a temporary withdrawal from court would help to allay rumours as to his own intentions.[2] It would be time enough when the new king's first parliament met, to attempt to dispel suspicions and to re-establish his proper influence. The date fixed

---

[1] N. B. Lewis ('The Continual Council in the Early Years of Richard II, 1377–80', *Eng. Hist. Rev.* xli (1926), 249) believes that Gaunt was offended by the preponderance of the Black Prince's dependents on this council, but no evidence is offered in support of the suggestion.

[2] Northumberland's resignation of the marshal's staff may also have been intended as a conciliatory move, though his plea of heavy responsibilities on the Border was certainly genuine.

for the opening of parliament was 13 October; and the offer of
two substantial loans in the preceding month could be read as
a good omen. Four wealthy Londoners—Walworth, Philipot,
Karlille, and Hadley—raised the sum of £10,000 and the city
itself advanced £5,000.[1] The success of the chancellor's oration
in February seemed to warrant another on the same lines and
parliament opened with a soundly royalist sermon from the
archbishop of Canterbury. Taking as his text, 'Behold, your
king cometh unto you', he drove home the lesson of the coronation, that the king was king 'not by election nor by any other
such way' but by lawful right of inheritance, so that subjects
were by nature bound to love and obey him. When the sermon
was finished and the commons had asked for a liaison committee
(which included Lancaster) to advise them on questions of
supply, the duke seized the opportunity to make a personal declaration. He

rose and . . . after kneeling before the king, prayed him humbly that
he would hear him a little on a weighty matter touching his own
person. The commons had chosen him to be one of the lords to consult with them but nothing could be done until he had been excused
of those things which had been evilly spoken of him. For, he said,
albeit unworthy, he was a king's son and one of the greatest lords in
the kingdom after the king: and what had been so evilly spoken of
him could rightly be called plain treason . . . and he said that none
of his ancestors on either side ever were traitors but good and loyal
men. And it would be a strange thing indeed were he to stray from
their path . . . for he had more to lose than any other man in the
realm. And if any man were so bold as to charge him with treason or
other disloyalty or with anything prejudicial to the realm, he was
ready to defend himself with his body as though he were the poorest
bachelor in the land.[2]

The effect of this speech was instantaneous. All the prelates and
barons rose to their feet and with one voice begged the duke to
desist from these words, since no one would wish to say such
things of him; while the commons exclaimed that the duke was
free from all blame or dishonour and that they had chosen him
to be their principal aid, comforter, and counsellor in this parliament. Gaunt replied to their protests that those guilty of

[1] Bird, p. 46; A. Steel, *The Receipt of the Exchequer 1377–1485* (1954), p. 38.
[2] *Rot. Parl.* iii. 5. It may have been Lancaster himself who arranged for a verbatim
report of his speech to be included in the roll.

spreading slanders were the traitors, for they were sowing discords which might destroy the whole kingdom; but for himself
he would forgive everything. There seems to be no compelling
reason for refusing to take his words at their face value, though
historians in general have been reluctant to do so. It was fully
consistent with his actions over the past year that Gaunt should
have wished to make public—even dramatic—affirmation of
his loyalty to the Crown and to quell the ugly rumours which, as
he rightly said, threatened the solidarity and well-being of the
realm. The argument that he had too much to lose to be a traitor was unanswerable and the repudiation of evil suspicions by
the lords and, still more, by the commons (whose Speaker was
again de la Mare) must have looked like the first stage in the
reassertion of his influence. But, as the session proceeded, it became clear that this was not to go unchallenged.

The first necessity was to make provision for the government
of the country while the king was under age. After some discussion a plan was adopted which seems to have been designed
with the object of preventing any one individual or clique from
gaining permanent control of policy.[1] Nine councillors were to
be appointed 'to sit continuously in the council for the needs of
the realm', with full authority to deal with all business of state
in the king's name; but they were to hold office only for a year
and were to be ineligible for re-election for a further two years.
None of the three royal uncles was a member of this first council, but they were made jointly responsible for seeing that the
councillors were not corrupted by bribes; and with his friend,
the bishop of St. David's, at the chancery Gaunt could feel con-
fident that his wishes would not be ignored. None the less, there
were danger-signs. The commons' generous grant of two fifteenths and two tenths was accompanied by a condition that
John Philipot and William Walworth, two of the London representatives in this parliament, should be made treasurers of the
funds to be collected and that these should be spent only on the
war. De la Mare made a bold, though unsuccessful, demand for
the nomination of the household officers in parliament and
reiterated the request that no parliamentary act should be repealed except in parliament. Still more distasteful to Lancaster
must have been the renewed publicity accorded to Alice

[1] The situation at the accession of Edward VI in 1547 affords an interesting
parallel.

Perrers. She had fled from court on the death of Edward III and within a month the government had laid hold of her property to the value of £2,626. 8s. 4d.[1] None the less, it was thought necessary to subject her to a formal trial and the commons clamoured that she should appear in person to hear her sentence.[2] Most sinister of all was the presentation among the common petitions of fourteen items taken almost verbatim from the Ordinances of 1311.[3] Whoever drafted these must have had a copy of the Ordinances before him. Someone, it seems, was concerned to remind those in power, and perhaps Lancaster in particular, that the history of Edward II had not been forgotten and that the prerogative of the Crown must not be pushed too far. Nor was much encouragement to be derived from the attitude of convocation; for, though the clergy granted two tenths for the war, the grant was preceded by the reading aloud of a formidable list of *gravamina*.

The second parliament of the reign met at Gloucester a year later, in an atmosphere of tense excitement which boded ill for Lancaster. Castilian galleys had made assaults on Cornwall and burnt the town of Fowey and, though an English garrison had gained control of Cherbourg, Gaunt himself had been repulsed, before St. Mâlo, a disaster which contrasted sharply with the success of John Philipot in capturing, with the aid of a fleet equipped at his own expense, a notorious Scottish pirate named John the Mercer—a bitter pill for the leaders of chivalry to swallow.[4] Worst of all was the scandal caused by a recent breach of sanctuary in Westminster abbey. Two squires named Robert Hawley and John Shakell had been imprisoned in the Tower pending investigation of their claim to the ransom of a valuable Spanish prisoner in which other powerful persons had interests.[5] In the summer of 1378 the squires escaped and fled to the abbey, pursued by the constable of the Tower, Sir Alan Buxhill, and a

[1] Steel, *Receipt of the Exchequer*, p. 37.

[2] *Rot. Parl.* iii. 15. The use of the word *impeachment* to describe this trial in which the steward of the household prosecuted at the command of the lords, and the commons had no share, may suggest an attempt on the part of the government to secure control of a new and dangerous procedure. See T. F. T. Plucknett, 'State Trials under Richard II', *Trans. Royal Hist. Soc.*, 5th ser. ii (1952), 160–1.

[3] J. G. Edwards, 'Some Common Petitions in Richard II's first Parliament', *Bull. Inst. Hist. Res.* xxvi (1953), 200–13.

[4] 'Proceres regni . . . illud aegre ferebant.' *Chron. Ang.*, p. 200.

[5] For details of the case, see E. Perroy, 'L'Affaire du Comte de Denia', *Mélanges* . . . *Louis Halphen* (1951), pp. 573–80.

chamber knight, Sir Ralph Ferrers, with a band of armed followers. Determined to recover their prisoners, they burst into
the church while mass was being said, fatally wounded a sacris-
tan who tried to intercept them, and murdered Hawley on the
steps of the altar. The resultant outcry shook the government to
its foundations. Gaunt was in Brittany at the time and can have
had no hand in the deed, but he was widely denounced as its
instigator and was unwise enough to allow himself to become
implicated. When Courtenay pronounced sentence of excommunication on all violators of sanctuary, with the too-pointed
exceptions of the king, the princess of Wales, and the duke of
Lancaster, Gaunt replied by denouncing him and, by the time
that parliament met, he had put himself in the impossible position
of defender of an indefensible outrage. The fact that the Spanish
prisoner's hostage had got into debt with some of the London
capitalists gave them an interest in the affair; and, as a crowning act of folly, Gaunt brought Wyclif into parliament to argue
the case against the abbey franchises. The occasion was hardly
propitious for the financial statement which the government
had perforce to make. The large subsidy granted twelve months
earlier had all been spent and more was needed. The Speaker,
Sir James Pickering (who may have been aware that a further
£4,000 had lately been raised by the sale of some of the
king's jewels), expressed his understandable astonishment at the
poverty of the Crown and demurred to requests for a further
subsidy; and it was with difficulty that Sir Richard Scrope who,
in the middle of the session, took over the seals from the bishop
of St. David's, extracted a small grant in return for a promise
that the names of the councillors should be reported as soon
as they were chosen.[1] Yet there were certain indications that
Gaunt still felt himself to have the upper hand. A provincial
parliament always put the Londoners at a disadvantage and the
enemies of the wealthy victuallers were quick to seize the opportunities offered them at Gloucester. A statute depriving the
fishmongers of their monopoly was pushed through; Philipot
and Walworth were not reappointed as treasurers; and Thomas

---

[1] The full number of councillors was not appointed in parliament, however,
because of its hasty dissolution (*Rot. Parl.* iii. 55). Difficulties of accommodation
often tended to curtail the sessions of parliaments and councils held in the provinces. At Gloucester the abbey was crowded out and the monks were forced,
despite the season, to bivouac in the orchard. *Hist. S. Petri Gloucestriae* (Rolls Ser.)
i. 52–54.

of Woodstock allowed himself a personal outburst against the late mayor, Nicholas Brembre. In their excitement at these sensational happenings, the chroniclers let pass almost without comment what was undoubtedly the most far-reaching act of the Gloucester parliament—the decision, which ensured the perpetuation of the Schism, to support the Roman pope, Urban VI, against his rival at Avignon.

With only a small subsidy in prospect, it was unlikely that the new councillors would succeed in extricating themselves from the financial morass. At a meeting of the great council in February 1379, the lords authorized the raising of loans and for the next two months special receivers were engaged in 'continuous daily borrowing' of small sums, amounting in all to £3,715.[1] This was quite inadequate and there was nothing for it but to call another parliament within six months of the dissolution of the last. Scrope apologized for this and expressed his readiness to concede inquiries into the accounts; his conciliatory attitude induced the commons to grant a graduated poll-tax, but its yield was disappointing. Another £5,000 had to be borrowed from London and more Crown jewels had to be sold. By the time that parliament met again, in January 1380, it was clear that new supplies could be had only at the price of administrative changes. The expedition organized by Sir John Arundel had ended in a disastrous shipwreck and the Speaker, Sir John Gildesburgh, demanded that all the members of the continual council be dismissed and that there be no more of them, 'seeing that the king is now of good discretion and fine stature'.[2] He proposed to return to traditional methods of government through the five great officers of state, whose names were to be given to the commons and who should not be removed before the next parliament. In addition, he demanded the appointment of what was, in effect, a committee of censure, to survey the royal household and the state of the national finances. Rebuffs to Gaunt were implicit in the inclusion of Walworth and Philipot among the members of the commission and in the removal of Scrope from the chancery, where his place was taken by the archbishop of Canterbury, Simon Sudbury. Yet Lancastrian influence was still powerful with the lords. Before the

[1] Steel, *Receipt of the Exchequer*, p. 39. In Jan. the Londoners had raised a loan of £600 to persuade 'the great lords of the realm' to return to London. Bird, p. 49.
[2] *Rot. Parl.* iii. 73.

dissolution they were persuaded to examine a London mercer named John Kirkeby, charged with murdering a Genoese merchant who was said to have suggested to the government that he could make Southampton the greatest port in Europe. As the Genoese was in England in an ambassadorial capacity and held the king's safe-conduct, his murder was adjudged an act of treason.[1] The details of the case are obscure, but the passions it aroused suggest that Lancaster may have been making a move to free the government from its financial dependence on the London capitalists; at all events, it was a sure sign of strained relations with the city that the parliament of November 1380 should have been summoned to Northampton, where Kirkeby could be executed without fear of retaliation and where the failure of Buckingham's expedition in the summer might be glossed over. During the session of this parliament Gaunt was on the Scottish border, but Walsingham none the less believed that it was his influence which secured the acquittal of Sir Ralph Ferrers (one of those implicated in the sanctuary case of 1378) on a charge of treasonable correspondence with France which the mayor of London had brought to the notice of the government. The question of supplies gave rise to heated debate, the Speaker pressing the chancellor for details of government expenditure and the commons attempting unsuccessfully to coerce the clergy, before the decision was reached to levy a poll-tax of three groats on all lay men and women over the age of fifteen.

This was the third poll-tax in four years and by far the most obnoxious. In 1377 a poll-tax at the rate of 4d. had occasioned much grumbling; the graduated poll-tax of 1379 was more equitable in principle but its yield, as has been seen, was disappointing and Walsingham denounces it as *nova taxa inaudita*; in 1380, beyond a vague stipulation that the rich ought to help the poor, no provision was made for grading and the rate was pushed up to 1s. The result was gross unfairness in the incidence of the tax. In villages where there was a resident man of means, his poorer neighbours were dependent on his generosity; in others, such as were common in East Anglia, where there was no rich man, the poorest labourer might find himself compelled to pay 2s. for himself and

[1] *Rot. Parl.* iii. 75. It is noteworthy that on this occasion reference was made to the Statute of 1352 which reserved judgement in doubtful cases of treason to the king in parliament.

his wife.[1] Most objectionable of all was the conduct of the collectors. A poll-tax, being based on something in the nature of a census, necessitated inquiries into personal circumstances which could hardly fail to provoke resentment; and the officials were not delicate in their methods. The result was evasion on a large scale. If we were to take the poll-tax returns as our evidence we should have to deduce a fall of one-third in the adult population between 1377 and 1381. The London collectors reported at the exchequer that they could not discover the names, rank, and condition of each person in the city and suburbs without causing dangerous agitation. The original plan had been to collect two-thirds of the tax by January 1381, the remaining one-third by June; but the money came in slowly and the government's urgent need of funds led to the appointment, in February, of collectors charged to levy the whole tax at once and to render account at the exchequer by 22 April. As the collectors were alleged to be negligent, further commissions were set up in March to make inquiries in certain counties. Rumours spread that a new tax was being levied without parliamentary authority and throughout the spring discontent mounted. By May the south-east of England was on the verge of revolution.[2]

The first open outbreak, in Essex, was provoked by the misdeeds of a commissioner sitting at Brentwood. Resistance by the three villages of Fobbing, Corringham, and Stanford-le-Hope soon developed into a general rising of the peasants of Essex against the tax-collectors. When Sir Robert Belknap, chief justice of the king's bench, was sent into the county with a commission of trailbaston, he was seized and forced to swear never to hold another such session. Houses and manors were sacked, among them Cressing Temple, the property of the treasurer, Sir Robert Hales, and attempts were made to raise the neighbouring shires. In Kent the activities of a royal serjeant-at-arms, John Legge, well known as a collector of unpopular taxes, seem to have given the signal for a general revolt; Walsingham tells us that the commons of this region demanded that there should be no more taxes levied, 'save the fifteenths which their fathers and forebears knew and accepted'.[3] Assembling in great numbers

---

[1] The figures for Suffolk published by E. Powell (*The Rising in East Anglia* (1896), pp. 67–119) afford ample illustration of the inequity of the tax.

[2] The literature of the Peasants' Revolt is discussed in the Bibliography below, pp. 559–60.　　　　　　　　　　　　　　　　　　　　　[3] *Hist. Ang.* i. 455.

at Dartford on 5 June, the Kentishmen took counsel and ordered
that all living within twelve leagues of the sea should remain at
home to guard the coasts; heavy and sustained taxation had not
freed the people of these exposed districts from fear of invasion.
Next day the Kentishmen moved to Rochester where they converged with some of the Essex men, forced Sir John Newton to
surrender the castle, and set free a runaway serf of Sir Simon
Burley's, who was imprisoned there. On 7 June they continued
their march south-eastward to Maidstone where they chose
Wat Tyler as their leader.

   Wat Tyler was the driving-force behind the Rising, but little
else is known of him. The jurors empanelled after his death
believed that he came from Essex, but the weight of the evidence
points to his being a man of Kent. Froissart describes him as a
disbanded soldier and there is a possibility that he may have
been an unruly cadet of the well-known Kentish family of Cul-
peper, whose estates lay on the Medway.[1] Whatever his origin,
he was undoubtedly a man of parts, 'vir versutus et magno sensu
praeditus', whose reputation soon spread far beyond the ranks
of his immediate followers.[2] Under his leadership the people
moved on to Canterbury, where they entered the cathedral
during mass and cried out to the monks to elect one of themselves as archbishop, for the present primate was a traitor and
would soon be beheaded. But it was not long before a more
acceptable candidate for the primacy presented himself to the
rebels in the person of John Ball, the vagrant priest whom they
released from the archbishop's prison. With Ball as their prophet and Tyler as their captain, they now swung round and on
Tuesday, 11 June, began their great march on London, Tyler's
energy driving them on to cover the seventy miles from Canterbury to Blackheath in the space of two days. When royal messengers intercepted them to inquire their purpose, they replied
that they were coming to destroy the traitors and to save the
king, with whom they desired to speak at Blackheath on the
following day. Richard II, who had been at Windsor, hurriedly
removed to London and took refuge in the Tower where many
members of the court and council were already assembled. The
situation was highly alarming, for the Essex men were moving
on London up the north bank of the river and by Wednesday,

[1] See Maude Clarke, *Fourteenth Century Studies*, pp. 95–97.
[2] *Hist. Ang.* i. 463.

12 June, they were encamped at Mile End, in the fields beyond Aldgate. Some of them probably joined the Kentishmen who were assembling in great numbers on Blackheath to await the king; and it may have been during this interval that John Ball took occasion to preach his famous sermon from the text,

> Whan Adam dalf, and Eve span,
> Wo was thanne a gentilman?[1]

Neither the government nor the city authorities seem to have made any preparation to meet the appalling danger which now threatened, though they had had several days in which to raise a force for the defence of London. This extraordinary inactivity is ascribed by the chroniclers to panic, but it is possible that the king's advisers were aware of the widespread sympathy for the rebels among the lower elements in the city and that they had already decided that their only hope lay in a policy of conciliation.[2] It must have been with some intention of offering terms to the insurgents that, on the morning of Thursday, 13 June, the king, with the chancellor, treasurer, and others of the council, set out by water from the Tower to Greenwich. But as the royal barge approached, those on board could see the frantic crowd gathered on the south bank of the river and it was clearly impossible to allow the king to land, though he seems to have been willing to do so.[3] If the ministers were alarmed, this was hardly surprising; for the rebels had already sent in a petition asking for the heads of the duke of Lancaster and the chief officers of state; but the return of the king's barge proved to be the signal for the great assault on London which it was now impossible to avert. Frustrated in his desire to speak with the king and aware that his men were running short of supplies, Tyler led them up to Southwark where they opened the Marshalsea prison and destroyed one of the marshal's houses before going on to Lambeth to burn the chancery records stored in the archbishop's manor there.[4] London Bridge was their next objective. At the

[1] Walsingham, who summarizes the sermon, says that it was preached 'ad le Blakheth', but gives no date. *Hist. Ang.* ii. 32.

[2] As suggested by Professor B. Wilkinson, 'The Peasants' Revolt of 1381', *Speculum*, xv (1940), 12–35.

[3] 'il respondist qil vodroit volunters, mes les ditz chanceller et tresorer luy conseillerount le reversee.' *Anon. Chron.*, p. 139.

[4] The *Anonimalle Chronicle* (p. 140) ascribes the burning of Lambeth to the Essex men but, although some of them may have been with Tyler, the main Essex body was still at Mile End.

subsequent inquiry, the London jurors ascribed the failure to keep the bridge to the treachery of two London aldermen, Horne and Sibill, but it is likely that their hands were forced by mobs in the city and suburbs and they were subsequently acquitted. Surging over the bridge, the invaders and their supporters took a westerly direction towards Fleet Street, where they opened the Fleet prison and then proceeded to the New Temple to burn the lawyers' rolls and a number of houses, including one belonging to the treasurer, Hales. Some then attacked Chester's Inn, the bishop of Lichfield's palace, where John Fordham, keeper of the privy seal, had his lodging, and refreshed themselves with wine from the episcopal cellars. Meanwhile, the Londoners had launched an assault on Gaunt's great palace of the Savoy where the Kentishmen joined them in an orgy of destruction. The splendid furnishings, the plate, jewels, and fine clothes were trampled under foot and burned to ashes or cast into the Thames; with the aid of gunpowder, the noblest house in the kingdom was reduced to a blackened ruin.[1] Probably it was Tyler who insisted that there should be no looting. A man found pocketing a silver piece was thrust with it into the flames, the rebels declaring that they were not thieves and robbers but zealots for truth and justice.[2] Elsewhere in the city and suburbs, the work of destruction went on apace. More prisons were opened and in Cheapside a number of lawyers, Flemings and other unpopular persons, among them Richard Lyons, were summarily beheaded; the hospital of the Knights of St. John in Clerkenwell burned for seven days.

From a vantage-point in the Tower, now virtually in a state of siege, Richard II could see the flames rising:

He called the lords who were around him . . . and asked their advice as to what should be done in such a crisis: and none of them knew, or was willing to advise him. Therefore the young king said that he would send word to the mayor of the city to order the sheriffs and aldermen to proclaim in their wards that, on pain of life and limb, all between the ages of fifteen and sixty should be at Mile End on the morrow, Friday, and meet him there at seven of the clock, so that all the commons threatening the Tower would raise the siege and come to see and hear him and those within the Tower could

---

[1] 'hospicium . . . cui nullum usquam in regno in pulchritudine et nobilitate potuit comparari.' *Hist. Ang.* i. 457.
[2] Knighton, ii. 135.

leave it and save themselves. But it was all in vain, for they lacked the grace to be saved.[1]

Meantime, the king made an effort to persuade the crowd on Tower Hill to disperse. Addressing them from one of the turrets he proclaimed that if they would return peaceably to their homes, all should be pardoned; as proof of his good faith, he had a charter of pardon written in his presence and sealed it with his signet before them all, sending it out by two of his knights to be read aloud. But the rebels declared the charter a mockery and some of them, returning to the city, ordered the execution of all lawyers and proceeded with the work of destruction. They had, however, accepted Richard's offer to go to Mile End as made in good faith and were beginning to stream there in great numbers. Whether or not the offer was made on the king's own initiative, his resolution to confront the insurgents in the open indicates a very high degree of personal courage in a boy not yet fifteen. Before setting out, he advised the archbishop, who had already surrendered the great seal, and others in the Tower to try to escape by water. Sudbury made the attempt, but he was recognized by *une malveys femme* and withdrew in confusion.

Thus, on Friday, 14 June, while a band of rebels led by one Jack Straw were setting fire to the treasurer's new manor of Highbury and others were attacking the house of the under-sheriff of Middlesex and making an abortive attempt to enter the exchequer, the king, accompanied by a handful of lords and knights and by William Walworth, mayor of London, approached Mile End. When the royal party appeared there, all the commons knelt, saying, 'Welcome, King Richard: we wish for no other king but you.' Through their spokesman, who was probably Thomas Faringdon,[2] they then demanded that the 'traitors' be handed over to them; to which Richard replied that they might work their will on any traitors who could be proved such by law. A petition was then presented for the abolition of villeinage, for labour services on a basis of free contract (probably an allusion to the Statute of Labourers), and for the right to rent land at fourpence an acre. The king replied that he would grant what they wished and that all should be free; and that they might go through the realm and seize traitors and

[1] *Anon. Chron.*, p. 143.
[2] The statement in the *Anonimalle Chronicle* that Tyler was at Mile End is not supported by any other authority.

bring them to him, so that justice might be done.[1] It is not clear
from the sources whether the tragedy at the Tower which followed was the result of this last concession, or whether it was
already being enacted while the king was at Mile End. The
failure of the garrison (consisting, it is said, of 600 men-at-arms
and 600 archers) to defend this great fortress has never been
satisfactorily explained; but it seems certain that the mob entered unopposed. Walsingham depicts them as bursting into the
Tower in jocular mood, pulling the beards of the guard and
inviting them to join their *societas*, invading the royal bed-
chambers and attempting familiarities with the Princess of
Wales, while not one of her knights ventured to intervene. We
learn from a different source that the privy wardrobe, where the
royal arms were stored, was thoroughly ransacked.[2] The archbishop, the treasurer, and other councillors were at their devo-
tions when they were discovered. Sudbury, Hales, John Legge,
and the duke of Lancaster's physician, a friar named William
Appleton, were dragged out on to Tower Hill and summarily
executed, their heads being taken in procession through the city
and set up on London Bridge. Some of her pages had meanwhile
smuggled the king's mother out of the Tower and escorted her
to the great wardrobe at Baynard castle where she was joined
by Richard who, extraordinary though it seems, apparently had
been wandering through the scenes of wild disorder in the city
and suburbs. It was probably here that the king delivered the
great seal to Arundel who was made 'chancellor for the day',
and arranged for clerks to write out charters in accordance with
the promises he had made at Mile End. Many of the Essex
rebels had begun to disperse; but Tyler and the Kentishmen had
still to be reckoned with.

The next day, Saturday, 15 June, opened ominously when
the keeper of the Marshalsea prison, Richard Imworth, who
had sought refuge in the abbey, was dragged from the shrine of
St. Edward and taken to Cheapside for execution. A little later
the king and some of his lords came to the shrine to pray.
Richard was determined to pursue the policy of conciliation

---

[1] According to the Monk of Evesham (pp. 27–28) the petitioners presented their
demands through a delegation appointed for the purpose and obtained letters
patent in confirmation. Adam of Usk (p. 2) adds that at Mile End the king conceded the liberation of all prisoners; and it seems clear from the city records that
many of the rebels believed that they had an unconditional permit to take traitors
wherever they could find them.   [2] See Tout, *Chapters*, iv. 460.

which had, at least, rid the city of some of the insurgents; and at Smithfield he faced another rebel host and ordered the mayor to proclaim that he wished to hear their demands. This was Tyler's opportunity. Eight days of unchallenged authority over the rebels had clearly gone to his head and he addressed the king with studied insolence, demanding a charter, written and sealed. Three of the concessions which he sought were comparable to those sought by the men of Essex—abolition of villeinage and of sentences of outlawry and 'no law but the law of Winchester', which may be interpreted as a protest against the labour legislation and a request for a return to the police law of Edward I. The remaining demands—that there be no more lordship save the lordship of the king, all other men to be equal; that the estates of the Church be confiscated and divided among the laity; and that all bishoprics save one be abolished—may well have had their origin in the fervid imagination of John Ball.[1] Richard replied easily that all should be granted and ordered Tyler to go home; but the rebel leader was suspicious and the little group around the king was growing restive. Words were exchanged and the mayor lost his temper. He pulled Tyler down from his horse and a squire named Standish finished him off. It was a moment of desperate peril; for the peasants, seeing their leader fall, set up a great cry and prepared to rush upon his assailants. But Richard kept his head. Spurring his horse forward, he rode out to confront the mob, crying, 'Sirs, will you shoot your king? I am your captain, follow me.' Tyler had disciplined his followers and they were accustomed to take orders; as the king rode slowly away to the north, they turned and followed. Walworth, meanwhile, had made off to the city where he hastily recruited a volunteer force, stiffened by some professional soldiers, which soon surrounded the rebels. Richard refused, however, to allow any violence to the sheeplike multitude.[2] He declared that he pardoned them all and that they should either take to flight, or go peaceably to their homes; his sole act of reprisal at this juncture was to send Tyler's head to replace Sudbury's on London Bridge. William Walworth, who, next to the king himself, had done most to save the situation,

---

[1] The *Anonimalle Chronicle* alone distinguishes clearly between the demands made at Mile End and at Smithfield, though the Westminster chronicler (Higden, ix. 5) suggests that the Mile End concessions were judged unsatisfactory and that what was sought at Smithfield was *emendatiorem . . . de libertate cartam.*

[2] 'come berbiz en caules.' *Anon. Chron.,* p. 149.

and two other loyal Londoners, John Philipot and Nicholas Brembre, were knighted by Richard in Clerkenwell Fields before he rode back to the wardrobe. The rising in London was at an end.

But the infection of revolt had spread and many other parts of England were in turmoil. Villages around London—Clapham, Croydon, Harrow, Hendon, Chiswick, Twickenham, and many others—had been plundered and burnt. In Bedfordshire the prior of Dunstable had been forced to concede a charter to his serfs; there were murders in Sussex; at Winchester the craftsmen rose against the urban authorities; and Knighton describes the panic caused in Leicester by a rumour that a band of rebels from London was advancing to destroy this Lancastrian stronghold. In the north of England there were at least three isolated outbreaks—at York, Scarborough, and Beverley—encouraged by rumours of a general collapse of the forces of law and order elsewhere. But the most serious risings outside London were at St. Albans and in the eastern counties.

Trouble began at St. Albans on Friday, 14 June—'illucescente die irae, die tribulationis et angustiae, die calamitatis et miseriae'—when news came to the abbey that the rebels had ordered the men of Barnet and St. Albans to arm themselves and hasten to London if they wished to save their towns from destruction.[1] It is somewhat hard to believe that such a message can really have been sent; but rumours of this kind served as a pretext for the departure of the Hertfordshire contingent to London. Coming in from the north, they reached the outskirts of the city in time to see rising from Highbury manor the flames kindled by Jack Straw, to whom they took oath to bear faith to king and commons. Moving on to the church of St. Mary Arches, they there discussed the terms of the charter which they hoped to obtain from the king. All their demands related to the seignorial rights of the abbot over his tenants and were purely local in content; neither the poll-tax nor the misdeeds of the government seem to have excited the men of St. Albans. But it is significant of the reputation already won by Tyler that some of them suggested that application should be made first to him, because they believed that there would soon be no authority but his throughout the land. Others, more cautious, advised

[1] *Hist. Ang.* i. 458 ff.

that application for a charter should first be made to the king. In the event, approaches were made to both. Under their leader, William Grindcob (whose treachery Walsingham finds hard to forgive because he had been nurtured in the abbey), they sought out the king, probably at Mile End, and obtained the desired charter; they then laid their grievances before Tyler who promised to come with 20,000 men and shave the monks' beards if the tenants met with any resistance. The assault on the abbey began the next day, when the townsmen broke down the gates of the abbot's park, drained his fishpond, killed his game, opened his prison, sacked the houses of his officials, burned the charters by virtue of which he enjoyed his manorial rights, and, displaying the king's charter, forced him to seal a deed of manumission concocted by themselves. Yet, in contrast to those of their fellows elsewhere, the doings of the Hertfordshire men were surprisingly moderate—a tribute, perhaps, to the strength and capacity of the great abbot, Thomas de la Mare, who kept his head throughout. Nobody, except a released prisoner, seems to have been killed and the abbey church and buildings suffered no serious damage.

The fate of the sister house of Bury St. Edmunds was a great deal worse. As the chapter was in the throes of a crisis over a disputed election, it fell to the prior, John of Cambridge, to defend the monastery. But he was not a man of much spirit[1] and when, on Friday, 14 June, a combined force of peasants and townsmen, led by a *sceleratissimus presbyter* named John Wrawe, advanced upon the abbey, he took to flight. Wrawe and his company had begun their operations two days earlier by burning Richard Lyons's manor of Overhall on the Suffolk–Essex border;[2] on their way to Bury they pillaged the church of Cavendish in the belfry of which the chief justice, Sir John Cavendish, had stored some valuable plate. His house in Bury was ransacked and he himself had the misfortune to fall into the hands of another group of rebels in the neighbourhood of Mildenhall; he was summarily executed and his head carried to Bury on the end of a pike. That same day (Friday, 14 June) the

[1] His gifts seem to have been musical: '. . . Orpheum Thracem, Neronem Romanum, Belgabred Britannum, vocis dulcedine, pariter et cantus scientia superantem'. *Hist. Ang.* ii. 2.
[2] Walsingham (*Hist. Ang.* ii. 2) states that Wrawe had been in consultation with Tyler; but Réville (*Soulèvement des travailleurs*, pp. 61–62) shows that the chronology of events makes this impossible.

fugitive prior of Bury was discovered in a wood near Newmarket; he was taken to Mildenhall, put through a mock trial before John Wrawe, and beheaded next morning, his body being left unburied for five days on Mildenhall heath and his head taken to Bury to be set up beside that of his friend, the judge. It was soon joined by a third, that of John Lakenheath, the monk charged with the collection of manorial dues and fines; for meanwhile, the rebels had been running amok through the great abbey. As at St. Albans, the monks were forced to hand over their deeds and muniments, their jewels and plate; and to issue a charter of manumission to the town under the seal of the sub-prior. But Wrawe was much more violent in his methods than Grindcob, and Bury was the only town excepted from the royal amnesty of December 1381.

In Cambridgeshire and Huntingdonshire the fury of the revolt was concentrated in one week-end, Saturday, 15 June, to Monday, 17 June. Though there had been isolated incidents earlier in the week, the signal for a general rising was given by one John Greystone of Bottisham, who had been at Mile End and claimed that he had obtained a charter empowering him to raise the people for the destruction of traitors. This was enough to start a campaign of looting and violence in north Cambridgeshire. In the south of the county another leader, John Hancchach (a landowner of some substance), led attacks on the Hospitallers' manor at Duxford; while in the Isle of Ely, one Adam Clymme rode around calling the peasantry to rebellion, and Richard of Leicester led his forces into the abbey church and reiterated the summons from the pulpit. On Monday this group seized and beheaded Sir Edmund Walsingham, a justice of the peace, broke open the bishop's prison and destroyed his rolls; only at Huntingdon and Ramsey abbey were the men of Ely repelled, thanks to the vigorous action of a minor chancery official.[1] Meanwhile, the scholars of Cambridge had been undergoing a severe ordeal. On Saturday, 15 June, the bell of Great St. Mary's gave the signal for revolt and a mob of peasants and townsmen attacked the house of the university bedel and sacked and gutted Corpus Christi college which was under the patronage of the dukes of Lancaster. Next morning they entered St. Mary's during mass and seized the chest containing the university archives and much of its plate and jewels; another chest full of parchments

---

[1] William Wightman, spigurnel of the chancery. See *Anon. Chron.*, p. 150.

was taken from the house of the Carmelites. Both chests (containing documents of priceless value to the historian of the medieval university) were publicly burned in Market Square, an old woman named Margery Starre crying, 'Away with the learning of clerks, away with it!' as she flung the parchments on the fire. The rebels then drew up a document whereby the university formally surrendered its privileges to the town and agreed to be governed by the municipal authorities. Before the revolt collapsed the mayor (who said later that he had acted under compulsion) had led an assault on the neighbouring priory of Barnwell.

In Norfolk trouble began on the southern border of the county on 14 June and soon spread rapidly; but there was no single leader in command and the outbreaks appear to have been sporadic and localized and concerned mainly with pillage. It was in east Norfolk that an outstanding leader appeared in the person of one Geoffrey Litster, who had the support of a knight, Sir Roger Bacon. Litster evidently valued upper-class support, for he tried also to secure the earl of Suffolk who was then in residence at one of his Norfolk manors; but the earl rose from his half-eaten dinner and fled across country to St. Albans. On Monday, 17 June, a great band of rebels, led by Litster, assembled on Mousehold Heath and forced the panic-stricken city authorities to open the gates of Norwich. Litster, whom his followers hailed as 'King of the Commons', established himself in the castle where he banqueted in state, waited on by some captive knights, while the mob plundered the city.[1] From Norwich detachments of rebels were sent out in different directions, one, under Roger Bacon, to Yarmouth, others to neighbouring villages and monasteries, where rolls were destroyed and several persons executed after mock trials. About 20 June, nearly a week after Tyler's death, Litster determined to send an embassy to London to request a general pardon and 'a charter more special than the charters granted to other counties'.[2] His emissaries were two knights—Sir William Morley and Sir John Brewes—and three peasants, who took with them a large sum of money levied from the citizens of Norwich. But the revolt was now

---

[1] These orgies may explain an item in the city accounts—'To Henry Lomynour (one of the acting treasurers) for wine at the time of the tumult, £10.' *Records of the City of Norwich*, ii. 46.

[2] *Hist. Ang.* ii. 6.

collapsing everywhere and they got no farther than Newmarket when they were intercepted by the militant bishop, Henry Despenser, who had already repelled an attack on Peterborough abbey and dealt vigorously with the rebels in Huntingdon and Cambridge. The two knights were spared but their peasant colleagues suffered immediate execution, and by 24 June the bishop was before the walls of his cathedral city. Litster and the main body of the rebels had taken to flight but Despenser pursued them. At North Walsham, where they had erected a primitive stronghold with carts and furniture, his cavalry surrounded them and, after shriving Litster, the bishop led him to the gallows. With Litster's death the rebellion in Norfolk came to an end. In Suffolk it melted away when the earl appeared in force at Bury on 23 June. John Wrawe was taken; and within a few days the revolt had petered out in nearly all the lesser centres of resistance.[1]

Meanwhile, the government had taken action. On 16 June the great seal had been transferred from Arundel to Sir Hugh Segrave, 'until the lord king could conveniently provide himself with another chancellor'. A commission of London aldermen was empowered to make provision for the safety of the city and its suburbs; another, of *oyer et terminer*, was set up under the chief justice of the common bench. The king himself went to Essex and on 23 June issued a proclamation warning his subjects against rumours that the deeds of the rebels had his approval. When a deputation of Essex men demanded ratification of the promises made at Mile End, he is said to have answered them harshly, 'Villeins ye are and villeins ye shall remain.' A last attempt at resistance near Billericay was crushed by Buckingham and Sir Thomas Percy, and on 2 July, from Chelmsford, Richard issued another proclamation announcing that by the

---

[1] There were still some rumblings in July, when the serfs of the abbot of Chester rose in the hundred of Wirral (Powell and Trevelyan, *The Peasants' Rising and the Lollards*, p. 13) and when the prior of Worcester excused himself from attendance at the General Chapter of his Order on the grounds that his tenants were trying to withdraw their services (W. A. Pantin, *Chapters of the English Black Monks*, iii. 205). It was about the same time that John Shirle of Nottingham was heard explaining to an appreciative audience in a tavern in Bridge Street, Cambridge, that John Ball was 'a true and upright man, prophesying things useful to the common people' and that the king, the justices, and many officials were more deserving of death than he (W. M. Palmer, 'Records of the Villein Insurrection in Cambridgeshire', *The East Anglian*, vi (1895–6), 136–7). There was another attempt to raise the villages of Cambridgeshire in Sept. 1382 (*Vict. County Hist. Cambridge*, ii. 402).

advice of his council he had revoked the letters patent of pardon 'lately granted in haste' to the rebels. The way was now open for full judicial inquiry and the new chief justice of the king's bench, Sir Robert Tresilian, began a tour of the disaffected districts with the king in his company. In Hertfordshire and Essex the severity shown moved even the monastic chroniclers to shocked protest; but, on the whole, the judicial proceedings reflect credit on the government. No mass reprisals were allowed; there were no tortures and very few attempts to convict without trial; and a surprisingly large number of persons whose guilt seems to have been clear, were either acquitted or punished with moderation. Tyler and Litster were already dead; Jack Straw was executed in London after confessing his guilt; John Wrawe was condemned to death after an elaborate trial in the course of which he tried to turn king's evidence; the Cambridgeshire captains were all hanged, as was William Grindcob; and John Ball was tried and executed at St. Albans. Certain Londoners and Sir Roger Bacon were acquitted. On 30 August the king ordered that all further arrests and executions should cease and all cases pending be transferred to the king's bench. The effect of this order was virtually to put an end to the capital sentences; and the parliament which met in the autumn, though it confirmed the king's revocation of his charters of manumission to the rebels, demanded also a general amnesty for all but a few specified offenders.

The history of this astonishing movement reveals something of the magnitude and diversity of the social grievances which underlay it.[1] Desire to be free of the disabilities of villeinage and of the hated labour laws was widespread, manifesting itself in the destruction of manorial records and in the attacks on judges and lawyers.[2] At St. Albans and at Bury St. Edmunds, however, the rebels were aiming, not merely at abolition of the abbot's seignorial rights but also at municipal charters for the towns which had grown up around the abbey walls; and the situation at Bury was further complicated by the dispute over the

---

[1] See above, pp. 331–8.
[2] In his introduction to Miss Putnam's *Proceedings before the Justices of the Peace in the Fourteenth and Fifteenth Centuries* (pp. cxxii–cxxiii), Professor Plucknett points out that (excluding Kent where the destruction of the records makes generalization impossible) the area of greatest intensity of the revolt coincides with that where there is definite evidence of the strictest enforcement of the labour laws.

abbatial election, in which some of the townsmen had become involved. At Cambridge the rising was made the occasion for working off old grudges between town and gown. At Yarmouth the main cause of violence seems to have been a feud of long standing over the burgesses' market monopoly and control of harbour dues. Some of the urban riots clearly arose from a movement of the poorer elements in the town against the rich burgesses who composed the governing body; it is evident that there was much disaffection of this kind in London. Hatred of foreigners was widespread, partly because of commercial jealousies, partly as a result of the sheer unreasoning prejudice which a long war is apt to engender. The lower clergy who played a considerable part in the revolt had their own grievances —heavy taxation, both royal and papal, and the inadequate stipends to which they were condemned by the labour laws.

The strength and vitality of the community of the shire is an aspect of the movement which has attracted little attention. Serfs and artisans who figure in manorial and urban records mainly as appendages of manor or gild, appear in 1381 as active and self-conscious members of much larger social groups. In the eastern and south-eastern counties, politically the most precocious regions of medieval England, we find ourselves confronted with a kind of tribalism almost reminiscent of the age of Bede. Manorial and urban arrangements which ignore or cut across this tribalism have not, it seems, radically affected it. For, whatever their tenurial or contractual commitments, the men of Kent—'totum illud germen Kentensium et Juttorum', as Walsingham, significantly, terms them—remain aware of themselves as a natural community, distinct from the community of their east Saxon neighbours, as from those of the North Folk and South Folk of the area of Anglian settlement. And, though most of the rebels were loyal to Richard II, some of their leaders may even have been contemplating the establishment of a many-headed monarchy, based upon the shires. According to 'Jack Straw's confession', Wat Tyler in Kent was to be the first of many county kings;[1] and Geoffrey Litster, banqueting in state in Norwich castle, no doubt felt himself to be enjoying a foretaste of his regal pomp. Himself a small artisan, he was *idolus Northfolkorum* and it was in search of a charter for his county that he sent forth his ill-fated emissaries. Fantastic as such pro-

---

[1] *Hist. Ang.* ii. 10.

jects must appear, they were not, perhaps, wholly retrograde; for they drew their strength from an ancient sense of common worth and a common will in the shires which was to survive the inevitable failure of the class-war of 1381 and the withering away of an alien feudalism, to remain an indispensable element in the political development of the English as a free people.

As to the communism of John Ball, this was both less original and less significant than the agitated chroniclers supposed. Dr. Owst has shown that such crude egalitarian doctrines were the common coin of contemporary preachers;[1] for they were of a kind to take root easily in the soil of medieval catholicism. The people were accustomed to hear from their clergy and to see depicted on the walls of their churches, lurid representations of the fate awaiting the rich, the proud, and the oppressor at the Judgement Day. They had been taught to reverence poverty in the person of

Iesu Cryst of heuene,
In a pore mannes apparaille,[2]

they knew that God is no respecter of persons and that rich men enter hardly into the kingdom of heaven. It cannot have been difficult for Ball to persuade an audience thus educated that the inequalities of wealth and status which they saw around them were contrary to the will of God, a violation of the divine order of creation. What made his teaching notorious was not his doctrine of equality, for which he and others might claim warrant in Holy Writ, but the discreditable use that he made of it to further his personal ambitions and to incite his hearers to violence. The sentiments of the people, so far as we can perceive them, were anti-clerical but not, for the most part, anti-Christian. Walsingham had heard of schoolmasters who were forbidden by the rebels to instruct their pupils in the Christian faith, but the majority seem to have assumed the continuing existence of some kind of Church. Even if all the prelates are to be slain, John Ball will be there to assume their functions; if the rich possessioners are removed, the mendicants will suffice for the administration of the sacraments. It was almost inevitable that monastic chroniclers should exaggerate the subversive nature of the rebels' designs on religion, that they should have

[1] G. R. Owst, *Literature and Pulpit in Medieval England* (1933), pp. 210 ff.
[2] *Piers Plowman*, B xi. 179–80.

been deaf to the note of Christian piety which is unmistakable in some of the vernacular literature of the revolt.

A rising of such dimensions presupposes some preparation. Unrest among the servile classes was of long standing and a petition presented to Richard II's first parliament shows that the landlords, mindful of the horrors of the Jacquerie, were already dreading the possibility of a similar movement among the English peasantry. Villeins, the petitioners alleged, were banding together to withdraw their services and had collected large sums of money in support of their cause.[1] The circulation of such strange allegorical letters as those preserved by Walsingham and Knighton, and the rebels' conception of themselves as a 'Great Society' (*Magna Societas*) show some kind of organized agitation at work.[2] Yet it remains very doubtful whether a general revolt would have resulted had not the nation been suffering from deep-seated political malaise. Sir Richard Waldegrave,[3] the Suffolk knight who acted as Speaker in the parliament of 1381, made this point in a speech which has been too little regarded. The causes of the Rising, he said, were the extravagance of the court and household, the burden of taxation, the weakness of the executive, and the inadequacy of the national defences.[4] It is evident that the commons in parliament were at one with the rebels in their condemnation of these evils; and dissatisfaction with the government may well afford partial explanation of the support given to the Rising by members of the landowning classes in Norfolk and elsewhere. Intensified by the levy of the poll-tax, this discontent served to light the fires of revolution, which put government and landlords at the mercy of the rebel leaders, who seized the opportunity thus offered them to ventilate their social as well as their political grievances. The Rising itself, however, had no perceptible effect on the disabilities of peasants or artisans nor (except that, here and there, it may have given some added impetus to the leasing of the demesne)[5] on the social and economic forces which, as has been shown in earlier chapters, were slowly transforming

[1] *Rot. Parl.* iii. 21.

[2] See the document printed in Powell, *Rising in East Anglia*, p. 127.

[3] He had been one of the commissioners appointed in Mar. 1381 to investigate evasions of the poll-tax in Essex.

[4] *Rot. Parl.* iii. 100.

[5] It is said to have been Wat Tyler's insurrection which persuaded Thomas IV, Lord Berkeley, to abandon cultivation of his demesne. *Lives of the Berkeleys*, ii. 5–6.

conditions of life and labour in town and countryside. Much of its importance lay in its revelation of the extent to which the government as a whole, and the duke of Lancaster in particular, had lost the confidence of the people. It came as the last of three successive indictments of the government within a decade; and its effects were felt in the sphere of political, rather than of social and economic history. The revolt did, indeed, kill the poll-tax; but it created an atmosphere of general nervousness which long outlasted its suppression. It encouraged the aristocracy to plunge once more into ill-considered enterprises abroad; and to Richard II, now growing to manhood, it had demonstrated the unique position accorded by popular sentiment to the person of the king and the almost mystical power inherent in his sovereign word.

XIV

RICHARD II, HIS FRIENDS AND HIS
ENEMIES (1381–8)

LITTLE is known of Richard of Bordeaux until he steps to the forefront of the political scene during the Peasants' Revolt. Born on the feast of the Epiphany 1367, he had been brought to England at the age of four, shortly after the death of his elder brother, Edward of Angoulême, and his early childhood was passed at the royal manors of Berkhamsted and Kennington, in the care of his mother, the princess Joan, and of the *magistri* chosen for him by his father. Despite her chequered matrimonial history, Joan of Kent is given a high character by her contemporaries. She was beautiful, gentle, and peace-loving, a devoted wife to the Black Prince and in no sense a political intriguer. It is certain that she must have trained her son to respect his father and to reverence his memory and all the evidence suggests that, when she became a widow, she looked for support to the duke of Lancaster, her husband's brother and most intimate friend. Richard's three *magistri* had all been knights in the service of the Black Prince. Guichard d'Angle, created earl of Huntingdon at the coronation, was merely an elderly figure-head; but Richard Abberbury remained in the king's household after his accession and later became chamberlain to Queen Anne; and in Simon Burley, Richard found an ally whose loyalty to the Crown was to cost him his life. Burley who, in 1377, was a man of about forty, had fought with the Black Prince in France and at Nájera; after the prince's death, he took office in the household of his widow. His reputation in the world of chivalry was high and he had some taste for books;[1] as acting-chamberlain of the household after Richard's accession, he had control of the group of chamber knights surrounding the king, upon whose favours they were exclusively dependent. Burley himself owed almost all his considerable fortune to court patronage;[2] and a knight of his reputation can hardly have

[1] The inventory of Burley's goods taken after his death shows that he possessed a considerable library. See Maude Clarke, *Fourteenth Century Studies*, pp. 120–1.

[2] For details see N. B. Lewis, 'Simon Burley and Baldwin of Raddington', *Eng. Hist. Rev.* lii (1937), 662–9, and C. D. Ross, 'Forfeiture for Treason in the reign of

failed to attempt the education of his pupil in the military and chivalric traditions proper to a grandson of Edward III. But Burley was too old to be Richard's closest confidant and companion. This place of intimacy was reserved for the hereditary chamberlain, Robert de Vere, earl of Oxford, who was the king's senior only by five years. Already a royal ward, de Vere was brought into the family circle by his marriage, in 1378, to Edward III's granddaughter, Philippa de Coucy. Henceforward, he was known as 'the king's kinsman' and Richard's partiality for him found expression in numerous grants of lands and privileges. Seemingly a man of neither talent nor judgement, though not wanting in personal courage, de Vere was foolish and irresponsible, rather than sinister or dangerous. But it was only to be expected that his friendship with the king should arouse jealousy and expose him to the kind of charges that had been levelled against Gaveston with whom, indeed, despite his ancient lineage, he seems to have had many characteristics in common. De Vere's folly, like Gaveston's, set his royal master's feet on the road to ruin and made his own destruction inevitable.

His mother, his favourite *magister*, and his young companion, with his uncle of Lancaster standing powerfully behind them, were the persons best placed to influence Richard's youth. During these early years he saw something of the pageantry of kingship and was encouraged to undertake certain formal duties. The effect of the coronation ceremony on an impressionable child of ten may well have been profound; and the king's customary appearances at the opening of each parliament would have served to keep alive memories of a drama in which his had been the leading part. No doubt his mentors tried to teach him that kingship implied responsibilities as well as privileges; but his whole environment in childhood and adolescence was such as to foster notions of himself as a unique personage; and such notions must have been strongly reinforced by the events of 1381. Richard's courage in the face of the rebel hosts is sufficient refutation of the calumny that he was by nature a coward or a weakling; but their astonishing readiness to follow him was heady flattery for his self-esteem. He alone, it seemed, could control them and it was for him alone to determine their fate.

Richard II', ibid. lxxi (1956), 564-5 and n. 4. Mr. Ross shows that there was much truth in Knighton's statement (ii. 294) that Burley had used his influence with the king to increase his hereditary patrimony from 20 marks to some 3,000.

He was young and we need not doubt that his impulses were generous; but the lavish promises of manumission and pardon must have brought an intoxicating consciousness of power. Educated as he had been, he can have understood little of the social and political grievances which had provoked the rising; but he understood, as never before, that he was king of England. It may well have been at Mile End and at Smithfield that the seeds of his tragic destiny were sown.

Something of this may have been apparent to Richard's subjects; for anxiety as to his environment and education is clearly audible in the parliament of November 1381. Complaints about the outrageous numbers preying on the resources of the royal household (with special reference to the king's confessor, Thomas Rushook), were linked with demands for a commission headed by the duke of Lancaster, to investigate both the household and the 'estate and governance' of the king's person; and before the dissolution it was announced that two mature and experienced statesmen, Richard, earl of Arundel, and Sir Michael de la Pole, had been appointed 'to attend the king in his household and to counsel and govern his person'.[1] Arundel, a prominent member of the older aristocracy, was then in his thirty-seventh year. He had been admiral of the west, a member of the council appointed in the Good Parliament and, after Sudbury's murder, temporary keeper of the great seal. He was a typical product of late fourteenth-century chivalry, aggressive in temper, militant in taste, and little disposed to humour the fitful impulses of an opinionated boy, even if that boy were his king. Richard never liked him; and Arundel was altogether incapable of the kind of *finesse* which might have enabled him to take advantage of his new position. Pole, too, had served as admiral and fought for many years in the French wars; he respected military prowess and was always anxious that Richard should prove himself as a soldier. But he came of a middle-class family which had achieved political influence by supplying the financial needs of the Crown; he was of a breed to feel more at home in king's palaces than in baronial circles, though his appointment in 1381 makes it clear that he was not at this time regarded as in any sense a royalist partisan. Almost certainly, Pole was the brains of what was to become the court party. He never behaved like an irresponsible favourite; and his subsequent decision to support the king

[1] *Rot. Parl.* iii. 100–4.

against his critics is easily explicable by the fact that he had
been one of Lancaster's retainers and by well-founded doubts
as to the character and motives of the opposition. His place in
Richard's affections was assured by the part he played in bringing about the king's marriage to Anne of Bohemia, which took
place in January 1382 and was intended by the pope to herald
the formation of a great Urbanist league against the Clementists.
The marriage was, at first, unpopular; for, although the new
queen was pious and well educated, she was also plain and poor
and the chroniclers considered that she had been bought at too
high a price.[1] As time passed, her virtues gained recognition,
though the Bohemians who accompanied her to England continued to enjoy the evil reputation commonly attaching to the
fellow-countrymen of a foreign queen.[2] We know little about
them except a few of their names. Richard employed certain of
them to buy horses for him in Bohemia, but the majority were
chaplains and ladies. None of them can be shown to have had
contacts with Wyclif or his followers; and it seems certain that
the queen herself can never have been a patron of the lollards
whom her husband regarded with open aversion.[3] For Anne
soon won Richard's passionate devotion and he would seldom
allow her to leave his side; but there is no evidence that she
sought to restrain his excesses and it is likely to have been her
docility which charmed him.

The king's marriage coincided with a period of confusion at
home and humiliation abroad. A state of public nervousness
had been engendered by the Peasant's Revolt and fears of a
recurrence of popular unrest were frequently expressed. London
was torn by faction and Oxford by heresy. There was general
agreement on the need for strong military and naval action but
no consensus of opinion as to what form such action should take.
John of Gaunt's unpopularity had diminished as a consequence

---

[1] *Hist. Ang.* ii. 46; Knighton, ii. 150; Higden, ix. 12. Queen Anne's dower was
fixed at £4,500. It was met by extensive grants of lands, rents, and prises and by
assignments on the customs of London and four other ports (*Cal. Pat. Rolls 1381–85*,
pp. 125–7). The dower was supplemented by a number of additional grants. The
Westminster chronicler rudely suggests that this was too high a price to pay *pro
tantilla carnis portione.* [2] *Hist. Ang.* ii. 119.
[3] See R. R. Betts, 'English and Čech Influences on the Husite Movement',
Trans. Royal Hist. Soc., 4th ser. xxi (1939), 71–102. Some of Wyclif's philosophical
works reached Bohemia in the last decade of the fourteenth century but it was not
until early in the fifteenth that his theological works became known there.

of the insurgents' attacks on his property and reputation and of
his success in negotiating a truce with the Scots. His appointment to the parliamentary commission of 1381 was indicative
of a new confidence in his judgement. He was now once more
turning his thoughts to Spain. Edmund of Cambridge was
already in Portugal whither he had been sent with a small
force in the hope of winning Portuguese support for an Anglo-
Aragonese alliance against Castile, the main assault to be
launched from the north under the command of Lancaster himself.[1] When parliament reassembled after the queen's marriage
and coronation, he felt strong enough to ask the commons to
guarantee a loan of £60,000 for the equipment of an expeditionary force. Some of the lords approved the scheme; for
Cambridge and his army were known to be in difficulties and
large-scale intervention in the Peninsula might put an end to the
raids of Castilian galleys on the English coast. Others protested,
however, at the danger of the duke leaving the country with a
large force so soon after the late rebellion; and the commons
were unenthusiastic. At a great council in March, Lancaster
proposed that the king in person should lead a large army to
France; once in Gascony, such a force could without difficulty
be diverted to Castile. But the parliament which met in May
rejected this scheme also and yielded nothing—partly, perhaps,
because of the persistence of rumours about the king's extravagance and the rapacity of the courtiers. Edmund Mortimer,
earl of March, had died in December leaving a seven-year-old
child as his heir; and Walsingham's complaint that the profits of
this great wardship which might have been used to reduce the
king's debts, were being dissipated in grants to his servants,
finds some support in the record of gifts of lands and offices
appertaining to it to knights and esquires of the household,
like Burley's nephew, Baldwin Raddington, Adam Ramsay and
Richard Hampton, and to yeomen of the chamber, like Matthew
Swettenham, John Horwode, and John Wimbush.[2] Many of the
lands of William Ufford, earl of Suffolk, who died without issue

---

[1] P. E. Russell, *The English Intervention in Spain and Portugal*, pp. 283 ff.

[2] *Hist. Ang.* ii. 68–69; *Cal. Pat. Rolls 1381-85*, pp. 88–94. At the end of 1383 a
great part of the Mortimer inheritance was put in the custody of a group of magnates headed by Arundel and Warwick; but a few months later the marriage of the
young earl was transferred from Arundel to Thomas Holland, earl of Kent, who
married him to his own daughter, Eleanor. See G. A. Holmes, *The Estates of the
Higher Nobility*, p. 18.

in February 1382, went to Pole. Burley himself was given a castle in Wales as a reward for his services in escorting the queen from Germany; a little later, he became justice of south Wales at an annual fee of £40, and under-chamberlain for life. Among his lesser perquisites was a shilling a day as master of the hawks and keeper of the mews for the king's falcons at 'Charryng by Westminster'.[1] The unpopular Thomas Rushook became archdeacon of Llandaff; and the value of the Crown revenues was being steadily diminished by a series of minor grants.[2] Royal borrowing on a considerable scale was the inescapable consequence of such extravagance. The crown itself, with other jewels, was pledged to the city of London for £2,000; Sir Robert Knollys received other Crown jewels as security for a loan of £2,779; four loans totalling over £5,000 came in on 25 March; in August £1,333. 6s. 8d. was borrowed from Brembre and in September £2,700 from a Lombard named Matthew Cheyne.[3] The praiseworthy efforts of the Lancastrian chancellor, Richard Scrope, to put a brake on Richard's ill-advised generosity resulted only in his dismissal from office; and Walsingham makes bitter comment on the malice of the courtiers and the folly of the young king who was their tool.[4]

Lancaster had not, however, abandoned hope of national support for his expedition to Castile; and when parliament met in October 1382, it was to listen to an able speech from his friend, the bishop of Hereford, who suggested that God had opened to the nation two ways of escape from the perils that beset her abroad. One was a general crusade to be launched under the direction of the bishop of Norwich against the Clementists, wherever they might be found. The other, also dignified with the title of crusade (for Juan I had declared for Clement VII), was Gaunt's projected excursion to Castile. Plenary remission had been promised by Urban VI to the supporters of either, and it was for the lords and commons to declare their will. Opinion was divided; although the bishop of Norwich, who had most of the commons behind him, took the bold step of

[1] Cal. Pat. Rolls 1381–85, pp. 161–2.
[2] e.g. £10 p.a. from the cloth subsidy in Berks. and Oxon to Adam de Colton; all the lands of John Wylkyn, one of the St. Albans rebels, to Hugh Martin; the goods of an Oxfordshire suicide to John Seymour (ibid., pp. 120, 125, 164).
[3] Steel, Receipt of the Exchequer, p. 46; Cal. Pat. Rolls 1381–85, p. 104.
[4] Hist. Ang. ii. 70. Scrope was succeeded at the chancery by the new bishop of London, Robert Braybroke, who had been the king's secretary.

publishing his bulls without waiting for authorization, Gaunt
had the support of the king and the lords. He had lowered his
price from £60,000 to £40,000 and when parliament was dissolved after a short session, a subsidy had been granted and it
looked as though he would have his way. Two disasters served
to shipwreck his hopes. The earl of Cambridge's expedition had
ended in failure, for Ferdinand of Portugal, tired of waiting for
the promised reinforcements from England, had made a secret
treaty with Juan of Castile (a set-back which had been carefully concealed from parliament); and, in Flanders, a large
French army secured control of Ypres. The Flemings were
routed at Roosebeke on 27 November and the leader of the
Gantois, Philip van Arteveldt, with whom the English had been
negotiating, was killed. Two days later Charles VI and the duke
of Burgundy entered Bruges in triumph, seized the goods of the
English merchants there, and ordered the cessation of commercial relations with England. As the season was far advanced,
Ghent was not invested; but French control of Bruges threatened
the wool staple at Calais and while arrangements were being
made for its transfer to Middelburg in Zeeland, the wool trade
was almost at a standstill and a main source of royal revenue
dried up. Unless Flanders was to become a base for an attack on
England, action against the French there had now become imperative; all that remained to be decided was the form the expedition should take. The arguments which had already been
stated in favour of the bishop of Norwich's crusade—that it
would be financed by the alms of the people and at small cost
to the government, that it would cement the Anglo-Flemish
alliance, and that, by diverting French troops to the north, it
would relieve pressure on the frontiers of Gascony and Spain[1]—
were now reinforced by others, even more cogent. But, whereas
Gaunt and the lords, fearing that Despenser would consider
only ecclesiastical interests, wished the expedition to be led by
the king or by some responsible layman, the commons expressed
nervousness at the prospect of either Richard or his uncles
leaving the country and, in the spring parliament of 1383, urged
acceptance of the bishop's offer. Their spokesmen were two of
the primate's brothers, Philip and Peter Courtenay, and they
had the backing of the clergy. When parliament was dissolved
on 6 March Despenser had won his point. The crusade was to

[1] *Rot. Parl.* iii. 140.

set forth under his leadership and, though he was to have the
assistance of some experienced captains, like Hugh Calveley and
William Elmham, he had rejected the suggestion that Arundel
should accompany him as king's lieutenant, and no position of
command was given to a lay magnate.

Such was the origin of the deplorable Norwich crusade.[1] It
was launched amid scenes of extraordinary enthusiasm, themselves significant of the prevailing mood of national hysteria.
The chroniclers bear witness to the avidity with which the
credulous of all classes, men, and more especially women,
sought to buy the plenary remissions offered to supporters of
the crusade. Gold and silver, jewels, rings, and precious stones—
'the secret treasure of the realm which is in the hands of its
women'—poured into the bishop's coffers. Knighton explains the
general enthusiasm by the papal offer to remit the sins, not only
of the living but also of the dead. Angels from heaven, the pardoners were saying, would descend at their bidding to bring
souls out of purgatory and waft them forthwith to the skies.[2]
Little wonder that Wyclif in his rectory at Lutterworth should
have been stirred to bitter protest at the use of such methods to
reinforce the common allurements of war.[3] Despenser himself
had seen service in Italy and his energy and resourcefulness had
been demonstrated in the summer of 1381; but he was no strategist and his enterprise was doomed from the start by fatal
delays, by the hostility or indifference of the lords and, above
all, by lack of firm direction and a clear-cut objective. Ignoring
a last-minute attempt by the king to recall him, Despenser embarked at Sandwich on 16 May, landed at Calais, and followed
the coastal road to Gravelines which he seized and sacked. The
inhabitants of Bourbourg submitted without a fight and the
bishop's vanguard pushed on to Dunkirk which was entered on
25 May, after a short encounter with a local force. The surrender of Nieuport, Dixmude, and other coastal towns soon followed
and a junction with the Flemish forces was achieved early in
June. To protests that the count of Flanders was Urbanist,
Despenser merely replied that effective power in Flanders now
rested, not with the count, but with the king of France and the

[1] See E. Perroy, *L'Angleterre et le schisme*, pp. 166–209, and F. Quicke, *Les Pays
Bas a la veille de la période Bourguignonne* (1947), pp. 336–55.
[2] Knighton, ii. 198–9. See also *Hist. Ang.* ii. 85.
[3] *Polemical Works*, ii. 592, 595, 603, 610 ff., 624.

duke of Burgundy. What was to be the next objective? The crusading zealots at home and in the army would have liked to see the bishop turn south in the direction of Artois with a view to chastizing the Clementists inside the borders of France itself; the realists who had supported the crusade in the hope of saving the wool trade, had expected that Bruges would be the first target; but the Gantois, thinking primarily in terms of defence against a renewed French attack, desired an assault on Ypres. Despenser yielded to their persuasions but with some reluctance. Ypres could be taken only by investing it and, though he had some primitive artillery,[1] his army was ill-equipped for a long siege. The new recruits, London apprentices and others, who poured out from England were an undisciplined rabble, interested only in loot, while others of the original force, wearied of the siege and deserted. When, at the beginning of August, news came through that Philip of Burgundy was approaching at the head of a French army, Despenser abandoned the siege and was himself abandoned by the indignant Gantois. He was now ready to listen to the arguments of those who, like Calveley, wished to see the crusaders turn their attention from Urbanist Flanders to Clementist France; but Elmham, Trivet, and the other captains refused to comply and wisely concentrated their attention on the now inevitable retreat. By the end of the month only Bourbourg and Gravelines still remained in English hands. Elmham accepted a French offer of 2,000 francs to evacuate the former; and, a few weeks later, Despenser himself, after a vain delay in the hope of receiving reinforcements, vented his rage in a sack of the town and evacuated Gravelines. When he returned with his disordered forces to England, there was no concealing the fact that the much-vaunted crusade had ended in total failure. Not a single schismatic had been converted; the French had not been chastised; the weavers of Ghent remained isolated and helpless;[2] the route from Calais to Bruges remained closed and access to the markets of Flanders was still denied to English traders. The bishop of Norwich, the Urbanist cause, and, indeed, the whole Church had been gravely discredited by the gross abuse of indulgences and by the exploits of an army of marauders

---

[1] The bishop is said to have had 'unam machinam magnam et unum trepget, cum una magna gunna, vocata gunna Cantuariensis'. Knighton, ii. 199.
[2] Lancaster, however, did insist on the inclusion of the Gantois in the subsequent truce of Lelighem (Jan. 1384).

masquerading as soldiers of the Cross. Retribution of some kind was unavoidable; and in the autumn parliament of 1383 Pole, who was now chancellor, undertook the 'impeachment' of Despenser on four charges, 'in the presence of the king himself and of my lord of Lancaster in full parliament'. Professor Plucknett (who has drawn attention to this remarkable instance of an impeachment in which the commons are not so much as mentioned, the proceedings are controlled by the chancellor, professedly acting under the king's orders, and the 'trial' takes the form of an altercation between him and the bishop, no witnesses being called nor written evidence produced), sees in it evidence of the desire of the government to capture the new weapon of impeachment and to use it as a procedure for prosecution by the Crown, instead of by the commons. The same procedure was used against Despenser's captains, who were accused of taking bribes from the French and defended themselves on the curious grounds that they had given value for the money, in the shape of good horses sold to the enemy.[1] All of them escaped more lightly than might have been expected. The bishop forfeited his temporalities but they were restored to him after two years and most of the captains suffered only a short term of imprisonment.[2]

Lancaster and his friends could now claim ample justification for their scruples about the crusade and he himself had by no means abandoned hope that his own alternative might yet be followed. When enthusiasm for the crusade was at its height, the commons had insisted on the opening of negotiations with Castile, which, had they come to fruition, would have made an end of Gaunt's ambitions there. But Pole arranged for instructions to be sent to the English envoys at Bayonne to delay their conversations, since the duke of Lancaster was determined to take the offensive in the spring of 1384. Within the Iberian peninsula itself, prospects had improved. The death of the vacillating Ferdinand of Portugal in October 1383 was followed by a dynastic revolution which smashed the Castile-Portugal alliance and brought Portugal back into the Urbanist fold. But the failure of the campaign in Flanders and the threat of

[1] T. F. T. Plucknett, 'State Trials under Richard II', *Trans. Royal Hist. Soc.*, 5th ser. ii, 161–4.

[2] Richard II is said to have 'consoled' the bishop after his sentence. It may have been grateful memories of this which led Despenser to attempt to put up a fight against Henry of Lancaster in 1399.

renewed trouble in the north made it impossible to contemplate another expensive expedition in 1384 and Gaunt had to content himself with concluding the truce of Lelighem with Charles VI (January 1384) and with a short excursion to Scotland. His brief spell of popularity was at an end and his relations with Richard were deteriorating. The king was growing up and de Vere was encouraging him to shake off the tutelage of his uncle; the depth of Richard's resentment became apparent during the stormy session of parliament at Salisbury in April 1384. He was in petulant mood and when Arundel began to complain of misgovernment, he lost his temper and told the earl to go to the devil. Lancaster intervened to restore the proprieties, but worse was to follow; for a Carmelite friar named John Latimer, after celebrating mass before the king in the earl of Oxford's chamber, informed Richard that his eldest uncle was plotting his murder. Richard seems to have been ready to believe him and to condemn Gaunt as a traitor without further inquiry; but the duke defended himself with dignity and the lords present in parliament persuaded the king to hold his hand and to have the friar committed to prison in the castle while his charge was investigated. On his way there, however, the friar was intercepted by a band of knights, the king's half-brother, John Holland and other Lancastrians among them, and after attempting to implicate Lord William Zouche, he was brutally tortured to death. No evidence has come to light to show what lies behind this extraordinary episode and attempts may well have been made to hush up details, if, as seems not unlikely, de Vere and the courtiers were in some way implicated.[1] However this may be, the friar's action occasioned not only the hasty dissolution of parliament, but also the first recorded breach between Richard and his uncle Thomas of Woodstock, who is said to have broken into the royal chamber in a rage and sworn that he would kill anybody, the king not excepted, who tried to impute treason to his brother of Lancaster.[2] Richard and his friends were temporarily silenced; but it is probable that the condemnation during the summer, of Gaunt's friend, John of Northampton, was not unconnected with what had happened at Salisbury.

[1] For a discussion of the episode see Tout, *Chapters*, iii. 392, n. 1. L. C. Hector has exploded the widely accepted view, based on a misreading of the passage in Higden, ix. 34 that Richard on this occasion threw his cloak and shoes out of the window. *Eng. Hist. Rev.* lxviii (1953), 62-65.

[2] *Hist. Ang.* ii. 114-15.

In the city of London John of Northampton had swum to power on the tide of reaction following the revolt of 1381. Elected mayor in October of that year, he had constituted himself the champion of the smaller masters and artisans of the gilds and in his attack on the monopoly of the fishmongers had enlisted the support of other elements as well.[1] His policy triumphed when a series of civic ordinances against the fishmongers was followed, in the October parliament of 1382, by a statute which granted complete freedom of trade to 'foreign' fishmongers, virtually excluded the London fishmongers from the trade, and forbade the holding of any judicial office in the city by a victualler. Walter Sibill, an alderman and a member of the fishmongers' gild, entered a protest in parliament, whereupon one of the city's representatives there accused him and another alderman of having assisted the rebels of 1381 in their assault on London Bridge. Northampton's confidence had been growing, for he had lately been re-elected to the mayoralty with the king's approval, and the fishmongers, who were believed to be forcing up prices, cheating the public, and seeking to extend the jurisdiction of their own court, were widely unpopular. But the statute of 1382 marked the turning-point in his career. Sibill was cleared of the charge against him[2] and the attack on the victualling gilds as a whole lost the mayor many supporters. Next year (1383) he was replaced by the grocer, Nicholas Brembre; and, though Northampton seems to have believed that the duke of Lancaster would use his influence to overset the election, Gaunt very properly refused to interfere. Northampton thereupon began to organize 'conventicles and confederacies' against the new mayor, there were riots in the city, and in January 1384 he was bound over to keep the peace. He continued, however, to act as before and in August he was tried before the king's council, his own clerk, Thomas Usk, bearing witness against him, and was sent to prison in Tintagel castle, after protesting against judgement being passed on him in the absence of 'his lord', the duke of Lancaster.[3] It was not until two years later that Gaunt was able to procure a mitigation of his sentence; and the origin and nature of the connexion between

[1] Sylvia L. Thrupp, *The Merchant Class of Medieval London*, pp. 77 ff.
[2] This seems to have been malicious. See Bird, *Turbulent London*, pp. 56–60.
[3] *Hist. Ang.* ii. 116; Higden, ix. 46. The Westminster chronicler records Richard's rage at the protest.

them remains obscure. We have no record of any contacts earlier than Northampton's appeal to Gaunt in October 1383, but the appeal itself suggests that by that date they must have been in some way linked. They may have had anti-clerical sentiments in common; for, though Northampton can hardly be described as a lollard, his attempts to suppress immorality in the city were tantamount to an attack on the jurisdiction of the ecclesiastical courts. Gaunt may have been hoping, with Northampton's help, to organize a Lancastrian group in the city to offset the hostility of Walworth, Philipot, and the other wealthy capitalists with whom he had long been at variance. He may, as Miss Clarke suggested, have been linked to Northampton through Pole, by his opposition to the Middelburg staple, which was favoured by Brembre.[1] At all events, Northampton, like Wyclif, evidently believed that he could look to Gaunt for support; and Gaunt's enemies may well have supposed that in striking at Northampton they were striking at him.[2] The duke's unpopularity was growing and the well-informed Westminster chronicle goes so far as to suggest that in the spring of 1385 Richard II was privy to a plot to have his uncle put to death. He resented Gaunt's attempts to make him take the field in France and his criticisms of the misgovernment of the realm; while the lords, says the chronicler, always feared the duke because of his mighty power, his foresight, and prudence. From the same authority we learn that Gaunt did not hesitate to speak strongly to his nephew about the evil counsellors who surrounded him, and to urge their dismissal.[3]

Here, at least, the duke spoke for many of his fellow-countrymen. Richard's character and the greed and unwisdom of his courtiers were giving rise to widespread anxiety.[4] The king was revealing himself as capricious, headstrong, and irresponsible. On the eve of the Norwich crusade he had attempted to recall the bishop; his response to the news that Gravelines was about to fall was to gallop through the night from Daventry to Westminster, pausing at St. Albans only to change horses; but when he had recovered from the ride, he lost interest and took no action. While the crusade was in progress the king and queen made

---

[1] *Fourteenth Century Studies*, pp. 41 ff.
[2] Brembre's election as mayor is said to have been *rege favente*. Higden, ix. 30.
[3] Ibid., ix. 55–56.
[4] See, for example, Gower, *Vox Clamantis* (probably written in 1383), vi, caps. 8–16.

a round of the abbeys of East Anglia coming, as Walsingham complains, not to offer gifts, but to prey upon the resources of the monks.[1] When Archbishop Courtenay ventured to remonstrate with the king on his choice of counsellors, Richard drew his sword and made to strike him through the heart and when Pole intervened, he too incurred the royal wrath. Walsingham speaks of Richard's habit of casting insults—*verba contumeliosa*—right and left in a manner most unbecoming to his dignity. Even to a not unfriendly observer like the Westminster chronicler, the king appeared 'greedy of glory', delighting to see the archbishop genuflect before him and other suppliants grovel on their knees. The unpopularity of the courtiers is reflected in numerous parliamentary petitions for reform and economy in the royal household and in complaints about illicit extensions of the jurisdiction of the household courts and the increased use of the signet.[2] When Pole made his first parliamentary speech as chancellor he was evidently on the defensive, forestalling criticism by explaining that he did not speak out of presumption or to advertise himself, but by the king's orders and because it was customary for the chancellor to open parliament.[3] In this same parliament of October 1383 evil counsellors are said to have been the cause of grave dissensions between the king and the lords.[4] Both Pole and de Vere had to sustain open attacks on their integrity. In the Salisbury parliament of 1384 a fishmonger named John Cavendish made a thinly veiled charge of bribery against the chancellor and was fined 1,000 marks for defamation; and a similar sentence was incurred by Walter Sibill who, six months later, charged de Vere with maintenance.[5] When Richard granted Queenborough castle to de Vere in March 1385, he thought it necessary to add an anathema to the deed of gift: 'The curse of God and St. Edward and the king on any who do or attempt aught against this grant.'[6] Walsingham tells us that when John of Northampton's property was confiscated, the courtiers fell upon it like harpies; and the charge seems to be well founded, for among those who profited by Northampton's fall were chamber knights, like John Golafre and John Beauchamp, yeomen of the chamber, like

---

[1] 'Non offerre sed auferre.' *Hist. Ang.* ii. 96; see also Higden, ix. 20.

[2] Under the secretaries, John Bacon (1381–5) and Richard Medford (1385–8), there was a marked increase in the number of warrants issued under the signet.

[3] *Rot. Parl.* iii. 149.   [4] Higden, ix. 26.

[5] *Rot. Parl.* iii. 168–70, 186.   [6] *Cal. Pat. Rolls 1381–85*, p. 542.

William Hert and John Elys, a Bohemian squire named Roger Siglem, and the king's chaplain, Nicholas Slake.[1] The same chronicler suggests that the circle at court, 'knights of Venus rather than of Bellona', were teaching the young king effeminate habits and discouraging him from undertaking, not only military exercises, but such kingly sports as hunting and hawking;[2] and, though this cannot have been true of either Burley or Pole, it may well be that among the younger courtiers, headed by de Vere, there was a tendency to indolence and softness. At a happier period of English history such weaknesses might have passed as the harmless foibles of youth; but the country had suffered two major disasters in the Rising of 1381 and the fiasco of the Norwich crusade and there had been many minor set-backs, both military and diplomatic. The more responsible among the magnates, the parliamentary knights, and the substantial merchants were keenly alive to the national humiliation, the damage to England's credit abroad and their own helplessness in the face of Richard's apparent indifference to public opinion and reliance on a closed circle at court. The only remedy seemed to be to encourage him to follow in the footsteps of his father and grandfather and learn the art of war; and it was with this end in view that the autumn parliament of 1384, 'considering that the lord king intends to fight his enemies in person' (a point which had been stressed by Pole) 'and that this is his first expedition', made a generous grant of two fifteenths.[3] An expedition to France was what the estates envisaged; but, instead, Richard found himself forced to lead an army to Scotland.

Gaunt had long been concerned with attempts to heal the running sore of the Scottish problem. He was personally well disposed towards the Scots, a number of whom were in his retinue; and a settlement on the northern border would not only strengthen and enrich the Crown but would also facilitate the prosecution of his own schemes in southern Europe. Lancastrian interference on the Border was not welcome, however, to the magnates of the north and the establishment of a Border commission with Gaunt at its head had involved him in a bitter quarrel with his former ally, Henry Percy earl of Northumberland, at the time of the Peasants' Revolt. None the less, a truce

---

[1] *Cal. Pat. Rolls 1381-85*, pp. 468, 493, 531, 569, 573, 581. The king had consulted Slake in the matter of the Carmelite friar. *Hist. Ang.* ii. 113.

[2] *Hist. Ang.* ii. 156.   [3] *Rot. Parl.* iii. 185.

with the Scots had been secured and such was Lancaster's popularity in Scotland that the Scottish lords even offered to lend him an army to lead against the insurgents. When, two years later, the truce expired and the Scots renewed their attacks, Lancaster went north once more; but he dealt with them as mildly as possible and on his return drew up an agreement with Northumberland entrusting him with the defence of the Border and the safety of the northern shires.[1] Such a reversal of Gaunt's former policy may well have been prompted by a desire to leave his hands free for other projects; and, despite the bishop of Winchester's opposition to the allocation of subsidies for the defence of the north,[2] large sums were assigned to both Northumberland and Neville in the course of the following year.[3] But an Anglo-Scottish settlement did not suit the French, eager to profit by England's domestic embarrassments; and Charles VI's advisers decided to send French troops to Scotland as a prelude to an invasion of England from the south. A small band which landed in 1384 was followed a year later by a substantial force under the command of the famous admiral, Jean de Vienne. Here was a challenge which could not be ignored; and preparations for the muster of an immense army were set on foot forthwith.[4] The size and splendour of his host, which included every magnate of consequence, delighted the young king; but the expedition was dogged with misfortune from the start and never had Richard shown to worse advantage than on this, his first military enterprise and the only great national effort of his reign. In a brawl near York, John Holland, whose violence had already been manifested in the affair of the Carmelite friar, killed the heir of the earl of Stafford, a young man who had been brought up in the queen's household. Holland fled into Lancashire and Richard, in a paroxysm of rage and grief, swore that his brother should be dealt with as a common murderer. The death of the princess Joan a few weeks later is said to have been precipitated by the bitterness between her sons. From York the great host moved north to Durham and Berwick; and when they were over the Border, the king judged the time propitious for the conferring of new titles on some of the leaders. His two younger uncles,

[1] See R. L. Storey, 'The Wardens of the Marches of England towards Scotland, 1377–1489', *Eng. Hist. Rev.* lxxii (1957), 595–8.   [2] *Hist. Ang.* ii. 108.
[3] e.g. £4,047 in Aug. 1384 and £2,000 in Dec. to Northumberland; over £2,700 to Neville in Jan. 1385. Steel, *Receipt of the Exchequer*, pp. 49–50.
[4] Above, p. 235.

Cambridge and Buckingham, became respectively dukes of York and Gloucester; and Michael de la Pole received the earldom of Suffolk, a title already foreshadowed by the grant to him of numerous Ufford lands. After burning the abbeys of Melrose and Newbattle, on the pretext that they were Clementist and were being put to secular uses by the enemy, the English entered Edinburgh, only to find that, though some of the Scots had fled to the north, their main army led by Jean de Vienne, had slipped down the western road to Carlisle. Gaunt advised a campaign of devastation north of the Forth; but Richard had wearied of the whole business and, goaded by de Vere, he turned on his uncle with rude reminders of his own military failures and announced that for himself, he was going home. The enemy had not been encountered and within about a fortnight of crossing the Border, the English army was back at Berwick, with nothing but the wasting of the lowlands to its credit.[1]

So inglorious an adventure was enough in itself to reinforce Lancaster's determination to pursue his ambitions in southern Europe where his prospects had still further improved. In response to the appeal of a Portuguese emissary, a small force of English volunteers had landed at Lisbon in April 1385; four months later the Portuguese won the resounding victory of Aljubarotta which freed Portugal finally from the dominance of Castile and ruined the hopes of its Clementists. Thus, when Gaunt approached parliament in the autumn with a fresh appeal for the 'way of Portugal' he found king, lords, and commons ready to listen to him. The outlook was promising; and it was suggested that the conquest of Castile by a Lancastrian army would force the king of France to treat for peace. Subsidies were granted for what was now accepted as a national enterprise; Gaunt dined with his old enemy Brembre and reconciled himself to the London capitalists; Urban VI's three-year-old bulls were brought out of storage and, though indulgences were rather less marketable than in 1383, considerable sums were raised. On 8 March 1386 Richard II in full council formally recognized his uncle as king of Castile. Gaunt was eager to be

[1] The view taken by Mr. Steel (*Richard II*, pp. 104-5, 108) that the Scottish expedition was a 'qualified triumph' was certainly not shared by contemporaries. See *Hist. Ang.* ii. 132-3. Jean de Vienne's return home, after wasting the villages of Cumberland, was due less to English activity than to French distaste for living conditions in Scotland (see Froissart, x. 333-9).

gone,[1] and Richard no less eager to see him go. So long as the earl of Oxford stood at the king's right hand, there was no hope that Gaunt could make his influence felt; and the security of de Vere's position had become plain to all when the parliament which confirmed the titles granted in Scotland acceded to his elevation to the rank of marquis of Dublin, an unprecedented title, reminiscent of that assumed by Roger Mortimer in 1328, and likewise designed to give him precedence over all the other earls. With the new title went a grant of all royal lands, jurisdictions, and authority in Ireland, in terms which pointed to the creation of a palatinate greater even than that of Lancaster. Hardly less ominous for the future had been the king's curt rejection, in the same parliament, of commons' petitions for an annual review of the royal household and for the names of its officers.[2] And a glaring example of royal recklessness was afforded by Richard's reception of Leo, titular king of Armenia, on whom he bestowed lavish gifts and entertainment and an annual pension of £1,000.[3] The king, says the Westminster chronicler, was so liberal that he gave to all who asked him, dissipating the revenues of his Crown so that he was forced to recoup himself by taxing his people.[4] There was no longer any place at court for the duke of Lancaster or for the policy which he had pursued consistently over the past ten years. Attempts to maintain the solidarity of the royal family, the dignity of the Crown, and the military reputation of the sovereign had broken down in the face of Richard II's obstinacy and instability. Much might have been hoped from Pole. But jealousy of his friendship with the king and of his elevation to the peerage and the suspicion that he was on the make, had lost him the confidence of the magnates and he was being driven, in self-defence, to identify himself, as others identified him, with the gang at court. Lancaster cannot have been blind to the possibility of a new parliamentary crisis and he had no wish to relive the mortifications of 1376. Thus, preparations for his departure were pushed ahead. As constable of his army which numbered some 7,000 men, he took the turbulent John Holland, now his son-in-law, and as admiral, one of the no less turbulent Percies; since his wife accompanied him, it was natural that he should take his daughters. His heir, Henry

---

[1] A point stressed by Froissart (xi. 325).　　[2] *Rot. Parl.* iii. 213.

[3] Leo V of Armenia was attempting to reconcile the western nations in the hope of persuading them to launch a crusade against the Turks.

[4] Higden, ix. 80–81.

Bolingbroke, remained to watch his private interests; and Katharine Swynford and the four young Beauforts were also left at home. Whatever he may have hoped for in Castile, Gaunt did not lack inducements to return. But when he sailed with his fleet out of Plymouth harbour on 9 July 1386, he left his country exposed to a situation as perilous as any that had confronted her since the crisis of the Peasants' Revolt.

Most immediately alarming was the threat of large-scale invasion from France. Philip of Burgundy had succeeded his father-in-law as count of Flanders and was urging upon Charles VI the preparation of a great expeditionary force. Taxes were imposed, stores assembled, and a great fleet mustered in the Channel ports; it was even proposed to transport across the narrow seas a number of wooden huts to serve as a base camp on English soil. Rumours of imminent invasion which had been rife since the beginning of May, revived again in the late summer and early autumn.[1] Gaunt was urged to hasten the return of his transports from Spain, the towns were ordered to see to the repair of their walls, arrangements were made for calling out levies and for raising loans. Possibilities of a counter-offensive were discussed in a great council at Oxford but it was obvious that no decision could be taken without reference to parliament; and it was primarily in order to discuss the war situation that the estates were summoned to Westminster for 1 October.[2] They met in an atmosphere of acute nervous tension, with the invasion scare at its height and large numbers of unpaid troops roaming through the suburbs and the home counties in search of food and plunder.[3] Public opinion, no less than private animosities, demanded a scapegoat; and the most obviously eligible victim was the principal minister of state, Michael de la Pole, earl of Suffolk. Richard was present as usual for the opening oration and heard the chancellor explain the king's desire to take the field in person and that rumours of his unwillingness to do so were wholly without foundation. At the conclusion of the speech the king withdrew to Eltham, leaving it to the lords and commons to decide whether or not they were willing to raise the enormous subsidy of four fifteenths which the chancellor had demanded. Knighton tells us that feelings were exacerbated by

[1] *Cal. Pat. Rolls 1385–89*, pp. 175, 177, 214, 217, 258–61.
[2] Chaucer represented Kent in this parliament.
[3] *Hist. Ang.* ii. 147–8; Knighton, ii. 213.

the announcement, a few days later, that the marquis of Dublin had been raised to the rank of duke of Ireland.[1] Such an extraordinarily ill-judged move was enough in itself to provoke a conflict and it soon became clear that whatever military action might be thought desirable, it would not be taken under the direction of the favourites whose loyalty to the national cause was already deeply suspect. Lords and commons sent word to the king at Eltham that they desired the removal of the chancellor and treasurer (John Fordham, bishop of Durham), because they were useless to the realm and because the estates had business to do with Suffolk which could not be done while he was still chancellor. Richard's reply was to order them to be silent, for he would not at their request remove a scullion from his kitchen. But the estates held to their point and after a short interval sent another message, demanding that the king should come in person and dismiss the chancellor. A little shaken by this, Richard intimated his readiness to receive a deputation from the commons. Such a body could easily have been intimidated; and no doubt it was the duke of Gloucester who saw to it that the wishes of parliament should be conveyed, not by a handful of knights and burgesses, but by himself and his friend, Thomas Arundel, bishop of Ely.

Both were young men in their early thirties and both had good reason to hate the favourites. Thomas of Woodstock, the youngest of Richard's uncles, was a man of choleric temper and militant tastes whose misfortune it was to have been born too late. Like his father and his eldest brother, he could have found a congenial outlet for his energies in the great campaigns of the forties and fifties. But he was in his cradle when Poitiers was fought and he had grown to manhood in a period of national decline. Too near in age to his nephew to share Gaunt's sense of paternal responsibility, he was no less eager to see the traditions of Edward III maintained and he had seized such opportunities for action as the times afforded—in the Channel, in Brittany, and in Scotland—while making the most of his position as president of the constable's court.[2] His resentment at Richard's treatment of Lancaster had already been shown at Salisbury; and any good that might have been done by his elevation to the

---

[1] 13 Oct. Knighton, ii. 215. (Knighton is our best authority for the conversations at Eltham and his chronicle provides an invaluable supplement to the meagre official roll of this parliament, *Rot. Parl.* iii. 215–24.)

[2] *Complete Peerage,* v. 719–28. As constable, Gloucester presided over the famous Scrope-Grosvenor case and at the trial of Sir John Annesley.

rank of duke of Gloucester was more than outweighed when the same honour was accorded to the supposedly effeminate de Vere. It was intolerable to him that the king should waste his time and his substance on the favourites and parasites at court while the tried soldiers of the aristocracy and the princes of the blood were being cold-shouldered and denied the stimulus of military action against the French. The brothers Arundel shared his feelings. The notorious ill will persisting between the king and earl Richard may have dictated Gloucester's preference for the bishop as his fellow-spokesman at Eltham; but both bishop and earl were the sons of a man, much of whose fortune had been made in the wars of Edward III and the cleric, no less than the soldier, had reason to fear the climbers at court. Gloucester and Ely were now determined to stand no nonsense from Richard II. In an address purporting to come from the whole parliament, they reminded the king that he was bound to summon parliament once a year and to be present at its meetings; if he withdrew himself, the lords and commons were entitled to disperse after forty days. Richard's answer could hardly have been more ill advised. Declaring that it was clear to him that his people were plotting rebellion, he said that he would seek help from his kinsman, the king of France; to which Gloucester retorted that if the king of France were to set foot in England, he would destroy the king, not assist him. In a speech instinct with all the passionate resentment of the chivalry against the unmilitant court of Richard II, the duke urged his nephew to remember the sufferings of his royal forebears and their subjects in the wars with France. Heavy taxation was producing a nation of paupers; evil counsellors were at the root of its miseries; let the king remember that 'by ancient statute and recent precedent' they have a remedy. A king who is guided by evil counsel, who withdraws himself from his people, neglects to maintain their laws, and is governed only by his own capricious impulses, can be removed from the throne and supplanted by some other member of the royal family. Adopting a rather more conciliatory tone, the emissaries then tried to appeal to Richard's better feelings. The great and famous realm of England, foremost in the world in the time of his grandfather, is now desolate and divided. Why is he so much under the influence of his counsellors that he dare not remove them? Nothing but ill can come to him from these divisions among the lords of his realm.

This speech, with its thinly veiled allusion to the fate of Edward II, broke Richard's resistance.[1] He returned to parliament, dismissed Suffolk from the chancery and Fordham from the treasury and replaced them by the bishops of Ely and Hereford.[2] The commons then, in the words of the official roll, 'came before the king, the prelates and the lords in the parliament chamber, complaining grievously of Michael de la Pole, earl of Suffolk, former chancellor of England, he being there present and accused him orally. . .'.[3] The articles of accusation which followed were, however, neither particularly remarkable nor very well substantiated. Three relating to the conduct of the war and the expenditure of supplies were subsequently dropped and the trial soon degenerated into an attack—reminiscent of the proceedings against Wykeham in 1376—on Pole's use of his powers as chancellor. He met his accusers with ability and spirit; and his brother-in-law, Sir Richard Scrope, himself an active critic of curial extravagance, reminded the estates of his long record of patriotic service. The name of the Speaker of the commons is not given; but the spectacular part played by them in Suffolk's trial seems to indicate a deliberate attempt—cynically directed, it may be, by Gloucester and Arundel—to follow the precedents of 1376 and to recapture the process of impeachment for the representatives of the shires and boroughs. It was never difficult to concoct charges of fraud against a medieval chancellor; but the case against Suffolk was manifestly weak and his sentence—forfeiture of all irregular grants made to him, payment of a fine, and imprisonment pending the king's pleasure—was relatively moderate. He was probably the least guilty of all the courtiers; but he was singled out for attack by reason of his official position and possibly, as Miss Clarke suggested, because Richard had made the immunity of his personal friends a condition of his dismissal of the chancellor and treasurer.[4]

Suffolk disposed of, the opposition leaders were concerned to

[1] There was, of course, no 'statute' authorizing the deposition of an unsatisfactory king; but the references in Knighton (ii. 219) and in the questions to the judges (below, pp. 448–9), suggest that the spokesmen must have had some actual document in mind—perhaps, the articles of Edward II's deposition or a record (since lost) of proceedings in the parliament of 1327.

[2] The new ministers were admitted to office on 24 Oct.; on the same date John Waltham replaced Walter Skirlaw as keeper of the privy seal.

[3] *Rot. Parl.* iii. 216.

[4] For the charges against Pole see *Fourteenth Century Studies*, pp. 49–52, and N. B. Lewis, *Eng. Hist. Rev.* xlii (1927), 402–7.

secure safeguards for the future. A 'great and continual council' with comprehensive powers over all matters of state and over the king's household and servants, was appointed to hold office for a year. In composition, it was reasonably unprovocative. The three principal ministers of state were members *ex officio* and the inclusion of the duke of Gloucester and the earl of Arundel was inevitable; but counsels of moderation could be expected from the two archbishops, the bishops of Winchester and Exeter and the abbot of Waltham, and from the duke of York and Sir Richard Scrope; while the two remaining members, John, Lord Cobham and Sir John Devereux, were both men of mature years and wide military and administrative experience. The tasks before the commissioners were to retrieve the king from the overweening influence of the favourites, to reform the administration and to take counter-measures against the enemy; and payment of part of the subsidies granted was made conditional on their being allowed freedom to do their work. Had Richard been prepared to co-operate with them, the majority of the commissioners would doubtless have been willing enough that he should assume his proper responsibilities when their year of office expired; but the king's minority had already been unduly protracted[1] and even before the dissolution he made his resentment clear, protesting that neither he nor his Crown should be in any way prejudiced by anything done in this parliament and that the prerogatives and liberties of the Crown should be saved and kept.[2] As a further gesture of defiance he appointed the courtier knight, Sir John Beauchamp, steward of the household; Suffolk's fine was remitted and he was sent to Windsor where Burley was his jailer and where the king joined them for Christmas. Meanwhile, the commissioners showed their intention of taking their duties seriously. They began sealing writs on their own authority and a number of minor officials were dismissed.[3] Much more important in the eyes of the public was the vigorous counter-offensive undertaken in the Channel by the earls of Arundel and Nottingham who gained a brilliant victory off Margate over a combined French and Spanish fleet,

---

[1] He was within a month of his twentieth birthday when this parliament was dissolved. Henry III's minority had ended when he was nineteen, Edward III's before he was eighteen.

[2] *Rot. Parl.* iii. 224.

[3] It seems to have been at this time that Geoffrey Chaucer lost his two posts in the customs. *Cal. Pat. Rolls 1385-89*, pp. 241, 248.

destroyed the fortifications of Brest, and returned home laden
with great quantities of wine which was sold off cheaply at
home—the one bright spot, so it seemed to Walsingham, in a
gloomy year.[1]

For the favourites were still, in Favent's words, buzzing around
the king and were widely believed to be plotting the murder of
Gloucester and his friends.[2] Since the commission was in control
at Westminster, the only means whereby the courtiers could
hope to retain their influence was by persuading Richard to leave
the capital. This he was ready enough to do; and on 9 February
1387 he embarked upon the ten months' 'gyration'[3] through the
midlands and the north, designed for the recruitment of a
private army and the consolidation of a royalist party based on
Cheshire and north Wales. Behind these regions lay Ireland,
now the palatinate of its new duke, de Vere; the earl of March
was a minor and the way lay open for the establishment of a new
power in Ireland and for the exploitation of its resources in the
interests of the Crown.[4] But for the present, the king was unwilling to part from his friend; and it is a measure of his infatuation
that he was even prepared to connive (and to persuade the
queen to connive) at de Vere's repudiation of his royal wife in
favour of a Bohemian woman named Agnes Lancecrona, and
to leave it to his uncles to express public indignation at the
annulment of the marriage which followed. De Vere was made
justice of Chester and north Wales; and a chronicler tells us
that while the king was in these parts, he enlisted archers and
other troops, giving his new retainers—the Cheshire archers and
Welsh pikemen who were to be his principal support for the
rest of the reign—badges of golden crowns.[5] Attempts to enlist
supporters elsewhere were less successful; for the agent who was
sent into the eastern counties was taken, and when the sheriffs
were asked what military forces they could raise in the shires and
whether they were prepared to manipulate the parliamentary

[1] *Hist. Ang.* ii. 170.

[2] 'circa regem glomorantes', 'Historia Mirabilis Parliamenti', p. 2 [*Camden
Miscellany* xiv].

[3] 'Revera non est auditum quod aliquis rex gyraverit fines regni in tam breve
tempore sicuti ille fecit.' Knighton, ii. 242.

[4] The Irish parliament in 1385 had asked for a visit from the king himself or
from the greatest and most trustworthy lord of England. See Curtis, *Medieval Ireland*, p. 255.

[5] Higden ix. 94. Cheshire mounted archers were employed as a royal bodyguard
at least as early as 1334 (*English Government at Work*, i. 341).

elections in the king's interests, they replied that they could do
nothing, since all the commons favoured the lords.[1] But the
Londoners protested their loyalty and Richard had other powerful friends. Alexander Neville, archbishop of York, had been
won over by the king's settlement of his long-standing dispute
with the canons of Beverley; and, besides a number of chamber
and household officers, the chief justice of the king's bench, Sir
Robert Tresilian, and other lawyers remained in Richard's
company, forming the nucleus of a private council to which on
occasions some of the magnates were summoned. No doubt it
was Tresilian who suggested the desirability of seeking a legal
basis for repudiation of the commission and reversal of the judgement on Suffolk and it is likely to have been he who arranged
for the two consultations with the judges, first at Shrewsbury
and then at Nottingham, in August 1387.[2]

As Professor Plucknett has pointed out, there was nothing in-
herently improper in the mere fact of the king consulting his
judges and other legal officers, whose function it was to advise
him on points of law.[3] Richard and his judges were blameworthy
only if he attempted to coerce them, and if they allowed themselves to be coerced, into giving opinions which were contrary
to the law as they knew it; and on these points the evidence is
far from clear. At their subsequent trial the judges pleaded compulsion and named the king's confessor, Thomas Rushook,
bishop of Chichester, as the man who had threatened them.[4]
He may well have done so. But most of the opinions to which
the judges subscribed at Nottingham were such as could have
been held in good faith by a body of men trained in the common-
law courts. It was arguable, to say the least, that the commission
of 1386 was damaging to the royal prerogative; it was historical
fact that the king might arrange the order of business in parliament and dissolve it at pleasure. The impeachment of the king's
servants had been repudiated as illegal in 1377 and to the judges
impeachment by the commons before the lords may well have

[1] These inquiries were made at Nottingham in August.
[2] No fully satisfactory explanation as to why two meetings were found necessary
has yet been offered. See S. B. Chrimes, 'Richard II's Questions to the Judges, 1387',
Law Qu. Rev. lxxii (1956), 365-90.
[3] *Trans. Royal Hist. Soc.*, 5th ser. ii (1952), 166.
[4] Though the judges' plea of compulsion was not accepted, the charge of having
threatened them was among the counts on which Rushook was impeached. *Rot.
Parl.* iii. 241.

looked like a dangerous innovation carrying a threat to their
own position. If impeachment was illegal, then the judgement
on Suffolk was *ipso facto* invalid; and to demand the production
in parliament of the 'statute' relating to the deposition of the
king's great-grandfather was undoubtedly an offence against
the king.[1] The judges gave it as their opinion that all those
responsible for certain of the acts named in the manifesto should
be punished 'as traitors'; and, though they refrained from saying in so many words that these acts were treasonable, there
can be no doubt that their replies held dangerous implications for the future. For the acts in question all came under the
head of 'accroaching the royal power', an offence not classed as
treason in the statute of 1352. This had reserved judgement in
doubtful cases to the king in parliament; and, as the history
of Edward II's reign had shown and as was soon to become
apparent again, charges of accroaching the royal power could
be brought against the king's friends as readily as against his
enemies.[2] The manifesto was weighty, for those who subscribed
it included the highest law officers of the realm—Sir Robert
Belknap, chief justice of the common bench, and three of his
colleagues, Sir John Holt, Sir William Burgh, and Sir Roger
Fulthorp, with John Lokton, a prominent serjeant-at-law and, of
course, Tresilian whose clerk, John Blake, had drafted the document. The witnesses to the judges' sealing were the archbishops
of York and Dublin, the bishops of Durham, Chichester, and
Bangor, the duke of Ireland, the earl of Suffolk, and two lesser
men—John Blake and John Ripon, who had acted for de Vere
at the Curia in the matter of his divorce. All were sworn to
secrecy, Richard's intention being to use the manifesto—like
the Cheshire archers and the Welsh pikemen—when the right
moment should come. He was recovering his confidence and
planning his return to the capital; the commission's term of
office was nearly at an end and he believed himself assured of
the support of London whose citizens had promised his representatives that they would stand by him in all that his royal
majesty demanded.[3] It is likely to have been about this time

---

[1] The questions to the judges, with their answers, are printed in *Rot. Parl.* iii.
233 and Higden, ix. 98–101.

[2] Professor Chrimes questions, however, whether the Statute of Treasons had
altogether superseded the common law of treason, under which 'accroaching the
royal power' had been deemed treason (*Law Qu. Rev.* lxxii. 382–4).

[3] Higden, ix. 104.

that he attempted to obtain letters apostolic against all persons threatening the rights of his Crown and his royal liberties.[1]

It may have been the attitude of the Londoners which led Richard so hopelessly astray in his estimate of the state of public feeling; for, when he entered the city on 10 November 1387, they escorted him in procession to St. Paul's and thence to Westminster as though he were returning victorious from a battlefield; no English king, says Knighton, had ever been received with so much honour in any English city, in time of peace.[2] But the Londoners were notoriously unstable as allies and Gloucester and Arundel were now thoroughly alarmed. The secret of the judges' manifesto had been betrayed to Gloucester, probably by Sir Roger Fulthorp; and both he and Arundel were convinced of Richard's intention to come to terms with the king of France. When, on 11 November, the king summoned them to his presence, they excused themselves on the grounds that he had surrounded himself with their enemies. Richard's reply was to issue a proclamation that no one in the city was to sell anything to the earl of Arundel;[3] whereupon he and Gloucester began to muster their forces in Harringay park, to the north of London. Here they were joined by Thomas Beauchamp, earl of Warwick, who, of all the appellants of 1388, has left the least definite impression on the pages of history. Yet the shrewd observer at Westminster clearly regarded him as the moving spirit and the most responsible—as he was the oldest— of the five.[4] From Harringay, the lords withdrew to Waltham Cross, whence they issued circular letters to the Londoners and to the principal religious houses, stating their case against the favourites and asking for support. Multitudes flocked to join them and news of this great combination of forces against him is said to have shaken the king. The favourites and, if Walsingham is to be believed, the archbishop of York in particular, were agog for a conflict;[5] but wiser counsels prevailed. The majority of the hated commissioners of 1386 were desirous of a settlement; and the delegation which went out from the king to

---

[1] *Diplomatic Correspondence of Richard II*, p. 52. (The letter is undated.)
[2] Knighton, ii. 241.
[3] A move ill received by the citizens who regarded him as one of the most valiant lords in the land. Higden, ix. 105.
[4] Ibid., pp. 107, 109-10.
[5] *Hist. Ang.* ii. 164. Neville may have been aware that he came high on Gloucester's list of public enemies.

the lords was composed of eight of the original fourteen members of this body.[1] Compromise of some sort was inevitable; for
the Londoners had already failed the king, saying that they
would fight his enemies but not his friends; and Sir Ralph Basset
who, while protesting his loyalty to Richard, declared that he
would not risk his neck for the duke of Ireland, probably spoke
for many.[2] At Waltham Cross, on 14 November, Gloucester,
Arundel, and Warwick presented the mediators with a formal
appeal (accusatio) against five of the leading king's friends—
Suffolk, de Vere, Tresilian, Brembre, and the archbishop of
York—accompanied by a demand that they should be put under
arrest. The mediators' reply was to urge the three lords to lay
aside their arms and come to Westminster where the king was
ready to receive them; and the lords accepted the invitation, no
doubt recognizing that to refuse it would be to put themselves
irretrievably in the wrong. Richard himself was now eager to
come to terms; for there can be little question that his main concern was to secure the safety of his friends and an opportunity to
rally his forces. Time was the chief requisite and dissimulation
necessary to gain it.

Thus, it was with elaborate courtesies that the appellant lords
were received when they appeared with their escorts before the
doors of Westminster Hall on 17 November. In a strictly formal
speech Sir Richard Scrope explained their fear of the favourites
(some of whom were present) and repeated the demand for their
arrest until the next parliament when they might be tried for
their offences under common law, if a procedure could be
found.[3] Richard accepted the proposal and by naming 3 February 1388 as the date for the meeting of parliament, allowed
himself eleven weeks in which to recover his position. It is unlikely that he had given much thought to what was to happen
when parliament met, for he hoped that by February, if his
enemies had not been put to confusion, at least his friends would
have made good their escape. As for the appellants, the words

---

[1] The archbishop of Canterbury, the bishops of Winchester and Ely, the duke of
York, John Waltham, John Cobham, Richard Scrope, and John Devereux. Higden, ix. 106.

[2] Knighton, ii. 244.

[3] '. . . et communi lege pro eorum maleficiis . . . persectarentur, si contingat
taliter inveniri' (Higden, ix. 108). The fact that Scrope spoke for the appellants does
not necessarily indicate that he had thrown in his lot with them. It may well have
been in the hope of effecting a compromise that he agreed to state their case.

of the well-informed Westminster chronicler show that they were still in some doubt as to the form the trial should take; but they too had time in which to seek advice and lay their plans. In short, a parliamentary hearing in February was satisfactory to both parties and the meeting ended with a great show of good-will when the king carried off the three lords to drink with him in his chamber.[1]

It soon became clear, however, that Richard's undertakings had not been given in good faith. No attempt was made to keep the accused in safe custody. The archbishop of York, abandoning his carriage and all his goods, fled to his diocese in the guise of a simple priest; de Vere disappeared to the north-west; Suffolk, disguised as a hawker, made for Calais where his brother was in command of the castle; but the captain of Calais[2] sent him back to Hull, whence he again escaped to the continent. Tresilian went into hiding in Westminster; and only Brembre remained to suffer arrest in the city of London. On 30 November Richard made another attempt to enlist armed help in the city; but the citizens replied that they were simple tradesmen, unversed in the art of war.[3] On 17 December the writs of summons to parliament were issued with a significant addition; 'indifference to recent disputes' was to be among the qualifications of those elected.[4] The Westminster chronicler tells us that when they became aware that he had played them false, some of the lords wished to depose the king, but that Warwick rejected this suggestion as disgraceful and proposed instead that a strong force should be sent against de Vere. Among the commanders of this army were Gaunt's son, Henry Bolingbroke, earl of Derby, and Thomas Mowbray, earl of Nottingham, both of whom had now openly taken part with the opposition. Derby may have been drawn to Gloucester by their common interest in the Bohun inheritance as well as by fear of the favourites; Mowbray, who was Richard's contemporary and formerly his

[1] Miss Clarke (*Fourteenth Century Studies*, pp. 134-5) rejected the Westminster chronicler's version in favour of Walsingham's which makes the suggestion of a parliamentary hearing come from the king. Gloucester, she believed, desired a hearing in the constable's court of which he was president. There seems to be no evidence, however, that such an idea was in his mind at this date. See B. Wilkinson, *Const. Hist.* ii. 254-6.

[2] He was Warwick's brother, Sir William Beauchamp, who, according to Knighton (ii. 244), had refused obedience to a royal order to cede Calais to the French.   [3] Higden, ix. 108.

[4] 'in modernis debatis magis indifferentes', *Lords' Report*, iv. 725.

friend,[1] had been alienated by de Vere's jealousy of the king's affection for him and aligned with the opposition through his own marriage to Arundel's daughter. De Vere was now known to be advancing from Chester whither he had gone with letters from the king, instructing the constable, Sir Thomas Molyneux, to raise a force with all possible speed. His scheme seems to have been to advance on London and join forces with the king; and when he learned that the lords had blocked the road through Northampton, he came down the Fosse Way to Stow-on-the-Wold hoping to slip past them on the west. By 19 December only two choices lay open to him—to proceed down the Fosse Way to Cirencester, which would take him still farther away from London, or to make a dash through Burford for Radcot Bridge and so put the Thames between himself and his enemies, who had blocked his line of retreat. He chose the second alternative, only to find himself trapped; for Derby was before him at Radcot and while de Vere was unfurling the royal standard and preparing to attack, Gloucester appeared in the rear. Judging the position to be hopeless, de Vere abandoned his army and fled downstream till he found a ford somewhere in the neighbourhood of Bablockhythe and escaped across it to be swallowed up in the December fog. He was the only one of the favourites to attempt to put up a fight for his master; and with him went Richard's last hope of defeating his enemies by show of arms.[2] According to the Westminster chronicle the two friends met once more before Richard smuggled him into Queenborough castle, whence he made his way overseas.

All hope of amicable compromise being now at an end, Richard withdrew for safety to the Tower. The appellants lingered in the Oxford region until after Christmas when they moved towards London and encamped in Clerkenwell fields. The mayor and aldermen had already refused the king leave to billet his troops in the city and, seldom eager to fight in a losing cause, they now went out to greet the victors of Radcot Bridge who were determined to have further speech with Richard. A parley was arranged by the archbishop of Canterbury and the earl of Northumberland and, taking elaborate precautions against treachery, the lords entered the Tower with 500 men,

[1] Richard appointed him marshal of England in June 1385.
[2] The campaign has been ably reconstructed by J. N. L. Myres in *Eng. Hist. Rev.* xlii (1927), 20–33.

while their supporters closed and guarded the gates. What
happened at this interview is obscure; but our best authority depicts the lords as speaking sternly to the king about his breach of
his coronation oath and his plots against themselves and reminding him that his heir was now of full age and could rule effectively; whereupon Richard, *stupefactus*, promised to be governed
by them, saving his Crown and his royal dignity.[1] Walsingham
describes Richard as vacillating and being forced to yield by the
appellants' threats that they would choose for themselves another king.[2] They may even have withdrawn their homage for
two or three days. At all events, there could be no doubt that
the king had now been brought to heel. On 1 January he was
forced to issue new writs of summons omitting the offending
phrase; and the sheriffs were ordered to see to the appearance
of the five appellees in parliament. Though their term of office
had expired, the commissioners began an elaborate purge of the
household, from the buttery (where they are said to have found
a hundred superfluous servants), up to the steward of the household himself. Burley was deprived of Dover castle and the
wardenship of the Cinque Ports and all the judges who had subscribed to the Nottingham declaration were removed from the
bench. Courtiers of all ranks, knights, clerks, bishops, and ladies,
were sent away.

On 3 February 1388 parliament assembled in Westminster
Hall. The king sat on his throne, the prelates on his right hand,
the lay magnates on his left, the crowd filling every corner of the
hall. Behind the king stood the chancellor, before him the five
appellants—Gloucester, Arundel, Warwick, Derby, and Nottingham—in golden surcoats. When the chancellor had opened
the session by a declaration, in very general terms, of the cause
of summons, Gloucester stepped forward and kneeling before
the king, disclaimed any intention to depose him and usurp his
place, offering to submit himself in this matter to the judgement
of his peers—a gesture which places it beyond doubt that he
was the 'heir of full age' referred to during the interview at the
Tower.[3] Richard declared that he held him blameless and the

---

[1] Higden, ix. 114-15.   [2] *Hist. Ang.* ii. 172.
[3] During the conversations at Eltham Gloucester seems to have repudiated the
law of primogeniture as applied to the succession to the Crown, in favour of the
doctrine which Hubert Walter had advanced at the coronation of John in 1199, viz.
that any suitable member of the royal family might be elected as king. (See Matt.
Paris, *Hist. Ang.* ii. 80.)

main business of parliament was then resumed. Geoffrey Martin, the clerk of the Crown, read aloud the articles of the appeal, a performance which occupied two hours. All hearts, says Favent, were shaken by the horrible revelations contained in the document and many were in tears. At the prayer of the appellants, the king then had it proclaimed within the hall and at the great door of the palace, that the accused should come to answer the appeal; when they failed to appear, the appellants demanded judgement by default. They may have been hoping to bluff their way through and secure a summary conviction, but they cannot have been altogether unprepared for the check that they now suffered when the king and the other lords withdrew to deliberate, not the guilt or innocence of the accused, but the much more fundamental question of the legality of the appeal. On this matter they took professional advice; and the common lawyers and civilians whom they consulted gave it as their unanimous opinion that the appeal was invalid, under either common or civil law.[1] This was a readily defensible opinion such as might have been expected from a body of jurists confronted with an unprecedented procedure; for, though the appeal, or *accusatio*, was an ancient and familiar process at common law, there was no precedent for its hearing in parliament. Gloucester and his friends must have been aware of this and must have had their answer ready; they had been scrutinizing documents touching 'the governance of the king's estate and of the realm';[2] and no doubt their supporters among the lords had been well coached; for it seems to have been without much further delay that they made their famous declaration. The civil law, they said, had no place at all, and the common law only a restricted place, in the realm, for above all lower courts was the law and course of parliament; and in cases of such high crimes, touching the person of the king and the estate of the realm, perpetrated by persons, some of whom were peers, it appertained to the lords

---

[1] The reference to the civil law is puzzling; but Miss Clarke's suggestion that it covers an allusion to the constable's court seems to strain the evidence too far. Professor Plucknett's view that the law—English, foreign, and fabulous—was being searched with anxious diligence (*Trans. Royal Hist. Soc.*, 5th ser. ii. 1952, 168–9) is to be preferred.

[2] Orders to send copies of such documents to Gloucester, Arundel, and Warwick were issued to the justices and the officers of chancery and exchequer on 18 Jan. (*Cal. Close Rolls 1385–89*, p. 394). By an order dated 12 Jan. (p. 387) the abbot of Reading was told to deliver a collection of king's bench memoranda to the appellants.

of parliament to be judges, with the king's assent.[1] In the allusion
to matters touching the person of the king and the estate of the
realm, we seem to catch an echo of the Statute of York of 1322;
and this may well have been among the records which Gloucester and his friends had searched. Yet the statute, with its emphasis on the pre-eminent position of the king, was not altogether to
their purpose and it could not conveniently be quoted; the doctrine that high crimes are matters for parliament alone had to be
combined with the well-established claim of peers to judge their
peers; and this the declaration skilfully achieved. Moreover,
though all the accused were charged with treason, the act of
1352 seems to have been carefully ignored, perhaps because it
also allowed the king too large a share in giving judgement,
perhaps because it would have revived in awkward fashion the
whole question of definition of treason. Thus, the famous declaration was framed with strictly practical ends in view; and the
claim to parliamentary supremacy is judicial, rather than legislative or political and is therefore less revolutionary than some
historians have supposed. Deriving from no clearly understood
principle, it established no fruitful precedent; and, their revenge-
ful purposes once attained, the lords were careful to disclaim intention to override the established laws and customs of the realm.

When the declaration had been made and the prelates had
withdrawn, the lords temporal devoted ten days to careful investigation of the articles of the appeal. The substance of the
thirty-nine charges against the accused is contained in the first
—that, seeing the youth of the king, they had accroached to
themselves royal power and had made the king follow their
counsel. Subsequent articles elaborated in detail the corrupt
methods whereby they had enriched themselves and their associates: their designs on the commission of 1386; their suborning
of the judges; the means they had used to persuade the king to
establish a great retinue and to give *signa*, such as had never
been given by former kings, to his retainers; their plots against
the lives of the appellants; the bargains which they were said to
have induced Richard to make with Charles VI; the oaths obtained from the Londoners; their culpability in neglecting the
defence of the realm. Certain articles relate only to individuals,
de Vere, for example, being charged in connexion with the
Radcot Bridge campaign, Brembre with unlawful executions in

[1] *Rot. Parl.* iii. 236.

the city of London. On 14 February the four absentees—Suffolk, de Vere, Tresilian, and the archbishop of York—were all pronounced guilty of treason and all, save the archbishop who was protected by his clergy, were condemned to its full penalties. Three days later Nicholas Brembre, the only one of the five appellees who was then accessible, was brought into the hall. He pleaded not guilty and offered to defend himself by battle 'as a knight should do'. It is at this point in the record that we find the first mention of the commons who intervened to declare that they had seen the articles of the appeal and would themselves have accused Brembre (presumably by way of impeachment) if the appellants had not forestalled them.[1] The next stage in the proceedings has to be pieced together from the Westminster chronicle, supplemented by Favent's *Historia*. The king seems to have intervened with a speech in Brembre's defence which roused a storm, the five appellants throwing down their gages and others following, 'like a fall of snow'. Trial by battle was ruled out by the lords, however, on the grounds that this lay only when there were no witnesses; and a committee of twelve peers was appointed to investigate the charges. They declared that they found Brembre guilty of nothing deserving of death— a judgement which affords striking evidence of the existence of a body of moderate opinion among the lords—and so drove the indignant appellants to attempt to do as the king had done in November and make use of the notorious factions in the city. Representatives of the gilds were summoned and asked if they thought Brembre guilty. Their reply was probably unsatisfactory, for the appellants then summoned the mayor, aldermen, and recorder and asked them the same question. They replied that he was more likely to be guilty than not and the recorder added that the man who concealed treason was worthy of death. Their opinion seems to have shaken the lords who now declared Brembre guilty and sentenced him to death. It is understandable that these unpalatable facts should have been omitted from the official record; and the chronicler's version is the more plausible because it serves to explain the three days' interval between Brembre's appearance in parliament on 17 February and his condemnation on 20 February.[2] Meanwhile, Tresilian

---

[1] *Rot. Parl.* iii. 238.

[2] The chronicler and the roll are in agreement on the dates (Higden, ix. 166–8; *Rot. Parl.* iii. 237–8).

had been discovered in hiding in Westminster; he was brought into parliament and, as he had already been condemned in absence, his plea that the process against him was illegal was ignored and his sentence carried out forthwith.[1] The former chief justice was dragged on a hurdle to Tyburn and brutally executed there; next day Brembre suffered the same fate.

The lesser offenders, none of whom was a peer, were dealt with by means of impeachment. Thomas Usk, under-sheriff of Middlesex and formerly clerk to John of Northampton, and John Blake were condemned and executed. The judges, who pleaded guilty but said that they had acted under constraint, were condemned to the full penalties of treason, but on the petition of the bishops their lives were spared. Thomas Rushook, who was accused of menacing them, was also impeached but his sentence was deferred; later, the pope translated him to the impoverished Irish see of Kilmore. On 12 March sixteen articles of impeachment were presented against four of the chamber knights—Burley, Beauchamp, Salisbury, and Berners—who were said to have taken advantage of the king's youth to set him against his proper counsellors and to have been the principal agents of the greater traitors. Burley was further charged with helping Suffolk to escape, encouraging the king to keep company with de Vere and to give extravagant presents to the Bohemians, and with making illicit use of the great seal. All the accused strenuously denied the charges and none of them was convicted on more than two counts. The case of Burley, who offered to defend himself with his body, excited much sympathy and his fate was not decided until after the adjournment from 20 March to 5 May. He was, says Favent, a knight of the Garter, powerful and humane, a gentleman of the king's court. The queen (who went on her knees to Gloucester), the duke of York, and even the two junior appellants, Derby and Nottingham, interceded for him; but Gloucester, Arundel, and Warwick, supported by the commons, refused to be moved; the most that they would concede to Burley, Beauchamp, and Berners was beheading in lieu of hanging. Sir John Salisbury, supposed to have been the principal agent in the negotiations with France, was made to pay the full penalty of treason. Apart from their determination to see Burley suffer, the appellants' interest in the proceedings seems to have diminished during the second session.

[1] Higden, ix. 167-8. Tresilian's protest finds no place on the official roll.

At the prayer of the commons, they had been awarded the enormous sum of £20,000 for their great expenses in saving the realm and destroying the traitors. Arundel went off to sea; and when the estates reassembled, more leniency was shown. A number of minor offenders were released under surety, Urban VI was persuaded to translate the archbishop of York to St. Andrews and to remove other royalist sympathizers among the bishops to less valuable sees. Provision was made for the dismissal of the Bohemians from the queen's household, for the nomination of the king's council, for a general review of the courts and offices of state, and for keeping unauthorized persons away from the king. To set the seal on the work of this parliament it was enacted that all who sought to reverse its judgements should be held as traitors and public enemies; but at the same time, it was laid down, on the petition of the commons, that its proceedings were not to be regarded as precedents and that in future trials for treason the act of 1352 was to stand. On 31 May Richard entertained the parliament at his manor of Kennington; and on 3 June there was an impressive ceremony in Westminster Abbey when, after mass, lords and commons renewed their oaths of allegiance and Richard promised to be 'a good king and lord' for the future.[1]

The Merciless Parliament well deserved its name.[2] Never before in our history, not even in the dark days of Edward II, had legal sanction been claimed for the destruction on such flimsy pretexts of so many men of gentle birth. It may readily be conceded that the courtiers, or most of them, were greedy, irresponsible, provocative, and wrong-headed; but none of them was a criminal and none was deserving of murder by act of parliament. The authors of these savage punishments must bear some responsibility for the long tale of violence and judicial murder which darkens so much of the parliamentary history, not only of the last years of Richard II, but also of the succeeding century. Yet some of Gloucester's fellow-peers were obviously troubled by doubts and scruples and it may be to their influence that we should ascribe the statute passed shortly before the dissolution—

[1] Favent's explanation (p. 24) of the renewal of oaths appears sufficient, i.e. that it was to allay scruples and because the king had taken his coronation oath as a minor.

[2] '... vocabatur parliamentum istud parliamentum sine misericordia.' Knighton, ii. 249.

which reaffirmed the immunity of estates tail from forfeiture and extended the same protection to the widow's inheritance and jointure—and the care taken to see that the statute was observed.[1] The complete success of the appellants in securing the verdicts they desired is not altogether easy to explain. A partial answer may be found in their wide territorial influence and in the strength of the bonds uniting them. Thomas of Woodstock was not only duke of Gloucester, he was also earl of Buckingham and Essex and, in right of his wife, of Hereford and Northampton; Richard Fitzalan was earl, not only of Arundel but also of Surrey; Thomas Mowbray was earl of Nottingham and heir-apparent to the great estates of the Bigods; and behind Henry Bolingbroke, earl of Derby, lay the vast resources of the house of Lancaster. Moreover, the duchess of Gloucester and the countess of Derby were sisters and Arundel was their uncle; the countess of Warwick and the countess of Arundel (who died in 1385) had been first cousins; and the countess of Nottingham was Arundel's daughter. Yet it does not seem that the appellants carried the northerners with them and it is unlikely that they could have attained their ends had they not enjoyed the full support and approbation of the commons in parliament and of a large section of public opinion outside parliament. None of our sources suggests that at any point in the proceedings did the commons raise a dissentient voice; and the lords' cause appears to have been widely popular in the country. Many Londoners were hostile to Brembre; Arundel had won renown by his exploits in the Channel; after Radcot Bridge, offers of help to the lords began to pour in from every side. Few of the commons are likely to have been inspired by personal jealousy of the courtiers; the causes of their discontent went deeper. All were suffering from the sense of national humiliation consequent on the failures abroad and from the nervous tensions engendered by social disturbances and invasion scares; and many must have been exasperated by the contrast between a luxurious and over-staffed court and an ill-governed and much-taxed people. Finally, the appellants served their own cause well by their decision to keep the king, so far as possible, out of the picture. No doubt, as Professor Plucknett has suggested, the procedure of the parliamen-

[1] See C. D. Ross, 'Forfeiture for Treason in the Reign of Richard II', *Eng. Hist. Rev.* lxxi (1956), 560-75. Mr. Ross shows that Miss Clarke was mistaken in her supposition (*Fourteenth Century Studies*, pp. 115-45) that the statute was ignored.

tary appeal commended itself to them as a criminal process over
which the king had no control; but it was not only because
Richard II might prove awkward that his role was reduced to a
formality and his youth and the evil counsel which had led him
astray, repeatedly stressed. None of the appellants is likely to
have been a revolutionary at heart. Even Gloucester, whatever
private ambitions he may have cherished, must have recognized that no scheme to dethrone the king in his own favour
would have been tolerated for a moment by his brother of Lancaster, by the bulk of the lords, or by public opinion generally.
Richard II was the lawful king; and the only hope for the future
lay in insistence that he had been misled and in allowing him,
in due course, the opportunity of a fresh start. In June 1388
many must have hoped, and some may even have believed, that
the worst troubles of the reign were over.

XV

THE RULE AND FALL OF RICHARD II
(1388-99)

THE king's innocence had been publicly proclaimed and his subjects had renewed their homage; but it was hardly to be expected that Richard II should assume the role of an active ruler forthwith. For nearly a year after the dissolution of the Merciless Parliament he seems to have been content to remain passive, stunned, it may be, by the fate that had befallen his friends. He performed the duties required of him but seldom took the initiative; and the only sign of emotion noted by the chroniclers was an outburst of rage on hearing of the victories of the Scots. Richard presided at meetings of the council and at the Cambridge parliament of September 1388; tournaments and joustings marked the Christmas celebrations at the royal palace of Eltham. Direction of public policy fell to the appellants and their friends. Gloucester took the lead on the council; Thomas Arundel, now archbishop of York, remained at the chancery, where he initiated some useful reforms;[1] his brother, the earl, resumed his naval activities and organized a raid on La Rochelle; Warwick was a member of the small committee appointed to advise and control the king.[2] It says something for the lords' administrative capacities that they succeeded in maintaining peace throughout most of the country,[3] and that in the Cambridge parliament an attempt should have been made to tackle social problems by means of a new statute of labourers and restrictions on the wearing of liveries.[4] It was in the turbulent country of the north that Gloucester and his supporters suffered their most damaging reverse. Encouraged by their knowledge of England's internal divisions, the Scots refused to

---

[1] Tout, *Chapters*, iii. 441 ff.

[2] The other members were the bishops of London and Winchester, John, Lord Cobham, and Sir Richard Scrope (Higden, ix. 178).

[3] Commissioners were sent down to Worcester in July to arrest certain individuals accused of blaspheming the king's person. *Issues of the Exchequer*, p. 239.

[4] In the absence of an official record, we are dependent for our knowledge of this parliament on the monk of Westminster (Higden, ix. 189-200); Knighton, ii. 299-308 and the statute roll (12 Ric. II).

renew the truce after 19 June 1388. Their armies soon appeared south of the Border and on 5 August, at Otterburne (Chevy Chase), on the road between Newcastle and Jedburgh, a Scottish force led by the earl of Douglas inflicted a crushing defeat on the English and captured the boldest of the northern captains, Henry Percy the younger. This disaster, which jeopardized the safety of the north for many years to come, had the effect of inclining some of the magnates to look more favourably on proposals for peace with France. An embassy led by the bishop of Durham to Calais at the end of the year was the first of several which produced a series of short truces culminating, in 1394, in an agreement to suspend hostilities for four years. Though there were still elements in the nation to whom this peace policy was distasteful, the days of the great *chevauchées* were temporarily at an end, and for the remainder of Richard's reign renewal of the war on a large scale was never seriously envisaged. Meanwhile, some of the old animosities at home seemed to be dying away. Though the appellants were busy creating a vested interest in what had been done in the Merciless Parliament by auctioning the lands there declared forfeit,[1] the king listened favourably to a proposal that de Vere's duchy of Ireland should be conferred upon Gloucester and supported the latter in a dispute with the monks of Westminster. John Holland was given the earldom of Huntingdon; and Nicholas Dagworth, who had been censured by the Merciless Parliament, was a member of the mission to Calais.

The king's sudden announcement, at a meeting of the council on 3 May 1389, of his intention to be free of tutelage and to rule as a monarch of full age, affords an interesting example of his taste for self-dramatization; but there is no reason to suppose that it proved seriously disconcerting to his hearers. It was obvious that Richard could not be kept in leading-strings for ever, nor the lords assume permanent responsibility for the government. Indeed, it may well have been with a sense of relief that Warwick withdrew to his estates, that Arundel planned to go crusading in the Holy Land and Derby in Prussia, whither Gloucester soon arranged to follow him. At a council in September the former appellants intimated their desire to see 'good love, unity, and agreement' established between the king and

---

[1] Lands to the value of nearly £10,000 had been granted in fee simple by midsummer 1389. See C. D. Ross, *Eng. Hist. Rev.* lxxi. 570–1.

his lords.[1] Richard had been carefully unprovocative. Though the great seal had been taken from the archbishop of York, it had been transferred to the venerable bishop of Winchester; the new treasurer, bishop Brantingham of Exeter, was likewise an experienced and elderly administrator; and Edmund Stafford, who became keeper of the privy seal, was a rising clerical cadet of a noble house. Five new judges were appointed and some of the appellants' friends were dismissed from the household; but there were no sensational acts of vengeance. The king's intentions had been announced to the nation in writs directing the sheriffs to proclaim that he had taken the government of the realm upon his own person, proposing, with the aid of his council, to rule more prosperously than before.[2] The chroniclers' comments make it clear that his action was generally approved; and the new régime opened auspiciously with the postponement of collection of part of the last subsidy. Richard was now twenty-two, king *de facto* as well as *de iure*; his capacities and his ultimate intentions remained to be revealed.

If the royal authority was to be restored on a basis of public confidence, the first essential was the recall of the duke of Lancaster; for the crisis through which he had lately passed must have awakened in Richard II some recognition of his eldest uncle's persistent care for the dignity of the Crown. Gaunt had withdrawn to Gascony in the autumn of 1387 and in the following May Gloucester, who was eager to keep him out of the country, procured his appointment as Lieutenant. The Spanish adventure had yielded little but private profit. Gaunt had struck bargains whereby his elder daughter, Philippa of Lancaster, was married to the king of Portugal and his younger daughter, Catalina, to the heir of Castille and he himself received an indemnity of £100,000 and an annual pension of £6,000, in exchange for renunciation of his claims to Castile and of the Urbanist cause there; but these arrangements were of no advantage to the nation.[3] Attempts to make them the basis of a permanent peace between England and the Spanish kingdoms had come to nothing; and Castile, Aragon, and Navarre had all been lost to the Clementists.[4] Urban VI showed his displeasure

---

[1] *Proc. and Ord. Privy Council*, i. 12.   [2] *Foedera*, vii. 618-19.

[3] For the campaigns in Spain and subsequent settlements, see P. E. Russell, *The English Intervention in Spain and Portugal*, pp. 421 ff.

[4] Perroy, *L'Angleterre et le schisme*, pp. 240-59.

by cancelling all the privileges he had conceded to the crusaders; but neither of the contending parties at home could now afford to quarrel with Lancaster. Gaunt's agent, Sir Thomas Percy, must have kept him fully informed of all that had occurred since his departure, but his reactions can only be surmised. It is likely that he deplored the fate of Burley and Pole and the humiliation suffered by Richard, while shedding few tears for de Vere; and that, since his eldest son's participation in the appeal made it difficult for him to intervene, he preferred to remain abroad. A peremptory order from Richard was needed to persuade him to change his mind, but he had no cause to regret the decision. The enthusiasm which greeted his arrival in November 1389 is a measure of the extent to which his reputation had been enhanced by the mere fact of his absence from the stormy scenes of the past three years. The chroniclers gloss over the Spanish fiasco and Walsingham explains his own change of front by a story of Lancaster's repentance for his former sins.[1] The king himself appeared determined to show his uncle all the respect he had denied him in the past. As Gaunt approached Reading, where the council was to meet on 9 December, Richard rode out two miles to meet him and give him the kiss of peace; an atmosphere of general goodwill pervaded both the council and the Hilary parliament which followed. Thereafter, the duke of Lancaster stood at the king's right hand, encouraging him to assert himself at home and to pursue peace abroad. His own fighting days were over and he had enlarged his fortune (though not his reputation) in the wars. He must have been aware by this date that Richard would never make a soldier; his role must be that of beneficent ruler of a realm at peace. With Hunting-don as chamberlain and Sir Thomas Percy as sub-chamberlain, two stout Lancastrians were well-entrenched at court; Gaunt himself was rewarded by the extension of his palatine powers in Lancashire to his heirs male and by the grant of the duchy of Aquitaine for life; and the duke of York's eldest son was given the earldom of Rutland. A great hunting-party at Leicester in August 1390, when Gaunt entertained the king and queen, to-gether with his brothers of York and Gloucester, his son-in-law, Huntingdon, the archbishop of York, the earl of Arundel, and many other bishops, lords, and ladies, seemed to betoken a general restoration of unity.[2]

[1] *Hist. Ang.* ii. 193–4.   [2] Knighton, ii. 313.

This atmosphere of harmony was never wholly lost during the next six years. It was a period of conciliar government during which the king accepted certain restraints on his freedom of action, notably on his right to make gifts; in some minor matters, the will of the council even prevailed against the king.[1] But neither in the ordinary council, about which the records now begin to yield fuller information,[2] nor in the occasional meetings of the great council, was there serious clash of interests. Indeed, it was contemplation of the *consilium ita magnum sicut parliamentum* which met at Stamford in the spring of 1392, that led an admiring foreigner, the duke of Gelderland, to exclaim that never in any realm had he seen a community so noble.[3] In the short parliaments of the period, lords and commons appeared more interested in economic and ecclesiastical questions than in criticism of the administration. The commons showed some nervousness about the expenditure of the supplies they granted and there were differences of opinion on the projected peace terms with France. But signs of hostility to the Crown are wanting and the prestige now enjoyed by Lancaster appears in the commons' request in 1391, that he should be one of the negotiators of the truce, 'because he is the most sufficient person in the realm'.[4] Foreign trade was booming; and it was not without justification that in his letters abroad, Richard referred with enthusiasm to his own prosperity and the peace of his realm.[5]

Yet Richard was not by temperament a peacemaker, nor was his mind readily forgetful of past ill will; and there is another side to the history of the so-called period of appeasement. It was soon after the feastings at Leicester that the king had told Gaunt that it was not in his power to pardon his former protégé, John of Northampton. When his uncle protested vigorously at the suggestion of any such curtailment of the royal prerogative of showing mercy—'This and greater things than this are in your power to do'—Richard had replied, 'If this be so, there are others who sustain great misery: now I understand what I may

---

[1] e.g. in 1393 the council quashed an attempt by the king to nominate a mayor for Northampton and ordered a free election. *Issues of the Exchequer*, p. 251.

[2] The minutes of council proceedings published by Sir Harris Nicolas begin in 1389; the journal kept by John Prophete, clerk (see Baldwin, *King's Council*, pp. 489–504), covers the years 1392–3.

[3] Higden, ix. 266.

[4] *Rot. Parl.* iii. 286.

[5] *Diplomatic Correspondence*, nos. 201, 216 (pp. 146, 158).

do for my friends in foreign parts.'[1] Pole died abroad in the autumn of 1389; but the magnates still lived in dread that Richard might seek to recall Neville or de Vere; and it was doubtless with the assent of Gaunt (who must have shared their dread) that in the council of February 1392, the king was made to promise never to pardon them, the lords swearing in return to stand by him in all things and to allow him full power to reign. Meanwhile, the unquiet ghost of Edward II was raising its head again. The autumn of 1390 saw Richard at Gloucester collecting evidence of his great-grandfather's alleged miracles; and while a succession of royal agents was pressing the suit for Edward's canonization at the Curia, the king was arranging with the abbot and convent for the maintenance of lights and ornaments around the shrine, and, perhaps about the same time, ordering the painting of white harts on the piers flanking the tomb.[2] In November 1391 the king secured from parliament a significant petition, that he should be 'as free in his regality, liberty and royal dignity . . . as any of his noble progenitors . . . notwithstanding any former statute or ordinance to the contrary, notably in the time of king Edward the Second who lies at Gloucester . . . and that if any statute was made in the time of the said king Edward, in derogation of the liberty and franchise of the Crown, that it be annulled'.[3] Richard's object in seeking this concession, or so his enemies later asserted, was to enable him to violate the laws of the land.[4] Next year his passionate temper betrayed him into an explosive quarrel with the Londoners. Riots between royal officers and citizens were made a pretext for 'taking the city into the king's hands', for the first time in seventy years. The mayor and sheriffs were removed from office, Sir Edward Dallingridge, a Lancastrian official and a member of the king's council, was installed as warden of the city, and a commission of inquiry was set up; the courts and the whole administration were ordered to remove to York. The chroniclers' story is that the citizens, after refusing a loan to the king, had expressed readiness to oblige a Lombard merchant and, though this may be mere gossip, it seems likely that money was at the root of the quarrel. Miss Bird has shown that loans

---

[1] Higden, ix. 238–9.

[2] Perroy, L'Angleterre et le schisme, p. 330; Issues of the Exchequer, pp. 247–8, 259; Cal. Pat. Rolls 1391–96. Richard is said to have assumed the badge of the white hart at the Smithfield tournament of 1390. Monk of Evesham, p. 122.

[3] Rot. Parl. iii. 286.

[4] Ibid. iii. 419.

from the city to the king fell away almost to nothing after June 1388; the fate of Brembre may well have made the citizens chary of financial dealings with the Crown, and the cessation of fighting in the Channel had removed the main inducement to generosity. An attempt to raise money in the city by distraint of knighthood had been a failure and the fact that recovery of civic liberties was put at the price of £100,000 strongly suggests that the king's object was to enrich himself at the Londoners' expense. In the end, £10,000 was accepted as a free gift; and the reconciliation between the king and the citizens was celebrated by a great procession through the city, reminiscent of the splendours of the coronation day.[1] None the less, it was thought expedient to hold the next parliament at Winchester; and Richard's rash action probably cost him the dubious loyalty of the Londoners. Seventeen of those who received the king's sentence in 1392 were members of the court of aldermen which welcomed Henry of Lancaster in 1399.[2] Walsingham may be right in suggesting that it was Gloucester who persuaded Richard to reduce the crushing fine originally imposed; and Gloucester's popularity in the city may well have had something to do with the king's sudden decision to forbid his projected expedition to Ireland.

The decision was unexpected; for by this date Gloucester appeared to be on the best of terms both with Richard and with Lancaster. Contrary winds had frustrated his intention to visit Prussia, but after his return he seemed willing to support the peace policy favoured by the court.[3] Richard appointed him lieutenant of Ireland in the spring of 1392 and within the next three weeks his 'treasurer of the wars' had obtained 9,500 marks as war wages, out of an agreed total of 34,000 for a period of three years.[4] When the king changed his mind and refused him licence to go to Ireland, Gloucester seems to have accepted the decision without protest; and the spring of 1393 saw him engaged in peace talks at Calais, in the company of Gaunt, Thomas

---

[1] Richard Maidstone, fellow of Merton, thought this occasion worthy of a Latin poem which runs to eighteen pages of print in the Rolls Series (*Polit. Poems*, i. 282-300). See also Helen Suggett, 'A Letter describing Richard II's Reconciliation with the City of London, 1392', *Eng. Hist. Rev.* lxii (1947), 209-13.

[2] Bird, *Turbulent London*, pp. 102-9.

[3] The Westminster chronicler (Higden ix. 282) suggests that Gaunt had bribed Woodstock to support the peace negotiations.

[4] Richardson and Sayles, *The Irish Parliament in the Middle Ages*, p. 151.

Percy, and the courtier knight, Sir Lewis Clifford. Nottingham, too, had long since been reconciled to Richard who made him warden of the east march of Scotland and then captain of Calais. Warwick and Derby (who was much abroad) had ceased to be active in politics. Of the former appellants, only Arundel maintained his old attitude of defiance and suspicion. He remained in sulky isolation, unable to indulge his taste for privateering because the post of admiral had gone to Huntingdon; he married an heiress without royal licence and was heavily fined; and when insurrection broke out on the northern estates of Lancaster and Gloucester during the early summer of 1393, he was believed to be secretly aiding the rebels. The causes of this rising are still obscure; but it seems that the peace negotiations with France were unwelcome to the bellicose population of Cheshire which had produced such war heroes as Hugh Calveley and Robert Knollys;[1] their leader, Sir Thomas Talbot, proclaimed it as his mission to save the king from his uncles of Lancaster and Gloucester. There was also trouble in Yorkshire where Sir William Beckwith, one of Lancaster's officials, was already engaged in a private war. Both princes found it necessary to return hurriedly from France to suppress the rebels. A rising in Cheshire against the duke of Gloucester might seem to carry sinister implications; and the king found it necessary to declare publicly that his men of Cheshire could look for no support from him in a movement directed against the magnates of the realm 'by whose *nobilitas*, the diadem of our dignity is principally sustained and honoured'.[2] It seems unlikely that Arundel was in any way responsible for the rising; but he certainly made no move to suppress it, although from his castle of Holt on the Dee he was in a good position to do so; and in the Hilary parliament of 1394, hostility between him and Lancaster flared up into an open quarrel. The duke accused the earl of assisting the rebels and the earl replied with angry protests against the duke's familiarity with the king, his overbearing manners in council and parliament, his acquisition of the duchy of Aquitaine, the cost to the nation of his Spanish adventure, and the peace treaty with France which he was engineering. Richard intervened with a spirited defence of his uncle and Arundel was forced to make grudging apology for words whereby the king and the lords had

[1] H. J. Hewitt, *Medieval Cheshire* (1929), pp. 150–60.
[2] *Foedera*, vii. 746.

been 'so mychel greved and displeisid'.[1] A little later, his uneasiness drove him to seek a special charter of pardon for his past misdeeds;[2] and when Queen Anne died in the summer, his resentment was still smouldering. He failed to join the funeral procession from Sheen and appeared late at the Abbey with a request for leave to withdraw, thereby provoking the king to strike him to the ground and pollute the sanctuary with his blood. The weeks which he spent in the Tower can hardly have mollified him and Richard, for his part, was determined to humiliate him to the uttermost. Arundel was made to appear before the king in person at the archbishop's manor of Lambeth and to take oath for his future good behaviour, his brother, the archbishop of York, the earls of Warwick, March, and Nottingham, and six others acting as his mainpernors for the enormous sum of £40,000.[3]

The four years' truce with France had been concluded in May, and there was nominal peace with the Scots. Opportunity now offered for a belated attempt to settle some of the problems of Ireland which had been ventilated in the previous parliament. At the root of the Irish question of the late fourteenth century was lack of effective lordship; and efforts to remedy this by legislation had proved quite ineffective. The ordinance of 1368 ordering all Irish landowners to return to their estates or, at the least, to make provision for their defence, was re-enacted in more stringent terms twelve years later; but there were no means of enforcing it and each year saw further Irish inroads into the lands of the English absentees. The situation was aggravated by the misfortunes of the house of March. Edmund Mortimer, the third earl, who had been appointed lieutenant in 1379, had indeed achieved much in a little time. He had secured nominal submissions from the leading Ulster chieftains and had pressed into the wild country of Tyrone, where few English deputies had been before him, and down the Shannon through the midlands to Munster; but he met his death crossing a ford in co. Cork and in 1381 his vast Irish estates fell to his seven-year-old son, Roger. With the Mortimer lands in the hands of a minor, the position of the Anglo-Irish became yet more desperate and there were repeated demands for a visit from the king in

---

[1] *Rot. Parl.* iii. 313-14; *Annales*, p. 166.   [2] *Cal. Pat. Rolls 1391-96*, p. 406.
[3] *Cal. Close Rolls 1392-96*, pp. 307, 368.

person or for the appointment of a leading English magnate as
lieutenant. But neither the earl of Oxford (whose commission as
duke of Ireland gave him almost royal rights), nor the duke of
Gloucester, appointed lieutenant in April 1392, was allowed to
go to Ireland; and since the Irish leader, Art McMurrough, was
penetrating Carlow, Kerry, and Kildare, it became increasingly
imperative that the king should go himself. His decision to do so
was generally approved and his visit was heralded by further
orders to the absentees to return home and by instructions to
Baldwin Raddington and others to find quarters for the entertainment of the king and his magnates.[1] On the surface,
Richard's motives in undertaking the expedition were obvious.
His lordship of Ireland had to be rescued and its revenues restored before it was lost to the rebellious natives; its clergy had
to be brought under discipline and protected from the overtures
of schismatics.[2] Moreover, the king himself was too much the
child of his age to be altogether blind to visions of the personal
renown which must attend success, could he achieve it. There is
no warrant in the evidence for suggestions that at this date he
was harbouring notions of making Ireland the basis of an independent royal dominion in the west.

Richard embarked from Haverfordwest towards the end of
September 1394, leaving Edmund of York as keeper of the
realm, John of Gaunt being bound for Gascony where he remained until the end of the following year. Generous subsidies
had been voted and large loans raised;[3] the imposing host which
accompanied the king included the duke of Gloucester, the earl
of Rutland (soon to be created earl of Cork as well),[4] the earl
of Nottingham, for whom the lordship of Carlow had been
revived, John Holland, earl of Huntingdon, the young earl of
March and Ulster (who was also lord of Connaught, Trim, and
Leix),[5] and many lesser barons. The army landed at Waterford

[1] Cal. Close Rolls 1392–96, pp. 295, 390; Cal. Pat. Rolls 1391–96, pp. 448, 451.

[2] See Perroy, L'Angleterre et le schisme, pp. 96–103.

[3] Mr. Steel puts the sum of these loans at well over £20,000. Thomas Arundel
contributed £1,000, his brother, the earl, £1,333. 6s. 8d. Receipt of the Exchequer,
p. 71.

[4] The title may have been conferred before he left England, see E. Curtis,
Richard II in Ireland (1927), p. 27, n. 1.

[5] For the support of his retinue March was allowed wardship of his lands in 1393,
two years before he was due to come of age. The Wigmore chronicler warmly
praised the guardians for their care of the estates; see G. A. Holmes, Estates of the
Higher Nobility, p. 19 and n. 2.

on 2 October, its immediate objective being the intimidation of McMurrough and the rebels of Leinster. But Richard showed little inclination for the unpleasant business of fighting. The earl of Ormonde was sent inland to deal with the rebels, while the king marched along the Barrow to Kilkenny and thence to Dublin, where he remained until after Christmas. In Dublin castle he and his advisers worked out a plan for the pacification of Ireland, which was to include a general pardon to the 'rebel English' who had allied themselves with the native movement, the creation of an 'English land', east of a line drawn from Dundalk to the Boyne and down the Barrow to Waterford, the expulsion of McMurrough and his supporters from Leinster, and confirmation of their lands to the other Irish chiefs, on con-dition of a double oath of allegiance, to the king and to the Anglo-Norman overlords. Richard was prepared, further, to take the bold step of admitting the native kings and chiefs to full legal status under the Crown. With the Anglo-Irish, or 'rebel English', he achieved little; but the concessions offered to the native princes resulted in the series of remarkable submis-sions which has been described as 'the most general recognition of the English Lordship in Ireland made between the reign of Henry II and 1541'.[1] Led by the O'Neill of Ulster, the heads of some of the greatest families—the O'Briens, O'Connors, and MacCarthys—were received by Richard in person during the early months of 1395. The name and presence of the king cast their inevitable spell and Richard was at pains to honour his strange guests, four of whom were knighted. The rebel McMur-rough had to be handled more severely; but even he, after he had surrendered his lands in Leinster in exchange for a vague promise of 'wages from the king to go and conquer other lands occupied by rebels against the king',[2] was graciously received and admitted to the honour of knighthood. Thus far, the policy of conquest and conciliation had been amply rewarding; but it was proving expensive and Gloucester had been sent back to England with instructions to raise further supplies. Ireland was the principal topic of debate in the Hilary parliament of 1395, where the chancellor (Thomas Arundel) 'rehearsed the manner of the king's passage to Ireland and the great zeal he had shown for the conquest of the rebels in those parts, and how honourably,

[1] Curtis, *Richard II in Ireland*, p. vi.
[2] Curtis, *Medieval Ireland*, p. 271.

with God's help, he had acquitted himself'.[1] Gloucester's appeal induced the reluctant commons to offer a tenth and fifteenth; but the letter of congratulation sent by the lords ended with a request to Richard to hasten his return. They thought it 'probable that you have conquered the greater part of that your land'; the Scots were threatening to break the truce; and other weighty matters needed the king's attention.[2] Richard was in no hurry to comply. He was enjoying himself in Ireland, where he had scored a personal success; and it was not until the middle of May, after the chancellor and the bishop of London had visited him in Dublin, that he embarked at Waterford, leaving it to March, whom he later appointed lieutenant, and the unwilling Anglo-Irish to do what they could to hold the native princes to their easy promises.[3]

Needless to say, Richard II had not solved the Irish problem by a few months' visit. But the submission of the chiefs represented his one substantial achievement since his dispersal of the rebels in 1381 and in the years 1395 and 1396 his credit stood higher than ever before. Opposition to the Crown seemed to have melted away. Richard had learned the value of family solidarity and his kinsmen massed in a strong phalanx around the throne. Foremost among them was, of course, John of Gaunt, his natural loyalty now reinforced by his enthusiasm for the French alliance and by his care for the interests of his third wife and her children. Constance of Castile had died in March 1394, and in January 1396 Gaunt married Katharine Swynford, their children being granted legitimacy in the first parliament to meet after the ceremony. John Beaufort became earl of Somerset and Henry Beaufort was given the deanery of Wells. Within his own territories, Lancaster's liberties were greatly enlarged;[4] and, since his heir, Henry of Derby, had returned from his foreign travels to take his place by his father's side, it seemed that Richard had now nothing to fear from the Lancastrians. His uncle of York had never been troublesome and York's son, Rutland, was among the king's closest confidants, as were also his half-brothers, the Hollands. Thomas Holland, earl of Kent,

---

[1] *Rot. Parl.* iii. 329.   [2] Curtis, *Richard II in Ireland*, pp. 217–19.
[3] There appears to be no foundation for Walsingham's assertion (*Annales*, p. 183) that it was the lollard manifesto to parliament which brought the king home. See *Rogeri Dymmok Liber*, ed. H. S. Cronin (1921), Introd., pp. xvii–xviii.
[4] See Somerville, *Hist. of the Duchy of Lancaster*, p. 65.

died in the spring of 1397 but his heir, another Thomas, remained in the court circle together with his uncle, John Holland,
earl of Huntingdon. The attitude of the duke of Gloucester was
more ambiguous. His support of the French alliance was lukewarm at best, and he was believed to share Arundel's jealousy
of Lancaster.[1] Yet, although Richard had cancelled his appointments as justice of Chester and lieutenant of Ireland, he had
not hesitated to entrust him with the appeal to parliament for
further supplies for the Irish expedition; and, presumably with
the object of keeping up appearances, the duke took an appropriate part in the ceremonies attendant on his nephew's second
marriage to Isabella of France. Outside the family circle few of
the magnates looked dangerous. Arundel was openly hostile and
might hope for some cautious support from Warwick, who had
a private quarrel with Nottingham over a piece of property in
Gower. Otherwise, his position was one of increasing isolation.
The elderly earl of Salisbury (once the husband of the king's
mother) was an adherent of the court, as was his heir-presumptive, Sir John Montagu; the earl of Nottingham now stood high
in the king's favour; the young earl of March had his hands full
in Ireland; and in Henry Percy, soon to be earl of Northumberland, Richard had a friend who had never taken part against
him. Among the bishops, too, the king had his allies. The death
of Courtenay in July 1396 removed a powerful moderating influence; but Richard raised no objection to the monks' choice of
Thomas Arundel as his successor, probably because this would
entail his removal from the chancery. The great seal was transferred in November to Edmund Stafford, now bishop of Exeter
and always the king's good servant. Robert Waldby, who replaced Arundel at York, had been a royal physician before his
elevation to the see of Dublin in 1391; so had Tideman of
Winchcombe, who became bishop of Worcester in 1395. Richard
Medford, translated from Chichester to Salisbury in the same year,
had been one of the king's secretaries; and the learned Thomas
Merke of Carlisle had served the king on diplomatic missions
and remained a loyalist to the end. Ministers of state and of the
household tended, as almost always, to stand by the king. Roger
Walden, treasurer from 1395, had done good service in the
organization of the signet office as a private royal secretariat.[2]
Northumberland's brother, the Lancastrian Sir Thomas Percy,

[1] Froissart, xv. 165.   [2] Tout, *Chapters*, v. 221-3.

was steward of the household, supported there by the chamberlain, Sir Baldwin Raddington, and by the sub-chamberlain, another prominent northerner, Sir William Scrope. Richard seems also to have made some effort to win support from members of the lesser territorial aristocracy, the politically influential class which furnished the knights of the shire. Prominent among them were Sir John Bussy, a Lancastrian official who first appears as Speaker in the parliament of 1394; Sir Henry Green, another Lancastrian retainer; and Sir William Bagot, a dependent of Nottingham's, who had followed his master into the royalist group.[1] All three had been several times members of parliament and all seem to have been to some degree implicated in the dark deeds of 1388; but they had made their peace with the king, and the pliancy of the commons in the parliaments of 1397–8 suggests that Richard must have had other friends among the knightly classes. He also had something in the nature of a private army. The Irish expedition had given him opportunity to recruit a large force of household troops; and all over England, but more especially in Cheshire, 'yeomen and archers of the Crown' had been enlisted, receiving the king's wages and liable to be called up in an emergency. These men wore the king's livery and his badge of the white hart and some of them received modest grants of land as retaining fees.[2]

The twenty-eight years' truce with France sealed on 9 March 1396, and the marriage alliance which symbolized it, looked like the culmination of Richard's hopes. He had long regarded the Anglo-French wars as intolerable,[3] and there was no mistaking his affection for the little queen who was formally delivered to him at Calais in October, amid scenes of splendid pageantry. Yet the French alliance was to prove the beginning of his troubles, a heavy liability rather than a political asset. It was unpopular in many quarters; the notorious clause in the

[1] Mr. K. B. McFarlane has pointed out (*Bull. Inst. Hist. Res.* xx (1947), 173–4) that both Green and Bagot were retained simultaneously by the king and by Gaunt, Bagot also holding a pension from Nottingham.
[2] The author of *Mum and the Sothsegger* (Early Eng. Text Soc. 1936, ii. 41–43) complains bitterly of the depredations committed by the king's liverymen and suggests that for every badge of the white hart which he bestowed, the king lost scores of faithful hearts.
[3] e.g. his speech to the Hilary parliament of 1397 with its reference to the 'tres grandes meschiefs & destructions de Guerre intolerables entre les deux Rôialmes'. *Rot. Parl.* iii. 338.

first draft of the treaty, whereby the French royal house pledged
its support for Richard 'against all manner of people who owe
him any obedience and also to aid and sustain him with all their
power against any of his subjects',[1] may have been intended
merely as a safeguard against the possibility of risings in Gascony or among the English peasantry; but the king's eagerness
for the alliance seemed to many to hold sinister implications.
There was widespread nervousness lest he and his friends should
make damaging concessions to France or commit England to
support of French continental enterprises, and general disappointment at the prospect of the prolonged uncertainty about
the succession which must result from marriage with a child.
Strange rumours began to circulate. It was said that when the
body of Robert de Vere, who had died at Louvain three years
before, was brought home for burial among his ancestors at
Colne in Essex, the king had caused the coffin to be opened and
had gazed long and earnestly upon the face of the embalmed
corpse, clasping the jewel-laden fingers.[2] And the French alliance cost Richard more than a decline in the reputation which
his Irish exploits had won him. With characteristic want of
prescience he seems to have failed to consider how frustrating
such an alliance was likely to prove to some of his other hopes
and schemes. It was the least of his embarrassments that he had
undertaken to send a force, led by Nottingham and Thomas
Holland, to support the Franco-Florentine war against Gian-
galeazzo Visconti of Milan. When the king himself expounded
this project in the first parliament of 1397, the commons, while
agreeing that he must honour his promises, disclaimed any
responsibility for financing them. What might have led to an
awkward clash of wills was luckily averted when the collapse of
the French scheme after the battle of Nicopolis enabled Richard
to withdraw his demand for a subsidy.[3] More serious was the
impossibility of reconciling the French alliance with Richard's
fantastic dream of winning the imperial Crown. Negotiations
with the German princes had been proceeding for some time
past, and two of them—the archbishop of Cologne and the
count palatine of the Rhine—had become his vassals in return

---

[1] *Foedera*, vii. 811.
[2] *Annales*, pp. 184-5. (Walsingham places this episode in Sept. 1395.)
[3] See D. M. Bueno de Mesquita, 'The Foreign Policy of Richard II in 1397:
Some Italian Letters', *Eng. Hist. Rev.* lvi (1941), 628-37.

for a pension. In the spring of 1397 the dean of Cologne came
to England and deluded Richard into supposing that the German electors were prepared to make him king of the Romans;
and by the end of the year he seems to have been prepared to
barter the French alliance for papal support of his claim. He
had made no serious effort to implement the policy of the 'way
of cession' which was by far the most damaging of his under-
takings to the French;[1] for even had there been any hope that
the English clergy would agree to withdraw their allegiance
from Boniface IX, Richard himself could not afford to quarrel
with him. Thus, the much vaunted Anglo-French alliance proved
stillborn; and when Henry of Lancaster was later driven into
exile, he found a ready welcome at the court of France.

With grandiose schemes of empire brewing in his mind,[2]
Richard was little disposed to tolerate criticisms from his own
subjects. The commons, who, in the Hilary parliament of 1397,
sponsored a bill which included a complaint of the extravagance
of the royal household and the multitudes of bishops and ladies
battening on the court, were forced to make abject apology. The
king consulted the lords who resolved that it was treason to
excite the commons or anyone else to reform the household; and
the author of the bill, one Thomas Haxey, proctor to the abbot
of Selby, was adjudged a traitor on these grounds. His clergy
saved him from the extreme penalty and he was shortly afterwards pardoned; but the implications of his case were sinister,
amounting to 'constructive treason', such as had been hinted at
in the judges' manifesto of 1387 and put into operation in the
Merciless Parliament of 1388. Haxey himself was clerk of the
court of common pleas and had received substantial rewards
from the king; his bill is likely to have been inspired, perhaps,
since he represented a northern abbot, by a group of disappointed prelates who had not gained access to the court, perhaps by general clerical hostility to the king's alignment with
the schismatic French. But the complaisance of the lords showed
Richard how little he had to fear from them; and, managed by
Bussy, the commons were ready to admit that their bill was
against his 'regality, royal estate and liberty'. Encouraged by
their subservience, Richard was able to secure the restoration of

[1] Above, p. 146.
[2] It is, perhaps, of some significance that in the roll of this parliament Richard
is described as 'entier Emperour' of his realm of England. *Rot. Parl.* iii. 343.

the judges condemned in 1388; but caution was still necessary, for at the same time the other acts of the Merciless Parliament were formally confirmed. Royal agents were busy in the spring and summer raising loans throughout the country, but the general picture continued to be one of prosperity and peace.[1] Westminster Hall was being reconstructed on a magnificent scale; distinguished foreign visitors came to court; the earl of Huntingdon secured from Boniface IX the resounding title of Gonfalonier of the Church and Vicar-General of the Patrimony of St. Peter, together with crusading privileges greater than any that had been accorded, either to the bishop of Norwich or to the duke of Lancaster.[2] Yet Arundel remained irreconcilable and Gloucester was once again drawing towards him. In February they had both enraged the king by ignoring his summons to a council; a few months later Gloucester broke out into angry complaints of the concessions being made to the French. There were rumours of a plot and the king's suspicions were aroused. He knew himself to be stronger than his enemies and he judged that the hour to strike had come.

His first move was to issue invitations to a banquet, compared by Walsingham to that at which Salome danced before Herod for the head of John the Baptist.[3] Gloucester made the excuse—apparently genuine—of ill health; Arundel, not unnaturally suspicious of his host's intentions, remained immured in his castle at Reigate. Warwick, who accepted the invitation, was received with great cordiality, the king taking his hand and telling him not to grieve over the lands in Gower which he had lately lost to Nottingham; only when the feast was over did the earl find himself being conducted to the Tower. A little later, Arundel was secured by treachery, Richard swearing a solemn oath to his brother, the archbishop, that he should suffer no bodily harm; when he appeared, he was put in Carisbrooke castle in the custody of Sir William Scrope. The king himself determined to apprehend Gloucester. Accompanied by the Hollands, Rutland, Nottingham, and a large number of household troops and Londoners, he rode down to Pleshey by night.

[1] *Annales*, pp. 199-200; *Eulogium*, iii. 372. Mr. Steel (*Receipt of the Exchequer*, p. 76) shows that considerable sums were raised during the summer, much coming from small lenders, though Wykeham produced £1,000 and the city of London over £6,000.   [2] *Annales*, pp. 200-1. See Perroy, *L'Angleterre et le schisme*, p. 343.
[3] *Annales*, p. 201.

When Gloucester came out to meet them, in procession with the priests and clerks of his newly founded collegiate church, his nephew told him that, since he would not heed his invitation, he had come to fetch him. To Gloucester's plea for mercy he replied that he should have just so much mercy as he himself had shown to Simon Burley, for whom the queen had interceded on her knees. Nottingham bore the duke off to Calais and his office of constable was given to Rutland. Unfavourable public reactions to the news of these arrests were countered by a proclamation that they were made, not on account of old offences, but for new ones committed since the king had issued his pardons to the appellants and that all would be made plain when parliament met.[1] A plausible story was also spread about that the king's moves against the magnates were all part of an elaborate game to persuade the visiting Germans that he was fully able to control his own barons.[2]

Plans for the parliament which was due to meet on 17 September 1397 were laid in a council at Nottingham where the judges were present. Eight lords—the king's kinsmen, Rutland, Kent, Huntingdon, and Somerset, together with Nottingham, Salisbury, Despenser, and Sir William Scrope—laid a bill of appeal against the prisoners, cautiously reserving the right to amend it. Following the precedent of 1387 the bill was first presented to the king and was referred by him to parliament. After the council it was publicly proclaimed that the imprisoned lords were detained on a charge of treason; and when parliament opened, an elaborate series of articles in defence of the appeal had been prepared. Since Westminster Hall was not available, the estates assembled in a temporary building in the palace yard. Armed forces were much in evidence; for those magnates whom the king felt able to trust had received licences to come with retinues for his protection and he had ordered all his own yeomen and retainers to attend him at Westminster. Within the building a throne of unusual height had been erected, with places on either side for the appellants and the accused and a wide space before it for the estates.[3] The proceedings opened with a sermon from the chancellor on the text *Rex unus erit omnibus*; the king, said the preacher, had caused this parliament to

[1] *Foedera*, viii. 6-7. [2] *Annales*, pp. 202-3.
[3] The most important sources for this parliament are the official roll (*Rot. Parl.* iii. 347-85); *Annales Ricardi Secundi*, pp. 209-25; *Eulogium*, iii. 373-9; and Adam of Usk's eye-witness account in his *Chronicon* (ed. E. Maunde Thompson), pp. 9-23.

be summoned in order to be informed if the rights of his Crown
had been in any way withdrawn or diminished. Bussy was then
presented as Speaker of the commons. The chronology of the subsequent proceedings (though not their nature) admits of some
doubt; but the first step was probably the revocation of the acts of
1386 and of the pardons granted to the members of the commission appointed in that year; from this revocation, however, all
the commissioners, save Gloucester, Warwick, and the two
Arundels, were at once exempted. The coupling of the archbishop
with his brother was unexpected; for Thomas Arundel had
taken his place as a trier of petitions and had attended the
opening meetings of parliament. It may be that he had opposed
the king's suggestion that the clergy should appoint a lay proctor
to act in their name. From Richard's standpoint, this was an
essential precaution; for the inevitable absence of the clergy
from processes involving judgement of blood might leave a
loophole for resuscitation of the argument used in 1322, when
the judgement on the Despensers had been declared invalid on
the grounds that the prelates, who were peers of the realm, had
not assented to it.[1] At all events, it was not until the archbishop
had withdrawn from parliament by the king's orders that the
clergy appointed Sir Thomas Percy as their proctor, with
plenary powers to act in their name. Immediately afterwards,
Bussy impeached the archbishop for his share in the proceedings
of 1386-8. He was allowed no answer to what were held to be
notorious facts and, on 25 September, he was sentenced to forfeiture of his temporalities and to perpetual banishment. Boniface IX was invited to translate him to a see in the obedience of
the anti-pope and machinery was set in motion for his replacement at Canterbury by the treasurer, Roger Walden.

Meanwhile, the appeal against the three lay magnates had
already begun. The appellants duly appeared, clad in robes of
red silk bordered with white and embroidered in gold, and set
forth their appeal against Gloucester, Arundel, and Warwick;
they asked that these persons be brought into parliament and
declared their readiness to prove the charges against them by
any means which the king should direct. Arundel was the first
to be produced. The duke of Lancaster, in his capacity of high

---

[1] The danger was real, since in this same parliament Thomas Despenser was
petitioning for reversal of the sentence on his ancestors and using the award of 1322
in support of his case. *Rot. Parl.* iii. 360-7.

steward of England, informed him of the appeal, adding (according to one version), 'Since parliament has accused you, you deserve according to your idea of law (*secundum legem tuam*) to be condemned without answer.'[1] Arundel's defence was to plead two pardons, one general and one particular, which he had received from the king; to this Lancaster replied that the pardons were given under constraint—'at the time when you were king' —and that they had been revoked by parliament. Arundel retorted that to grant pardon was the king's prerogative and that if the steward denied this, he was more guilty than himself of damaging it. Bussy broke in on these wranglings and, with his fellow-commons, demanded judgement on Arundel as a traitor. Arundel turned on him—'Ye are all liars! I am no traitor!'[2] The appellants then threw down their gloves and Lancaster asked the accused if he wished to say anything more; when he remained silent, the chief justice of the king's bench was directed to explain to him the consequences of a refusal to speak. His second attempt to plead the charter of pardon was brushed aside and the appellants prayed the king to pronounce sentence on the earl as convicted of all the points in their appeal. By command of the king, the lords temporal, and the proctor of the clergy, Lancaster then declared him guilty of treason and pronounced the customary sentence, which the king mitigated forthwith by substituting the axe for hanging. Escorted by the Cheshire archers, Arundel was taken to Tower Hill, where he was beheaded in the presence of the two Hollands and of his own son-in-law, the earl of Nottingham. Gloucester's case was taken next. In reply to an order to produce him his jailer, the earl marshal, announced that he was already dead. If this were so, the appeal manifestly failed;[3] so, in order to ensure forfeiture of his property, the appellants prayed that none the less he be adjudged a traitor, as one who had waged war against the king's person; and the commons again intervened to stress the notoriety of his appearance in warlike array at Harringay and of his levying war against the king. Thus, it was by looking back once again to what Professor Plucknett calls 'primitive legal notions of conviction by notoriety' that the lords found him guilty and sentenced him, as they had sentenced the archbishop, to

---

[1] *Eulogium*, iii. 374. [2] Usk, p. 14.
[3] Under an act of 1361, forfeitures for treason (other than forfeitures of war) could not be exacted posthumously.

forfeiture of his estates.[1] On the following day, 25 September, probably in the hope of forestalling criticism of this sentence, a justice named William Rickhill appeared at the appellants' request and read aloud a full confession of guilt which, he said, he had obtained from Gloucester at Calais, the day before his death. The eccentricity of the proceedings against Gloucester is almost certainly to be explained by the need to observe reticence on the circumstances of his death; for, though some of the details still remain obscure, there can be little doubt that he had been murdered—Richard not daring to have him produced in parliament—and that Nottingham was party to the crime.[2] Warwick was then brought in. The same procedure was adopted against him as against Arundel, but he made no attempt to defend himself. Instead, 'like a wretched old woman, he made confession of all contained therein, wailing and weeping and whining that he had done all, traitor that he was, submitting himself in all things to the king's grace'.[3] His sentence was the full penalty of treason; but afterwards the king, 'moved by pity . . . and by the prayer of the appellants and commons and all the other lords',[4] commuted it to one of perpetual banishment to the Isle of Man. Satisfaction, rather than pity, is likely to have prompted the concession, for Walsingham reports Richard as saying that Warwick's confession was dearer to him than all the lands of Arundel and Gloucester.[5]

Formal appeals were directed only against the three magnates; but some lesser persons were named in the attached articles. Sir Thomas Mortimer, a kinsman of the earl of March, was impeached in his absence by the commons, with the king's leave, as a fugitive guilty of notorious treasons and was ordered to present himself within three months, or take the consequences; and John Lord Cobham, a member of the commission which had condemned Burley, was likewise impeached and sentenced to exile in Guernsey.

The judicial proceedings of 1397 owed much to earlier state trials and Professor Plucknett has demonstrated the ingenuity

[1] 'Impeachment and Attainder', *Trans. Royal Hist. Soc.*, 5th ser. iii (1953), 151.

[2] For a discussion of this fascinating problem in criminal detection, see J. Tait in *Owens College Historical Essays* (1902), pp. 193-216; A. E. Stamp in *Eng. Hist. Rev.* xxxviii. 249-51 and xlvii. 453; R. L. Atkinson, ibid., xxxviii. 563-4: and H. G. Wright, ibid., xlvii. 276-80.

[3] Usk, p. 16.

[4] *Rot. Parl.* iii. 380.

[5] *Annales*, p. 220.

with which the king's legal advisers helped him to circumvent
the obstacles in his path. The necessary consent of the spiritual
peers was secured by their appointment of a lay proctor. Arundel's defence was answered in advance by the repeal of the pardons granted to the appellants of 1388. The king's control of
impeachment was admitted by the formality with which Bussy,
on each occasion, sought his leave to proceed; and king, lords,
and commons combined to redefine the law of treason and to
extend its forfeitures so as to cover, not only entailed property,
but also all corrodies, fees, and pensions held by the retainers
and dependents of the condemned traitors, their widows alone
being allowed some means of subsistence.[1] For all his military
resources and the high claims made for his regality, Richard II
yet found it necessary to pay some attention to the forms of law.
In the circumstances, this was not easy. For the principle, applied under Edward II, that the king's record was sufficient to
convict a man of treason, had since been rejected, and the lords,
as peers of the realm, had acquired a prescriptive right to pronounce judgement on their peers. Moreover, subsequent state
trials, in 1376, 1386, and 1388, had been the work of opponents
of the Crown and both impeachment and appeal had been devised for use against the king's favourites and friends, not against
his enemies. No direct action in parliament was therefore open
to the king; and it was as much legal necessity as dramatic instinct which drove Richard to parody the arrangements of 1388
by organizing a group of magnates to present an appeal and by
securing a subservient Speaker to initiate impeachments. Parliament, in short, could be manipulated, but the forms of
legality had to be respected. Only outside its walls could a
would-be tyrant allow free rein to his unrestricted sovereign
will.

The king's enemies having been punished, it now remained
for him to reward his friends. Five of them became dukes. The
title of Hereford was conferred on Henry Bolingbroke, that of
Norfolk on Thomas Mowbray; the king's half-brother, John
Holland, became duke of Exeter, his nephew, Thomas Holland,
duke of Surrey, his cousin, Edward of Rutland, duke of Aumale.
John Beaufort became marquis of Dorset; and the earldoms
of Gloucester, Westmorland, Worcester, and Wiltshire were
severally bestowed on Thomas Despenser, Ralph Neville,

[1] See C. D. Ross in *Eng. Hist. Rev.* lxxi. 574-5.

Thomas Percy, and William Scrope. The common people, says Walsingham, spoke mockingly of the king's new *duketti*; for, though all were of gentle birth, the highest rank in the peerage had been jealously guarded hitherto. Nor was the king neglectful of his own interests. The county of Chester, enlarged by inclusion of some of the Arundel property on the marches of Wales, was declared a principality annexed to the Crown; and the hereditary sheriffdom of Worcester, formerly the perquisite of the earls of Warwick, was similarly transferred. On the last day of the session (30 September) high mass was celebrated in the Abbey and at its conclusion a number of lords spiritual and temporal swore before the shrine of the Confessor (whose supposed arms Richard had already added to his shield), to uphold and maintain all the acts of the present parliament. The lay proctor for the clergy, and the parliamentary knights were likewise sworn, before parliament adjourned until 27 January 1398.

That Richard should have been prepared to allow a four months' recess at such a juncture indicates his confidence in the strength of his position; and too much weight need not be attached to Walsingham's tales of how the king's sleep was broken by Arundel's ghost, how he dared not retire to rest without a guard of 300 Cheshire archers, how he had Arundel's body exhumed by night and ordered the Austin friars with whom it rested, to move the tomb so that the people should not venerate the late earl as a martyr.[1] Yet there was uneasiness abroad. Adam of Usk tells us that when the young earl of March appeared at Shrewsbury for the second session, the people met him joyfully, wearing his colours and hoping through him, the heir-presumptive, for deliverance from the grievous evil of such a king.[2] And the choice of Shrewsbury as the meeting-place of parliament suggests that Richard may have felt it wise to be within easy reach of his western appanage. Moreover, he showed no inclination for a long sesssion. Four days proved enough for the business he had in hand.

The first act of the Shrewsbury session was the chancellor's request for supplies for the defence of the realm and of the king's possessions overseas. Then, at the instance of Bussy and the commons, the earl of March (who had not been present at Westminster) took the oath already taken by the other peers.

[1] *Annales*, pp. 218-19.   [2] Usk, pp. 18-19.

Seven of the eight appellants next entered a request, which was granted, that all the judgements and acts of the Merciless Parliament be annulled as 'done without authority and against the will and liberty of the king and the right of his Crown'.[1] The questions put to the judges in 1387 were rehearsed and the serjeants-at-law gave it as their opinion that they had counselled the king loyally and properly. Led by William Thirning, chief justice of the common pleas, the judges of 1398 did muster courage, however, to put in a word for the Statute of Treasons. Declaration of a treason not hitherto defined, said Thirning, appertains to parliament; but he saved the king's face and probably his own skin as well, by adding that, had he been a lord or a peer of parliament, his opinion would have concurred with that given in 1387. The earldom of Suffolk was then restored to the heirs of Michael de la Pole; the king received the wool and leather customs for life, and a liberal subsidy; and fresh oaths were sworn to maintain the acts of this parliament, attempts to reverse them being declared treasonable. Richard was anxious about the future, and with good reason; for a new danger was threatening.

Only seven of the eight appellants of 1397 had appeared at Shrewsbury. The duke of Norfolk was missing; and the reason for his absence became clear to all when, on 30 January, Henry duke of Hereford appeared in parliament and, directed by the king, related what had passed recently between himself and his former associate. Norfolk, so Hereford alleged, had warned him that they were both about to be destroyed, 'by reason of Radcot Bridge', and that, although they now seemed to enjoy the king's favour, he was minded to do with them what he had already done with Arundel, Gloucester, and Warwick.[2] Hereford had confided in his father who advised him to repeat the conversation to the king; Richard had ordered him to put it in writing and to bring the record into parliament. But it was not the king's intention to risk trouble by having the scandal thrashed out there. Hoping to secure a settlement with less publicity, he persuaded parliament to agree to the appointment of a committee of eighteen persons, including the magnates nearest to him, Worcester, as proctor for the clergy, Bussy, Green, and four other knights, to discuss and settle the matters raised in Hereford's

[1] *Rot. Parl.* iii. 357.
[2] Ibid. 360.

complaint. At the same time, a body almost identical in constitution,[1] was set up on the petition of the commons 'to examine
and answer petitions still outstanding'. The appointment of this
committee led Stubbs to designate the Shrewsbury parliament
as 'suicidal'; but it has long since been demonstrated that the
terms of reference of the two committees (which, for all practical
purposes, could be regarded as one) were strictly limited to the
two matters of the Hereford-Norfolk dispute and the termination
of petitions.[2] Moreover, the committee of 1398 was little different in principle from that set up at the end of the Merciless Parliament to deal with petitions which could not be answered for
lack of time. It was not until some months later that Richard
conceived the plan of enlarging the committee's powers for his
own purposes. For the present he was content to wind up the
session with the issue of a general pardon, subject to the significant exception of 'all who had risen or ridden' with the appellants against him in 1387–8, and to write to Boniface IX for
letters apostolic confirming the acts of this parliament. He had
achieved his immediate purpose by destroying his enemies,
clearing the reputation of his friends both living and dead,
quelling lay and clerical opposition, and securing a large measure
of financial independence for himself.

The complaisance of the commons in Richard II's last parliament has led several historians to accept at their face value
the charge of interference with elections laid against him in the
articles of his deposition and Walsingham's assertion that the
knights in 1397 were elected 'not by the community as custom
demands, but by the royal will'.[3] Evidence on such matters is
hard to come by; but Walsingham is a biased witness and no
other chronicler supports him.[4] The famous words of Arundel—
'Where be those faithful commons? . . . the faithful commons are
not here'—were the words of a desperate man on trial for his
life and should not be construed as anything more.[5] Whatever
private negotiations Richard may have had with the sheriffs,
their names will bear inspection as those of the normal type of

---

[1] The arrangements for a quorum were slightly different.
[2] See J. G. Edwards, 'The Parliamentary Committee of 1398', *Eng. Hist. Rev.* xl
(1925), 321–33.   [3] *Annales*, p. 209.
[4] The charge is reiterated, however, in *Mum and the Sothsegger*, written after 1399.
This poem preserves, albeit in obscure allegorical language, what appears to be an
eyewitness account of the commons' debates in the Shrewsbury session (iv. 31–93).
[5] Usk, p. 14.

country gentleman; the appointment of a number of subservient sheriffs did not occur until October 1397, too late to affect the elections.[1] Nor was there anything abnormal in the status or qualifications of the knights returned in September; if parliament was in any sense 'packed', it was certainly not packed with Cheshire men or other accomplices of ill fame. The most that can plausibly be suggested is that Richard may have tried to ensure the loyalty of many of the country gentry by grants from the Crown made at various dates in the preceding years. Bussy and his friends were skilful managers. Even so, the elaborate military precautions taken by the king indicate that he was none too sure of the acquiescence of the estates in the measures to be proposed.

Richard's movements after the dissolution at Shrewsbury suggest uncertainty of purpose. He made no immediate attempt to return to London, but wandered through the midlands and the west, accompanied always by his Cheshire guard and adding constantly to the number of his retainers. The issue between Hereford and Norfolk was still unresolved and the king was present when the committee of eighteen held its first meeting at Bristol on 19 March. Five petitions outstanding from the late parliament were dealt with before the committee proceeded to consider the accusations laid by the duke of Hereford. It was decided that if sufficient proofs were not forthcoming, the matter should be tried, as was customary in such cases, by battle. Six weeks later, both disputants appeared before the committee at Windsor and, as proofs were still wanting, it was arranged that there should be a duel at Coventry on 16 September. So long a postponement is probably to be explained by Richard's hope that a settlement of some kind might be reached in the interval. Disappointed in this, he chose instead to order the disputants to lay down their arms as soon as they had entered the lists, causing it to be publicly proclaimed that the duke of Hereford had fulfilled his debt, so far as in him lay.[2] Immediately afterwards, and without giving any reason, Richard declared Hereford banished for ten years and Norfolk for life, heavy penalties

[1] See A. Steel, 'Sheriffs of Cambs. and Hunts. in the Reign of Richard II', *Proc. Camb. Ant. Soc.* xxxvi (1934), 1–34, and H. G. Richardson, 'The parliamentary representation of Lancashire', *Bull. John Rylands Lib.* xxii (1938), 175–222.

threatening anyone who should attempt to intervene for either. It may well be, as Mr. Steel has suggested, that he could not bear to see either of them win.[1] A victory for Norfolk would be taken as proof of the justice of the charges against the king; a victory for Hereford would add to the growing renown of a man whom the king had never liked and whom, by this date, he almost certainly feared. But whatever the reason for his impulsive gesture, it is clear that Richard's mind was riddled with suspicions and that his self-control was fast deserting him. With reckless indifference to public opinion, he embarked on a policy of pillaging his people. All who had been implicated with the appellants in 1387-8 had been ordered to seek individual pardons by midsummer, those who had been at Harringay or Radcot Bridge being reserved for special treatment; and it may have been shortly after the scene at Coventry that he devised the scheme of making large financial profits out of these pardons. Exceptions to them were made to cover the seventeen counties which had supported the king's enemies, the price of recovery of the royal favour—the notorious *pleasaunce*—being either £1,000 or 1,000 marks for each county. It seems, too, that the king exacted from the accredited representatives of these counties 'blank charters' under their seals which he kept for some unknown purpose of his own.[2] And applications for loans—from churches, communities, and individuals—continued unremittingly. The object of it all was probably very much as Richard's enemies later defined it—'ostentation, pomp and vainglory'.[3] Reckless expenditure on building, tournaments, furniture, food and dress, lavish presents, and the desire to live magnificently, all demanded resources far in excess of any to be had through the ordinary channels; the king's grandfather had recognized that such things could be paid for only by the profits of a victorious war. Even more significant of Richard's strange mentality was his persistent use of the special oath. Not content with the solemn swearings at Westminster and Shrewsbury, he forced eighteen prelates, sixteen lay peers, and four commoners to renew their oaths, with an addition covering the banishment of

---

[1] *Richard II*, p. 252.

[2] The rumour repeated by Walsingham (*Annales*, p. 236) that Richard intended to use them to negotiate the sale of Calais to Charles VI is wholly incredible in view of their strained relations at this date. See J. G. Dickinson, 'Blanks and Black Charters in the Fourteenth and Fifteenth Centuries', *Eng. Hist. Rev.* lxvi (1951), 384-5.   [3] *Rot. Parl.* iii. 419. (From the articles of the deposition.)

Hereford and Norfolk; and writs were issued to all subjects, demanding special pledges to maintain the acts of the late parliament. There is good reason to believe that the king was influencing the appointment of sheriffs; and he was certainly overriding the common law. Critics of the régime were brought before the courts of the constable and marshal and accused persons of all ages were forced to defend themselves by battle. Papal support was bought by the concordat of November 1398, in which Richard II yielded almost all the defensive positions against the Curia formerly acquired by his grandfather.[1]

Three months later, on 3 February 1399, John of Gaunt died. After the dissolution of the Shrewsbury parliament, he had taken little part in public life and we know nothing of his reactions to the sentence on Hereford. He made no recorded protest. No doubt he was reluctant to prejudice the prospects of his duchess and their children by risking a quarrel with the king; but it is possible also that the treasonable activities of his eldest son in 1387–8 had weighed upon his conscience, arousing fears lest in earl Henry the sinister traditions of earl Thomas be revived and the house of Lancaster once again invite ruin by assuming the leadership of an opposition movement. Gaunt may have been little disposed to criticize the king's decision. Hereford had been licensed to appoint attorneys to receive any property which might fall to him during his exile; and there were grounds for hope that on the death of his father he might be recalled and reinstated. At all events, Gaunt allowed his will, with its substantial bequests to the king, to stand;[2] and the eldest of the Beauforts remained loyal to Richard throughout the revolution. This was in the paternal tradition; for John of Gaunt was nothing if not a loyalist. Professor Russell's recent study of the Spanish enterprise reveals Gaunt as deficient 'in almost every quality required of a commander in the field', as preferring his private profit to a settlement which would eliminate Castile from the war, and as prepared to compromise the independence of Portugal in order to gain his ends.[3] Yet at home Gaunt had shown himself a faithful friend, even to embarrassing allies like Wyclif and John of Northampton; and throughout his life he spared no effort to maintain the solidarity of the royal family

---

[1] Above, pp. 282–3.
[2] For the full text of this remarkable will, see Armitage-Smith, *John of Gaunt,* pp. 420–36.   [3] *English Intervention in Spain and Portugal,* pp. 479, 495 ff.

and the prestige of the Crown. He gave wholehearted support to his father, to his elder brother, and, so far as circumstances allowed, to his incalculable nephew; and it was his death at the age of fifty-eight that sealed the fate of Richard II.

For, at a stroke, the death of Gaunt removed the strongest pillar of the monarchy and brought the king face to face with the problem of Henry of Hereford. It was commonly expected that he would be recalled; but this was a step that Richard either feared or disdained to take. On 18 March the committee of eighteen was summoned, Hereford's licence to appoint attorneys was revoked, and his sentence was extended to one of banishment for life. The estate of the duchess Katharine was not seriously disturbed and the annuities granted by Gaunt to various dependents were confirmed; but the bulk of the Lancastrian lands were divided among the king's partisans, Exeter, Aumale, and Surrey;[1] and the sentence on Hereford made it plain to every property owner in the country that here was a king in whose hands the most indefeasible of all rights, the right of inheritance, was no longer safe. Richard had already found means to enlarge the powers of the committee of eighteen by altering the parliament roll to include the phrase, 'and all other matters moved in the presence of the king and all things arising therefrom'. His intention was probably to use the committee as an instrument of vengeance rather than as a substitute for parliament; and, the matter of Hereford once disposed of, Sir Robert Plessyngton, sometime chief baron of the exchequer, was posthumously condemned as a traitor because he had spoken for the appellants in 1388; at the last recorded meeting, on 23 April, Henry Bowet, archdeacon of Lincoln, who had supported Hereford's plea for attorneys, was likewise condemned.[2] Such perversions of justice were sufficiently disquieting; and all the time, the king's megalomania grew. The continuator of the *Eulogium* tells how on feast-days he would have prepared a lofty throne where he would sit from dinnertime to vespers, speaking to no one, but watching them all. Any man who caught his eye was required to genuflect forthwith.[3] Walsingham speaks of the king's dependence on soothsayers and other charlatans who persuaded him that he should be greatest among the princes of the earth; of the flatterers who told him that he was worthy to be

---

[1] R. Somerville, *History of the Duchy of Lancaster*, pp. 134-5.
[2] *Rot. Parl.* iii. 384-5.
[3] *Eulogium*, iii. 378.

named a mighty conqueror because he had vanquished all his
enemies without a battle; of the Cheshire guard swarming
always about him to protect him from the fury of the people.[1]
Into this world of vain delusions broke the shattering news that
Art McMurrough, whom the king himself had admitted to
knighthood, had seized the opportunity offered by the death of
the fourth earl of March and was once again in revolt. Richard
determined to castigate him. Ignoring the danger of leaving
England at such a juncture, he pushed ahead with plans for
a second Irish expedition. Surrey was commissioned as lieutenant and sent to Ireland in advance, while the king went on
pilgrimage to Canterbury, increased his exactions, and held his
last Garter feast at Windsor, where he said farewell to the young
queen. Once again the leading earls and barons of the court
circle accompanied him; and it may have been someone with
more foresight than Richard who saw to it that the sons of the
late duke of Gloucester and of the duke of Hereford were not left
behind. The removal of the Crown jewels and of the royal treasure
was later made part of the case for the deposition. On 1 June
the army landed at Waterford; but McMurrough could not be
brought to battle and, claiming that he alone was the rightful
king of Ireland, he refused to consider terms of submission.
Richard decided to go in person to Leinster to punish him; but as
he was about to set forth, he received the news which forced
him to change his plans.

Edmund of York had been left as keeper of the realm, with
the support of the three principal officers of state (Stafford,
Wiltshire, and Clifford) and of a council in which Bussy, Bagot,
and Green were prominent. But the king's departure gave rise
to widespread alarm and rumour fed on rumour. Richard intended, it was said, to crown the duke of Surrey as king of Ireland and to settle there permanently himself, exploiting the
wealth of England from the west; many nobles were to be murdered and their estates farmed for the king's use by the earl of
Wiltshire. Panic mounted and the shire courts ceased to meet.
The opportunity was ripe for a bold invader and Henry of Lancaster seized it.

At the time of his exile nine months before, Henry had gone
to Paris, where he had been joined by Archbishop Arundel and

[1] *Annales*, pp. 233–7.

his nephew, the son of the late earl. Richard II now had few friends at the French court and the exiles were well placed to gather news.[1] Towards the end of June, when Richard was known to be safely in Ireland, Bolingbroke set out for Boulogne with a small company; offers of French help had been wisely rejected. After a brief landing at Pevensey (where a greater conqueror had been before him) he sailed north to the area of Lancastrian influence. Early in July he disembarked at Ravenspur on the lower Humber and directed his way towards his ancestral stronghold of Pontefract, where 'all the people of the north country' came out to join him. Among them were the greatest of the northern lords—the earl of Northumberland, with his son, Harry Hotspur, and Ralph Neville, earl of Westmorland; as he moved southwards, men streamed to enlist under his banner, and only the stalwart bishop of Norwich made any attempt to muster a force against him. The duke of York, as always, was pitifully ineffective. Fearing the hostility of the Londoners he had transferred the government to St. Albans, whence belated orders were issued for the raising of troops. Urging the king to return post haste from Ireland, the councillors then moved westward in the hope of joining him, making first for Oxford and then for Wallingford, where the queen was lodged. It had been expected that Bolingbroke would march on London; but he too deflected his forces to the west, in the hope of driving a wedge between Richard and the councillors. Wiltshire, Bussy, and Green, who seem to have been anxious to put up a fight, showed good sense in hastening to Bristol; but York and the remainder of the council took refuge at Berkeley castle in Gloucestershire. Here, Bolingbroke secured their submission; and bearing York on with him to Bristol, forced him to order the surrender of its castle. Bussy and Green were taken and executed; Bagot had already fled to Cheshire where he was soon captured.

The executions at Bristol took place on 29 July, two days after Richard had left Ireland.[2] His departure had been delayed by the treasonable schemes of his cousin Aumale, who had now resolved to abandon him. It was Aumale who persuaded the

[1] The garrisoning of Kenilworth for Henry from 2 June shows that preparations for his return were being planned as soon as Richard left the country. See Somerville, pp. 136-7.

[2] For a detailed reconstruction of Richard's movements between his landing and his capture, see Clarke, *Fourteenth Century Studies*, pp. 66-77.

king to take the fatal step of dividing his army. Salisbury was
sent ahead with a strong detachment and instructions to raise
troops in north Wales, while Richard himself with the remainder
of the army landed at Haverfordwest. But the troops were tired
and out of heart and efforts to raise men in Glamorgan met with
no success. Advised by Aumale and by Worcester (whom he
had long trusted), Richard decided to disband his own force
and to stake everything on that which Salisbury was believed to
be raising in north Wales. These vacillations gave Bolingbroke
the time he needed. From Bristol he marched up the Welsh
border to Chester which, he rightly guessed, would be Richard's
objective, and entered it on 9 August. Richard took the longer
route by the coast and arrived at Conway two days later. Aumale and Worcester had deserted him for Henry; and Salisbury, believing a rumour that the king was dead, had allowed
his army to scatter and had withdrawn to Conway with barely
a hundred men. In the strongly fortified castle of Conway, with
shipping in the bay, Richard's position was not, perhaps, altogether hopeless; though by this time it must have been hard for
him to know whom he could trust. When archbishop Arundel
and the earl of Northumberland appeared with an offer of
terms, he gave them audience; and, as the terms were not un-
reasonable, he agreed to accept them, no doubt hoping to find
a way later to take his revenge. He was to restore the Lancastrian
inheritance, submit Henry's claim to the hereditary stewardship to a full parliament, and surrender five of his councillors
(unnamed) for trial there. Northumberland swore on the Host
that the king should retain his royal dignity and power and that
Henry would observe the terms as he had stated them.[1] Richard
then left the castle, only to fall into an ambush which had been
laid for him. He was taken to Flint castle and thence to meet
Henry at Chester. Writs for a parliament to meet at Westminster on the last day of September were issued from Chester in his
name on 19 August; and the cousins then moved southwards

[1] It may be significant that it was Northumberland and not Arundel who took
the oath. Miss Clarke did not consider the possibility that Northumberland himself may have been deceived by Henry. But, although it is not altogether easy to
believe him guilty of such deliberate perjury, the fact that on 2 Aug. he had
accepted appointment as warden of Carlisle and the west march by Henry's commission under the seal of the duchy of Lancaster, points to his having been a party
to the usurpation. See R. L. Storey, 'The Warden of the Marches towards Scotland,
1377–1489', *Eng. Hist. Rev.* lxii (1957), 603.

towards London, where Richard was placed in the Tower. Nothing was to be feared from the Londoners who had already sent an official deputation to Henry, commending the city to him and withdrawing allegiance from Richard.[1]

Thus, once again after seventy years, a body of revolutionaries faced the task of deposing a king. They enjoyed an advantage over their predecessors in having a precedent to work from; and this was followed, in principle, with remarkable fidelity. First, the king must abdicate by what purported to be a voluntary act. Richard proving less easy to coerce than Edward II, Lancastrian propaganda overreached itself with tales of a voluntary offer to resign made at Conway by the king, *in sua libertate existens*, of his confirmation of this offer in the Tower, 'with cheerful countenance', and of his expressed desire that Henry of Lancaster should succeed him on the throne.[2] The demand which he almost certainly made for a hearing in parliament, and the speeches of the bishop of Carlisle and others in support of it, find no place in the official version.[3] What could pass for a voluntary cession of the Crown had been secured and this completed the first stage of the rebels' task. The second was to convey the abdication to some body representative of the nation, to have it accepted, and to fortify it with an act of deposition. Here again, the precedents of 1327 proved useful. An assembly of the estates of the realm, already summoned for a parliament but, on this occasion, not sitting *in forma parliamenti*, supported by the clamour of a *populus* composed mainly of Londoners, would give the rebels what they needed. Awkward questions as to the effect of the cession of the Crown on the validity of the parliamentary writs, or of the capacity of parliament to judge the king, might thereby be evaded. After much preliminary discussion it was to such an assembly that Richard's act of resignation was read, first in Latin and then in English; and the *status et populus* accepted it. But more was needed. Either, as the official roll states, to remove all doubts and scruples, or, as is more likely, to silence the bishop of Carlisle, a case for the king's deposition was put to the assembly in the form of a schedule of thirty-three articles setting forth his *plura crimina et defectus*. When this had been heard, estates and people agreed unanimously that the

---

[1] Usk, p. 28.
[2] *Fourteenth Century Studies*, pp. 76 ff.
[3] *Rot. Parl.* iii. 416.

facts were notorious, that the king was worthy to be deprived, and that he be deprived. A commission comprising representatives of the spiritual and temporal peers and of the knights, together with William Thirning, chief justice of the common bench, was appointed to convey the decision of the assembly to Richard in the Tower; his removal could now be held to be complete. It may have been to guard against such ambiguities as still surrounded the deposition of Edward II that those responsible for the proceedings arranged for a full and detailed record to be entered on the parliament roll.

Meanwhile, answer was being sought to a yet more delicate question. In 1327 it had been enough to remove the existing king and to substitute his heir-apparent; in 1399 it was necessary to agree as to the terms on which a usurper might be allowed to take the Crown. For Henry himself, the most obviously satisfactory solution would have been the admission that he was no usurper, but the legitimate heir. His most dangerous potential rival, Roger Mortimer, earl of March, had met his death in Ireland in July 1398 and a plausible case might have been concocted against the claim of his eight-year-old heir; but Henry, for reasons of his own, did not attempt this. He preferred to resurrect the preposterous legend of the seniority of Edmund Crouchback of Lancaster to his brother, Edward I—a legend which, if accepted as true, must invalidate the claims of all three Edwards, as well as of Richard II. When his supporters rejected it, Henry suggested that the facts should be faced and that he should claim by conquest, an even less welcome proposal; for a conqueror, as chief justice Thirning reminded him, was under no obligation to respect the laws, lives, and property of his subjects. Henry was in a difficult position, faced, on the one hand, with archbishop Arundel's desire to make him accept, if not a parliamentary title, at least one deriving from some form of national consent, and, on the other, with the reluctance of some of the magnates to see him king on any terms. Delay might well be fatal; and the ambiguous terms in which the claim was finally stated were in the nature of a compromise. After Richard's resignation had been accepted, and in the same assembly which accepted it, Henry challenged the Crown, on the grounds of his descent from Henry III 'and through the right that God of his grace hath sent me, with the help of my kin and of my friends to recover it'. The estates and people having accepted the claim,

Henry displayed Richard's signet, given him, so he said, in token of the late king's approval of his claim. Arundel placed him on the vacant throne and preached a sermon from the text, *vir dominabitur populo.* Henry then thanked the lords and all the estates and disclaimed intention to deprive any man of his heritage, franchises, or other rights, 'except those persons that have been against the good purpose and common profit of the realm'. The precise nature of his claim has remained ambiguous ever since; and the law of succession to the Crown still remained undefined. But the practical objectives of the revolution had been attained. Richard II had been supplanted by Henry of Lancaster, whose reign was held to begin from the day on which he challenged the Crown, 30 September 1399.[1]

A cloud of romantic illusion has gathered round the name of Richard II, and the student of the revolution of 1399 does well to remind himself of its essential causes. In the last two years of his reign, Richard had deprived men of their property and inheritances and had driven them into exile by arbitrary act; he had tampered with the parliament roll in order to have his enemies condemned as traitors by a tribunal furnished with no judicial authority; his vindictiveness towards his enemies exceeded even theirs towards his friends. He had imprisoned men without trial and denied them access to the courts of common law; he had exacted unprecedented oaths and unparliamentary taxes; and he had made use of a private army to terrorize his people. Whether or not he ever said that the laws were in his own mouth and in his own breast and that the lives and property of his subjects were at his disposal absolutely,[2] it was on these assumptions that he acted. He rode roughshod over common right; and the nation at last repudiated him for the tyrant that he was.[3] Whatever may be thought of the manner of his removal and however its consequences be judged, it is certain that Richard II had to be removed; for, as a near-contemporary

---

[1] The complex issues raised by the proceedings described above have been very fully debated in recent years. See the articles by G. T. Lapsley, H. G. Richardson, and B. Wilkinson in *Eng. Hist. Rev.* xlix (1934), 423-9, 577-606; lii (1937), 39-47; liv (1939), 215-39; S. B. Chrimes, *Constitutional Ideas of the Fifteenth Century* (1936), pp. 106-17; B. Wilkinson, *Constitutional History,* ii (1952), 284-304.

[2] *Rot. Parl.*i ii. 419, 420.

[3] For the criticisms of a not unsympathetic contemporary observer, see *Mum and the Sothsegger,* i. 1-19.

justly remarked, 'whenever kingship approaches tyranny it is near its end, for by this it becomes ripe for division, change of dynasty, or total destruction, especially in a temperate climate . . . where men are habitually, morally and naturally free'.[1] Richard's actions, as the articles of his deposition repeatedly insist, were in flagrant violation of the oath sworn at his coronation to maintain the laws and customs of the realm. They threatened such security as his subjects had painfully achieved in the course of many generations, a security always precarious and seldom more so than in his time when, despite many signs of mounting prosperity, men's lives and property were still menaced by pestilence and famine, local feuds, over-mighty lords, undisciplined retainers, by all the flotsam and jetsam of half a century of war. From time immemorial the king had been conceived of as a bulwark against injustice and a refuge from oppression. Strong and sagacious monarchs were the greatest need of the age and much might be forgiven to an autocrat like Edward III since, under him, the forms of law were on the whole preserved. The reign of Richard II itself affords ample illustration of the extent to which parliament could be manipulated under a cloak of legality. Richard's supreme blunder in his final years was to cast the cloak of legality away and leave tyranny revealed in all its nakedness. In the end, the case for deposing him looked stronger than the case for deposing his great-grandfather. Edward II had been a weakling and a fool; but Richard II had become dangerous, perhaps dangerously mad.

His final breakdown is the more tragic because, although too little of a realist to be consistently effective as a statesman, either at home or abroad, Richard had been pursuing an intelligible and sufficiently intelligent policy between 1389 and 1397. A settlement with France would put an end to coastal raids and invasion scares, make trade recovery possible, and permit diversion of some of the national resources from the arts of war to the arts of peace. Internal stability would be maintained by a king, supported by a brilliant court and by men pledged to his service in the shires, yet set high above popular and partisan clamour by the *mystique* of monarchy which Richard was making peculiarly his own. It is to this period of the reign that we must assign

[1] Nicholas of Oresme (c. 1320–82), *De Moneta*, ed. C. Johnson (1956), pp. 42–43, (Nicholas was a master of the university of Paris, canon and dean of Rouen, chaplain to Charles V, and finally (1377–82) bishop of Lisieux.)

two and, probably, three, distinguished works of art, each in its way symbolic of 'the divinity that doth hedge a king'. The new Westminster Hall, with its great roof of timber arches, hammer beams and supporting angels, and its statues of kings at the south end, was begun in 1393; the Westminster Abbey portrait of Richard II, aloof in majesty, may have been painted in 1390; and the Wilton Diptych, the loveliest English painting of the century, between 1394 and 1399.[1] To the same period of the nineties belong Richard's most strenuous efforts to remove the blot on the family scutcheon by securing the canonization of Edward II; and his impalement of the arms of Edward the Confessor with his own. By the opening of the year 1397 the régime had gone far to justify itself. The war was over, the country was prospering, the king had the Irish expedition to his credit and, so far as can be judged, he was not unpopular. It is clear that he had not forgotten his former friends; but the evidence affords little warrant for supposing that elaborate schemes of vengeance had been brewing in his mind since 1388. Rumours that Gloucester was organizing a plot may have been enough to persuade him that he must rid himself, once and for all, of the former appellants. But the malice and cunning with which he carried through his acts of revenge, his mounting recklessness, his dark suspicions, and the evident disquiet aroused in the minds of many who had been his friends, all suggest a sudden loss of control, the onset of a mental malaise. If Richard was sane from 1397 onwards, it was with the sanity of a man who pulls his own house about his ears. His is, indeed, a tragic history, the history of a prince who came to the throne too young, who was held in tutelage too long, and whose memories were bitter. An exemplary husband, a loyal friend, a generous patron, he was devout in his religious observances and sensitive to beauty in many forms. Something of his charm has survived the centuries. But it would have taken an abler and a stronger man to impose *dominium regale* on the magnates and people of a land where parliament and the common law had struck their roots; and Richard badly misjudged his age. His failure was a personal tragedy; but his success would have been the tragedy of a nation.

[1] The dating of the Diptych is a highly controversial question; but in a paper read to the Society of Antiquaries of London on 7 Feb. 1957, Mr. John Harvey made a strong case for assigning it to this period. I am much indebted to Mr. Harvey for his courtesy in allowing me to see a draft of his paper. (See p. 566 below.)

XVI

LEARNING, LOLLARDY, AND LITERATURE

WHEN Chaucer's 'litel clergeon, seven yeer of age' first went to school, it was as a matter of course that he learned—

> Swich manere doctrine as men used there,
> This is to seyn, to syngen and to rede,
> As smale children doon in hire childhede,[1]

for reading and chanting were the foundations of a clerkly education. At a more advanced stage, in the schools intended primarily for choristers, though some grammar was a necessary adjunct, the study of plainsong and the psalter naturally took first place: and the same was probably true of the almonry schools which emerge in the later Middle Ages as adjuncts to some of the monasteries. A school of this type is found at Ely in 1314, another at St. Albans about 1330.[2] But the expansion of the universities in the thirteenth century and the multiplication of chantries in the fourteenth, stimulated the development of a different type of school—the grammar school, where grammar (or Latin) was the principal subject of study. All schools were under the control of the Church. None might be opened in any diocese without the consent of the cathedral chancellor and almost every cathedral and some collegiate churches assumed direct responsibility for the maintenance of both a grammar school and a song school. Other grammar schools were normally under the control of the archdeacons; they might be kept by the parish clergy or by schoolmasters whom the clergy were encouraged to admit to their houses. In London we find schools in charge of masters attached, not only to St. Paul's, but also to great parish churches, like St. Martin-le-Grand and St. Mary-le-Bow.[3] Founders of chantries, whether they were individuals,

---

[1] *Canterbury Tales* (Prioress's Tale), ll. 499–501.
[2] Knowles, *Religious Orders*, ii. 295.
[3] It is thought that Chaucer may have attended one of these schools. See J. M. Manly, *New Light on Chaucer* (1926), pp. 5–6.

like Sir John Mounteney of Chelmsford, or gilds, like that of St. Lawrence at Ashburton in Devon, often made provision for the mass-priest to teach a grammar school.[1] In the university towns grammar schools took root easily; and although at Cambridge the *magister glomeriae*, who was in charge of them, long remained subject to the archdeacon, at Oxford control passed to the university, in the person of two regent masters, early in the fourteenth century.[2] Walter de Merton had attached a grammar school to his college and his example was followed by the founder of Queen's (1341) who provided for the maintenance of poor boys. Their principal function was to serve as choristers but they were also to receive a careful training in grammar and logic and were to dispute with the fellows of the college, as they sat at meals at high table.[3] Only twelve boys were provided for in the original foundation and Eglesfield's plan for an expansion of these numbers up to seventy-two, the number of Christ's disciples, never was realized. In founding the hall that was to become Exeter College, Bishop Stapledon had also projected a grammar school in connexion with it, though not in Oxford; the bishop's school was to be attached to the hospital of St. John the Baptist by the east gate of his cathedral city of Exeter. William of Wykeham had these precedents to guide him when he came to found Winchester in 1382; and, like the founders of Merton, Exeter, and Queen's, the bishop intended his school to serve as a nursery for his college. But in three respects the foundation of the college at Winchester marked a departure from precedent. It was by far the largest venture of its kind so far executed, its total complement of ninety-six scholars, choristers, and commoners being about eight times as big as the schools of Merton and Queen's; entry to the sister college at Oxford (New College) was confined to its members; and its status as a self-governing independent body was far superior to that enjoyed by any other existing school.[4] Wykeham's foundation, which was later to afford a model for the royal foundation at Eton, may thus fairly be said to mark an epoch in the development of English education.

A few of the pupils from the grammar schools might pass on, as did Chaucer, to service in a royal or noble household; others

[1] A. F. Leach, *The Schools of Medieval England* (1916), pp. 197, 211.

[2] M. D. Lobel, 'The Grammar Schools of the Medieval University', *Vict. County Hist. Oxford*, iii (1954), 40.

[3] J. R. Magrath, *The Queen's College*, 2 vols. (1921), i. 46.

[4] A. F. Leach, *A History of Winchester College* (1899), pp. 84–90.

to study in the schools of common law concentrated in and around the London Temple; others into the world of trade and industry; but the majority were destined for the Church. Of these, some might receive further education at one of the internal schools maintained by monks, canons, and friars for those intending to enter their orders; some might assist the parish clergy as acolytes, or obtain minor secretarial posts in large households until they were ripe for further orders. A few of the bishops, notably Grandisson at Exeter (1327–69) and Ralph of Shrewsbury at Wells (1329–63), were insisting that their cathedral chancellors should fulfil their proper function of lecturing and teaching and a boy might pursue his studies beyond the grammar-school level at one or other of these cathedral schools.[1] But if he were ambitious and intelligent, it was likely, by this date, that he would seek to enter one or other of the universities.

It remains something of a puzzle why medieval England, once so rich in cathedral and other schools, should have given birth to no more than two universities; and, indeed, the secession of Oxford masters to Stamford (already a flourishing centre of learning) suggests at least the possibility that even so late as 1334, a third *studium generale* might have established itself. But on this, as on other occasions, Oxford and Cambridge showed themselves determined to suppress all potential rivals:[2] and long before the century ended, these two, with Oxford well in the lead, had won an exclusive and unassailable position in the world of learning. Already by 1322 Oxford was claiming to be older than Paris and no whit inferior in dignity; the legend of the Greek philosophers who came with Brutus to Cricklade (Crekelade) and removed thence to Oxford, 'propter ampnium pratorum et nemorum adiacencium amenitatem', was current by 1350.[3] The wars made communication with Paris difficult and increased the self-confidence of the Oxford masters; a royal proclamation from Carfax in 1369, ordering all French students to leave the kingdom, may be read as an ominous sign of changing times.[4] The period is one of crucial importance in the

---

[1] Pantin, *English Church in the Fourteenth Century*, pp. 113–16.

[2] For correspondence relating to the Stamford schism, see *Collectanea*, i, ed. C. R. L. Fletcher (Oxf. Hist. Soc., 1885), pp. 8–39, and *Oxford Formularies*, ed. H. E. Salter, W. A. Pantin, and H. G. Richardson (ibid. 1942), i. 85 ff.

[3] *Statuta Antiqua Universitatis Oxoniensis*, ed. S. Gibson (1931), p. 17.

[4] A. G. Little, *The Grey Friars in Oxford* (Oxf. Hist. Soc., 1892), p. 86.

development of the university, administratively as well as intellectually; and it was not until the next century that the evil consequences of academic isolationism became fully evident.

By the end of the fourteenth century the constitutional development of the university was far advanced. The chancellor, elected by the regent masters from among the doctors of theology and canon law, had gained full independence of his diocesan, the bishop of Lincoln. He was the principal executive officer of the university, presiding over convocation and congregation, conferring licences (degrees), and enjoying wide jurisdiction in both criminal and civil actions where members of the university were implicated.[1] Many of his duties were undertaken by a deputy, known as a commissary, not yet as a vice-chancellor, and much public business was entrusted to the two proctors elected annually by the regents, the one to represent the northern, the other the southern 'nation'.[2] Convocation, which included the regents and non-regents of all faculties, was the supreme legislative assembly of the university, with power to enact, repeal, and amend statutes.[3] Congregation consisted of regents only, that is, of younger men, and dealt with formal university business. There was also a third assembly, less well defined, known as the black congregation of the faculty of arts, which claimed the right of deliberating all measures to be brought before convocation. The predominance of the faculty of arts—described in 1339 as *fons et origo ceteris*[4]—over the higher faculties of medicine, civil and canon law, and theology, was the outstanding characteristic of all these assemblies and was in itself a principal cause of the bitter quarrel with the friars which disrupted the university in the early years of the century. It meant, in effect, that control of university policy lay in the hands of very young men—the masters of arts who formed a majority in congregation and were thus in a position to outvote their seniors in the higher faculties and to thwart the designs of the

[1] The chancellor (or his deputy) was bound to be present in person only for cases between regents; or other suits, the doctors of canon and civil law nominated, and the chancellor and proctors elected, the judges.

[2] The northern nation comprised the Scots and the northern English; the southern, the Irish, Welsh, and southern English.

[3] Every M.A., after inception, had to dispute and lecture for at least two years as a regent master, after which he might remain as a 'regent at will', teaching for pay. When he ceased to lecture he became a non-regent. See S. Gibson, 'The University of Oxford', *Vict. County Hist. Oxford*, iii. 10.

[4] *Stat. Ant.*, p. 142.

religious orders. Since the minimum age for the mastership was twenty and inception at twenty-one was normal, it is hardly surprising that when the university resisted his visitation, archbishop Arundel should have complained to Henry IV that he was being opposed by a pack of juveniles.[1]

The quarrel between the university and the friars was only one aspect of the much wider conflict between seculars and mendicants. University hostility to them undoubtedly owed much to jealousy—of their buildings, their influence, and their privileges.[2] Until about 1320, when Thomas Cobham began to build the old congregation house, the university had no public buildings, but was forced to hold its meetings in the churches of St. Mary and St. Mildred and to lecture in hired schools, which the friars, with their relatively spacious houses and halls, were in a good position to supply.[3] The many eminent scholars among them exercised an inevitable and enviable influence over the young, whom they were suspected of enticing to join their ranks. And, although the university did not lack powerful friends, particularly among the bishops, the mendicant orders stood high in the favour of the monarchy, of noble families, like the Bohuns and the Valences, and of many Oxford citizens.[4] Against this background, the mendicant claim to a position of privilege within the university itself became increasingly hard to tolerate. By virtue of a well-established custom, the friars received graces to omit the normal arts course and to proceed directly to the theological degrees with which alone they were concerned.[5] The granting of these graces was a matter for congregation, that is, for the regents in arts who, at the opening of our period, were beginning to show a disposition to question them; whereas the friars, for their part, resented the necessity of seeking *de gratia* from this body what they thought should be theirs by right. The

[1] *Snappe's Formulary*, ed. H. E. Salter (Oxf. Hist. Soc. 1924), pp. 103–4, 172–3. See also H. Rashdall, *The Universities of Europe in the Middle Ages*, ed. Powicke and Emden (1936), III. xviii.

[2] See *Oxford Formularies*, i. 1–79.

[3] When the news of his condemnation by a university committee was brought to Wyclif, he is said to have been sitting in his master's chair in the hall of the Austin Friars. *Fasciculi Zizaniorum* (Rolls Ser.), p. 113.

[4] In the second half of the fourteenth century one-third of the wills of Oxford citizens included bequests to Franciscans. Little, *Grey Friars*, p. 101.

[5] The arts course comprised the seven liberal arts of antiquity, modified to suit medieval requirements, viz. grammar, rhetoric, logic, arithmetic, music, geometry, and astronomy, supplemented by natural, moral, and metaphysical philosophy. See L. J. Paetow, *The Arts Course at Medieval Universities* (1910).

Dominicans, in particular, wished to obtain in Oxford the privilege which their order already enjoyed in Paris, namely, that the chancellor should be allowed by the pope to confer theological degrees after examination, without reference to the faculty; but the regents in arts were determined that their discipline should not thus be evaded. A protracted dispute, in the course of which both parties appealed to the Curia, ended in a victory for the university on the main issue, the masters promising, however, not to withhold graces maliciously. In 1320 the university informed the bishop of Carlisle that the Dominicans, having asked pardon on their knees in full congregation, were restored to grace and favour.[1] But resentment continued to smoulder and we hear of further quarrels in the middle of the century, when the friars were charged with preaching heresy and reviling the faculty of arts. It was an eminent Oxford secular, Richard Fitzralph, archbishop of Armagh (*Armachanus*), who, in 1357, accused the friars before the Curia of enticing children into their orders; and next year the university passed a statute forbidding them to admit boys under eighteen. But the friars had powerful friends and by laying their grievances before the king in parliament they were able to procure the repeal of this enactment.[2] Richard II, like his predecessors, was a patron of the mendicants and he wrote more than once to the chancellor on their behalf when it was thought that graces were being refused maliciously. Altogether, the conflict was exacerbating, expensive, and, in the main, inconclusive; but it had at least one useful consequence in that it prompted the university authorities to put their privileges and statutes on record. The oldest portion of the chancellor's book (Registrum A) which embodies the first attempt at codification cannot be later than 1350 and may be as early as 1325.[3]

Quarrels between the university and the town of Oxford were less far-reaching, but they could be dangerous. The main point at issue between them was the scope of the chancellor's jurisdiction over the townsmen. A statute of 1290 had allowed him jurisdiction in all cases of crime—except homicide and maiming —committed in Oxford, and over all contracts arising there,

[1] *Oxford Formularies*, i. 72–73.
[2] A fact which was noted with satisfaction by the Franciscan annalist at Lynn. See Antonia Gransden, 'A Fourteenth Century Chronicle from the Grey Friars at Lynn', *Eng. Hist. Rev.* lxxii (1957), 276–7.
[3] *Stat. Ant.*, pp. xii ff.

when one of the parties was a scholar. In 1328 Edward III granted custody of the assize of bread and ale to the chancellor, jointly with the mayor; a series of lesser disputes arising in the ensuing years were settled mainly in favour of the university. The town had been suffering, meanwhile, from the decline in the local woollen industry;[1] and there must have been much accumulated resentment behind the affair of St. Scholastica's Day (10 February 1355) which, starting as a tavern brawl, developed into a general conflagration costing many lives. Most of the scholars fled and the town was placed under interdict. Heavy damages were afterwards exacted by the university, an annual penance was imposed on the burgesses,[2] and the chancellor's jurisdiction was extended, so as to give him sole custody of the assizes of bread and ale and of weights and measures, together with other privileges which had the effect of placing the town virtually under the jurisdiction of the university. All this was undoubtedly galling for the townsmen; but there is no reason to suppose that they suffered materially or that the university abused its powers. It may even have been memories of St. Scholastica's Day and its consequences which helped to preserve Oxford from violence during the Rising of 1381.

Cambridge was less fortunate; and the wanton destruction of the university archives by the rebels is a principal cause of the relative obscurity of her early history.[3] Her constitutional development seems, however, to have been comparable to that of Oxford, except that there is no trace of anything similar to the black congregation of artists. Rashdall's dismissal of fourteenth-century Cambridge as a third-rate university[4] was certainly unwarranted, for a number of eminent scholars, including Duns Scotus, Robert Holcot, and John Bromyard, are now known to have studied or taught there. But she was younger and smaller than Oxford and her emancipation from episcopal control was

[1] Above, p. 364.

[2] Every year, on St. Scholastica's Day, the mayor, bailiffs, and sixty burgesses had to attend mass at St. Mary's for the souls of the murdered scholars, each of the townsmen offering a penny at the high altar. A proportion of the sum thus collected was distributed by the proctors among poor scholars.

[3] Much of this obscurity is likely to be dispelled by the forthcoming publication (announced by Professor Knowles in *The Times* of 23 Dec. 1957) of a collection o. the early statutes of Cambridge made between 1235 and 1272, which has lately been discovered in a library in Rome by Dr. Benedict Hackett, O.S.A.

[4] H. Rashdall, *The Universities of Europe in the Middle Ages*, ed. Powicke and Emden, 3 vols. (1936), iii. 284.

more hardly won. At Oxford confirmation of the chancellor's election by the bishop had been reduced to a mere formality by 1350 and was finally abandoned in 1367; but it was not until 1374 that the Cambridge chancellor seems even to have questioned his obligation to take an oath of canonical obedience to the bishop of Ely who, in 1400, was still maintaining his right of confirmation; and it was not until 1383 that he obtained the criminal jurisdiction which the Oxford chancellor had then enjoyed for nearly a century. Yet Cambridge attracted notable patrons; and in the first half of the fourteenth century she could boast seven new collegiate foundations for seculars as against Oxford's three.[1]

Recent work has corrected an earlier tendency to exaggerate the importance of the colleges in the medieval universities. Until the Reformation the great majority of undergraduates and teachers lived in private halls or in lodgings. The main (though not the sole) purpose of the colleges was to afford opportunity for a few men of promise to proceed to the higher degrees which would equip them for posts of responsibility in the Church; and the numbers envisaged by the founders were very small—thirteen at Exeter, ten at Oriel, twelve at Queen's (apart from chaplains and boys), six at Michaelhouse, twenty 'when funds permitted' at Clare, twenty-four at Pembroke. William of Wykeham was the first to project a sizeable community of seventy; but even after the foundation of New College, it is unlikely that the seven Oxford colleges then in existence housed more than 150 out of a total university population of at least 1,200. If all the colleges had been swept away in 1400, it has been argued, the blow to the university would not have been crushing.[2] This may well be true; and Dr. Salter's suggestion rightly underlines the contrast between the medieval and the modern university. Yet, even if the colleges were in a sense external to the university they were not in themselves negligible; late fourteenth-century Merton has even been described as 'the most distinguished house of learning in England';[3] and it was the colleges rather than the

---

[1] Fourteenth-century foundations at Cambridge were King's Hall, c. 1316; Michaelhouse, 1324 (both of these were subsequently absorbed in Trinity); Clare, 1326; Pembroke, 1347; Gonville Hall, 1349; Trinity Hall, 1350; Corpus Christi, 1352. The Oxford foundations were Exeter, 1314; Oriel, 1324–6; Queen's, 1341; and New College, 1379.

[2] H. E. Salter, *Medieval Oxford* (1936), 97.

[3] H. W. Garrod, 'Merton College', *Vict. County Hist. Oxford*, iii. 105.

universities which attracted distinguished patronage. Among
their founders and benefactors were Edward II, Edward III,
and Queen Philippa, great aristocrats like Henry, duke of Lancaster, and the ladies of Clare and Pembroke, bishops like
Stapledon, Bateman, and Wykeham, a judge (Hervey de Stan-
ton, founder of Michaelhouse), lesser civil servants like the
chancery clerk, Adam de Brome, or the king's clerk, Robert de
Eglesfield, knights like Sir Philip de Somervyle who largely re-
modelled Balliol in this century, rectors of churches, like Philip
of Beverley who made it possible for University College to endow two new fellowships, and the Cambridge gilds of Corpus
Christi and Blessed Mary. This flow of wealth and property to
the colleges was to prove of the utmost importance for the future.
For (though contemporary critics of ecclesiastical endowments
were by no means disposed to exempt the universities and colleges from their general censures) when disaster in the end over-
took the monks and friars, it was the academic societies which
alone survived to carry over into the modern world the medieval
educational ideal of true religion rooted in sound learning.

Meanwhile, the monks themselves were not indifferent to the
advantages of university education. Rewley abbey, founded
about 1280 by Edmund of Cornwall, was intended to serve as a
centre for Cistercian studies at Oxford;[1] and Gloucester Hall to
fulfil the same purpose for the Benedictines. But economic difficulties and the inveterate individualism of the old black monk
houses proved insuperable and Gloucester remained 'a collection of *camerae* belonging each to a different monastery' rather
than a true college.[2] The hall colonized from Durham early in
the century languished until Bishop Hatfield undertook its en-
dowment, and, even then, his project for a college of eight monks
with eight seculars attached in a subordinate capacity, was not
completed until after his death in 1381. At Islip's foundation of
Canterbury Hall there were difficulties and dissensions until
1371 when the seculars, whom Islip had wished to include, were
finally expelled and the monks left with undivided control.[3]

The main purpose of these monastic colleges was to afford a
training in canon law or theology to some of the abler monks,

[1] *Oxford Formularies*, ii. 286–327.
[2] V. H. Galbraith, 'New Documents about Gloucester College', *Snappe's Formulary*, p. 341.
[3] H. S. Cronin, 'John Wycliffe the Reformer and Canterbury Hall, Oxford',
Trans. Roval Hist. Soc., 3rd ser. viii. 55–76.

who might thereby qualify for high office in their houses. Similar practical ends directed the studies of the great majority of seculars. It may have been the mathematicians and philosophers who gave Oxford her reputation in the outside world and it is they whose fame has survived the centuries; but we should not be misled into supposing them to have been anything but exceptions. Then, as now, the average Oxford or Cambridge man contented himself with the degree in arts which gave him a basic education; and those who aimed higher were commonly in pursuit of the degree in canon or civil law which would equip them for administrative work in Church or State, or of the degree in medicine which might win them posts in noble households, or of the theological degree which might lead to high ecclesiastical preferment. The disinterested thinkers were always the few.

Among them we must reckon the mathematicians and astronomers. Men accustomed to look to the sky as clock and calendar, to think of the Milky Way as a familiar thoroughfare,[1] and to credit the stars with determination of their destinies, did not need to be taught to respect astronomers. Oxford owed her reputation as a kind of medieval Greenwich to scholars like John Maudit, who compiled a famous astronomical table—'mirabiliter inventus in civitate Oxon. MCCCCX'—or Richard Wallingford, afterwards abbot of St. Albans, whose astronomical clock was said to be without rival in Europe.[2] Wallingford, who was the son of a blacksmith, may have owed something to inherited skill, for the instruments then available to the practical mathematician—a straight edge, a square, and a pair of compasses—would often have been little, if any, better than the blacksmith's tools. Most of the mathematicians and astronomers seem to have been fellows of Merton. Such were John Ashenden, Simon Bredon, William Rede (who built Merton library), and the great Thomas Bradwardine, generally regarded as the most eminent English mathematician of the century. But two of the mendicants—the Franciscan, John Somer, who wrote an

---

[1]   'Se yonder, loo, the Galaxie,
                          Which men clepeth the Milky Wey,
                          For hit ys whit (and somme, parfey,
                          Kallen hyt Watlynge Strete'.).
                               Chaucer, *House of Fame*, ll. 936-9.
[2] R. T. Gunther, *Early Science in Oxford* (Oxf. Hist. Soc.), 2 vols. (1923), i. 96-98; ii. 44-65. Maudit and Wallingford are said to have been the initiators of western trigonometry. See A. C. Crombie, *Augustine to Galileo* (1952), p. 68.

astronomical calendar for Joan of Kent, and the Carmelite, Nicholas of Lynn, who at Gaunt's request compiled a calendar of the latitude and longitude of Oxford, which was used by Chaucer for his treatise on the astrolabe, also enjoyed high reputations; and William of Wykeham ordained that two of the fellows of New College should be students of astronomy.

The most powerful philosophical influence at work in early fourteenth-century Oxford was that of William of Ockham, *doctor invincibilis*, a Franciscan thinker of profound originality, whose opinions were condemned, first at Oxford and then, in 1326, at Avignon, whence he fled to find refuge at the court of the emperor Lewis IV. In Ockham's predecessor, Duns Scotus, who died in 1308, Oxford had known a philosopher whose speculations, extravagant though some of them were, are yet recognizable as a development from the thought of St. Thomas, in so far as Scotus was reluctant to admit a fundamental divorce between reason and faith, between philosophy and religion, but sought to embrace both in a single intellectual system. Ockham, who carried nominalism to its logical conclusion, abandoned all such attempts at reconciliation between human understanding and the mysteries of God. Relegated thus to the sphere of the incomprehensible, God became a remote and mysterious arbiter of the destiny of creation, an absolute Will, bound by no human concepts of reason or justice. Speculation on the nature and attributes of such a Being was futile, for God and his ways belonged to a realm where reason had no place. Man's capacity for fruitful investigation was therefore limited to the sphere of his own experience. Since Ockham never presumed to question the truth of the Christian revelation, he came dangerously near to denial of the fundamental principle of all human thinking, that reality may not contradict itself. His teaching opened the door, however, to the layman and the scientist, for it pointed the way to free inquiry, untrammelled by theological presuppositions, into human history and the world of sight and sense; and it was consistent with the trend of his thought that at the court of Lewis IV he should have developed a fully secularist theory of the State. For religion, Ockham's teaching was deadly. To dismiss God from the sphere of human thought, was to jettison the painfully garnered wisdom of the theologians, and to reduce faith to superstition. Thomas Bradwardine, *doctor profundus*, an adherent of the Augustinian theology of grace, produced a counterblast to Ockham in his *De Causa Dei contra Pelagianos*; but this, like the work of the Dominican Robert Holcot, seems to have made little impression at the time; and when Wyclif began to teach at Oxford in the sixties, the brand of nominalism favoured by Ockham was still the prevailing philosophical trend.

John Wyclif, a Yorkshireman from the Lancastrian honour of Richmond, was a junior fellow of Merton in 1356, a master at Balliol in 1360, and a doctor of divinity by 1372, when he was probably just over forty years of age. Though his stature as a philosopher was almost certainly less than Ockham's or Brad-wardine's, there can be no doubt that he dominated the Oxford of his own day; he was *flos Oxonie*, 'holden of full many men the greatest Clerk that they knew then living'.[1] The training of the schools was exacting and Wyclif must have been a very able man. His cast of mind was radical, like Ockham's, but his philosophical standpoint was very different. For Wyclif was a realist, who held that reality consisted only in ideas which, for him, meant archetypes in the mind of God.[2] The divine mysteries, so far from lying outside the sphere of human reason, were fit subjects for intellectual wrestling; *hoc investigari non potest* he regarded as unchristian counsel.[3] Up to the time when he took his doctor's degree there was little, except his intellectual eminence, to distinguish Wyclif from other scholars in the higher faculties. He drew the bulk of his income from two rural parishes which he held in succession, mainly as an absentee—Fillingham in Lincolnshire (1361–8) and Ludgershall in Buckinghamshire (1368–74)—and from the prebend of Aust in the collegiate church of Westbury-on-Trym which Urban V granted him in 1362. For a few months he was warden of Canterbury Hall, but the reassertion of monastic control there soon led to his expulsion. He occupied lodgings in Queen's (though he was never a member of that college) and it is to this period of his life that we must assign such philosophical works as the *De Logica*, the *De Compositione Hominis*, and the *Summa de Ente*. His first important theological treatise, the *De Incarnatione Verbi*, or *De Benedicta*

---

[1] *Eulogium* (Rolls Ser.), iii. 345; A. W. Pollard, *Fifteenth Century Prose and Verse* (1903), pp. 118–19. See Knighton, ii. 151, 'In philosophia nulli reputabatur secundus, in scholasticis disciplinis incomparabilis.'

[2] S. H. Thomson, 'The Philosophical Basis of Wyclif's Theology', *Journal of Religion*, xi (1936), 96.

[3] *De Eucharistia*, ed. J. Loserth (1892), p. 286.

Incarnatione, which almost certainly embodies a course of lectures delivered in the school of theology, may date from 1370.[1]

It is likely to have been shortly before he obtained his doctorate in 1372 that Wyclif entered the service of the Crown as *peculiaris clericus*. This was not a full-time appointment; the new clerk's services were, so to say, 'retained', to be called on when needed. No doubt Wyclif's reasons for seeking such employment were partly financial. He had been disappointed of more than one hope of preferment and the exchange of Fillingham for the much less valuable Ludgershall looks like a manœuvre to raise some ready cash.[2] But his presence as a spectator in the parliament of 1371, when the Church was under fire from mendicants as well as from the laity,[3] suggests that he may also have welcomed the prospect of addressing himself to a wider public than could be reached at Oxford. The government, for its part, could always find room for a well-trained university man and the arrangement worked out fairly satisfactorily for both. The king presented his clerk to the rectory of Lutterworth in Leicestershire, which was equal in value to Fillingham, and three months later (July 1374) sent him on a diplomatic mission to Bruges, where it may have been hoped that his academic training would enable him to counter the subtle arguments of the papal agents. For this undertaking Wyclif received £60 in wages and expenses. But he was no diplomat; and by September he was back in Oxford where already he was deeply embroiled in more than one academic controversy—with the Carmelite, John Kenningham, who questioned the theological implications of his ultra-realist metaphysic, and with the Benedictines, William Binham and John Uhtred of Boldon, the Cistercian, Henry Crumpe, and the Franciscan, William Wodeforde, all of whom criticized his theories of lordship. For it is likely that before he went to Bruges, Wyclif had written the second part of his famous *Determinatio*, an imaginary conversation in which seven lay lords argue the case against the papal tribute. This pointed the way to the two massive treatises, *De Dominio Divino* and *De Dominio Civili*, which were the outcome of lectures delivered at Oxford between 1374 and 1376. The thought owes much to Fitzralph and is not markedly original; but at a time when anti-clerical sentiment

---

[1] A. Gwynn, *The English Austin Friars in the Time of Wyclif* (1940), pp. 211–12.
[2] K. B. McFarlane, *John Wycliffe* (1952), p. 29.
[3] Above, p. 290.

was running high, the books attracted considerable attention—
though not, we may surmise, many readers—outside university
circles. Wyclif's main contention that the exercise of all human
lordship depends upon grace, or, in other words, that the sinful
man has no right to authority or property, led him on to the
specific conclusions that, if an ecclesiastic abuse his property, the
secular power may deprive him; that pope and cardinals alike
may err; that neither is essential to the true government of
the Church; and that a worldly pope is a heretic who ought to
be deposed. Such notions proved widely attractive—to John of
Gaunt, who was at odds with the bishops and whose interest in
Wyclif first becomes evident soon after the Good Parliament
of 1376, to many Londoners, and to some of the mendicants, at
both Oxford and Cambridge.[1] Since Wyclif did not scruple to
publish his opinions from the London pulpits, he began to seem
dangerous; and after the failure of Courtenay's first attempt to
have him silenced,[2] a higher authority intervened. Some of
Wyclif's opponents had sent a list of his conclusions to the Curia
and from these Gregory XI selected eighteen for censure and, in
a series of bulls drafted in May 1377, ordered the archbishop
and the bishop of London to present the offender before him
within three months. Publication of these bulls in England was
somewhat inexplicably delayed and in the meantime Wyclif
took opportunity to present Richard II's first parliament with a
tract maintaining that in case of necessity the English government might lawfully detain money due to the pope. When,
towards the end of the year, the contents of the bulls were made
known, they were not well received by the university, whose
officers contented themselves with putting Wyclif under formal
arrest pending further inquiry. He continued to lecture; and his
treatise on the Bible, the *De Veritate Sacrae Scripturae*, was published in the spring of 1378. About the same time he was cited
before an ecclesiastical court at Lambeth, but the intervention
of the princess of Wales enabled him to escape with no more
than a formal reproof; and in the autumn Gaunt brought him
into the Gloucester parliament to deliver arguments against the

---

[1] At Cambridge an Austin friar named Adam Stocton was copying passages from
the works of Wyclif whom he describes as *venerabilis doctor*; later, Stocton erased
these words and substituted *execrabilis seductor*. Gwynn, *Austin Friars*, pp. 237–9.

[2] Above, p. 396. It is noteworthy that on this occasion, Gaunt was able to produce representatives of all the four orders of friars to defend Wyclif. *Chron. Ang.*,
p. 118.

privilege of sanctuary, in the Hawley-Shakell case. Wyclif was still an influential and widely respected figure; and, though his *De Ecclesia* (1378) and *De Potestate Papae* (1379) probably went beyond what most of his contemporaries, even among the anticlericals, were prepared to accept, he had not yet finally repudiated the papacy, as such. But his innate radicalism was fast driving him beyond the limits of prudence; from attacks on the authority of the priesthood he now passed to an attack on the main source of priestly influence and power. In 1379 he was delivering the lectures soon to be embodied in his *De Eucharistia*.

Debates of the type to be found in this book were as old as the schools themselves. Philosophers and theologians of many generations had discussed the relation of accidents to substance, the significance of the words of consecration, the nature of Christ's presence in the sacrament of the altar. There was nothing to call for comment in the mere fact that yet another Oxford schoolman found himself in difficulties over the annihilation of substance.[1] The much more eccentric theological opinions of John Uhtred of Boldon, for example, aroused little attention because expression of them was confined to the precincts of the schools and because their author was prepared to bow to ecclesiastical authority.[2] What made Wyclif's denial of the doctrine of transubstantiation a cause of scandal was his position as a public figure, a well-known critic of abuses in the Church and a man with very powerful friends. Worst of all was his refusal to recant and the apparent readiness of some of his academic colleagues, not only to listen to him with respect, but even to subscribe his teaching. It was by a very narrow majority that a council of Oxford theologians, convened by the chancellor, William Barton, condemned as erroneous and contrary to catholic teaching twelve propositions which Wyclif was alleged to have defended in his lectures on the eucharist. When the news of this sentence was brought to him, he is said to have appeared confused; but he stoutly declared that neither the chancellor nor anyone else could persuade him to alter his opinions; and it was consistent with his long-standing bias against ecclesiastical authority that he should have decided to appeal against the condemnation, not to any academic or ecclesiastical body, but to the king—

---

[1] Wyclif had already denied the possibility of such annihilation in his early treatise, *Logice Continuacio*, see S. H. Thomson, p. 8.

[2] See Knowles, *Religious Orders*, ii. 83 ff.

'like a heretic, adhering to the secular power in defence of his error'.[1] Since the king was a child, it must have been to the duke of Lancaster and to the princess of Wales that Wyclif was looking for support; but it was no part of Lancaster's anti-clerical programme to offer open challenge to the catholic faith. So far as is known, Wyclif's appeal received no formal answer; but Gaunt is said to have approached him privately and forbidden him to speak further on such matters. Wyclif was not deterred, however, from publishing in May 1381 his openly heretical *Confessio*.

It was peculiarly unfortunate for him that these interchanges should have coincided so nearly with the Peasants' Revolt; for subverters of all kinds became suspect at a time of general subversion. The shocking murder of Archbishop Sudbury could be read as a judgement on the pastor who, confronted with dangerous heresy, had shown himself so negligent a guardian of his flock; and there were many ready to believe that John Ball had learned his tricks from Wyclif. Towards the end of the year he found it prudent to withdraw to Lutterworth; and Courtenay, who succeeded Sudbury at Canterbury, was able to ride to victory on the tide of reaction following the revolt. In May 1382 a synod of eminent theologians, heavily weighted with mendicants, met at the London convent of the black friars to consider twenty-four propositions extracted from Wyclif's writings, touching the eucharist, confession, papal jurisdiction, lordship, and grace. Parliament was in session and Wyclif had already submitted to it a memorandum on ecclesiastical endowments, payments to Rome, and other topics likely to gain a sympathetic hearing from the commons. We do not know what reception was accorded to this document; but the Blackfriars synod, after four days' deliberation, interrupted by an earthquake (variously interpreted as indicative of the divine reaction to its proceedings), unanimously condemned as either heretical or erroneous all the opinions submitted to it. Wyclif himself, however, was not named. For some reason, probably because Lancaster spoke for him behind the scenes, it was decided not to submit him to formal trial.[3] But as a suspected heretic, he was forbidden to preach and teach in his old university and he may

---

[1] *Fasciculi Zizaniorum* (Rolls Ser.), p. 114.
[2] In one of his later tracts Wyclif refers to Gaunt as the friend of poor priests. *Polemical Works*, i. 95.

even have given some kind of undertaking to refrain from preaching and teaching elsewhere.[1] Apart from an abortive attempt to bring him to Rome, he was left to linger on at Lutterworth suffering, it is thought, from the effects of a paralytic stroke. But his pen was never more active. Helped by his friend and secretary, John Purvey, he poured forth a stream of pamphlets, sermons, and treatises, the most important of which, the *Trialogus*, makes it plain that his zeal had not abated nor his opinions changed. He died on the last day of 1384, still in communion with the Church of Rome, and his bones were left to lie in peace at Lutterworth for another forty years.

Appraisement of a figure at once so controversial and so inaccessible to posterity must necessarily be hazardous. Most of the English works formerly ascribed to Wyclif are now known to be of doubtful authorship.[2] His Latin works—bulky, discursive, and repetitive—make difficult and, for the most part, unattractive reading; and little of the man himself comes through to us from any of them. Moreover, there has been inevitable reaction against centuries of protestant hagiology; many modern historians tend to minimize Wyclif's importance and to question his motives and methods, some of which seem, indeed, to have been far from admirable. Determined pursuit of his own material interests comes out unmistakably in such matters as the wardenship of Canterbury Hall, or the Lincoln canonry which he failed to secure.[3] Inconsistency between precept and practice may be detected in his absences from his cures, his too frequent exchanges of benefices and his neglect of his duties at Westbury-on-Trym; and it is sad defence for one claiming the title of reformer to urge that these were the common practices of the age. Little in the way of personal heroism was called for in the man whose friendship with the great ensured his immunity from danger; and, so far from seeking martyrdom, Wyclif seems to have been at some pains to avoid it. Not even his enemies could charge him with the grosser sins; but it is hard to believe that he was amiable. No coterie of devoted disciples, but only the faithful Purvey, sustained his declining years; and the harsh

[1] As suggested by K. B. McFarlane, pp. 115–16. See also J. H. Dahmus, *The Prosecution of John Wyclyf* (1952), pp. 129–57.

[2] See H. B. Workman, *John Wyclif*, i, app. C.

[3] H. S. Cronin, *Trans. Royal Hist. Soc.*, 3rd ser. viii, pp. 55–76; Miss M. E. H. Lloyd, 'John Wyclif and the Prebend of Lincoln', *Eng. Hist. Rev.* lxi (1946), 388–94.

censoriousness of much of his writing reflects unpleasantly upon
the author. His love for Oxford must, indeed, be set to his credit.[1]
But his narrowly theological outlook, his distaste for music, art,
romance—'jeestis of battles and fals cronycles'—and for all
secular learning, is none the less repellent for having been shared
by many of his orthodox contemporaries. As a practical reformer, Wyclif showed himself conspicuously maladroit. By
temper and conviction an adherent of the secular power, and
fortunate in his lay patrons, he deliberately threw away the advantages he enjoyed, his theological extravagances and, above
all, his attack on the sacramental system of the Church, making
it impossible for those who had been willing to support him to
continue to do so. Infinitely less sensitive than Hus to the climate
of contemporary opinion, Wyclif seems to have been incapable
of effective organization of his resources.[2] His movement did
little to promote reform in England and may even have tended
to delay it.

Is his reputation, then, entirely spurious? Such a conclusion
is possible only if we take no account of his religious convictions;
for not even his prolix Latinity can altogether conceal the
strength and sincerity of his faith. The driving-force of all his
thought and teaching was his devotion to the Incarnate Christ,
his longing to bring men face to face with him, as *verissime fratrem
nostrum, homo cum aliis.*[3] Almost everything in contemporary religious practice that offended him, from ecclesiastical endowments
to the doctrine of transubstantiation, offended him because it
seemed to blur or distort this vision of the human Jesus.[4] More
and more he himself found the Christ he was seeking in the gos-
pels and missed him in the Church. Though never a schismatic
in intention, his aim being to purify the Church, not to divide it,
his remarkable sense of history persuaded him of the novelty of
many of the Roman claims; and his search for what he saw as
the inward truth of his religion drove him inexorably on to the
conclusion, 'quod ecclesia Romana potest errare in articulis

---

[1] '... locus amenus fertilis et optimus ... domus Dei et porta celi congrue voci-
tata.' *Opera Minora*, p. 18.
[2] R. R. Betts, 'English and Čzech influences on the Husite Movement', *Trans.
Royal Hist. Soc.*, 4th ser. xxi (1939), 98 ff.
[3] *De Benedicta Incarnatione*, ed. E. Harris, pp. 28, 91.
[4] For example, in the *De Eucharistia* Wyclif argues that because men take the
consecrated Host as their God, they cannot rightly conceive the doctrines of the
Trinity and the Incarnation.

fidei'[1]—and so to cross a bridge between two worlds. Wyclif's
reputation among protestants has never really rested either on
his personal character or on his positive achievement. It was
over 300 years ago that Thomas Fuller wrote of him, 'I intend
neither to deny, dissemble, nor excuse, any of his faults. We have
this treasure (saith the Apostle) in earthen vessels; and he that
shall endeavour to prove a pitcher of clay to be a pot of gold will
take great pains to small purpose.'[2] Fuller and his fellow-pro-
testants revered the medieval *doctor evangelicus* as a prophet of the
Reformation; and not without good reason, for such, indeed,
he was. By his repudiation of papal and ecclesiastical authority,
his confidence in the sole sufficiency of Holy Scripture, both as a
revelation of the Christ and as a rule of life for every man, and his
demand for a Bible open to the people, Wyclif anticipated the
most fundamental protestant convictions. For good or for ill
these convictions were to mould the character and help to determine the destiny of the English-speaking peoples throughout the
world. The man who first set them forth was in a very real sense
the spiritual ancestor of Bunyan, and Baxter, and Whitefield.
Whatever his faults and whatever his limitations, he has a title
to his countrymen's respect.

The lollards, as Wyclif's followers came to be called,[3] may be
divided roughly into three main groups—the scholars who had
supported him at Oxford; a number of humbler disciples, both
clerical and lay; and some of the landed gentry. Oxford lollardy
was given its death-blow by the archbishop's determination to
enforce the decrees of the Blackfriars council of May 1382 within
the university. Six days before the council was due to meet, the
university chancellor—a fellow of Merton, named Robert Rigg
—had allowed Nicholas Hereford, a prominent Wyclifite, to
preach an inflammatory sermon at St. Frideswide's on Ascen-
sion Day; and he had invited another, Philip Repingdon, to
deliver the university sermon on 5 June, the feast of Corpus
Christi. When Courtenay ordered him to publish the Blackfriars decrees, the chancellor chose to interpret this as a violation

---

[1] *De Eucharistia*, p. 32.   [2] *Church History of Britain* (1655), p. 129.
[3] The word *lollard* probably derived from the Middle Dutch *lollaerd*, mumbler
(of prayers) which had already been applied to the followers of certain religious
movements; it was deliberately confused with the Middle English *loller*, a loafer,
and, by way of a pun, with the Latin *lolia*, tares. See *Oxford English Dictionary* and
Workman, *John Wyclif*, i, app. A.

of academic privilege. No bishop or archbishop, he quite erroneously declared, had any jurisdiction over the scholars of Oxford, not even in matters of heresy; and he underlined his defiance by going in procession to St. Frideswide's to hear Repingdon defend Wyclif's opinions and declare publicly that the duke of Lancaster would stand by those who professed them. Courtenay made short work of this recalcitrance. Summoning the chancellor and proctors to Lambeth, he brought them before a small synod which found them guilty of condoning heresy and error in the persons of Wyclif, Hereford, and Repingdon. Rigg was also found guilty of contempt of the archbishop's mandate and had to ask pardon on his knees, the magnanimous Wykeham interceding for him. When he protested that he dared not publish the Blackfriars decrees in Oxford, Rigg was summoned before the king's council and sharply ordered to obey his metropolitan. His fears were not unfounded, however; for the publication of the mandate on the following day caused an academic uproar. Hereford and Repingdon betook themselves to Tottenhall to seek out Gaunt who seems to have been ready to listen to them; when a body of orthodox doctors appeared, they were given a cool reception. But after hearing both sides, the duke swung round completely and stigmatized the Wyclifites as demons.[1] A few days later the opinions of Hereford and Repingdon were formally condemned at Canterbury and they themselves pronounced contumacious and excommunicate.

These strong measures had the desired effect; and Courtenay drove the lesson home by arranging to hold the convocation of Canterbury at Oxford, in November. Here Repingdon and another Oxford lollard, John Aston, made full submission to authority. Repingdon's repentance seems to have been genuine. He returned to the fold and lived to become, successively, abbot of his own house of St. Mary-of-the-Meadows at Leicester, chancellor of his university and, finally, bishop of Lincoln. Aston, however, relapsed and a little later is found disseminating lollard doctrines along the borders of south Wales; but it is possible that he too conformed in the end.[2] Meanwhile, Hereford, with extraordinary temerity, had gone to Rome to pursue his appeal to the pope in person. He was thrown into prison by

---

[1] *Fasciculi Zizaniorum*, p. 318.
[2] Aston may have resumed his fellowship at Merton in 1391–2. See A. B. Emden. *A Biographical Register of the University of Oxford to A.D. 1500* (1957), i. 67.

Urban VI but was lucky enough to escape and, returning to England, he joined forces with Aston; in the dioceses of Hereford and Worcester they and their friends gave the authorities plenty of trouble. But, early in 1387, Hereford was taken at Nottingham and, at some date before 1391, he too recanted finally. He became chancellor of Hereford cathedral and a zealous hammer of heretics, and died at a ripe old age in the Carthusian house of St. Anne at Coventry. With the flight or recantation of the leading Wyclifites, Oxford was lost to lollardy and such hopes as Wyclif may have entertained of winning the Church to share his views were finally defeated. Some Oxford scholars continued to interest themselves in the project for a vernacular Bible; but the movement as a whole had been cut off from its intellectual roots and deprived of the opportunity to influence men destined for high ecclesiastical office. It could now hardly fail to become the preserve of semi-literate clerical enthusiasts and their lay sympathizers. Moreover, the orthodox party within the university hastened to fill the breach left by the routing of the Wyclifites, old opponents of Wyclif, like Uhtred of Boldon and William Wodeforde, pouring forth many tracts and treatises in the last two decades of the century. It may be of some significance, however, that the bishop of Worcester should have submitted the heresies of William Swinderby to a body of Cambridge doctors and that it should have been the chancellor of Cambridge who presided over the learned committee that condemned the eccentric opinions of Walter Brute.

Walsingham's assertion that Wyclif dispatched a number of missionaries from Oxford is now largely discredited.[1] There can be no doubt that he attached the utmost importance to preaching; and there is abundant evidence that towards the end of the century unlicensed preachers, variously described as 'poor priests', 'poor clerks', *idiotae et simplices*, and so forth, were perambulating much of the country. The vernacular sermons, suitable for delivery on village greens, which have come down to us show, moreover, that many of these preachers were disseminating Wyclif's ideas. But there is no reason to suppose that he had any direct responsibility for them and few, if any, can have been his personal disciples. His intimates were scholars, like Nicholas Hereford and John Aston; and, though both of these were active preachers, their missionary endeavours are

[1] *Hist. Ang.* ii. 53.

more likely to have been a consequence of their enforced departure from Oxford than to have formed part of a large-scale missionary enterprise devised by Wyclif himself. Had Wyclif been
actively concerned in anything of the kind, Lutterworth would
have been its natural focus.

William Swinderby is the most remarkable figure among the
non-academic lollards of the first generation.[1] Beginning in
Leicester as the leader of a small group of heretics which early
established itself in the chapel of St. John the Baptist near the
leper hospital, Swinderby was condemned by his bishop, in July
1382, to make public recantation. But he soon departed to the
country beyond Severn, and in August 1387 he was one of five
lollards whom the bishop of Worcester prohibited from preaching in his diocese.[2] Swinderby seems to have been popular with
the local gentry; for it was at their request that the bishop gave
him a safe-conduct when he appeared at Kington, in June 1391,
to hear the charges against him. Excommunicated by his bishop
and his arrest ordered by the king, he made good his escape into
the marches of Wales and was never brought to justice. Swinderby had been responsible for making many converts in the
part of the country where John Oldcastle, the son of a Hereford-
shire squire, was then growing up; and, though he was better
educated and certainly abler than some of them, he was the
forerunner of the semi-literate unbeneficed clerks who were to
be the typical lollards of the next generation. Associated with
him in the same area were the Oxford clerks, John Purvey (who
was probably not a graduate), a layman named Stephen Bell,
and a loquacious Welshman, Walter Brute, who was later to
join Oldcastle. This south Welsh march was the most important
field of lollard activities outside Oxford; but some of the midland boroughs were also troublesome to the bishops. Lollardy
did not die at Leicester with Swinderby's enforced recantation;
seven years later, when Courtenay visited it, eight persons were
denounced as teachers of heresy and excommunicated forthwith.
At Northampton the mayor, one John Fox, was said to have
given asylum to a strange assortment of heretics, including a
former archdeacon of Sudbury, a runaway London apprentice,
and an Oxfordshire squire named Thomas Compworth.

[1] For Swinderby see especially K. B. McFarlane, *Wycliffe*, pp. 103–4, 121–5,
129–35.
[2] The others were Hereford, Aston, Purvey, and John Parker, of whom nothing
else is known.

Compworth was representative of a class from which the lollards clearly hoped much. Among the parliamentary knights, there were always some anti-clericals attracted by schemes of disendowment and Swinderby, as has been seen, received some protection from the gentry. Yet it is not easy to be sure which, or how many, of the knight class had fully committed themselves to Wyclifite opinions. Walsingham and Knighton between them supply the names of ten knights believed to have been guilty of heresy;[1] but Sir Thomas Latimer is the only one of these for whose lollardy we have firm evidence. A kinsman of Lord Latimer and, like him, with Lancastrian affinities, Sir Thomas was brought before the king's council in 1388 on a charge of possessing heretical books; and his seat at Braybrooke in Northamptonshire became a centre for lollard teaching. Sir John Montagu, whom Walsingham considered the worst of the lollard knights, is said to have removed the images from his chapel and harboured lollard preachers, Nicholas Hereford among them. But Montagu's lollardy, if such it was, must have been only a temporary aberration, for he is found crusading in Lithuania by 1391. Clifford and Stury, who were members of the household of the princess of Wales, may have been interested in Wyclif when he was her protégé;[2] and both Clifford and Cheyne were closely connected with Latimer. But Walsingham's allegation that the lollard manifesto affixed to the doors of Westminster Hall during the parliament of 1395 was the work of Clifford, Stury, Latimer, and Montagu, has been shown to be unfounded;[3] and, except for Latimer, the careers of most of the knights named are hard to reconcile with open adherence to heresy.

There is no evidence that the lollard broadsheet of 1395 was ever discussed in parliament; but it probably contributed to the mounting alarm of the ecclesiastical authorities who were beginning to regard the existing machinery for the repression of heresy as altogether inadequate. Before the rise of the lollards, if a bishop found himself unable to deal with any isolated case

[1] Thomas Latimer, John Trussell, Lewis Clifford, John Pecche, Richard Stury, Reginald Hilton (Knighton, ii. 181); William Neville, John Clanvowe, John Montagu (*Hist. Ang.* ii. 159); John Cheyne (*Annales*, p. 290). See W. T. Waugh, 'The Lollard Knights', *Scottish Hist. Rev.* xi (1913), 55–92.

[2] In his will, dated 17 Sept. 1404, Sir Lewis Clifford described himself as false, and a traitor, unworthy to be called a Christian man. N. H. Nicolas, *Testamenta Vetusta*, i. 164.

[3] *Rogeri Dymmok Liber*, ed. H. S. Cronin (1921), Introd., pp. ix ff.

of heresy that might arise, he could apply for the writ *significavit*
which instructed the sheriff to keep the offender in prison until
he had made his peace with the Church. But this procedure was
not well adapted for dealing with heretics in large numbers, particularly when they moved rapidly from shire to shire, and in
1382 the sheriffs had been authorized to arrest and imprison
unlicensed preachers, pending their examination by an ecclesiastical court. Six years later the Merciless Parliament authorized new commissions, designed to secure the seizure of heretical
writings; to teach or maintain lollard doctrines became an
offence at common law rendering the offender liable to imprisonment and forfeiture.[1] It is likely to have been the demonstration
of 1395 which convinced the bishops that yet sterner measures
were needed. A copy of the lollard manifesto was sent to Boniface IX and it was after his reply had been received that, early
in 1397, the bishops of both provinces came out into the open,
asking parliament to authorize the death penalty for heresy in
England, 'as in other realms subject to the Christian religion'.
But for this they had to wait until Richard II had been deposed
and Archbishop Arundel's support of Henry of Lancaster had
put him in a position to make his own terms with the new king.

The lollard Bible was the outcome of Wyclif's conviction
that the text of Holy Scripture should be accessible to all. He
and his learned colleagues knew very well, of course, that much
of the Bible is difficult; but they believed that the New Testament was 'opyn to the undirstanding of simple men, as to the
poyntis that be most nedeful to saluacioun'.[2] Only the better
educated among the clergy and a few of the literate laity could
read the Latin Vulgate;[3] the Anglo-Norman Bible and the revised version of Jean de Sy (1355) were rare in England and
would not have been intelligible to many outside the ranks of
the aristocracy;[4] and, by this date, the old West Saxon gos-
pels had become obsolete. No doubt the average layman was

[1] H. G. Richardson, 'Heresy and the Lay Power under Richard II', *Eng. Hist. Rev.* li (1936), 1–28.
[2] From the General Prologue to the second lollard Bible.
[3] Margaret Deanesly, *The Lollard Bible* (1920), pp. 156 ff. The Lady of Clare possessed a Vulgate, though whether she could read it is another matter. Both Chaucer and Gower, however, knew the Vulgate well.
[4] Thomas of Lancaster is known to have borrowed a French Bible from a clerk of York. See R. L. Atkinson, 'A French Bible in England about the year 1322', *Eng. Hist. Rev.* xxxviii (1923), 248–9.

sufficiently familiar with the main outlines of the Bible story, epi-
sodes from which were represented on the walls, windows, and
screens of many churches, retailed by preachers, and re-enacted
in the popular miracle plays. Versified English paraphrases of
Genesis and Exodus had been composed about 1250; and the
Cursor Mundi offered an encyclopaedia of scriptural story in
24,000 lines, to him who 'na French can'. But Bible-reading by
the laity was not encouraged; and, in any event, none of these
poetic or pictorial versions would have satisfied Wyclif's demand for the literal text as the key to right understanding of
Holy Writ. Earlier fourteenth-century translations of the Psalter
and parts of the New Testament suggest that there may have
been an incipient movement in favour of vernacular scriptures
before Wyclif's time; but the lollards were the first to plan and
execute an English translation of the whole Bible. Wyclif himself is now thought to have taken little if any part in the actual
work of translation; but he may have supervised, and he certainly inspired, the earlier of the two versions that has come
down to us. This is a strictly literal rendering probably intended
as a key to the Vulgate for those with little Latin, that is, for the
inferior clergy and for laymen of some education. Since there is
good manuscript authority for believing the Old Testament,
down to Baruch iii. 20, to have been the work of Nicholas Hereford, it is likely that this first version was an Oxford enterprise;
but the translator, or translators, of the remainder of the Old
Testament and of the New are unknown to us. Whoever they
were, their work made it plain to the lollards that word-for-
word rendering of the Vulgate into English was inadequate to
convey the true sense of the original; a second version, freer,
more idiomatic, and more readable was needed; this was begun
in the eighties and completed probably about 1396. The translator, generally believed to have been John Purvey (though the
evidence for his authorship is not conclusive), describes himself
in his prologue as a 'simple creature' and tells us that he sought
many helpers in his attempt to solve the four main problems
confronting any translator of the Bible in this period—to establish a satisfactory text of the Vulgate; to unravel the sense of the
text with the aid of the glossators; to find apt English equivalents
for hard words and hard sentences; and to produce a lucid rendering. Allusions to Oxford in the prologue suggest that these
helpers may have been Oxford scholars; and the result was a

translation which was widely copied in the fifteenth century and, shorn of its outspokenly lollard prologue, remained the best English version until the time of Tyndale and Coverdale.[1] If the association of the vernacular Bible with unorthodoxy was in some ways unfortunate, it is none the less evident that the lollards had met a demand which extended far beyond the circle of their adherents.[2] The decline of French as the language of educated society and the great resurgence of English as a literary language in the second half of the fourteenth century, meant that, despite persistent discouragement of Bible-reading by authority, an English Bible had become a necessity, long before the Reformation.

Evidence of this development of the vernacular meets us in many quarters. Writing in 1385 Trevisa observes that some forty years earlier, two Oxford grammar masters, John Cornwall and Richard Pencrich, had first decreed that boys should construe their Latin, not into French, but into English, and that this custom had since become general.[3] It was in 1362 that the statute ordering the use of English in the law-courts was enacted and in 1363 that the chancellor first opened parliament in his native tongue, a precedent which was commonly, though not invariably followed.[4] French was still in use at court until the end of the century, though, when Froissart was received by Richard II in 1395, he thought it worthy of comment that the king could speak and read French very well.[5] The historians in the monastic *scriptoria* continued, for the most part, to write in Latin; but the *Brut*, the oldest prose chronicle in Middle English, was translated from Norman-French between 1350 and 1380. John Trevisa, who had been Wyclif's contemporary at Oxford, finished his translation of Higden's *Polychronicon* in 1387 and of the great encyclopaedia of Bartholomew the Englishman (*De Proprietatibus Rerum*) in 1398. North of the Border, John Barbour, archdeacon of Aberdeen, had completed his *Bruce* in 1375. A poet like Gower, who was at home in all three languages

[1] Sir W. A. Craigie, 'The English Versions (to Wyclif)', *Ancient and English Versions of the Bible*, ed. H. Wheeler Robinson (1940), pp. 137–45.

[2] Forshall and Madden examined nearly 150 manuscripts for their edition of the lollard Bible (1850) and others have since come to light.

[3] *Polychronicon*, ii. 161.

[4] *Statutes* 36 Edw. III, st. 1, c. 15 (see above, p. 197); *Rot. Parl.* ii. 275, 283. Parliament seems to have opened in French in 1377.   [5] Froissart, xv. 167.

and used them with equal facility, is becoming something of an exception by 1400, and it is noteworthy that Gower chose English for his last important work (c. 1390). For by this date English had invaded the realms of lyric and romance, of comedy and tragedy, of allegory and drama, of religion and education.[1] It had become the language, not of a conquered, but of a conquering people.

The geography of this fourteenth-century revival is interesting. Most of the literature was provincial and much of it was north-midland and north-western, springing from regions which had been almost completely silent for over 500 years. In sharp contrast to the metrical romances composed in the south under French influence (and satirized by Chaucer in *Sir Thopas*), the northern poems are written in unrhymed, alliterative verse, an ancient form which may have survived, without record, since before the Norman Conquest. But the revival was not merely antiquarian; for the structure of the language had been subtly modified by the passage of centuries, and the influence of France, though weakened by time and distance, is unmistakable in the most distinguished of the northern alliterative productions—four poems, all contained in a single manuscript and likely to have been the work of a single author.[2] Three of them—*Pearl* (a moving allegory on a dead child in which the author uses the vision convention of the romances), *Patience*, and *Purity* —present familiar medieval moralizing in attractive dress. The fourth and most impressive—*Sir Gawayne and the Grene Knight*— is an Arthurian romance set in wintry and mountainous country by an artist of real imaginative power, whose vocabulary includes many words of Scandinavian origin, not to be found in the southern writers. In the same alliterative tradition are such poems as *The Parlement of the Thre Ages* (Youth, Middle Age, and Old Age), and *Wynnere and Wastoure* (Winner and Waster). All the authors are anonymous. But this strange flowering of alliterative allegories and romances in the age of Chaucer, who 'writes as if they had no existence and would have written no differently had he known them',[3] may serve to remind the historian how much still remains to be discovered of the knightly

[1] See J. E. Wells, *A Manual of the Writings in Middle English* (1916). (Supplement in progress.)

[2] Brit. Mus. MS. Cott. Nero A. X. See *Pearl*, ed. E. V. Gordon (1953), pp. xli–xliv.

[3] G. Sampson, *The Concise Cambridge History of English Literature* (1941), p. 46.

and aristocratic households of the north and midlands where such poets must have found their audiences.

Further evidence of the literary vitality of these regions may be found in the miracle plays and in devotional literature. The three almost complete cycles of plays which have come down to us—the York, Wakefield (or Towneley),[1] and Chester cycles—all derive from this part of the country; and, though the manuscripts date from the fifteenth century, the plays were almost certainly taking shape during our period. Among religious writers, Richard Rolle (c. 1300–49), of Thornton-le-Dale in Yorkshire, turned his back on Oxford and withdrew to his native county to live as a hermit at Hampole, where he translated the Psalter into English prose and composed a number of religious lyrics and devotional tracts in both Latin and English.[2] Robert Mannyng who, in the 12,000 octosyllabic lines of his *Handlyng Synne*, made an heroic effort to distract his readers from secular romances, came from Bourne in Lincolnshire.[3] Walter Hilton (d. 1396), author of the widely read *Scale of Perfection*, was a canon of Thurgarton, near Newark; and the *Cursor Mundi* (c. 1300–25) is also of northern origin. In a very different genre the patriotic verses of Lawrence Minot, composed between 1333 and 1352, are likewise traceable to the north.

Yet the greatest of all the alliterative poems is of southern origin. *Piers Plowman*, which survives in three distinct versions (the so-called A-text, composed 1362–3; the B-text, 1377; and the C-text, c. 1393), has presented its critics with many problems.[4] Expert opinion tends, on the whole, to favour belief in a single author, William Langland, who may have been born about 1332, had his schooling at Great Malvern priory, and taken minor orders before, having completed the A-text, he removed to London to live on Cornhill with his wife Kitte and his daughter, Calotte. In the B-text Langland altered and greatly expanded his original poem; in the C-text he undertook a final revision, not generally regarded as an improvement. The book is

---

[1] So called from the family which long owned the manuscript.

[2] See Hope Emily Allen, *Writings ascribed to Richard Rolle* (1927).

[3] This work is a version of William of Waddington's *Manuel des Péchiez*. See *Robert of Brunne's Handlyng Synne (1303) and its French Original*, ed. F. J. Furnivall (Early Eng. Text Soc. cxix, cxxiii).

[4] The standard edition by W. W. Skeat, 2 vols. (1886), has been reprinted (1954) with a bibliographical note by J. A. W. Bennett.

divided into two main parts—the 'Vision of William concerning Piers the Plowman' and the 'Life of Dowel, Dobet and Dobest' (Do Well, Do Better and Do Best). Though little of it is easy reading, the persistent reader will not lack reward; for the poem is of compelling sincerity and power, its sombre landscape lit with gleams of beauty. Langland is a bitter critic of the whole top-heavy ecclesiastical system—'we han so manye maistres'—and of the rich and proud; but he is not a demagogue, still less a revolutionary. He is learned, austere, a seeker after truth and a teacher of righteousness. In his lovely lines on Charity he evokes an ideal which has nothing in common with the rantings of John Ball:

'Charite,' quod he, 'ne chaffareth nouȝte · ne chalengeth, ne craueth.
As proude of a peny · as of a pounde of golde,
And is as gladde of a goune · of a graye russet
As of a tunicle of Tarse · or of trye scarlet.
He is gladde with alle gladde · and good tyl alle wykked,
And leueth and loueth alle · that owre lorde made.
Curseth he no creature · ne he can bere no wratthe,
Ne no lykynge hath to lye · ne laughe men to scorne.
Al that men seith, he let it soth · and in solace taketh,
And alle manere meschiefs · in myldenesse he suffreth;
Coueiteth he none erthly good · but heuene-riche blisse.'[1]

Piers Plowman offers us a congested canvas of late fourteenth-century society. The value of the poem as a source for social history has been widely recognized, more widely, perhaps, than its poetic quality. Yet, in contrast with *Troilus*, or the *Canterbury Tales*, or even with *Confessio Amantis*, the figures are two-dimensional, we do not see them in the round; we seem to look at a medieval fresco, rather than to mingle with a medieval crowd. Perhaps by reason of its form—for the Old English alliterative line was burning itself out—the whole poem has an archaic air. It is none the less impressive for that. But, whereas Gower presents us with the conventions of a polite society, and Chaucer with human nature as we know it, Langland speaks to us from a forgotten world, drowned, mysterious, irrecoverable.

John Gower, who came of a family of Kentish squires, was a Londoner by adoption, a friend of Chaucer's, and a generous benefactor of the priory of St. Mary Overy in Southwark (now the cathedral), where his effigy may still be seen, the head

[1] B xv, ll. 159–70.

resting on his three principal works—the French, *Speculum Meditantis* (or *Mirour de l'Homme*), the Latin, *Vox Clamantis*, and the English, *Confessio Amantis*. Gower's will was proved in 1408, but we know neither the date of his birth nor much about his life, though this may be presumed to have been leisurely. There are 10,000 lines in the Latin book, over 29,000 in the French, 33,000 in the English—and some minor works as well. Gower was essentially a stylist, whom the ordinary reader probably underrates. He is distinguished principally by his 'correctness' and by the ease and lucidity of his French and English verse.[1] But he was also a perceptive and skilful narrator and, though inordinately addicted to moralizing, he does not lack humour. For historians, much of the interest of his work lies in what he tells us of his own reactions to contemporary history. In the first book of *Vox Clamantis*, for example, Gower succeeds, despite (or, perhaps, by reason of) what W. P. Ker called the 'detestable verse', in conveying an idea of the horror and panic aroused in the minds of the gentry by the rising of 1381:

> Quidam sternutant asinorum more ferino,
> Mugitus quidam personuere boum;
> Quidam porcorum grunnitus horridiores
> Emittunt, que suo murmure terra tremit.[2]

The same poem, believed to have been composed in 1382 or 1383,[3] also reveals something of the poet's attitude to Richard II. Having begun by declaring the king too young to be held responsible for the corruption of his court, Gower, in a slightly later revision of the poem, takes it upon himself to read Richard a lecture on his kingly duties, urging him to be worthy of his famous father and reminding him that the God-given beauty of his person ought to be matched by the virtue of his soul. It says much for Richard's forbearance that when, two or three years later, he met his surly critic on the Thames, he should have invited him into the royal barge and encouraged him to write something more palatable:

> He hath this charge upon me leid,
> And bad me doo my besynesse
> That to his hihe worthinesse

---

[1] W. P. Ker, *Essays on Medieval Literature* (1905), pp. 104–7; see also C. S. Lewis, *The Allegory of Love* (1938), pp. 198–222.　　[2] *Vox Clamantis*, I. xi, ll. 799–802.

[3] G. C. Macaulay, *The Works of John Gower*, iv (1902), p. xxx. Many of the dates of Gower's works are conjectural.

Some newe thing I scholde boke,
That he himself it mihte loke
After the forme of my writynge.[1]

Gower was naturally flattered: and it seems likely that his
laudable decision to write in English and to season edification
with love-stories was taken on the king's advice. The first recen-
sion of *Confessio Amantis*, completed in 1390, opens with a prologue which describes it as

A bok for king Richardes sake,
To whom belongeth my ligeance
With al myn hertes obeissance
In al that evere a liege man
Unto his king may doon or can.

Oddly enough, it seems not to have been the events of 1386–8 which
shook Gower's loyalty to Richard, but something that occurred in
or after 1390. By 1391 he had cut out the compliments to Richard
and, by 1393 at latest, had substituted a dedication to Henry of
Lancaster. Whatever brought about this metamorphosis, it can
hardly have been the gift from Henry of an inexpensive collar,
valued at 26s. 8d.; all that is clear is that henceforward Gower
became increasingly critical of Richard II. The *Cronica Tripertita*,
which he added to his *Vox Clamantis* early in the next reign, recapi-
tulates the story of Richard's last years from the Lancastrian
standpoint; and one of the last poems that he ever wrote—perhaps the last, for he went blind shortly afterwards—is addressed
'In Praise of Peace' to Henry:

God hath the chose in comfort of ous alle.

Though he seems to have had no particular attachment to
Gaunt, Gower must have had Lancastrian affiliations before
1399. It is interesting to find that *Confessio Amantis* was translated into Portuguese by an English canon of Lisbon, probably
in Gower's own lifetime, when the queen of Portugal was
Henry's sister, Philippa of Lancaster.

We come, at last, to Chaucer, the glory of the age and its epi-
tome. Little in his time escaped him; he was familiar with almost
every aspect of lay society touched on in this book. Born, probably in 1343 or 1344, the son and grandson of London vintners,

[1] *Confessio Amantis*, Prologue, ll. 48–52.

and from 1374–86 controller of the petty custom on wine and merchandise in the port of London, he knew the world of trade and business. His service as a page in the household of the duchess of Clarence in the fifties, and as valet and afterwards esquire to Edward III in the sixties and seventies, his marriage to Philippa Roet, one of the queen's ladies and sister to Katharine Swynford, brought him into the innermost circles of courtly society. He saw active service in the campaign of 1359–60, when he was taken prisoner and ransomed, and in 1369, probably with Lancaster in Picardy.[1] He was sent abroad on several diplomatic missions, to France in 1368, 1377, and 1387, and to Italy in 1372–3, when he visited Genoa and Florence, and again in 1378, when he went to Lombardy. He was justice of the peace in the county of Kent and represented it in the momentous parliament of 1386. As clerk of the king's works from 1389–91, he carried a heavy burden of responsibility; and when he relinquished this appointment, it was to accept another as deputy-forester in the royal forest of North Petherton in Somerset. Both in Somerset and in Kent, where much of his time was spent, he lived the life of a country gentleman; but London was his chosen home. For many years he occupied a house above Aldgate; and only a few months before his death, on 25 October 1400, he had leased another, in the garden of Westminster Abbey. He never went to a university, but he understood the issues which the clerks were debating in the schools; and whether or not he had a formal legal training, he knew the language of the law.[2] His activities would have sufficed to fill the life and absorb the whole attention of an ordinary man; but Chaucer, who was not ordinary, returned nightly to his books. The speech of the eagle in the *House of Fame* allows us a precious glimpse of the poet as reader:

> For when thy labour doon al ys,
> And hast mad alle thy rekenynges,
> In stede of reste and newe thynges,
> Thou goost hom to thy hous anoon;
> And, also domb as any stoon,
> Thou sittest at another book
> Tyl fully daswed is thy look,
> And lyvest thus as an heremyte,
> Although thyn abstynence ys lyte.[3]

[1] There is a blank in the records of Chaucer's life from 1360 to 1367.
[2] It is now thought likely that Chaucer was a member of one of the inns of court. See *The Works of Geoffrey Chaucer*, ed. F. N. Robinson (2nd ed. 1957), Introd., p. xxv.   [3] *House of Fame*, ll. 652–9.

Elsewhere, Chaucer pays moving tribute to the books that have been his solace:

> And yf that olde bokes were aweye,
> Yloren were of remembraunce the keye.
> Wel ought us thanne honouren and beleve
> These bokes, there we han noon other preve.
> And as for me, though that I konne but lyte,
> On bokes for to rede I me delyte,
> And to hem yive I feyth and ful credence,
> And in myn herte have hem in reverence
> So hertely, that ther is game noon
> That fro my bokes maketh me to goon,
> But yt be seldom on the holyday,
> Save certeynly, whan that the month of May
> Is comen. . . .[1]

The range of his reading is astonishing. He knew the Vulgate thoroughly and the hymns and services of the Church; he knew Virgil, Ovid, Statius, Claudian, and other classical authors; and he knew well such early medieval versions of the classical myths and histories as the *Roman de Troie* and *Li Hystore de Julius Caesar*. He knew Jerome and, of course, Boethius; and he was widely read in the multifarious Latin literature of the Middle Ages, in its poetry, its history, its philosophy, its science. The *Treatise on the Astrolabe* which he compiled for his young son, was accepted as the standard English textbook on the subject.[2] It goes without saying that he knew the *Roman de la Rose* of Guillaume de Lorris and Jean de Meun, and the work of his French contemporaries, Guillaume de Machaut, Eustace Deschamps, and Froissart. After his first visit to Italy he added the Italians to his store, Dante, Petrarch, and Boccaccio.

Though few of his works can be dated precisely, Chaucer scholarship has established the sequence of most of them. Of the major works, the *Book of the Duchess*, a lament for Blanche of Lancaster ('goode faire White'), who died in 1369, belongs to the earliest period when Chaucer was writing mainly under French influence. Between 1372 and 1380, after his first visit to Italy, come the transitional works, partly in the French tradition but showing the effects of his Italian reading—the *House of Fame* and, possibly, the *Parliament of Fowls.* Between 1380 and

---

[1] *Legend of Good Women*, ll. 25–37.
[2] R. T. Gunther, *Early Science in Oxford*, ii. 62–63.

1386 when the Italian influence had been fully assimilated, there appeared *Troilus* and the *Legend of Good Women*. The *Prologue*, the earlier *Canterbury Tales*, and the *Astrolabe* were written between 1387 and 1392 and the later *Tales* between 1393 and 1400.[1] The *Canterbury Tales* stand supreme in their own kind; but the reader who confines his attention to them misses some of the best in Chaucer—the delicate comedy of the dialogue between the poet and the eagle in the *House of Fame*, for example, the brilliance and vivacity of the *Parliament of Fowls*, above all, the matchless beauty of *Troilus and Criseyde*. Each is the work of a great artist, a man of poetic genius and of noble heart and mind. Chaucer did not invent his tales; but he told them as they had never been told before, and in his hands the puppets of the *fabliaux* become living men and women. Some of the humour is broad, but none of it is insensitive, 'for pitee renneth soone in gentil herte', and Chaucer's infinite compassion embraces all his creatures. He knew his way about the world and nothing that was human was beyond his understanding—neither bawdy, farce, nor wit; neither love nor passion; neither simple virtue nor high adventure; neither pain, nor fear, nor age, nor death; nor yet the immortal longings of the heart. His speech may sound strangely in our ears, as the speech of the ancient world sounded strangely in his own—

> Ye knowe ek that in forme of speche is chaunge
> Withinne a thousand yeer, and wordes tho
> That hadden pris, now wonder nyce and straunge
> Us thinketh hem, and yet thei spake hem so,
> And spedde as wel in love as men now do;[2]

—but his language is the language of all time; and, more than any other writer of his age, he opens the gate to comprehension of a vanished world. For with Chaucer, as with Shakespeare, whose universal quality he shares, genius is sufficient to persuade us that the children of his imagination were of like passions with ourselves and were bred in a land we know.

---

[1] Chaucer's authorship of the recently discovered *Equatorie of the Planetis*, ed. D. J. Price (1955), has not been conclusively established.
[2] *Troilus and Criseyde*, book ii, ll. 22–28.

BIBLIOGRAPHY

Note. The following list of authorities (which is not intended to be in any sense complete) is arranged on a plan similar to those adopted in the two preceding volumes of the present History. Articles and specialized studies referred to in the footnotes, which are generally not included here, may be found in the Index under the names of their authors. *E.H.R.* = *English Historical Review*; R.S. = Rolls Series; E.E.T.S. = Early English Text Society.

1. Bibliographies and Books of Reference.
2. Charters, Records, and other Documents.
3. Ecclesiastical Records, Letters, and Wills.
4. Narrative Sources.
5. General and English Political History.
6. Ecclesiastical History.
7. Law and Institutions.
8. Social and Economic History.
9. Scotland.
10. Wales.
11. Ireland.
12. Literature and Art.

1. BIBLIOGRAPHIES AND BOOKS OF REFERENCE

The best general guide to the printed sources and historical literature of the period, C. Gross, *The Sources and Literature of English History from the earliest times to about 1485* (2nd ed., London, 1915), is now nearly half a century out of date but a new edition is projected. The *Annual Bulletins of Historical Literature* published by the Historical Association may be consulted for historical literature published since 1915. A new edition of A. Potthast, *Bibliotheca Historica Medii Aevi* (2nd ed., Berlin, 1896), is in course of preparation. Manuscripts of literary sources are described in T. Duffus Hardy's *Descriptive Catalogue of Materials relating to the History of Great Britain and Ireland*, 3 vols. (R.S., 1862–71), which includes an appendix on sources in print at that date. For French narrative sources, which contain much material for the English history of this period, see the critical bibliography by A. Molinier, *Les Sources de l'histoire de France* (Paris, 1901–6). H. Maxwell Lyte, *Historical Notes on the use of the Great Seal in*

England (Stationery Office, 1926), and V. H. Galbraith, *An Introduction to the use of the Public Records* (Oxford, 1934), are useful guides. A new edition of the authoritative survey by M. S. Giuseppi, *Guide to the Manuscripts preserved in the Public Record Office* (Stationery Office, 1923–4), is in preparation. The bibliographies appended to the relevant chapters of vols. vii and viii of the *Cambridge Medieval History* (1932, 1936) will be found helpful for both original and secondary sources.

The *Dictionary of National Biography*, which includes lives of all the more important persons of the period with summaries of the biographical material, is an indispensable work of reference. For biographies of the nobility and the descent of noble families, see G. E. Cokayne, *The Complete Peerage*; the new edition begun in Hardy (Oxford, 1854; new edition in preparation). A list of medieval parliaments with details relating to election and attendance of the commons will be found in the *Interim Report of the Committee on House of Commons Personnel and Politics 1264–1832* (1932). Lists of kings and ministers of state, bishops, dukes, earls, ecclesiastical councils, and parliaments, together with much chronological material are assembled in the *Handbook of British Chronology*, ed. F. M. Powicke and others (Royal Hist. Soc., 1939); the Society is also responsible for the very useful *Handbook of Dates*, ed. C. R. Cheney (1945). *L'Art de vérifier les dates*, 18 vols. (Paris, 1818–19), gives valuable brief accounts of the royal families and the higher nobility of Europe. For pedigrees of reigning houses, see *Genealogical Tables* by H. B. George, 6th ed. by J. R. H. Weaver (Oxford, 1930). The *Victoria History of the Counties of England*, still in process of publication, is a mine of information on the descent of manors and on local history in general.

2. CHARTERS, RECORDS, AND OTHER DOCUMENTS

The *Calendars of Charter Rolls*, vols. iii–v (Stationery Office, 1908–16), include such royal charters and confirmations as were enrolled in the chancery. A valuable catalogue of medieval cartularies has been compiled by G. R. C. Davis, *Medieval Cartularies of Great Britain* (London, 1958); the majority are ecclesiastical but the list includes such important secular cartularies as those of the Beauchamp earls of Warwick and the 'Liber Niger 1910 was completed in 1959. Lists of dignitaries of the Church are given in J. le Neve, *Fasti Ecclesiae Anglicanae*, ed. T. Duffus

de Wigmore'. A number of private charters deriving from this period are printed *in extenso* or summarized in *Sir Christopher Hatton's Book of Seals*, ed. L. C. Loyd and D. M. Stenton (Oxford, 1950). For summaries of the urban charters of the fourteenth century, see M. Weinbaum, *British Borough Charters, 1307–1660* (Cambridge, 1943).

Rymer's *Foedera*, as prepared for the Record Commission (vols. i–iii, 1816–19, and vol. iv, printed 1833, published 1869), extend only to 1383; for the remaining years of the century it is necessary to refer to vols. vii and viii of the first edition (20 vols. London, 1704–35). The *Foedera* comprise a large miscellany of documents drawn from diverse sources and including many treaties and letters. A syllabus of the contents was published by T. Duffus Hardy, 3 vols. (R.S., 1869–85).

Good examples of collections of documents drawn from various classes of records to afford material for a particular theme are *The Diplomatic Correspondence of Richard II*, ed. E. Perroy (Camden, 3rd series, xlviii, 1933), and *The War of Saint-Sardos, 1323–25*, ed. P. Chaplais (ibid. lxxxvii, 1954).

Chancery. The *Calendars of Patent Rolls* (so called because the letters were delivered open with the great seal pendant at the bottom and were addressed 'to all to whom these presents shall come') are complete for the period 1307–99 in 27 vols. (Stationery Office, 1894–1916). So, too, are the *Calendars of Close Rolls*, 24 vols. (Stationery Office, 1892–1927); these comprise abstracts of letters close which, being normally mandates addressed to one or more persons, were closed and sealed on the outside. The *Calendar of Chancery Rolls (Various) 1277–1326* (Stationery Office, 1912) includes summaries of the scutage rolls, 1285–1324, and some supplementary close rolls. A single volume, the *Calendar of Chancery Warrants, 1244–1326* (Stationery Office, 1927), summarizes warrants issued under the king's privy seal authorizing the issue of letters under the great seal. Vols. ii–xi of the *Calendars of Fine Rolls* (Stationery Office, 1912–29) cover the period 1307–99: they record payments made to the king for various favours, such as licences, pardons, and exemptions from military service. A beginning has been made with the publication of the *Treaty Rolls*, vol. i, 1234–1325, ed. P. Chaplais (Stationery Office, 1955).

Exchequer. Thomas Madox brought his *History and Antiquities of the Exchequer* (2nd ed., London, 1769)—the classic commentary which includes numerous extracts from the Pipe Rolls and other records—to a close with the reign of Edward II, but it is none the less indispensable to students of the fourteenth century as a whole. *Issues of the Exchequer,* ed. F. Devon (London, 1837), contains extracts, in translation, from the liberate, issue, memoranda, and household rolls: Devon also edited a translation of the *Issue Roll of Thomas de Brantingham, 1370* (London, 1835). Extracts from the liberate rolls relative to the repayment of loans by Italian merchants to the Crown in the thirteenth and fourteenth centuries were published by the Society of Antiquaries of London in *Archaeologia,* xxviii (1840), 207–326. The Society has also published summaries of the wardrobe accounts of 10, 11, and 14 Edward II (ibid. xxvi (1836), 318–45) and the wardrobe accounts of 1344–9 (ibid. xxxi (1846), 5–103) and of 1392–3 (ibid. lxii, pt. ii (1911), 497–514). The six volumes of *Inquisitions and Assessments relating to Feudal Aids* (Stationery Office, 1899–1920) are arranged topographically. Published records of taxation include the assessment of the ninth of 1340, *Nonarum Inquisitiones* (Rec. Com., 1807), and a summary of the subsidy roll of 51 Edward III (the poll-tax of 1377), *Archaeologia,* vii (1785), 337–47. Subsidy rolls have been published by many local record societies (for examples, see Gross, pp. 428–35). Vol. iii of *The Red Book of the Exchequer,* ed. H. Hall, 3 vols. (R.S., 1896), includes the Cowick ordinances of 1323, a household ordinance of 1324, and a fourteenth-century treatise on the mint. The *Rotulorum originalium in curia scaccarii abbreviatio,* 2 vols. (Rec. Com., 1805–10), is an abstract of the *originalia* rolls of the lord treasurer's remembrancer which consisted of extracts from the chancery rolls by means of which the exchequer was kept informed of all writs, letters, and charters on which fines were payable. These rolls, as Professor Galbraith has pointed out (*The Public Records,* p. 43), afford an interesting example of co-operation between the two great departments: and the same is true of the *Inquisitiones post Mortem,* inquiries into the land held by a tenant-in-chief at his death, which were compiled in duplicate in the chancery, the second copy being forwarded to the exchequer. These inquests have been calendared for the reigns of Edward II and Edward III, vols. v–xiv (Stationery Office, 1908–52).

Parliament and Council. For the parliament rolls—*Rotuli Parliamentorum* (1272–1503), ed. J. Strachey and others, 6 vols. (1767);

Index (Rec. Com. 1832),—see above, p. 182. The two great volumes produced by F. Palgrave, *Parliamentary Writs and Writs of Military Summons, Edward I and Edward II* (Rec. Com., 1827–34), are likewise indispensable. The parliament roll for 12 Edward II is printed in H. Cole, *Documents Illustrative of English History in the Thirteenth and Fourteenth Centuries* (Rec. Com., 1844), and some supplementary rolls for the years 1279–1373 by H. G. Richardson and G. O. Sayles, *Rotuli parliamentorum anglie hactenus inediti* (Camden, 3rd series, li, 1935). William Prynne, *A brief Register . . . of Parliamentary Writs*, 4 pts. (London, 1659–64), contains many expenses writs and returns. William Dugdale, *A perfect copy of all summons of the nobility to the Great Councils and Parliaments of the Realm* (London, 1685) and the first three volumes of *Reports from the Lords' Committees . . . touching the Dignity of a Peer*, 5 vols. (London, 1820–9), contain much material of value for the study of the medieval parliament. The *Statutes of the Realm*, 11 vols. (Rec. Com., 1810–28), is the most nearly complete collection of the early statutes. Vol. i of *Proceedings and Ordinances of the Privy Council*, ed. N. H. Nicolas, 7 vols. (Rec. Com., 1834–7), contains memoranda of council proceedings in the last decade of the century. Lists of the names of parliamentary knights and burgesses will be found in *Parliamentary Papers*, vol. lxii, pt. i (1878). These lists were compiled from the writs *de expensis* and many more names have since come to light, mainly in borough records, and scattered for the most part in the publications of local record societies, pending publication of the medieval section of the forthcoming official History of Parliament. For the *Modus tenendi parliamentum* see above, pp. 183–4.

Legal. Certain cases heard by the council, or by the council in parliament, are recorded in the Parliament Rolls. See also *Select Cases before the King's Council*, ed. L. G. Leadam and J. F. Baldwin (Selden Soc., 1918). Other aspects of royal justice are illustrated in the following publications of the Selden Society: *Select Pleas in the Court of Admiralty*, ed. R. G. Marsden (1894); *Select Cases in Chancery*, ed. W. P. Baildon (1896); *Select Pleas from the Coroners' Rolls*, ed. C. Gross (1896); *Select Pleas of the Forest*, ed. J. G. Turner (1901); *Select Cases concerning the Law Merchant*, ed. C. Gross and H. Hall, 3 vols. (1908–32); *Select Bills in Eyre*, ed. W. C. Bolland (1914); *Public Works in Medieval Law*, ed. C. T. Flower, 2 vols. (1915–23); *Select Cases in the Exchequer of Pleas*, ed. H. Jenkinson and B. Fermoy (1932); and in the *Calendar of Inquisitions, Miscellaneous* (Chancery), vols. ii–iv, 1307–88 (Stationery Office, 1916–57). Material of value to the historian of the law is also to be found in two older volumes—the *Placita de Quo Warranto, Edward I–Edward III*, ed. W. Illingworth (Rec. Com., 1818), and the *Placitorum Abbreviatio, Richard I–Edward II* (ibid., 1811), the latter being an abstract of miscellaneous pleas compiled by the indefatigable Elizabethan antiquary, Arthur Agarde. Proceedings before the justices of the peace are illustrated in *Some Sessions of the Peace in Lincolnshire, 1360–75*, ed. R. Sillem (Lincoln Record Soc. xxx, 1937), and in a similar volume for 1381–96, ed. E. Kimball (ibid. xlix, 1955). *The Scrope and Grosvenor Controversy, 1385–90*, ed. N. H. Nicolas, 2 vols. (London, 1832), records a *cause célèbre* in the court of chivalry.

Points of law debated in the courts and the opinions of the judges are preserved in the Year Books, on which see W. C. Bolland, *Manual of Year Book Studies* (Cambridge, 1925), and R. V. Rogers, 'Law reporting and the multiplication of law reports in the fourteenth century', *E.H.R.* lxvi (1951), 481–506. Numerous Year Books for the reign of Edward II have been published by the Selden Society (Year Book Series, ed. F. W. Maitland and others, 1903–51). Year Books of 11–20 Edward III have been edited by A. J. Horwood and L. O. Pike, 15 vols. (R.S., 1883–1911). Those of Richard II's reign are to be completed in 8 volumes (Ames Foundation publications), 3 of which have already appeared—12 Ric. II, 1388–9, ed. G. F. Deiser (1914); 13 Ric. II, 1389–90, ed. T. F. T. Plucknett (1929); 11 Ric. II, ed. I. D. Thornley and T. F. T. Plucknett (1937). The Selden Society has published for the Historical Manuscripts Commission a *Preliminary Edition of the Register of MSS of Year Books extant*, prepared by Jennifer Nicholson (1956).

In contrast to the preceding age, the fourteenth century was not prolific of legal tracts or works on jurisprudence. The *Mirror of Justices*, possibly the work of Andrew Horn, one time chamberlain of London, may date from the reign of Edward II but Maitland's preference for an earlier date is supported by N. Denholm-Young (*Collected Papers* (Oxford, 1946), p. 79, n. 1). Otherwise, only *Novae Narrationes* (London, 1561), a French tract dealing with the method of pleading and *Olde Teners newly Corrected* (London, 1525), both anonymous and both assigned to the reign of Edward III, have come down to us.

The Army and Navy. Many writs of military summons are entered on the Close Rolls and the Treaty Rolls; commissions of array and other documents relating to military levies will be found in Palgrave's *Parliamentary Writs and Writs of Military Summons. Crecy and Calais*, ed. G. Wrottesley (W. Salt Archaeol. Soc., *Collections*, xviii, pt. ii, London, 1897), includes translated extracts from the French and memoranda rolls, the Norman roll and the Calais roll and from the account of the treasurer of the household. For the study of maritime law the principal source is the *Black Book of the Admiralty*, ed. with translation by Sir Travers Twiss, 4 vols. (R.S., 1871–6).

Private and Local. Notable collections relating to two great honours are the *Register of Edward the Black Prince*, 4 pts. (Stationery Office, 1930–3), and *John of Gaunt's Register, 1372–76*, ed. S. Armitage-Smith, 2 vols., Camden 3rd series, xx–xxi (1911), and *1379–83*, ed. E. C. Lodge and R. Somerville, ibid. lvi–lvii (1937). Manorial and urban records are too numerous for detailed specification here but many have been printed or used as the basis of special studies (see the references given in Chapters XI and XII above). Mention must be made, however, of the very valuable collections of London records which survive from this period. The most important publications are the *Calendar of Letter-Books of the City of London* (C to H), ed. R. R. Sharpe (London, 1901–7), which include recognizances of debts and miscellaneous civic regulations; the *Calendar of Letters from the Mayor and Corporation of the City of London, 1350–70*, ed. R. R. Sharpe (London, 1885); the *Calendar of Coroners' Rolls, 1300–78*, ed. R. R. Sharpe (London, 1914); the *Calendar of Plea and Memoranda Rolls of the City of London, 1323–1412*, ed. A. H. Thomas, 3 vols. (London, 1926–32); and the *Liber Custumarum* (c. 1320), ed. H. T. Riley, in *Munimenta Gildhallae Londoniensis* (R.S., 1859–62), vol. ii. For other urban collections, see C. Gross, *A Bibliography of British Municipal History* (Cambridge, 1915). Extracts from many urban, ecclesiastical, and private records are embodied in the reports and appendixes of the Historical Manuscripts Commission (1870 ff.).

3. ECCLESIASTICAL RECORDS, LETTERS, AND WILLS

General. The essential guide to the mass of material for British and Irish history preserved in the papal archives is the *Calendar of entries in the papal registers relating to Great Britain and Ireland:*

Papal Letters, vols. i–v, ed. W. H. Bliss and others (Stationery Office, 1893–1904), and *Petitions to the Pope*, vol. i, ed. W. H. Bliss (ibid., 1896). A great part of the ecclesiastical legislation of the period is to be found in *Concilia Magnae Britanniae et Hiberniae*, ed. D. Wilkins, 4 vols. (London, 1737); a much-needed new edition of this important work is in preparation. The *Nonarum Inquisitiones* of 14–15 Edward III (above, p. 536) specify the value of every ecclesiastical benefice at that date, in relation to the *Taxatio* of Nicholas IV, 1291. (See Rose Graham, *English Ecclesiastical Studies*, pp. 271–301.)

Episcopal Registers. The following registers have been published, or are in preparation: *Canterbury*: Robert Winchelsey, 1294–1308, ed. Rose Graham (Cant. and York Soc., 1917–51); Simon Langham, 1366–8, ed. A. C. Wood (ibid., 1947–54). *Bath and Wells*: John Drokensford, 1309–29, ed. E. Hobhouse (Somerset Record Soc. 1887); Ralph of Shrewsbury, 1329–63, ed. T. S. Holmes (ibid., 1896). *Chichester*: Robert Rede, 1397–1415, ed. C. Deedes (Sussex Record Soc., 1908–10). *Coventry and Lichfield*: Abstracts of the registers of Roger Northburgh, 1322–59, ed. E. Hobhouse (W. Salt Archaeol. Soc., 1881), Robert Stretton, 1360–85, and the *sede vacante* register, ed. R. A. Wilson (ibid. 1907, 1910). *Ely*: Extracts from the registers of Simon Montacute, 1337–45, Thomas de Lisle, 1345–61, Thomas Arundel, 1374–88, and John Fordham, 1388–1425, are printed in the *Ely Diocesan Remembrancer* (1889–1914). *Exeter*: Indexes of the contents of the registers of Walter Stapledon, 1307–26, John Grandisson, 1327–69, Thomas Brantingham, 1370–94, and Edmund Stafford, 1395–1419, with illustrative extracts have been edited by F. C. Hingeston-Randolph (London, 1886, 1906). *Hereford*: All the fourteenth-century registers have been issued jointly by the Cantilupe Society (Hereford) and the Cant. and York Soc. Richard Swinfield, 1283–1317, ed. W. W. Capes (1909); Adam Orleton, 1317–27, ed. A. T. Bannister (1908); Thomas Charlton, 1327–44, ed. W. W. Capes (1912); John Trilleck, 1344–61, ed. J. H. Parry (1910, 1912); Lewis Charlton, 1361–9, ed. J. H. Parry (1913); William Courtenay, 1370–5, ed. W. W. Capes (1913); John Gilbert, 1375–89, ed. J. H. Parry (1913); John Trefnant, 1389–1404, ed. W. W. Capes (1914–15). *Lincoln*: None of the fourteenth-century registers has been published but the effect of the Black Death on the clergy of the diocese was examined by A. Hamilton Thompson, 'Registers of John Gynewell, Bishop of Lincoln for the years 1347–1350', *Archaeol. Journal*, lxviii (1911), 301–60. *London*: Register of Ralph Baldock, Gilbert Segrave, Richard Newport, and Stephen Gravesend, 1304–38, ed. R. C. Fowler (Cant. and York Soc., 1911); Simon Sudbury, 1362–75, ed. R. C. Fowler and C. Jenkins, (ibid., 1927–38). *Rochester*: Hamo de Hethe, 1319–52, ed. C. Johnson (ibid., 1914–48); an edition of T. Brinton, 1373–89, by Sister Mary Devlin is in preparation. *Salisbury*: Simon of Ghent, 1297–1315, ed. C. T. Flower and M. C. B. Dawes (Cant. and York Soc., 1914–34); an edition of Roger Mortival, 1315–30, by Kathleen Edwards is in preparation. *Winchester*: Henry Woodlock, 1305–16, ed. A. W. Goodman (Cant. and York Soc., 1940–1); John Sandale and Rigaud de Asserio, 1316–23, ed. F. J. Baigent (Hants Record Soc., 1897); William of Wykeham, 1366–1404, ed. T. F. Kirby (ibid., 1896–9). *Worcester*: *Sede vacante*, 1301–1435, ed. J. W. Willis Bund (Worcs. Hist. Soc., 1893–7); Walter Reynolds, 1308–13, ed. R. A. Wilson (Dugdale Soc., 1928); Thomas Cobham, 1317–27, ed. E. H. Pearce (Worcs. Hist. Soc., 1930). *York*: William Greenfield, 1306–15, ed. W. Brown and A. Hamilton Thompson (Surtees Soc., 1931–8). Copious extracts from the York registers are printed in the three volumes edited by J. Raine, *Historical Letters and Papers from the Northern Registers* (R.S., 1873) and *Historians of the Church of York and its Archbishops*, 2 vols. (R.S., 1879–94). See also A. Hamilton Thompson, 'Some letters from the register of William Zouche, archbishop of York', *Historical Essays in honour of James Tait* (Manchester, 1933), pp. 327–43. *Carlisle*: John de Halton, 1292–1324, ed. W. N. Thompson (Cant. and York Soc., 1913). *Durham. Records of Antony Bek*, 1283–1311 (whose register has perished or is lost), ed. Constance M. Fraser (Surtees Soc., 1953), assembles material from many sources to illustrate the administration of the liberty and the diocese. *Registrum palatinum Dunelmense*, ed. T. Duffus Hardy, 4 vols. (R.S., 1873–8), includes the register of Richard Kellaw, 1311–16, and part of that of Richard of Bury, 1333–45. Extracts from the Carlisle and Durham registers are included in *Letters and Papers from the Northern Registers*.

Religious Orders. William Dugdale's indispensable *Monasticon Anglicanum*, ed. J. Caley and others, 6 vols. (London, 1846), includes many monastic charters. For a short catalogue of extant cartularies of religious houses, see G. R. C. Davis, *Medieval Cartularies of Great Britain*, pp. 2–137. The legislation and *acta* of the

Chapters of the English Black Monks, 1215–1540 have been edited by W. A. Pantin, 3 vols., Camden 3rd series, xlv, xlvii, liv (1931, 1933, 1937). Comparable material for other orders will be found in *Collectanea Anglo-Premonstratensia*, ed. F. A. Gasquet, 3 vols., Camden 3rd series, vi, x, xi (1904, 1906), and in *Chapters of the Augustinian Canons*, ed. H. E. Salter (Cant. and York Soc., 1921–2). Proceedings relating to the Templars are printed in Dugdale's *Monasticon*, vi. 813–54, and in Cole's *Documents Illustrative of English History*, pp. 139–230. L. B. Larking, *Knights Hospitallers in England* (Camden Soc., 1857), comprises a survey of manorial balance-sheets drawn up in 1338 for submission to the grand master of the order. For the most important printed records of individual houses, see the valuable bibliography in M. D. Knowles, *The Religious Orders in England*, ii. 376–81. *Literae Cantuarienses*, ed. J. B. Sheppard, 3 vols. (R.S., 1887–9), contains long and valuable extracts from the letter-books (or registers) of Christ Church, Canterbury, and is of especial interest for this period. For this class of document in general and for a list of other surviving monastic letter-books, W. A. Pantin, 'English Monastic Letter-Books' in *Historical Essays in Honour of James Tait*, pp. 213–22, should be consulted.

Wills. For the wills of Edward III, Richard II, the Black Prince, and other persons of importance, see J. Nichols, *A Collection of the Wills of the Kings and Queens of England* (London, 1780). Wills preserved in the registers of the archbishops of Canterbury are listed in *Testamenta Lambethana*, ed. A. C. Ducarel (Middle Hill Press, 1854). Translated extracts from many wills chosen with a view to their value as evidence for social history were collected by N. H. Nicolas for his *Testamenta Vetusta*, 2 vols. (London, 1826). See also *Calendar of Wills proved and enrolled in the Court of Husting, London, 1258–1688*, ed. R. R. Sharpe, 2 vols. (London, 1889–90). Three late fourteenth-century wills are printed in *The Fifty Earliest English Wills in the court of probate, London, 1387–1439, 1454*, ed. F. J. Furnivall (E.E.T.S., 1882).

Schools and Universities. Material of great value for the history of the medieval university of Oxford has been published by the Oxford Historical Society, notably *Collectanea*, ed. C. R. L. Fletcher and others, 4 vols. (1885–1905), *Medieval Archives of the University of Oxford*, 2 vols. (1917–19), and *Statuta antiqua univer-*

sitatis Oxoniensis, ed. Strickland Gibson (1931). The older publication, *Munimenta Academica*, ed. H. Anstey, 2 vols. (R.S., 1868), is still valuable but relates mainly to a later period. *Snappe's Formulary*, ed. H. E. Salter (Oxford Hist. Soc., 1923), and *Oxford Formularies*, ed. H. E. Salter, W. A. Pantin, and H. G. Richardson (ibid., 1942), contain much miscellaneous material about Oxford, including confirmations of the university chancellors by the bishops of Lincoln and letters of Thomas Sampson and other Oxford *dictatores*. The *Liber Epistolaris* of Richard de Bury, ed. N. Denholm-Young (Roxburghe Club, 1950), also contains a section of formularies relating to Oxford. Publications of the Oxford Historical Society of value for the history of the colleges include *Oxford Deeds of Balliol College*, ed. H. E. Salter (1913), *Oriel Records*, ed. C. L. Shadwell and H. E. Salter (1926), and *Canterbury College*, ed. W. A. Pantin, 3 vols. (1946–50). A. B. Emden, *A Biographical Register of the University of Oxford to A.D. 1500*, vols. i–ii, A–O (Oxford, 1957–8), when completed will do for Oxford what has already been done for Cambridge by J. and J. A. Venn, *Alumni Cantabrigienses* (Cambridge, 1922–7). Ancient statutes both of the university and of the colleges of Cambridge are printed in *Documents relating to the University and Colleges of Cambridge* (London, 1852). For full bibliographies see vol. iii of Rashdall's *Universities of Europe*, ed. Powicke and Emden (1936), iii. 1–5, 274–6.

A few documents relating to the education of children in this period are included in A. F. Leach, *Educational Charters and Documents A.D. 598–1909* (Cambridge, 1911).

Wyclif and the Lollards. For a complete list of Wyclif's Latin writings and of the English works ascribed to him, see *Cambridge Medieval History*, vii. 900–4. For the rise and decline of lollardy in Oxford and for the history of Wyclif generally, the documents linked by a sparse narrative which comprise *Fasciculi Zizaniorum Magistri Johannis Wyclif*, ed. W. W. Shirley (R.S., 1858), are of fundamental importance. Relevant matter will also be found in Foxe's *Acts and Monuments* and in Snappe's *Formulary*.

4. NARRATIVE SOURCES
Works by Regulars

(a) *Black monks.* The tradition of historical writing established at St. Albans in the thirteenth century was maintained, albeit at a lower level, in the first half of the fourteenth, in the *Annales* ascribed to John Trokelowe and Henry Blaneforde (ed. H. T. Riley, R.S., 1866), though both may be the work of William Rishanger; they cover the years 1307–26. From this date until about 1376, when Thomas Walsingham began to write history, there was a suspension of literary activity. Walsingham endeavoured to fill the gap by a large-scale history, linking Matthew Paris with a history of his own times. He was the author of *Historia Anglicana*, 1272–1422, ed. H. T. Riley, 2 vols. (R.S., 1863–4), *Chronicon Angliae*, 1322–88, ed. E. M. Thompson (R.S., 1874), and *Annales Ricardi Secundi*, ed. H. T. Riley (R.S., 1866). The interrelation of the numerous manuscripts and the proofs of Walsingham's authorship of these and other St. Albans chronicles are discussed by V. H. Galbraith, 'Thomas Walsingham and the St. Albans Chronicle, 1272–1422', *E.H.R.* xlvii (1932), 12–30, and in *The St. Albans Chronicle 1406–20* (Oxford, 1937), Introd., pp. ix–lxxi. Westminster was another Benedictine house with a good historiographical tradition. A Westminster monk named Robert of Reading composed the last section (1306–25) of a continuation of Matthew Paris, compiled at Westminster under the title *Flores Historiarum*, ed. H. R. Luard, 3 vols. (R.S., 1890). This is a full and valuable chronicle packed with detail on public events: the author's sympathies are strongly pro-Lancastrian. A scant and inaccurate compilation links his work with that of John of Reading whose chronicle covers the period 1346–67 (ed. J. Tait, Manchester, 1914). Though by no means a first-rate historian John of Reading has to be reckoned one of the principal narrative sources for a decade (1356–66) when these are meagre. Also from Westminster comes the remarkable chronicle, wrongly attributed to John of Malvern, which is printed as a continuation of the *Polychronicon*, vol. ix, ed. J. R. Lumby (R.S., 1886). (See J. Armitage Robinson, 'An unrecognized Westminster Chronicle, 1381–94', *Proc. of the British Academy*, iii (1907), 61–77.) The author writes with first-hand knowledge of many of the main events of the period and his judgements are more dispassionate than Walsingham's. The *Polychronicon* of Ranulph Higden, a monk of St. Werburgh's, Chester, extending to 1340, was the standard work on general history, planned on the grand scale. The narrative is scrappy and repetitive but includes some lively character-sketches and some interesting reflections of public opinion. Vol. viii is a continuation to 1381, written by a monk of Worcester named John of Malvern. The Rolls Series edition, ed. C. Babington and J. R. Lumby (1865–86), includes Trevisa's English translation completed in 1387.

The literary activity of the scriptorium of Christ Church, Canterbury, is represented by the *Vitae Archiepiscoporum Cantuariensium* (ed. H. Wharton, *Anglia Sacra* (1691), i. 1–48), wrongly ascribed to Stephen Birchington, which is of particular value for the crisis of 1340–1; and possibly by the continuation of the *Eulogium Historiarum* (ed. F. S. Haydon, 3 vols., R.S., 1858–63). The first part of this general survey of English history was probably written by a monk of Malmesbury who completed the work about 1367: the continuation from 1362 to 1413 was thought by the editor to be by a Canterbury monk: but E. J. Jones has argued a case for the authorship of John Trevor, bishop of St. Asaph (*Speculum*, xii, 1937). From St. Mary's, York, comes the *Anonimalle Chronicle, 1333–81*, ed. V. H. Galbraith (Manchester, 1927): its unique importance derives from two long interpolations dealing, the one with the Good Parliament of 1376, the other with the Rising of 1381. (For a discussion of its authorship see A. F. Pollard in *E.H.R.* liii (1938), 577–605.) The *Historia Vitae et Regni Ricardi Secundi*, ed. T. Hearne (1729), commonly ascribed to a monk of Evesham, is an independent authority from 1390 to 1399. *Historia de statu ecclesiae Dunelmensis, 1214–1336* (Surtees Soc. (1839), pp. 33–123) contains some material of interest for ecclesiastical history but is not always reliable. The remaining Benedictine chronicles are valuable chiefly for the internal affairs of their respective houses. Such are the *Historia Sancti Petri Gloucestriae*, ed. W. H. Hart, 3 vols. (R.S., 1863–7); Abbot Burton's *Chronica monasterii de Melsa* (Meaux) to 1406, ed. E. A. Bond, 3 vols. (R.S., 1866–8); the *Chronicon abbatiae de Evesham* to 1418, ed. W. D. Macray (R.S., 1863); William Thorne's *Chronica de rebus gestis abbatum Sancti Augustini Cantuariae, 578–1397*, ed. R. Twysden, *Scriptores X* (1652), pp. 1753–2202, English translation, ed. A. H. Davis (Oxford, 1934); and the *Gesta Abbatum S. Albani* of Walsingham, ed. H. T. Riley, 3 vols. (R.S., 1867–69). Vol. II of *Historians of the Church of York*, ed. J. Raine, 3 vols. (R.S., 1879–94), contains *Chronica Pontificum Ecclesiae Eboracensis*, the section from 1147–1373 being the work of a fourteenth-century author, Thomas Stubbs. Conflicts between the abbot and the burgesses of Bury St. Edmunds are described in vol. ii of *Memorials of*

St. Edmunds Abbey, ed. T. Arnold, 3 vols. (R.S., 1890–6), pp. 327–61.

(*b*) *Austin Canons*. The important chronicle of Henry Knighton (or Cnitthon), ed. J. R. Lumby, 2 vols. (R.S., 1889–95), comes from the Augustinian house of St. Mary-of-the-Meadows at Leicester. It is now known that the last section (1377–95) of this chronicle was written first (see V. H. Galbraith, 'The Chronicle of Henry Knighton', *Fritz Saxl 1890–1948*, ed. D. J. Gordon (London, 1957), pp. 136–45). Knighton is a source of primary importance for the reign of Richard II: the earlier section of his chronicle, though largely derived from well-known sources, owes some interesting local detail to a chronicle of his own house, now lost. The house of Austin canons at Guisborough produced the chronicler long known as Walter Hemingford, or Hemingburgh. His chronicle (ed. H. Rothwell, Camden 3rd series, lxxxix, 1957) extends only to 1312 but is of high quality. A continuation from 1327–46 has been shown by Professor Rothwell to have only a very tenuous connexion with Guisborough's chronicle; it appears to have been based on the (unpublished) *Historia Aurea* of John, vicar of Tynemouth. (See V. H. Galbraith, in *Essays in History presented to R. L. Poole*, ed. H. W. C. Davis (Oxford, 1927), pp. 379–98, and *E.H.R.* xliii (1928), 203–17.) The *Gesta Edwardi de Carnarvan* by a canon of the Austin priory of Bridlington (ed. W. Stubbs in *Chronicles of the Reigns of Edward I and Edward II* (R.S., 1882–3), ii. 25–151) is an important source for the reign of Edward II, particularly for events in the north of England. In the Austin priory of Lanercost, Cumberland, the canons produced a composite work, based on a lost Franciscan original, which extended first to 1296 and was continued by another hand to 1346, ed. J. Stevenson (Edinburgh, 1839) and translated for the years 1272–1346 by H. Maxwell, with an introduction by J. Wilson (Glasgow, 1913). For an analysis of the structure of this chronicle, see A. G. Little, in *E.H.R.* xxxi (1916), 269–79.

(*c*) *Cistercian*. Very slight, but by no means devoid of interest for the history of Richard II are the *chronicula* emanating from the Cistercian houses of Dieulacres (Staffs.), ed. M. V. Clarke and V. H. Galbraith, *Bull. J. Rylands Lib.* xiv (1930), and Kirkstall (Yorks.), ed. with English translation, by J. Taylor, *Thoresby Society*, xlii (1952).

Works by Secular Clerks. Those emanating from St. Paul's cathedral are among the most important. The *Annales Paulini* (*Chronicles of Edward I and Edward II*, i. 253–370) cover the period 1307–41. The author, who may have been a canon of St. Paul's, was in close contact with another canon, Adam Murimuth, whose *Continuatio Chronicarum*, 1303–47, ed. E. M. Thompson (R.S., 1889), relates to the same period. Both writers tell us much of events in London and Murimuth is an important source for ecclesiastical affairs, having been several times employed on diplomatic missions to the papal curia. Letters relating to military and diplomatic affairs are included in the *De gestis mirabilibus regis Edwardi tertii* of Robert of Avesbury (ed. E. M. Thompson, R.S., 1889) who, though he was registrar of the court of Canterbury, writes almost exclusively as a military historian. The *Historia Roffensis* (1314–50) of William Dene (ed. H. Wharton, *Anglia Sacra*, i. 356–83) is the work of a notary public and is especially valuable for the latter years of Edward II and his deposition. Of primary importance for this reign (to 1325) is the anonymous *Vita Edwardi Secundi* (ed. N. Denholm-Young, 1957), wrongly ascribed in the past to a monk of Malmesbury and thought by its latest editor to be the work of John Walwayn, D.C.L., a dependent of the earl of Hereford. Whoever the author, this is a contemporary biography by a shrewd and well-informed critic of men and events. It contrasts sharply with the *Chronicon* of Geoffrey le Baker of Swinbrook, 1303–56 (ed. E. M. Thompson, Oxford, 1889), a country clerk who began to write after 1341 at the request of his patron, Sir Thomas de la More. Though not in general reliable, Baker supplies some valuable material for the latter years of Edward II, his patron having been present at Kenilworth in the train of the bishop of Winchester. For the deposition of Richard II, Adam of Usk, whose *Chronicon* written in the reign of Henry V has been edited with a translation by E. M. Thompson (London, 1904), is an important authority.

Works by Lay Writers. Though the lay historian is still a fairly rare phenomenon, he is beginning to emerge in this century: the capital and the army offered congenial soil for his development. *Annales Londonienses*, 1194–1330 (*Chronicles of Edward I and Edward II*, i. 1–251), the work of a London citizen who had access to the corporation records, is a valuable source for the first half of Edward II's reign; the author is careful and impartial and includes a number of documents in his narrative. By contrast, the *Croniques de London*, 1260–1344, ed. G. J. Aungier (Camden Soc., 1844), is unsophisticated, a typical civic chronicle, purely annalistic in method, but valuable for events in London. The *Scalacronica* of Sir Thomas Gray of Heton, ed. J. Stevenson (Edinburgh, 1836, English translation by H. Maxwell, Glasgow, 1907), is the work of a soldier, written in Edinburgh jail in 1355; it is important for the history of the Border. The herald who served the famous Sir John Chandos was another soldier with a taste for history; his metrical *Life of the Black Prince*, written in French (ed. with a prose translation by M. K. Pope and E. C. Lodge, Oxford, 1910), preserves an eyewitness account of the Castilian campaign of 1366–7.

Foreign Authors. Among the foreign authors who concern themselves with English internal affairs Jean Froissart stands pre-eminent. Much of the earlier part of his Chronicle (which covers the period 1307–1400) is derived from that of Jean le Bel, 1272–1361 (ed. J. Viard and E. Déprez, Soc. de l'histoire de France, 2 vols., Paris, 1904–5). The standard editions of Kervyn de Lettenhove (25 vols., Brussels, 1867–77) and of S. Luce and G. Raynaud, Soc. de l'histoire de France (11 vols. to 1385, Paris, 1869–99) are both indispensable; the former includes an elaborate index and many documents. Lord Berners's delightful English translation, first published 1523–5, is edited by W. P. Ker in 6 vols. (*Tudor Translations*, 1901–3, reissued in 8 vols. 1927–8). Froissart derived much of his material from oral testimony and is often bewilderingly inaccurate and inconsistent. But his writing has concreteness, spontaneity, and an eye for visual detail and is far more satisfying artistically than that of any other historian of the period.

The importance of the group of interrelated French narratives which cover the fall of Richard II is now generally recognized. These are: the *Chronique du religieux de St. Denis*, 1380–1422, ed. with French translation of Latin text by L. Bellaguet, *Documents inédits*, 6 vols. (Paris, 1839–52); Jean le Beau, *Chronique de Richard II*, 1377–99, ed. J. A. C. Buchon, *Collection des chroniques françaises*, xxv, supplement ii (Paris, 1826); the *Chronique de la traïson et mort de Richard II*, 1397–1400, ed. with English translation by B. Williams (Eng. Hist. Soc., 1846); and Jean Creton, *Histoire du roy d'Angleterre Richard*, ed. J. A. C. Buchon, *Collection des chroniques françaises*, xxiv (Paris, 1826), 321–466.

The Brut. This popular chronicle, which stands in a class by itself, has been unduly neglected by historians. The English version to 1333 is merely a translation from a French original but there are independent English continuations for the remainder of the fourteenth and much of the fifteenth centuries. The standard edition, which includes a full examination of the manuscripts is that of F. Brie (E.E.T.S. 1906, 1908). See also C. L. Kingsford, *English Historical Literature in the Fifteenth Century* (Oxford, 1913), pp. 113–39, and J. Taylor, 'The French "Brut" and the Reign of Edward II', *E.H.R.* lxxii (1957), 423–37.

Occasional Pieces. Vol. I of *Political Poems and Songs*, ed. T. Wright, 2 vols. (R.S., 1859–61), comprises an anthology of topical verses in Latin, French, and English, relating to the reigns of Edward III and Richard II. *The Treatise of Walter de Milemete*, an exhortation to Edward III on the occasion of his accession to the Crown, has been published in facsimile, ed. M. R. James (Roxburghe Club, 1913). The *De Speculo regis Edwardi III* (*c.* 1330), ed. J. Moisant (Paris, 1891), is a remonstrance addressed to Edward III on the subject of purveyance. Thomas Favent's *Historia mirabilis parliamenti*, ed. M. McKisack, Camden Miscellany, xiv (1926), is a political pamphlet written in justification of the lords appellant, probably soon after the dissolution of the parliament of 1388. Much valuable historical material lies embedded in the numerous publications of the Early English Text Society. Examples are *Mum and the Sothsegger*, ed. M. Day and R. Steele (1936), and Trevisa's *Dialogus inter Militem et Clericum*, ed. A. J. Perry (1924).

5. GENERAL AND ENGLISH POLITICAL HISTORY

General. The general history of the period may conveniently be studied in the following: *Cambridge Medieval History*, vii (1932); H. Pirenne and others, *La Fin du moyen âge*, pt. i; *La Désagrégation du monde médiéval* (1285–1453), in the series 'Peuples et Civilisations' vii (Paris, 1931); *L'Europe occidentale de 1270 à 1380*, pt. i (1270–1328) by R. Fawtier (Paris, 1940), pt. ii by A. Coville (Paris, 1941); *L'Europe occidentale de la fin du xive siècle aux guerres d'Italie*, pt. i by E. Déprez, *La France et l'Angleterre en conflit* (Paris, 1937), the last three forming part of the 'Histoire générale' directed by G. Glotz. E. Lavisse, *L'Histoire de France*, vol. iii, pt. ii (1226–1328), and vol. iv, pt. i (1328–1422), by

A. Coville (Paris, 1911), is still useful. The only full-scale biography of a fourteenth-century king of France is R. Delachenal, *Charles V*, 5 vols. (Paris, 1909–31); it covers also some aspects of the reign of John the Good. G. Mollat, *Les Papes d'Avignon* (1305–78), 6th ed. (Paris, 1930); L. Salembier, *Le Grand Schisme d'Occident*, 5th ed. (Paris, 1922); and N. Valois, *La France et le grand schisme d'occident*, 4 vols. (Paris, 1896–1902), cover the general history of the Church and papal policy in this century.

England. Sir James Ramsay's *Genesis of Lancaster, 1307–99*, 2 vols. (Oxford, 1913), is a serviceable factual study. T. F. Tout, *The Political History of England*, vol. iii, 1307–77 (London, 1905), is still well worth reading though it needs to be supplemented by Tout's later work, particularly by *The Place of the Reign of Edward II in English History*, 2nd ed. revised, Hilda Johnstone (Manchester, 1936), and by vols. iii and iv, pp. 1–68, of *Chapters in the Administrative History of Medieval England* (Manchester, 1928) which constitute the best general survey of fourteenth-century administration and politics. A. R. Myers, *England in the late Middle Ages, 1307–1536* (Pelican History, 1952), is an interesting introduction to the period. There is no good modern biography of Edward III; the best is still Joshua Barnes, *History of Edward III* (Cambridge, 1688). H. Wallon, *Richard II*, 2 vols. (Paris, 1864), is good on foreign affairs and has not been altogether supplanted by the compact and readable study of A. Steel, *Richard II* (Cambridge, 1941); for a lively comment on the latter see V. H. Galbraith, 'A New Life of Richard II', *History*, n.s. xxvii (1942), 223–39. The only other substantial biography of note is S. Armitage Smith's *John of Gaunt* (London, 1904), a standard work, though now inevitably somewhat out of date on particular points. M. V. Clarke, *Fourteenth Century Studies*, ed. L. S. Sutherland and M. McKisack (Oxford, 1937), includes some important contributions to the general history of the period and is the work of a fine scholar.

The Hundred Years War. The literature of the war is voluminous. The most recent general survey, that of E. Perroy, *La Guerre de Cent ans* (Paris, 1945), English translation by W. B. Wells, with introduction by D. C. Douglas (London, 1951), has a good select bibliography. For a more elaborate bibliography see *The Chronicle of Jean de Venette*, ed. R. A. Newhall (New York, 1953),

pp. 315–33. On the diplomatic origins of the war the fundamental work is still E. Déprez, *Les Préliminaires de la guerre de
Cent ans* (Paris, 1902), but G. P. Cuttino, *English Diplomatic Administration, 1259–1339* (Oxford, 1940), throws new light on the
mechanics of diplomacy in the pre-war years. H. S. Lucas in
The Low Countries and the Hundred Years War, 1326–47 (Ann
Arbor, 1929) unravels the complex relations of Edward III and
his allies in the first decade of the war. His book should be read
in conjunction with those of J. de Sturler, *Les Relations politiques
et les échanges commerciaux entre le duché de Brabant et l'Angleterre au
moyen âge* (Paris, 1936), and F. Quicke, *Les Pays Bas à la veille de
la période bourguignonne* (Paris, 1947). Helen Jenkins, *Papal Efforts
for Peace under Benedict XII, 1334–42* (London, 1933), is useful and
G. Templeman makes an interesting contribution to a much-
debated topic in his 'Edward III and the beginnings of the Hundred Years War', *Trans. Royal Hist. Soc.* 5th ser. ii (1952), 69–88.
For Gascony and the effects of the war there, see C. Bémont, *La
Guienne pendant la domination anglaise, 1152–1453* (London, 1920);
E. C. Lodge, *Gascony under English Rule* (London, 1926); and
the important study by R. Boutrouche, *La Crise d'une société:
seigneurs et paysans du Bordelais pendant la guerre de Cent ans* (Paris,
1947). The work of P. E. Russell, *The English Intervention in Spain
and Portugal in the time of Edward III and Richard II* (Oxford, 1955),
supplies a long-felt want, for it deals admirably, not only with
a neglected theatre of war but also with the largely neglected
period between the treaty of 1360 and the truce of 1396. The
effects of the war in France may be studied in H. Dénifle, *La
Désolation des églises, monastères et hôpitaux en France pendant la
guerre de Cent ans*, i (1337–84) (Paris, 1899). On the military
aspect of the war, the standard work of Sir Charles Oman, *The
Art of War in the Middle Ages*, 2nd ed., 2 vols. (London, 1924),
has been supplemented by that of F. Lot, *L'Art militaire et les
armées*, 2 vols. (Paris, 1946). Both are useful but, for the student
of a limited period, their range and the consequent absence of
adequate documentation renders them somewhat unreliable
on points of detail. By contrast, *The Crecy War* of A. H. Burne
(London, 1955), is the work of a professional soldier whose interests are concentrated on the first phase of the Hundred Years
War. The author's doctrine of 'inherent military probability'
should be received with caution and he is not altogether at ease
with the political and diplomatic background to his subject.

None the less, this is an original and stimulating book. For the war at sea and the navy generally the standard authorities are N. H. Nicolas, *A History of the Royal Navy*, 2 vols. (London, 1847), and M. Oppenheim, *A History of the Administration of the Royal Navy* (London, 1896). See also the introduction to *Select Pleas in the Court of Admiralty*, ed. R. G. Marsden (1894).

6. ECCLESIASTICAL HISTORY

For the organization of the Church in England, F. Makower, *The Constitutional History and Constitution of the Church of England* (English transln., London, 1895), is still useful. Three important further studies are those of Irene Churchill, *Canterbury Administration*, 2 vols. (London, 1933); A. Hamilton Thompson, *The English Clergy and their Organization in the later Middle Ages* (Oxford, 1947); and Kathleen Edwards, *The English Secular Cathedrals in the Middle Ages* (Manchester, 1949). *Convocation of the Clergy* (London, 1937) by Dorothy B. Weske includes some useful lists. On Anglo-papal relations, F. W. Maitland, *Roman Canon Law in the Church of England* (Cambridge, 1898) is fundamental but should be read in conjunction with such more recent studies as G. Mollat, *La Collation des bénéfices ecclésiastiques sous les papes d'Avignon* (Paris, 1921); E. Perroy, *L'Angleterre et le grand schisme d'occident 1378–99* (Paris, 1933); G. Barraclough, *Papal Provisions* (Oxford, 1935); W. E. Lunt, *Financial Relations of the Papacy with England to 1327* (Cambridge, Mass., 1939); C. Davis, 'The Statute of Provisors of 1351', *History*, xxxviii (1953), 116–33; W. A. Pantin, *The English Church in the XIV century* (Cambridge, 1955); and E. W. Kemp, *Introduction to Canon Law in the Church of England* (London, 1957).

M. D. Knowles, *The Religious Orders in England*, vol. ii (Cambridge, 1955), now the authoritative general survey, includes a good bibliography. In collaboration with R. N. Hadcock, Professor Knowles has published an invaluable work of reference, *Medieval Religious Houses* (London, 1953),[1] and with J. K. St. Joseph, *Monastic Sites from the Air* (Cambridge, 1952). Eileen Power's *Medieval English Nunneries* (Cambridge, 1922), though less distinguished than some of her later work, is the best conspectus of the female communities. R. H. Snape has examined *English Monastic Finances in the later Middle Ages* (Cambridge, 1926). E. H. Thompson's is the standard work on *The Carthusian*

[1] For some additions and corrections, see *E.H.R.* lxxii (1957), 60–87.

Order in England (London, 1930); the history of the Premonstratensians has been written by H. M. Colvin, *The White Canons in England* (Oxford, 1951). For the Franciscans, the work of A. G. Little, exemplified by his *Grey Friars in Oxford* (Oxford Hist. Soc., 1892), and *Franciscan Papers, Lists and Documents* (Manchester, 1943), is indispensable. J. R. H. Moorman, *The Grey Friars in Cambridge* (Cambridge, 1952), and A. Gwynn, *The English Austin Friars* (Oxford, 1940), are both good studies.

Brief biographies of individual bishops will be found in the introductions to the printed registers (see above, pp. 540–1). Not many fourteenth-century prelates have been the subject of full-scale studies but attention may be drawn to two lives of William of Wykeham by R. Lowth (3rd ed., Oxford, 1777) and G. H. Moberley (2nd ed., London, 1893), to C. M. Fraser's *History of Antony Bek, Bishop of Durham, 1283–1311* (Oxford, 1957) and to Gordon Leff's *Bradwardine and the Pelagians* (Cambridge, 1957). Kathleen Edwards in *Church Quar. Rev.* cxxxviii (1944), 57–86, surveys 'Bishops and Learning in the Reign of Edward II'. Sister Mary Devlin has edited the *Sermons of Thomas Brinton, Bishop of Rochester*, 2 vols. (*Royal Hist. Soc.*, 1954); and sermon literature is discussed with a wealth of illustrative detail by G. R. Owst in *Preaching in Medieval England* (Cambridge, 1926) and *Literature and Pulpit in Medieval England* (Cambridge, 1933). Kathleen Wood-Legh has written of the chantries in *Church Life under Edward III* (Cambridge, 1934) and in *Trans. Royal Hist. Soc.*, 4th ser. xxvii (1946), 47–60. For the recluses, see R. M. Clay, *The Hermits and Anchorites of England* (London, 1914). Agreeable books dealing with the popular religion are E. L. Cutts, *Parish Priests and their People in the Middle Ages in England* (London, 1898); B. L. Manning, *The People's Faith in the time of Wyclif* (Cambridge, 1919); and H. Maynard Smith, *Pre-Reformation England* (London, 1938).

For a good sketch of education in the later Middle Ages see the chapter by G. R. Potter in *Cambridge Med. Hist.* viii (1936), 688–717, and the bibliography (p. 985), which includes a list of the numerous studies of medieval schools by A. F. Leach. The monastic schools are described by M. D. Knowles, *Religious Orders*, ii. 294–97. J. Hastings Rashdall, *The Universities of Europe in the Middle Ages*, rev. ed. (Oxford, 1936), covers the English universities in vol. iii and is the authoritative work. Rashdall owed much to Sir H. C. Maxwell-Lyte's *History of the*

University of Oxford (London, 1886), an important pioneer study. For an eminently readable history of the medieval university, its colleges and halls, see Sir Charles Mallet, *History of the University of Oxford*, vol. i (London, 1954), and for an up-to-date survey, the *Victoria County History of Oxford*, vol. iii, *The University of Oxford* (1954). Pending publication of the corresponding volume for Cambridge,[1] the student is dependent on J. B. Mullinger, *History of the University of Cambridge to 1535* (Cambridge, 1873), which is unsatisfactory for the Middle Ages.

For Wyclif and the lollards, see the bibliography in *Cambridge Med. Hist.* vii. 900–7. The standard biography is by H. B. Workman, *John Wyclif*, 2 vols. (Oxford, 1926). K. B. McFarlane's *John Wycliffe and the beginnings of English Nonconformity* (London, 1952) is a brilliant sketch of the history of the lollard movement up to the fall of Oldcastle. Older but still useful studies are those of G. V. Lechler, *Johann von Wiclif und die Vorgeschichte der Reformation*, 2 vols. (Leipzig, 1873), English (abridged) transln. by P. Lorimer, *John Wycliffe and his English Precursors* (London, 1884); R. Buddensieg, *Johann Wiclif und seine Zeit* (Gotha, 1885); J. Gairdner, *Lollardy and the Reformation in England*, vol. i (London, 1908); R. L. Poole, *Wycliffe and Movements for Reform* (London, 1911); *Illustrations of the History of Medieval Thought and Learning* (Chap. x, 'Wycliffe's Doctrine of Dominion'), 2nd ed. (London, 1920). *The Lollard Bible* by M. Deanesly (Cambridge, 1920) is the indispensable introduction to its study. See also Miss Deanesly's Ethel M. Wood lecture, *The Significance of the Lollard Bible* (London, 1951) and Sir William Craigie's essay on 'The English Versions (to Wyclif)' in *The Bible in its Ancient and English Versions*, ed. H. Wheeler Robinson (Oxford, 1940).

7. LAW AND INSTITUTIONS

The basic works are vols. ii and iii of W. Stubbs, *Constitutional History of England*, first published 3 vols. (Oxford, 1874–78), 4th ed. of vol. ii reprinted 1906, 5th ed. of vol. iii (1903); F. W. Maitland, *Constitutional History of England* (Cambridge, 1908) and *Collected Papers*, ed. H. A. L. Fisher, 3 vols. (Cambridge, 1911). A selection of the last which includes Maitland's obituary notice of Stubbs has been edited by Helen Cam, *Selected Historical Essays of F. W. Maitland* (Cambridge, 1957). See also her

---

[1] Published (1959) after this volume had gone to press.

'Stubbs Seventy Years After', *Camb. Hist. Journal*, ix (1948), 129–47, and J. G. Edwards, *William Stubbs* (Hist. Assocn., 1952). T. P. Taswell-Langmead, *English Constitutional History* (London, 1875), has been revised by T. F. T. Plucknett (1949). W. H. Holdsworth, *History of English Law*, vol. i, revised by A. L. Goodhart, and H. G. Hanbury (London, 1956), vols. ii and iii (3rd ed., 1922–3), is valuable. J. E. A. Joliffe, in *The Constitutional History of Medieval England*, has written a stimulating book; B. Wilkinson, *The Constitutional History of Medieval England*, vols. ii and iii (London, 1952, 1958), includes select documents and useful bibliographies together with informed comment. An interesting essay on the Ordinances of 1311 will be found in the same author's *Studies in the Constitutional History of the 13th and 14th Centuries* (Manchester, 1937). Important essays by the late G. T. Lapsley are assembled in *Crown, Community and Parliament in the Later Middle Ages*, ed. Helen Cam and G. Barraclough (Oxford, 1951). P. E. Schramm, *A History of the English Coronation* (English transln. L. G. Wickham Legg, Oxford, 1937), is useful for the whole medieval period. The significance of the oath sworn by Edward II at his coronation (see above, pp. 4–6) is discussed in the following: B. Wilkinson, 'The Coronation Oath of Edward II', *Historical Essays in honour of James Tait* (Manchester, 1933), pp. 405–16; 'The Coronation Oath of Edward II and the Statute of York', *Speculum*, xix (1944), 445–69; *Constitutional History of Medieval England*, ii. 86–111; H. G. Richardson, 'The English Coronation Oath', *Trans. Royal Hist. Soc.*, 4th ser. xxiii (1941), 129–58; 'The Annales Paulini', *Speculum*, xxiii (1948), 630–40; 'The English Coronation Oath', ibid. xxiv (1949), 44–75; R. S. Hoyt, 'The Coronation Oath of 1308', *Traditio*, xi (1955), 235–57; 'The Coronation Oath of 1308', *E.H.R.* lxxi (1956), 353–83.

J. F. Baldwin, *The King's Council in England during the Middle Ages*, is a useful study which may be supplemented by his essay in *The English Government at Work, 1327–36*, ed. J. F. Willard and W. A. Morris (Medieval Academy of America, 1940), i. 129–61. In *The High Court of Parliament and its Supremacy* (New Haven, 1910), C. H. McIlwain followed up and somewhat exaggerated Maitland's emphasis on the curial character of the assembly. A. F. Pollard's brilliant study, *The Evolution of Parliament*, 2nd ed. (London, 1926), must be read with caution so far as it relates to the Middle Ages; the author was insufficiently appreciative of the genius of Stubbs. T. F. T. Plucknett, 'Parliament' (*The English Government at Work* i. 82–128), though it covers only a single decade, is probably the best analysis to be produced since Stubbs; Professor Plucknett's other articles, referred to above (see Index), are among the most important recent contributions to medieval parliamentary history. Ludwig Riess, *English Electoral Law in the Middle Ages* (English transln. ed. K. Wood-Legh, Cambridge, 1940), is a good monograph. In addition to those already cited in the text, the following articles may be read with profit: J. G. Edwards, 'The Personnel of the Commons in Parliament under Edward I and Edward II', *Essays presented to T. F. Tout* (Manchester, 1925), pp. 197–215; 'Re-election and the medieval Parliament', *History*, xi (1926), 204–10; *The Commons in Medieval English Parliaments* (Creighton Lecture, 1957); H. G. Richardson and G. O. Sayles, 'The Parliaments of Edward III', *Bulletin Inst. Hist. Research*, vii (1930–1), 65–82, ix (1931–2), 1–18; T. F. Tout, 'The English Parliament and Public Opinion', *Collected Papers* (Manchester, 1932), 173–90; L. C. Latham, 'Collection of Wages of Knights of the Shire in the 14th and 15th centuries', *E.H.R.* xlviii (1933), 455–64; N. B. Lewis, 'Re-election to Parliament in the Reign of Richard II' (ibid., pp. 364–94); H. M. Cam, 'The Relation of English Members of Parliament to their constituencies in the 14th Century', in *Liberties and Communities in Medieval England* (Cambridge, 1944), pp. 223–35; 'The Community of the Shire and the Payment of its Representatives in Parliament' (ibid., pp. 236–50); 'The Legislators of Medieval England', *Proc. British Academy*, xxxi (1947), 127–50. For the parliamentary subsidies, see J. F. Willard, *Parliamentary Taxes on Personal Property 1290 to 1334* (Medieval Academy of America, 1934).

The Statute of York of 1322 (see above, pp. 71–73) has been discussed in the following: G. T. Lapsley, 'The Commons and the Statute of York', *E.H.R.* xxviii (1913), 118–24; 'The Interpretation of the Statute of York', ibid. lvi (1941), 22–51, 411–46 (reprinted in *Crown, Community and Parliament*, pp. 153–230); G. L. Haskins, *The Statute of York and the Interest of the Commons* (Cambridge, Mass., 1935); 'A Draft of the Statute of York', *E.H.R.* lii (1937), 74–77; M. V. Clarke, *Medieval Representation and Consent* (1936); J. R. Strayer, 'Statute of York and Community of the Realm', *Amer. Hist. Rev.* xlvii (1941), 1–23; W. A. Morris, 'Magnates and Community of the Realm in Parliament, 1264–1327', *Medievalia et Humanistica*, i (1943), 58–94; B. Wilkinson, *Constitutional History*, ii. 134–56; Gaines Post, 'The Two Laws and the Statute of York', *Speculum*, xxix (1954), 417–32; and J. H. Trueman, 'The Statute of York and the Ordinances of 1311', *Medievalia et Humanistica*, x (1956), 64–81.

T. F. Tout has written the classic work on the administration. His *Chapters in Administrative History*, 6 vols. (Manchester, 1923–35), is essential for all students of the fourteenth century, in dealing with which he reached his greatest heights. J. Conway Davies, *The Baronial Opposition to Edward II* (Cambridge, 1918), was a pioneer venture in administrative history and though some of its general conclusions are open to question, the material is valuable. S. B. Chrimes, *An Introduction to the Administrative History of Medieval England* (Oxford, 1952), will be found helpful, not only to beginners. For the chancery, see B. Wilkinson, *The Chancery under Edward III* (Manchester, 1929) and his essay in *The English Government at Work*, i. 162–205. The exchequer and financial administration in general are dealt with in vol. ii of this work, ed. W. A. Morris and J. R. Strayer (1947). A. Steel, in *The Receipt of the Exchequer, 1377–1485*, provides some useful statistics relating to royal revenues and disbursements under Richard II, though these should not be regarded as complete. See also his articles on finance in *E.H.R.* xliii (1928), 172–80, li (1936), 29–51, and in *History*, xii (1927), 298–309.

Valuable essays on the administration of justice in the royal courts will be found in the introductions to the Selden Society publications referred to above, p. 537. An essay on the court of common pleas was contributed by Nellie Neilson to *The English Government at Work*, vol. iii, ed. J. F. Willard, W. A. Morris, and W. H. Dunham (1950), pp. 259–85. See also L. Ehrlich, *Proceedings against the Crown, 1216–1377* (Oxford, 1921). T. F. T. Plucknett, *Statutes and their Interpretation in the first half of the 14th Century* (Cambridge, 1922), illustrates the slow process of reconciliation between written and customary law. *The Place in Legal History of Sir William Shareshull* has been investigated by Bertha Putnam (Cambridge, 1950); that of *Sir Robert Parvyng* by J. R. M'Grath (Kendal, 1919); E. L. G. Stones has an essay on Sir Geoffrey le Scrope in *E.H.R.* lxix (1954), 1–17.

Some aspects of the jurisdiction of local courts are dealt with in vol. iii of *The English Government at Work*, where H. M. Cam writes on 'Coroners, Constables and Bailiffs' and B. H. Putnam on 'Keepers of the Peace and Justices of the Peace'. An earlier and useful work on the justices is that of C. H. Beard, *The Office of Justice of the Peace in England* (New York, 1904). H. Jenkinson and Mabel Mills (*E.H.R.* xliii (1928), 21–32) have discussed the sheriffs' records as illustrated by some Bedfordshire survivals which have since been edited by G. H. Fowler for the Beds. Historical Society (Aspley Guise, 1929). For private and palatinate jurisdictions see especially, G. T. Lapsley, *The County Palatine of Durham* (New York, 1900), and R. Somerville, *The Duchy of Lancaster I, 1265–1603* (London, 1953). Though it deals mainly with earlier periods, J. Tait, *The Medieval English Borough* (Manchester, 1936), is still the best constitutional history of the boroughs.

8. SOCIAL AND ECONOMIC HISTORY

General. Good general surveys are to be found in the first two volumes of the *Cambridge Economic History of Europe* (1941, 1952). See especially the sections by Nellie Neilson in vol. i and by M. Postan, E. Carus-Wilson, and G. P. Jones in vol. ii and the bibliographies. Sir J. Clapham, *A Concise Economic History of Britain to 1750*, ed. J. Saltmarsh (Cambridge, 1949) and E. Lipson, *Economic History of England*, vol. i, 11th ed. (London, 1956), will be found useful for England. *The Historical Geography of England before 1800*, ed. H. C. Darby (Cambridge, 1936), includes essays by R. A. Pelham and D. T. Williams relevant to the period and some good maps. J. J. Jusserand, *English Wayfaring Life in the Middle Ages*, transld. L. Toulmin Smith (4th ed. London, 1950), is attractively written and still valuable. L. F. Salzman, *English Life in the Middle Ages*, is also good. J. C. Russell, *British Medieval Population* (Albuquerque, 1948), is a stimulating essay in demography but needs cautious handling.

Barons and Knights. W. Dugdale's *Baronage of England* (London, 1675) (though many of the pedigrees are incorrect), is memorable as pointing the way to subsequent studies, notably those of J. H. Round, *Studies in Peerage and Family History* (Westminster, 1901) and *Peerage and Pedigree* (London, 1910). Sir Harris Nicolas, *History of the Orders of Knighthood* (London, 1841–2), is a work of great erudition. For the Order of the Garter, see p. 251 above, and W. H. St. John Hope, *The Stall Plates of the Knights of the Order of the Garter, 1348–1485* (London, 1901). The best introduction to the complex subject of heraldry is by A. R. Wagner, *Heralds and Heraldry in the Middle Ages* (Oxford, 1939). C. N. Elvin, *Dictionary of Heraldry* (London, 1889), and J. W. Papworth, *An Alphabetical Dictionary of Coats of Arms*, 2 vols. (London, 1858–74), are helpful for the identification of particular coats of arms. Among studies of noble families, *The Lives of the Berkeleys* by J. Smyth, ed. J. MacLean, 2 vols. (Gloucester, 1883–4), is of unique interest as Smyth (1567–1640) was steward of Berkeley and had access to the family papers.

Agrarian Society. Air-photography has done much in recent years to elucidate the history of field-systems and the movement of rural and urban populations. *Medieval England, an Aerial Survey* by M. W. Beresford and J. K. St. Joseph (Cambridge, 1958) comprises a fine collection of photographs with learned commentaries and bibliographies. See also M. W. Beresford, *The Lost Villages of England* (London, 1954), and W. G. Hoskins, *The Making of the English Landscape* (London, 1955). Among older works H. L. Gray, *English Field Systems* (Cambridge, Mass., 1915), is a useful comparative study, and *The Open Fields* by C. S. and C. S. Orwin (2nd ed. Oxford, 1954) has become a classic. For the peasantry, Sir Paul Vinogradoff's *Villainage in England* (Oxford, 1892) is basic, but needs to be supplemented by such works as B. H. Putnam, *The Enforcement of the Statute of Labourers* (New York, 1908), G. C. Coulton, *The Medieval Village* (Cambridge, 1925), H. S. Bennett, *Life on the English Manor, 1150–1400* (Cambridge, 1937), E. A. Kosminsky, *Studies in the Agrarian History of England*, ed. R. H. Hilton, transld. Ruth Kisch (Oxford, 1956), and by numerous local studies some of which are referred to above (Chapter XI).[1] Though many of its conclusions need modification, J. E. T. Rogers, *A History of Agriculture and Prices in England*, 7 vols. (Oxford, 1866–1902), is a pioneer work of fundamental importance. On corn and prices see also N. S. B. Gras, *The Evolution of the English Corn Market* (Cambridge, Mass., 1915); W. J. Ashley, *The Bread of our Forefathers* (Oxford, 1928); and Sir William Beveridge's papers in *Econ. Hist. Review*, ii (1929), vii (1936), and 2nd series viii (1955).

Rising of 1381. The most important contribution to the history of the revolt is that of A. Réville, *Le Soulèvement des travailleurs*

[1] J. A. Raftis, *The Estates of Ramsey Abbey* (Pontifical Institute of Medieval Studies, Toronto, 1957), did not become available until after the present volume had gone to press.

d'Angleterre en 1381 (Paris, 1898), which supplies a detailed history of the movement in Hertfordshire and East Anglia with a valuable appendix of documents relating to these and other districts. An English translation of the introduction by C. Petit-Dutaillis, 'The Causes and General Characteristics of the Rising of 1381', is included in vol. ii of his *Studies Supplementary to Stubbs's Constitutional History* (Manchester, 1915), pp. 252–304. The notable work by D. M. Petruševski, *Vozstanie Vota Tailera* (Moscow, 1927), is not available in English but A. Savine's review of the first edition in *E.H.R.* xvii (1902), 780–2, may be consulted. Readable general surveys of the rising are those of Sir C. Oman, *The Great Revolt of 1381* (Oxford, 1906), and A. Steel, *Richard II*, pp. 58–91. The account of the rising given by G. M. Trevelyan, *England in the Age of Wycliffe*, is supplemented by a collection of documents ed. G. M. Trevelyan and E. Powell, *The Peasants' Rising and the Lollards* (London, 1899). The Suffolk poll-tax lists are printed as an appendix to E. Powell, *The Rising in East Anglia* (Cambridge, 1896). G. Kriehn's article 'Studies in the Sources of the Social Revolt of 1381' is interesting, if not always convincing and the same is true of *The English Rising of 1381* (London, 1950), by R. H. Hilton and H. Fagan, an attempt at a Marxist interpretation. Excellent accounts of the local risings will be found in the *Victoria History*, notably in vol. ii of *Cambridge and the Isle of Ely* (1948) by H. C. Darby and E. Miller.

Commercial, Industrial, and Municipal History. L. F. Salzman, *English Trade in the Middle Ages* (Oxford, 1931), and *English Industries of the Middle Ages*, 2nd ed. (Oxford, 1923), are good introductions; and the same author's *Building in England down to 1540* (Oxford, 1952), is a distinguished addition to the history of a key industry. W. J. Ashley, *The Early History of the English Woollen Industry* (Baltimore, 1887), should be read in conjunction with the work of E. Carus-Wilson, referred to above, Chapter XII. *Finance and Trade under Edward III*, ed. G. Unwin (Manchester, 1918), includes the editor's valuable essay on 'The Estate of Merchants' which deals mainly with their relation to the Crown. Eileen Power's brilliant Ford Lectures, *The Wool Trade in English Medieval History* (Oxford, 1941), were intended to break the ground for the larger work she did not live to write. The customs figures have been analysed by N. S. B. Gras, *The Early English Customs System* (Cambridge, Mass., 1918).

C. Gross, *The Gild Merchant*, 2 vols. (Oxford, 1890), is fundamental. The history of the London gilds has been written by G. Unwin, *The Gilds and Companies of London* (London, 1908); among numerous studies of individual gilds mention may be made of Frances Consitt, *The London Weavers' Company*, vol. i (Oxford, 1933), and Sylvia Thrupp, *A Short History of the Worshipful Company of Bakers* (London, 1933). Miss Thrupp's *Merchant Class of Medieval London, 1300–1500* (Chicago, 1948) includes much interesting biographical data. On a smaller scale but very useful is Ruth Bird's *Turbulent London of Richard II* (London, 1948). The history of London under Edward II is dealt with by M. Weinbaum, *London unter Eduard I and II*, 2 vols. (Stuttgart, 1933). Town histories are too numerous to be listed here but M. D. Lobel, *The Borough of Bury St. Edmunds* (Oxford, 1935), and J. W. F. Hill, *Medieval Lincoln* (Cambridge, 1948), may be cited as good examples of recent work in this field.

9. SCOTLAND

Many documents relating to Scotland are printed in Rymer's *Foedera*, some of them included in the *Calendar of Documents relating to Scotland preserved in the Public Record Office*, ed. J. Bain, 4 vols. (Edinburgh, 1881–8); vols. iii and iv cover the fourteenth century. The *Rotuli Scotiae*, vol. i (Rec. Com., 1814), also contains many fourteenth-century texts, mandates, letters, and safe-conducts. Scotland is generally poor in medieval records but *Accounts of the Great Chamberlains . . . rendered at the Exchequer* (1326–1453), ed. T. Thomson (Edinburgh, 1817–36); *Acts of the Parliaments of Scotland* (1124–1567), ed. T. Thomson and C. Innes, 2 vols. (Edinburgh, 1814); the *Registrum magni sigilli regum Scotorum*, vol. i (1306–24), ed. J. M. Thomson (Edinburgh, 1912); and the *Exchequer Rolls*, ed. J. Stuart and G. Burnett (Edinburgh, 1878) contain much fourteenth-century material. The *Calendar of Papal Letters*, vol. i, and the *Statuta Ecclesiae Scoticanae*, ed. J. Robertson, 2 vols. (Edinburgh, 1866), are essential for the history of the Church. *Medieval Religious Houses: Scotland*, by D. E. Easson (London, 1957), does for Scotland what M. D. Knowles and R. N. Hadcock have done for England and includes excellent bibliographical notes. English chronicles of the period, notably the *Scalacronica* and those written in the north of England, are valuable for the affairs of the Border and Anglo-Scottish relations generally. The only important literary sources of Scottish provenance are the *Scotichronicon* of John Fordun, ed. W. Goodall, 2 vols. (Edinburgh, 1759), and the vernacular poem, *The Bruce* by John Barbour, archdeacon of Aberdeen, ed. W. W. Skeat (E.E.T.S. 1870–89) and W. M. Mackenzie (London, 1909).

There is no really satisfactory history of medieval Scotland but A. Lang's *History of Scotland*, vol. i (Edinburgh, 1900), is well written and P. Hume Brown, *Short History of Scotland*, 2nd ed. (Edinburgh, 1951), is a repository of useful information. R. Rait, *An Outline of the Relations between England and Scotland, 500–1707* (London, 1901), is an admirable sketch. A. R. MacEwen, *History of the Church in Scotland*, vol. i (1913), and J. Dowden, *The Medieval Church in Scotland* (Glasgow, 1910), are the best of their kind. For the literature of Bannockburn, see above, pp. 35–36.

10. WALES

Printed records relating to Wales include a *Survey of the Honour of Denbigh*, 1334, ed. P. Vinogradoff and F. Morgan (British Academy, 1914), and *Flintshire Ministers' Accounts, 1301–28*, ed. and translated A. Jones (Prestatyn, 1913). Charters and other muniments relating to the lordship of Glamorgan (441–1721) have been edited by G. L. Clark, 6 vols. (Cardiff, 1910). The so-called *Record of Carnarvon* (Record Com., 1838) includes extents of manors, *quo warranto* proceedings, a taxation of the clergy and other documents, mostly of the time of Edward III. Many local documents have been described and published by the Cambrian Archaeological Association and the Society oi Cymmrodorion (see Gross, pp. 582–5).

For narrative sources, the historian of this period is dependent almost exclusively on Latin authors, such as Adam of Usk and the Wigmore chronicle (published in Dugdale's *Monasticon* vi, pt. i, 344–55). (See M. E. Giffin, 'A Wigmore Manuscript at the University of Chicago', *National Library of Wales Journal* (1952).)

There is no good recent history of late medieval Wales: but J. Conway Davies' 'The Despenser War in Glamorgan', *Trans. Royal Hist. Soc.*, 3rd ser. ix (1915), 21–64, is important for the reign of Edward II; and W. Rees, in *South Wales and the March* (Oxford, 1924), has achieved a fruitful agrarian study. He has also published an excellent map of this area.

11. IRELAND

The stormy history of the Celtic fringe has not been conducive to the preservation of records and many of the most important Irish documents had already disappeared before the fires of 1922. The reports of the Irish Record Commissioners (1815–25) and of the Deputy Keeper of the Public Records (from 1869), together with H. Wood's *Guide to the Records* (Dublin, 1919), give some idea of what has been lost and themselves embody much valuable material, e.g. the appendixes to the 35th and subsequent reports of the Deputy Keeper (Dublin, 1903 ff.) contain an inventory of the Irish pipe rolls from Henry III to 15 Edward III; and transcripts of numerous charters and privileges to towns and other corporate bodies were published by the Record Commissioners (Dublin, 1889). *Historic and Municipal Documents of Ireland, 1172–1320*, ed. J. T. Gilbert (R.S., 1870), is valuable for the early part of this period; vol. i of *Statutes and Ordinances of the Parliaments of Ireland* (R.S., Dublin, 1907) preserves a few legislative acts and representations to the king. The *Roll of the Proceedings of the King's Council in Ireland, 1392–93* was edited by J. Graves for the Rolls Series (1877) from one of the Ormonde MSS.: the Ormonde deeds, 1172–1350, have been calendared for the Irish MSS. Commission (Dublin, 1932) by E. Curtis who also printed texts relating to the submissions of the Irish chiefs in his *Richard II in Ireland, 1394–95* (Oxford, 1927) and unpublished letters from Richard II in Ireland in *Proc. Royal Irish Academy*, xlii (1927), 276–303. The register of Archbishop Sweetman of Armagh (1362–80) has been calendared by H. J. Lawlor in *Proc. Royal Irish Academy*, xxix (1911). For documents relating to the period of Windsor's administration (ed. M. V. Clarke) and for a calendar of the muniments of Edmund Mortimer, 3rd earl of March, touching his liberty of Trim (ed. H. Wood), see ibid. xl (1932). Other local documents are listed by Gross, pp. 539–42. In their *Parliaments and Councils of Medieval Ireland*, vol. i (Irish MSS. Commn., 1947), H. G. Richardson and G. O. Sayles have published an important selection of documents drawn mainly from the Public Record Office in London. Both Rymer's *Foedera* and the calendars of papal registers contain material of importance for the history of Ireland and the Irish Church. Translations of a few of the more important fourteenth-century texts will be found in *Irish Historical*

Documents, 1172–1922, ed. E. Curtis and R. B. McDowell (London, 1943).

Several contemporary annals are in print, among the most important being the 'Annales Hiberniae' of the Franciscan, John Clyn, a contemporary authority for the years 1315–49, in *Cartularies of St. Mary's Abbey, Dublin*, ed. J. T. Gilbert (R.S., 1884), ii. 303–98. Native annals which have been published in translation include the *Annals of Loch Cé, 1014–1590*, 2 vols. (R.S., 1871), the *Annals of Ulster, 431–1541*, 4 vols. (Dublin, 1887–1901), both edited by W. M. Hennessy; the *Annals of the Four Masters*—a digest of ancient annals compiled in the Franciscan house in Donegal in the seventeenth century—ed. J. O'Donovan, 7 vols. (Dublin, 1848–51); the *Annals of Clonmacnoise to 1408* (which survive only in a seventeenth-century translation), ed. D. Murphy (Dublin, 1896); and Henry of Marlborough's *Chronicle of Ireland*, 1285–1421, translated by James Ware in *Ancient Irish Histories*, vol. ii (Dublin, 1809).

E. Curtis has written a good *History of Medieval Ireland* (Dublin, 1923) and G. H. Orpen, *Ireland under the Normans, 1169–1333*, 4 vols. (Oxford, 1911–20), is valuable for the early fourteenth century. The best list of chief governors and deputies is that of H. Wood, published in *Proc. Royal Irish Academy*, xxxvi (1923), 206–38. *The Irish Parliament in the Middle Ages* (Philadelphia, 1952) by H. G. Richardson and G. O. Sayles is likely to prove the definitive work on the subject. Vol. ii of the *History of the Church of Ireland*, ed. W. A. Phillips (London, 1934), includes two useful chapters on the medieval Church by St. J. D. Seymour.

For the literature of the Bruce invasion see above, p. 43.

12. LITERATURE AND ART

Literature. The authoritative guides are the *Cambridge Bibliography of English Literature*, 4 vols. (1940), vol. v (supplement) (1957), and J. E. Wells, *A Manual of the Writings in Middle English* (New Haven, 1916), with supplements 1919–52. A useful short introduction is provided by W. L. Renwick and H. Orton, *The Beginnings of English Literature, to Skelton* (London, 1939). W. P. Ker, *English Literature: Medieval* (Home Univ. Library, 1912; 1955), is the work of a distinguished critic; and the *Concise Cambridge History of English Literature* (Cambridge, 1941) by George Sampson is much more than a mere digest of the larger enterprise. For the early drama, see E. K. Chambers, *The Medieval Stage*, 2 vols. (Oxford, 1903); for the Arthurian legend and the romances generally, the same author's *Arthur of Britain* (London, 1927) and C. S. Lewis, *The Allegory of Love*, 2nd ed. (Oxford, 1938); for religious literature and mystical writers, M. D. Knowles, *The English Mystics* (London, 1927), and W. A. Pantin, *The English Church in the Fourteenth Century* (Cambridge, 1955), pp. 189–280. Good anthologies are K. Sisam, *Fourteenth Century Verse and Prose* (Oxford, 1921); Carleton Brown, *Religious Lyrics of the Fourteenth Century* (Oxford, 1924); and A. W. Pollard, *English Miracle Plays, Moralities and Interludes*, rev. ed. (Oxford, 1927). *The Age of Chaucer*, ed. Boris Ford (Pelican Books, 1954), includes some well-chosen passages.

Piers Plowman is edited by W. W. Skeat, 2 vols. (Oxford, 1886), reprinted with a bibliographical note by J. A. W. Bennett (1954). Nevill Coghill has produced a modernized version of the B-text (with some omissions), *Visions from Piers Plowman* (London, 1949). The complete works of Gower have been edited by G. C. Macaulay, 4 vols. (Oxford, 1899–1902). The standard edition of Chaucer is by W. W. Skeat, 7 vols. (Oxford, 1897), the most convenient working edition that of F. N. Robinson (Harvard, 1933, rev. ed. 1957). *The Canterbury Tales* have been translated by Nevill Coghill into modern English (Penguin Books, 1951). For editions of particular works and for Chaucer criticism, see the bibliographies in H. S. Bennett, *Chaucer and the Fifteenth Century* (Oxford, 1947), pp. 273–82, and *Cambridge Bibliography*, v. 130–45. For biographical material, see *Life Records of Chaucer* (Chaucer Society, 2nd ser. xii, xiv, xxi, xxii, 1875–1900), ed. R. E. G. Kirk and others, and J. M. Manly, *Some New Light on Chaucer* (New York, 1926).

Art. The reader is referred to the excellent bibliography on pp. 232–49 of Joan Evans, *English Art, 1307–1461* (Oxford, 1949). Important publications since 1949 include three volumes in the Pelican History of Art: M. Rickert, *Painting in Britain: the Middle Ages* (1954); L. Stone, *Sculpture in Britain: the Middle Ages* (1955); and G. Webb, *Architecture in Britain: the Middle Ages* (1956). E. W. Tristram, *English Wall Painting of the Fourteenth Century* (London, 1955), is fundamental for the art history of the period. *The Holkham Bible Picture Book*, ed. W. D. Hassall (London, 1954), is of great interest, having been made for the use of a Dominican preacher, about 1330, probably in London. To the literature of the Wilton Diptych cited by Dr. Evans (notes to pp. 102–4) there may be added her own essay, 'The Wilton Diptych Reconsidered', *Archaeological Journal*, cv (1948), 1–5; Margaret Galway, 'The Wilton Diptych: a Postscript', ibid. cvii (1950), 9–14; and F. Wormald, 'The Wilton Diptych', *Journ. of Warburg and Courtauld Institutes*, xvii (1954), 191–203. (See also p. 498 n. above.) J. Harvey, *English Medieval Architects, a biographical dictionary to 1500* (London, 1954), is a valuable work of reference.

Note. The following books, which have appeared since this volume was completed, should be added to the bibliographies in the appropriate sections:

E. L. C. MULLINS, *Texts and Calendars* (Royal Hist. Soc. Guides and Handbooks, 1958).

G. D. SQUIBB, *The High Court of Chivalry* (Oxford, 1959).

INDEX

(*Books and articles* not included in the Bibliography *are indexed here by single references under the authors' names*)

OXFORD

MORE OXFORD PAPERBACKS

This book is just one of nearly 1000 Oxford Paperbacks currently in print. If you would like details of other Oxford Paperbacks, including titles in the World's Classics, Oxford Reference, Oxford Books, OPUS, Past Masters, Oxford Authors, and Oxford Shakespeare series, please write to:

UK and Europe: Oxford Paperbacks Publicity Manager, Arts and Reference Publicity Department, Oxford University Press, Walton Street, Oxford OX2 6DP.

Customers in UK and Europe will find Oxford Paperbacks available in all good bookshops. But in case of difficulty please send orders to the Cash-with-Order Department, Oxford University Press Distribution Services, Saxon Way West, Corby, Northants NN18 9ES. Tel: 0536 741519; Fax: 0536 746337. Please send a cheque for the total cost of the books, plus £1.75 postage and packing for orders under £20; £2.75 for orders over £20. Customers outside the UK should add 10% of the cost of the books for postage and packing.

USA: Oxford Paperbacks Marketing Manager, Oxford University Press, Inc., 200 Madison Avenue, New York, N.Y. 10016.

Canada: Trade Department, Oxford University Press, 70 Wynford Drive, Don Mills, Ontario M3C 1J9.

Australia: Trade Marketing Manager, Oxford University Press, G.P.O. Box 2784Y, Melbourne 3001, Victoria.

South Africa: Oxford University Press, P.O. Box 1141, Cape Town 8000.

THE OXFORD AUTHORS

General Editor: Frank Kermode

THE OXFORD AUTHORS is a series of authoritative editions of major English writers. Aimed at both students and general readers, each volume contains a generous selection of the best writings—poetry, prose, and letters—to give the essence of a writer's work and thinking. All the texts are complemented by essential notes, an introduction, chronology, and suggestions for further reading.

Matthew Arnold
William Blake
Lord Byron
John Clare
Samuel Taylor Coleridge
John Donne
John Dryden
Ralph Waldo Emerson
Thomas Hardy
George Herbert and Henry Vaughan
Gerard Manley Hopkins
Samuel Johnson
Ben Jonson
John Keats
Andrew Marvell
John Milton
Alexander Pope
Sir Philip Sidney
Oscar Wilde
William Wordsworth

HISTORY IN OXFORD PAPERBACKS

THE STRUGGLE FOR
THE MASTERY OF EUROPE 1848–1918

A. J. P. Taylor

The fall of Metternich in the revolutions of 1848 heralded an era of unprecedented nationalism in Europe, culminating in the collapse of the Hapsburg, Romanov, and Hohenzollern dynasties at the end of the First World War. In the intervening seventy years the boundaries of Europe changed dramatically from those established at Vienna in 1815. Cavour championed the cause of *Risorgimento* in Italy; Bismarck's three wars brought about the unification of Germany; Serbia and Bulgaria gained their independence courtesy of the decline of Turkey—'the sick man of Europe'; while the great powers scrambled for places in the sun in Africa. However, with America's entry into the war and President Wilson's adherence to idealistic internationalist principles, Europe ceased to be the centre of the world, although its problems, still primarily revolving around nationalist aspirations, were to smash the Treaty of Versailles and plunge the world into war once more.

A. J. P. Taylor has drawn the material for his account of this turbulent period from the many volumes of diplomatic documents which have been published in the five major European languages. By using vivid language and forceful characterization, he has produced a book that is as much a work of literature as a contribution to scientific history.

'One of the glories of twentieth-century writing.' *Observer*

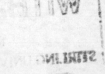

OXFORD REFERENCE

THE CONCISE OXFORD COMPANION TO ENGLISH LITERATURE

Edited by Margaret Drabble and Jenny Stringer

Based on the immensely popular fifth edition of the *Oxford Companion to English Literature* this is an indispensable, compact guide to the central matter of English literature.

There are more than 5,000 entries on the lives and works of authors, poets, playwrights, essayists, philosophers, and historians; plot summaries of novels and plays; literary movements; fictional characters; legends; theatres; periodicals; and much more.

The book's sharpened focus on the English literature of the British Isles makes it especially convenient to use, but there is still generous coverage of the literature of other countries and of other disciplines which have influenced or been influenced by English literature.

From reviews of *The Oxford Companion to English Literature*:

'a book which one turns to with constant pleasure ... a book with much style and little prejudice' Iain Gilchrist, *TLS*

'it is quite difficult to imagine, in this genre, a more useful publication' Frank Kermode, *London Review of Books*

'incarnates a living sense of tradition ... sensitive not to fashion merely but to the spirit of the age' Christopher Ricks, *Sunday Times*

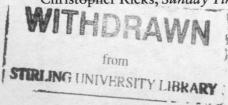